Introduction to

Vector Analysis

Seventh Edition

Harry F. Davis
University of Waterloo, Canada

Arthur David Snyder
University of South Florida

Book Team

Developmental Editor *Daryl Bluflodt*
Production Editor *Kay Driscoll*
Art Editor *Tina Flanagan*

Cover design and illustration by Schneck-De Pippo Graphics

Hawkes Publishing
James Hawkes, President

Library of Congress Catalog Card Number: 94-71618

ISBN 0-697-16099-8

Printed in the United States of America by **Rose Printing Company, Inc.,**
2503 Jackson Bluff Road, Tallahassee, Florida 32304

Contents

Preface

Current trends in mathematics continue to emphasize its applications to engineering and physics. Of course, this is nothing new for vector analysis, which has virtually been the *language* of mechanics and electromagnetism since the beginning of the century. The previous editions of *Introduction to Vector Analysis* have recognized this aspect of the subject, and in preparing the seventh edition we have continued to emphasize the treatment of applications in the hopes of initiating a broader and more diverse group of readers into the realms of vector techniques.

As in the earlier editions, we stress the geometric–analytic duality inherent in the subject. The concepts and theorems are first visualized and understood heuristically, and then they are reduced to an algebra–calculus framework for computation or mathematical scrutiny. A mastery of the geometric aspects of the subject is essential for all who wish to use it, and with it the rigorous details take on more of the flavor of "confirmations" rather than "proofs." This geometric–analytic dualism pervades the text, beginning with the basic concept of a vector and continuing with the physical significance of the vector differential operators and the advanced integral theorems. Table summaries are inserted at key points in the development to help the reader maintain perspective.

For the seventh edition we have rearranged some of the topics to improve the pedagogical development. Most significant is the early introduction of curvilinear coordinate expressions for the vector operators. This tactic provides immediate reinforcement of the coordinate-free interpretation of the operators as well as additional flexibility in the choice of examples illustrating the integral theorems. In its revised form the book's first four chapters now constitute a compact, one-semester exposition of the methods and applications of vector analysis. Deeper mathematical insights are available in chapter five.

The versatility of the text for use at several levels is enhanced by the utilization of optional reading sections and appendices dealing with specialized topics. These include introductory tensor concepts, the Frenet formulas, dyadics and three-dimensional Taylor polynomials, the Kuhn–Tucker conditions, transport theorems, the vector aspects of matrix algebra, the vector equations of classical mechanics and electromagnetism, the equations of potential theory, the physical significance of the laplacian and the vector potential, and the computation of the potentials from the field sources. The equations of fluid mechanics are developed in the exercises.

In closing let us state that it has been our continuing goal to design a textbook that will stand the engineer and scientist in good stead for his or her professional needs, to give the aspiring mathematician a firm grasp of the three-dimensional versions of the theorems of higher geometry and their applications, and to continue to serve the needs of every reader for some time after the successful completion of formal training.

The authors wish to thank the following teachers and colleagues who—directly or indirectly—have influenced our perspective of the subject of vector analysis: Francis Low, Richard Hutchinson, Edward O'Neil, Samuel Poss, and John Wyatt. We also wish to thank the following reviewers for their valuable recommendations: Thomas Metzger, University of Pittsburgh; Judith Baxter, University of Illinois; Selim Haïdar, Grand Valley State University; Maurice Ngo, Chabot College; Adel Boules, University of North Florida; and Ethan Devinatz, University of Washington. Particular thanks go to Mathew Sadiku of Temple University, whose review lent an important engineering perspective to the revisions.

1 Vector Algebra

1.1 Definitions

The vector concept is closely related to the geometrical idea of a *directed line segment.* Roughly speaking a vector is a quantity that has direction as well as magnitude. It is represented by an arrow of length equal to its magnitude, pointing in the appropriate direction. Two vectors* **A** and **B** are said to be equal, $\mathbf{A} = \mathbf{B}$, if they have the same length and direction.

This description of a vector conveys the intuitive concept, but as a definition it suffers from a lack of precision. Let us go back to basics and see if we can formulate this idea more carefully and unambiguously.

Consider two points P and Q in space. If P and Q are distinct points, there will exist one and only one line passing through them both. That part of the line between P and Q, including both P and Q as endpoints, is called a *line segment.* A line segment is said to be *directed* when the endpoints are given a definite order. The same line segment determines two directed line segments, one denoted PQ and the other QP (or $-PQ$). If P and Q coincide, PQ is said to be *degenerate,* and the line segment is a point.

---

* In this book, vectors are represented by boldfaced letters such as **A, B, C**,. . . . Since you cannot conveniently imitate this, the authors suggest that you either underline the letter, $\underline{A}$, orput an arrow above it, $\vec{A}$. Be sure to distinguish between the number 0 and the vector **0.**

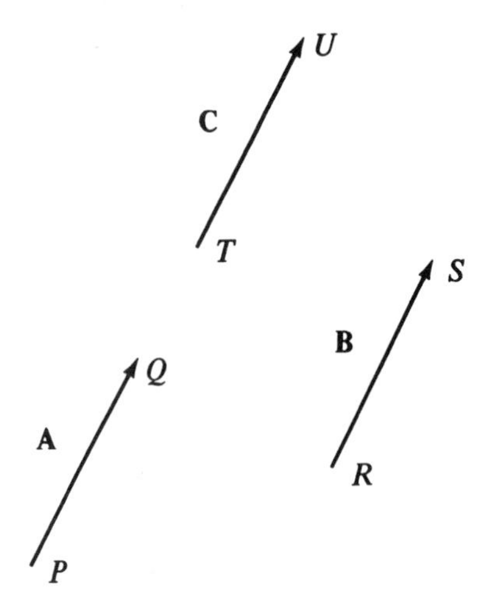

Figure 1.1

Now a directed line segment is a quantity with magnitude (the distance between P and Q) and direction (one exception: the degenerate segment, or point!). Historically, vectors were defined to be directed line segments. Experience has taught us, however, that it is convenient to consider two directed line segments as representing the *same* vector if they are parallel translates of each other; that is, if they point in the same direction and have the same length. Thus in figure 1.1 we see that PQ, RS, and TU are all equivalent and they represent the same vector. Being careful to consider all the possibilities, we can now formulate the following definitions. Two directed line segments PQ and RS are said to be *equivalent* if PQ and RS have the same length and are parallel, and also PR and QS have the same length and are parallel (the last condition ensures that RS is not directed opposite to PQ—draw a sketch to see this). *A vector is defined to be a collection of equivalent directed line segments.*

We may represent a vector by any one of the directed line segments in the collection. Thus, we may represent a vector by giving a particular directed line segment PQ, but it is understood that the vector itself is the set of all directed line segments that are equivalent to PQ.

In this book, boldface letters are usually used to denote vectors. In the diagrams, a single directed line segment will often be drawn to represent a vector and will be labeled by a boldface letter to denote the vector it represents. In figure 1.1, PQ is labeled **A**, RS is labeled **B**, and TU is labeled **C**. Since these all represent the same vector, we can write $\mathbf{A} = \mathbf{B} = \mathbf{C}$. Notice, however, that PQ and RS are not the same directed line segment since they occupy different positions in space, so we would not write $PQ = RS$.

To summarize, $\mathbf{A} = \mathbf{B}$ implies that PQ is parallel to RS, that PQ and RS have the same directed sense, and that the distance between P and Q is the same as the distance between R and S. This common distance is called the *magnitude* of the vector. Any point (degenerate line segment) represents the zero vector $\mathbf{0}$. This vector has zero magnitude and no direction; it is the exception to the intuitive characterization of "vector" given in the first paragraph.

Many of the quantities of physics have magnitude and direction, and thus are conveniently represented by vectors. As examples we mention force, displacement, velocity, acceleration, and magnetic field intensity. Such quantities are represented graphically by arrows pointing in the appropriate direction, of length proportional to the magnitude of the quantity. Thus the *magnitude* of a force vector may be 5 newtons, and it might be represented by an arrow of *length* 2 centimeters if we agree that each centimeter represents 2.5 newtons. Of course, we are disinclined to say that 5 newtons is the "length of the force."

In some books, what we call *directed line segments* are called *bound vectors,* and what we simply call *vectors* are called *free vectors.* The idea is that a "free vector" can be moved freely through space; provided it is always kept parallel to its initial position, and is never allowed to reverse its sense or to vary in magnitude, it does not really "change," whereas a "bound vector" could not be moved about in space. The distinction creates logical difficulties for both the pure mathematician and the physicist. For the pure mathematician it is difficult to accept such loose terminology as "moving freely through space" in the definition of a quantity that does not fundamentally involve the idea of time or motion at all. For the physicist the difficulty is in determining whether *force* is a bound or a free vector. In many cases the effect produced by a force acting on a body depends not only on its magnitude and direction but also on its point of application. Hence force might well be regarded as a bound vector; but in deeper theoretical work this becomes extremely awkward. Most physicists regard force as a vector quantity (i.e., a "free vector"), recognizing nevertheless that the *effect* of a force may depend on the point where it is applied (which, itself, can be located by a position vector).

In this book, the word *scalar* is used as a synonym for *number.* Those quantities of physics that are characterized by numerical magnitude alone (and have nothing to do with direction) are called *scalars* or *scalar quantities*. Examples are mass, time, density, distance, temperature, and speed (as read from a speedometer).

Loosely speaking, you can think of a vector as simply an arrow, but recognize that two arrows are considered equal, from a vector viewpoint, provided they are parallel, have the same directed sense, and the same magnitude.

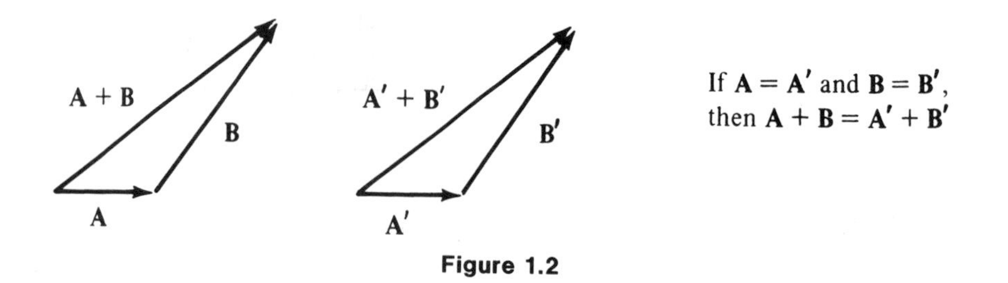

Figure 1.2

Suppose you are sitting at a desk with a horizontal surface. How many vectors are there that are perpendicular to this surface, are directed upward, and have magnitude of 3 centimeters? *Only one.* There are an infinite number of directed line segments with these properties, but they are *identical* as vectors.

1.2 Addition and Subtraction

The *sum* $\mathbf{A} + \mathbf{B}$ of two vectors may be defined in the following way. Let the vectors be represented so that the terminal point, or tip, of **A** coincides with the initial point, or tail, of **B**. Then $\mathbf{A} + \mathbf{B}$ is represented by the arrow extending from the tail of **A** to the tip of **B** (fig. 1.2). Reasoning with congruent triangles, we easily see that this definition of addition is compatible with the notion of equivalence; that is, if $\mathbf{A} = \mathbf{A}'$ and $\mathbf{B} = \mathbf{B}'$, then $\mathbf{A} + \mathbf{B} = \mathbf{A}' + \mathbf{B}'$. It is also commutative (fig. 1.3):

$$\mathbf{A} + \mathbf{B} = \mathbf{B} + \mathbf{A}$$

From figure 1.4 we see that vector addition is associative:

$$(\mathbf{A} + \mathbf{B}) + \mathbf{C} = \mathbf{A} + (\mathbf{B} + \mathbf{C})$$

and so no ambiguity arises from writing $\mathbf{A} + \mathbf{B} + \mathbf{C}$ without parentheses.

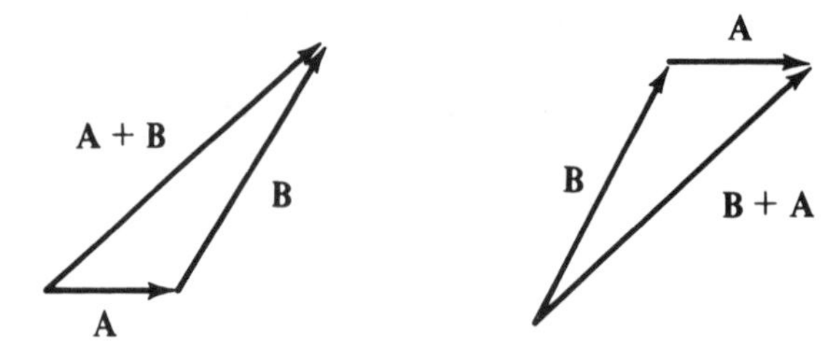

Figure 1.3

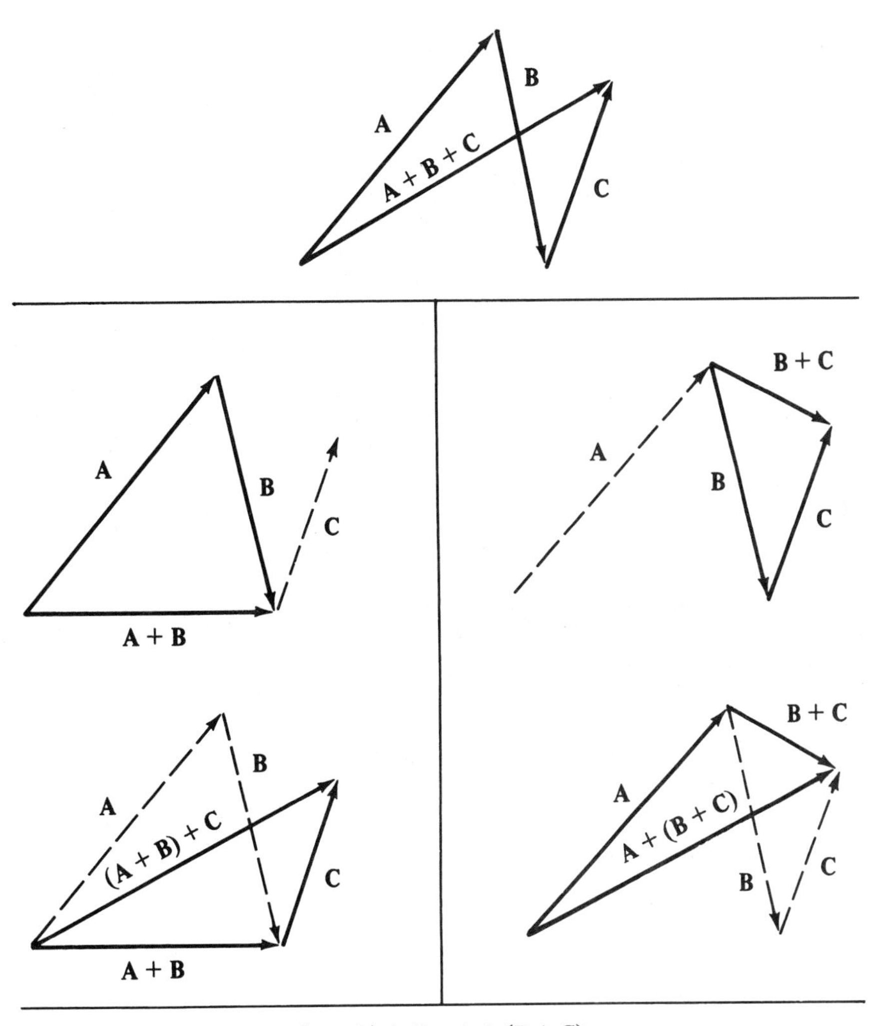

$(\mathbf{A} + \mathbf{B}) + \mathbf{C} = \mathbf{A} + (\mathbf{B} + \mathbf{C})$

Figure 1.4

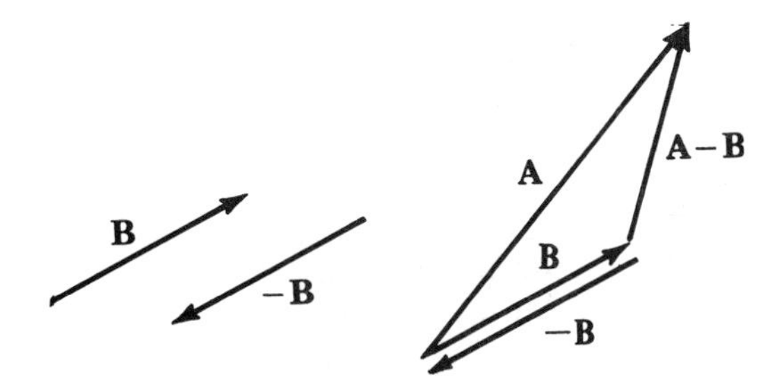

Figure 1.5

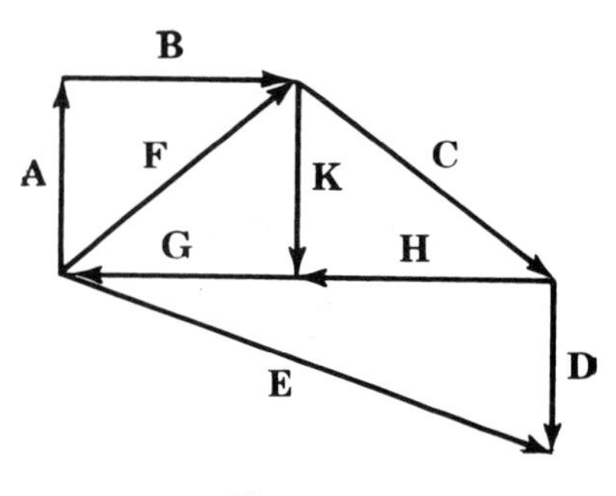

Figure 1.6

If **B** is a vector, $-\mathbf{B}$ is defined to be the vector with the same magnitude as **B** but opposite direction (fig. 1.5). Subtraction of vectors is defined by adding the negative:

$$\mathbf{A} - \mathbf{B} = \mathbf{A} + (-\mathbf{B})$$

The student who ignores this definition and simply memorizes figure 1.5 will inevitably confuse $\mathbf{A} - \mathbf{B}$ with $\mathbf{B} - \mathbf{A}$, which has the opposite direction. A good way of avoiding confusion is to keep in mind that $\mathbf{A} - \mathbf{B}$ is, algebraically, the vector that must be added to **B** to produce **A**; hence it runs from the tip of **B** to the tip of **A**, when **A** and **B** share a common tail.

The above definitions apply to the vector **0** if it is represented by a degenerate line segment. We have $\mathbf{0} = -\mathbf{0}$, $\mathbf{A} - \mathbf{A} = \mathbf{0}$, $\mathbf{A} + \mathbf{0} = \mathbf{A}$, and $\mathbf{0} + \mathbf{A} = \mathbf{A}$ for every vector **A.** The zero vector (which should be distinguished from the zero scalar) does not have a well-defined direction.

EXERCISES

The first four problems refer to figure 1.6.

1. Write **C** in terms of **E, D, F.**
2. Write **G** in terms of **C, D, E, K.**
3. Solve for $\mathbf{x} : \mathbf{x} + \mathbf{B} = \mathbf{F}$.
4. Solve for $\mathbf{x} : \mathbf{x} + \mathbf{H} = \mathbf{D} - \mathbf{E}$.
5. If **A** and **B** are represented by arrows whose initial points coincide, what arrow represents $\mathbf{A} + \mathbf{B}$?
6. By drawing a diagram, show that if $\mathbf{A} + \mathbf{B} = \mathbf{C}$, then $\mathbf{B} = \mathbf{C} - \mathbf{A}$.
7. Is the following statement correct? If **A, B, C,** and **D** are distinct nonzero vectors represented by arrows from the origin to the points *A, B, C,* and *D* respectively, and if $\mathbf{B} - \mathbf{A} = \mathbf{C} - \mathbf{D}$, then *ABCD* is a parallelogram.

8. Let the sides of a regular hexagon be drawn as arrows, with the terminal point of each arrow at the initial point of the next.
   (a) If **A** and **B** are vectors represented by consecutive sides, find the other four vectors in terms of **A** and **B.**
   (b) What is the vector sum of all six vectors?

1.3 Multiplication of Vectors by Numbers

The symbol $|\mathbf{A}|$ denotes the *magnitude* of the vector **A.** Although it should not be confused with $|s|$, which denotes (as usual) the absolute value of a number s, it does have many properties that are quite similar. For example, $|\mathbf{A}|$ is never negative, and $|\mathbf{A}| = 0$ if and only if $\mathbf{A} = \mathbf{0}$. Since **A** and $-\mathbf{A}$ have the same magnitude, we can always write $|\mathbf{A}| = |-\mathbf{A}|$ and $|\mathbf{A} - \mathbf{B}| = |\mathbf{B} - \mathbf{A}|$. The "triangle inequality"

$$|\mathbf{A} + \mathbf{B}| \leq |\mathbf{A}| + |\mathbf{B}|$$

is the vector expression of the fact that any side of a triangle does not exceed, in length, the sum of the lengths of the other two sides (fig. 1.7).

If s is a number and **A** is a vector, $s\mathbf{A}$ is defined to be the vector having magnitude $|s|$ times that of **A** and pointing in the same direction if s is positive or in the opposite direction if s is negative. Any vector $s\mathbf{A}$ is called a *scalar multiple* of **A** (fig. 1.8).

Here are the fundamental properties of the operation of multiplying vectors by numbers:

$$0\mathbf{A} = \mathbf{0} \qquad 1\mathbf{A} = \mathbf{A} \qquad (-1)\mathbf{A} = -\mathbf{A} \tag{1.1}$$

$$(s + t)\mathbf{A} = s\mathbf{A} + t\mathbf{A} \tag{1.2}$$

$$s(\mathbf{A} + \mathbf{B}) = s\mathbf{A} + s\mathbf{B} \tag{1.3}$$

$$s(t\mathbf{A}) = (st)\mathbf{A} \tag{1.4}$$

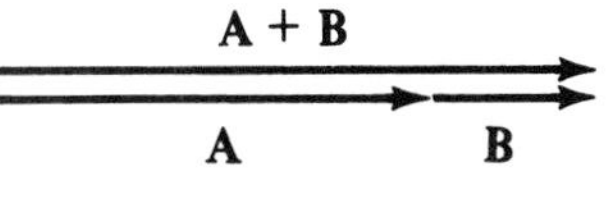

Figure 1.7

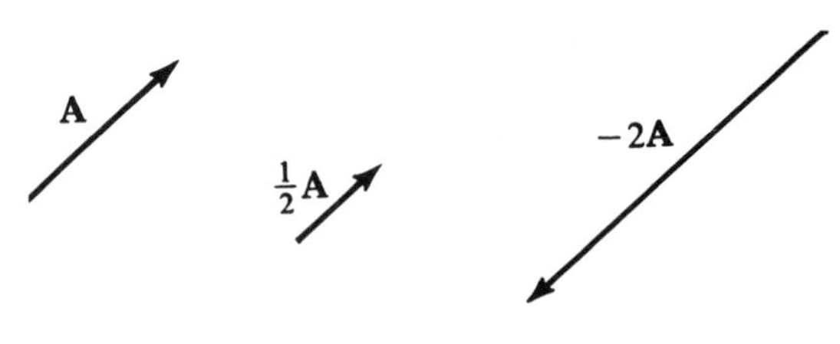

Figure 1.8

A vector whose magnitude is 1 is called a *unit vector.* To get a unit vector in the direction of **A**, divide **A** by $|\mathbf{A}|$ (equivalently, multiply **A** by $|\mathbf{A}|^{-1}$):

$$\left|\frac{\mathbf{A}}{|\mathbf{A}|}\right| = \frac{|\mathbf{A}|}{|\mathbf{A}|} = 1$$

EXERCISES

1. Is it ever possible to have $|\mathbf{A}| < 0$?
2. If $|\mathbf{A}| = 3$, what is $|4\mathbf{A}|$? What is $|-2\mathbf{A}|$? What can you say about $|s\mathbf{A}|$ if you know that $-2 \leq s \leq 1$?
3. If **A** is a nonzero vector, and if $s = |\mathbf{A}|^{-1}$, what is $|-s\mathbf{A}|$?
4. If **B** is a nonzero vector, and $s = |\mathbf{A}|/|\mathbf{B}|$, what can you say about $|s\mathbf{B}|$?
5. If **A** is a scalar multiple of **B**, is **B** necessarily a scalar multiple of **A?**
6. If $\mathbf{A} - \mathbf{B} = \mathbf{0}$, is it necessarily true that $\mathbf{A} = \mathbf{B}$?
7. If $|\mathbf{A}| = |\mathbf{B}|$, is it necessarily true that $\mathbf{A} = \mathbf{B}$?
8. You are given a plane in space. How many distinct vectors of unit magnitude are perpendicular to this plane?
9. How many distinct vectors exist, all having unit magnitude, perpendicular to a given line in space?
10. If **A** is a nonzero vector, how many distinct scalar multiples of **A** will have unit magnitude?
11. Let **A** and **B** be nonzero vectors represented by arrows with the same initial point to points A and B respectively. Let **C** denote the vector represented by an arrow from this same initial point to the midpoint of the line segment AB. Write **C** in terms of **A** and **B.**
12. Prove that $|\mathbf{A} - \mathbf{B}| \geq |\mathbf{A}| - |\mathbf{B}|$.
13. Find nonzero scalars a, b, and c such that $a\mathbf{A} + b(\mathbf{A} - \mathbf{B}) + c(\mathbf{A} + \mathbf{B}) = \mathbf{0}$ for every pair of vectors **A** and **B.**
14. Derive a formula for a vector that bisects the angle between two vectors **A** and **B.**

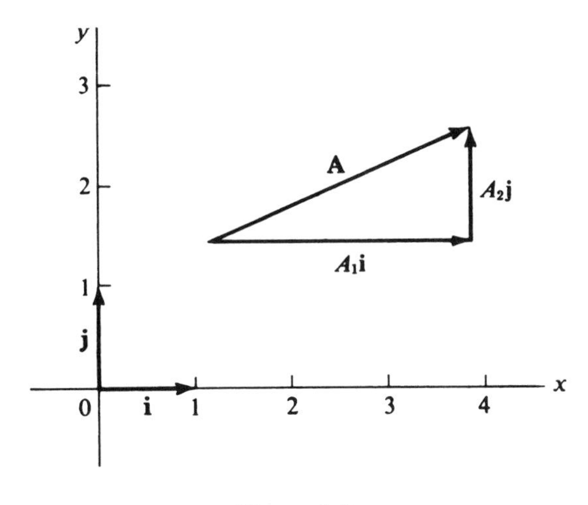

Figure 1.9

1.4 Cartesian Coordinates

Let us consider a cartesian coordinate system in the plane, obtained by introducing two mutually perpendicular axes, labeled x and y, with the same unit of length on both axes (fig. 1.9). We assume that the reader is already familiar with this construction, which sets up a one-to-one correspondence between points in the plane and ordered pairs (x,y) of numbers.

Let **i** denote the unit vector parallel to the x axis, in the positive x direction, and **j** the unit vector in the positive y direction. Every vector in the plane can be written uniquely in the form

$$\mathbf{A} = A_1\mathbf{i} + A_2\mathbf{j}$$

for a suitable choice of numbers A_1 and A_2. These numbers are called the *components* of **A** in the x direction and y direction, respectively; the component of a vector in a given direction is the orthogonal projection of the vector in that direction.

The magnitude of **A** can be determined from its components by using the pythagorean theorem (fig. 1.9):

$$|\mathbf{A}| = \sqrt{A_1^2 + A_2^2}$$

To determine the components of a vector, *any* directed line segment representing the vector can be used. Thus, if $P_1(x_1,y_1)$ and $P_2(x_2,y_2)$ are points in the xy plane, the vector represented by the directed line segment P_1P_2 (initial point P_1, terminal point P_2) is $(x_2 - x_1)\mathbf{i} + (y_2 - y_1)\mathbf{j}$. Any other directed line segment equivalent to P_1P_2 would give the same components.

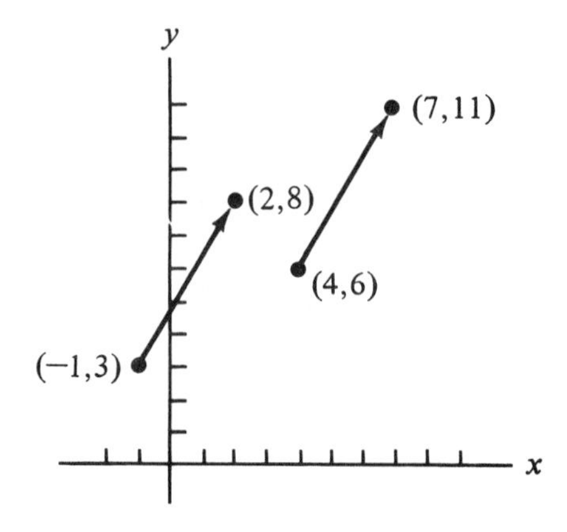

Figure 1.10

Example 1.1 The directed line segment extending from (4,6) to (7,11) is equivalent to the directed line segment extending from $(-1,3)$ to (2,8) because both of these directed line segments represent the vector $3\mathbf{i} + 5\mathbf{j}$ (fig. 1.10).

EXERCISES

1. What is the x component of **i**?
2. What is the x component of **j**?
3. What is the magnitude of $\mathbf{i} + \mathbf{j}$?
4. What is the magnitude of $3\mathbf{i} - 4\mathbf{j}$?
5. With the axes in conventional position (fig. 1.9), directions may be specified in geographical terms. What is the unit vector pointing west? south? northeast?
6. Vector **A** is represented by an arrow with initial point (4,2) and terminal point $(5,-1)$. Write **A** in terms of **i** and **j**.
7. The direction of a nonzero vector in the plane can be described by giving the angle θ it makes with the positive x direction (see fig. 1.11). This angle is conventionally taken to be positive in the counterclockwise sense. Write A_1 and A_2 in terms of $|\mathbf{A}|$ and this angle θ.
8. In figure 1.11, if $|\mathbf{A}| = 6$ and $\theta = 30°$, determine A_1 and A_2.
9. In terms of **i** and **j**, determine
   (a) the unit vector at positive angle 60° with the x axis.
   (b) the unit vector with $\theta = -30°$ (θ as in exercise 7).
   (c) the unit vector having the same direction as $3\mathbf{i} + 4\mathbf{j}$.
   (d) the unit vectors having x components equal to $\frac{1}{2}$.
   (e) the unit vectors perpendicular to the line $x + y = 0$.
10. Determine $|6\mathbf{i} + 8\mathbf{j}|$, $|-3\mathbf{i}|$, $|\mathbf{i} + s\mathbf{j}|$, $|(\cos\theta)\mathbf{i} + (\sin\theta)\mathbf{j}|$.
11. In terms of **i** and **j**, determine the vector represented by the arrow extending from the origin to the midpoint of the line segment joining (1,4) with (3,8).

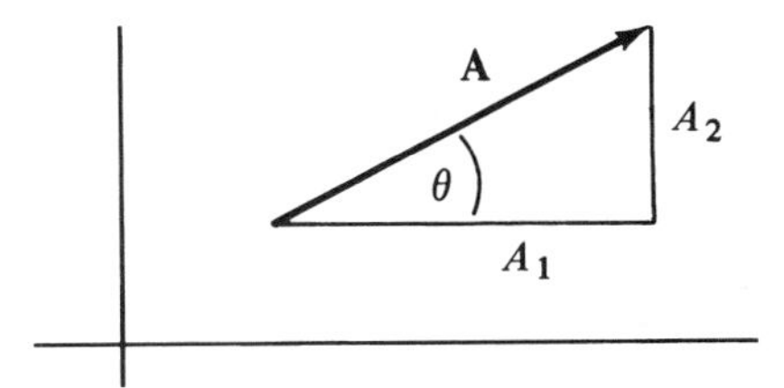

Figure 1.11

12. If the vector $\mathbf{V} = 2\mathbf{i} + 3\mathbf{j}$ represents the segment AB, and the midpoint of AB is $(2,1)$, find A and B.

13. If $\mathbf{V}$ is a unit vector in the xy plane making an angle of 30° with the positive y axis, express $\mathbf{V}$ in terms of $\mathbf{i}$ and $\mathbf{j}$ (*two solutions*).

1.5 Space Vectors

Throughout most of this book, we shall be concerned with vectors in three-dimensional space. By the introduction of three mutually perpendicular axes, with the same unit of length along all three axes, we obtain the usual cartesian coordinate system. The conventional orientation of axes is shown in figure 1.12. Every vector can be expressed in the form $\mathbf{A} = A_1\mathbf{i} + A_2\mathbf{j} + A_3\mathbf{k}$, where $\mathbf{i}$, $\mathbf{j}$, and $\mathbf{k}$ are unit

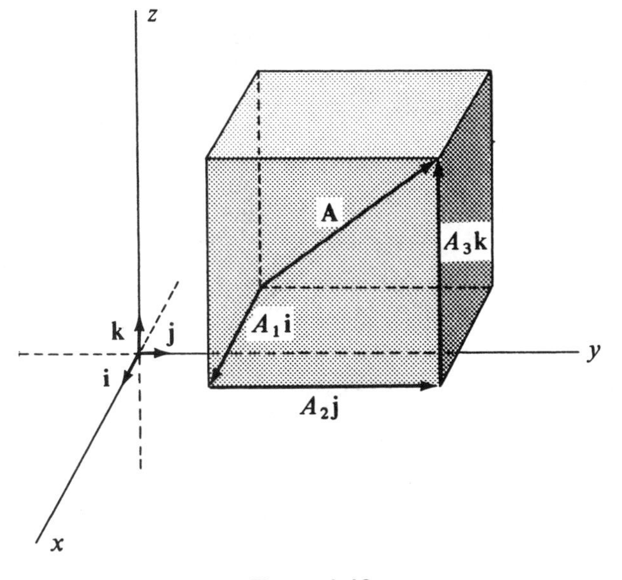

Figure 1.12

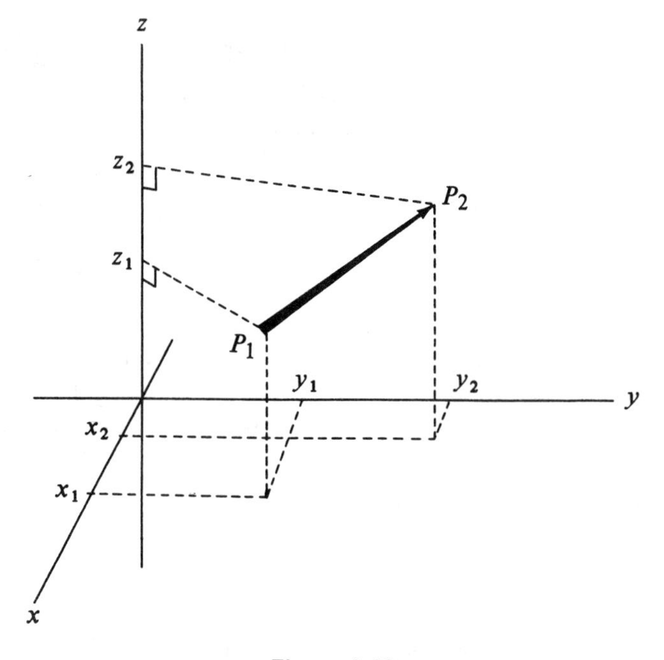

Figure 1.13

vectors in the positive x, y, and z directions, respectively. The numbers A_1, A_2, and A_3 are the *components*, or *orthogonal projections*, of **A** in the x, y, and z directions, respectively.

If $P_1(x_1,y_1,z_1)$, $P_2(x_2,y_2,z_2)$, and $P_3(x_3,y_3,z_3)$ are points in space, the vector represented by P_1P_2 is (fig. 1.13)

$$(x_2 - x_1)\mathbf{i} + (y_2 - y_1)\mathbf{j} + (z_2 - z_1)\mathbf{k}$$

Similarly, P_2P_3 is represented by

$$(x_3 - x_2)\mathbf{i} + (y_3 - y_2)\mathbf{j} + (z_3 - z_2)\mathbf{k}$$

and P_1P_3 by

$$(x_3 - x_1)\mathbf{i} + (y_3 - y_1)\mathbf{j} + (z_3 - z_1)\mathbf{k}$$

Observe that the components of P_1P_3 are given by the sums of the corresponding components of P_1P_2 and P_2P_3; for example, in the x direction we have

$$x_3 - x_1 = (x_2 - x_1) + (x_3 - x_2)$$

Since, furthermore, P_1P_3 represents the *vector* sum of P_1P_2 and P_2P_3, we have shown that *vector addition proceeds componentwise;* that is,

$$\mathbf{A} + \mathbf{B} = (A_1 + B_1)\mathbf{i} + (A_2 + B_2)\mathbf{j} + (A_3 + B_3)\mathbf{k}$$

Similar reasoning for multiplication by a scalar shows that, in terms of components,

$$s\mathbf{A} = (sA_1)\mathbf{i} + (sA_2)\mathbf{j} + (sA_3)\mathbf{k}$$

The commutative and associative laws of addition, as given in section 1.2, are valid for space vectors; one simply interprets figures 1.3 and 1.4 as three-dimensional. Alternatively, they become very obvious statements when expressed componentwise (see exercise 19).

By a double application of the pythagorean theorem, we obtain (fig. 1.12)

$$|\mathbf{A}| = \sqrt{A_1^2 + A_2^2 + A_3^2}$$

An alternative description of a vector in space is obtained by giving its magnitude and direction. We can specify the direction by prescribing the three *direction angles* α, β, and γ between the vector and the positive x, y, and z directions, respectively (see fig. 1.14). Sometimes it is more convenient to prescribe $\cos\alpha$, $\cos\beta$, and $\cos\gamma$, the direction cosines, because they are given in terms of the components by the following simple formulas:

$$\cos\alpha = \frac{A_1}{|\mathbf{A}|} \qquad \cos\beta = \frac{A_2}{|\mathbf{A}|} \qquad \cos\gamma = \frac{A_3}{|\mathbf{A}|}$$

(compare figs. 1.12 and 1.14). It is easy to verify that the direction cosines are related by

$$\cos^2\alpha + \cos^2\beta + \cos^2\gamma = 1 \tag{1.5}$$

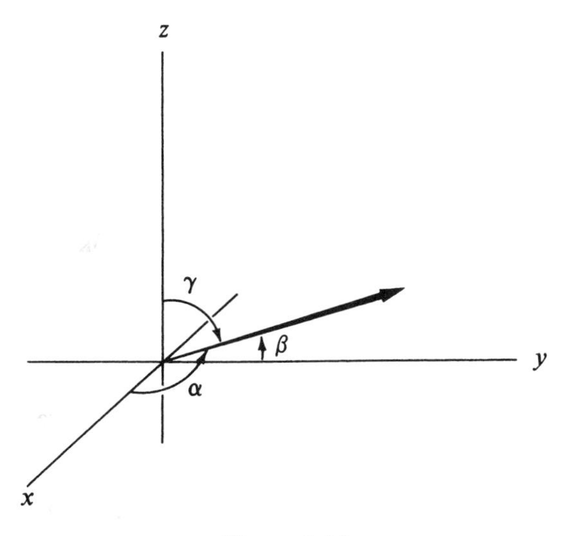

Figure 1.14

Thus we are not free to assign values to α, β, and γ arbitrarily. (See exercises 20–22.)

There is no way of telling from the direction cosines what the magnitude of the vector may be; the magnitude must be specified separately. For example, *any* vector parallel to the yz plane and making an angle of 45° with the positive y and z directions has direction cosines

$$\cos\alpha = 0 \qquad \cos\beta = \frac{\sqrt{2}}{2} \qquad \cos\gamma = \frac{\sqrt{2}}{2}$$

EXERCISES

In the first seven problems, let $\mathbf{A} = 3\mathbf{i} + 4\mathbf{j}$, $\mathbf{B} = 2\mathbf{i} + 2\mathbf{j} - \mathbf{k}$, and $\mathbf{C} = 3\mathbf{i} - 4\mathbf{k}$.

1. Find $|\mathbf{A}|$, $|\mathbf{B}|$, and $|\mathbf{C}|$.
2. Find $\mathbf{A} + \mathbf{B}$ and $\mathbf{A} - \mathbf{C}$.
3. Determine $|\mathbf{A} - \mathbf{C}|$.
4. For what values of s is $|s\mathbf{B}| = 1$?
5. Find the unit vector having the same direction as $\mathbf{A}$.
6. Let $\mathbf{A}$ and $\mathbf{C}$ be represented by arrows extending from the origin.
   (a) Find the length of the line segment joining their endpoints.
   (b) This line segment is parallel to one of the coordinate planes. Which one?
7. Let α denote the angle between $\mathbf{A}$ and the positive x direction. Determine $\cos\alpha$.
8. Determine all unit vectors perpendicular to the xz plane.
9. Compute $|\mathbf{i} + \mathbf{j} + \mathbf{k}|$.
10. Write the vector represented by P_1P_2 in terms of $\mathbf{i}$, $\mathbf{j}$, and $\mathbf{k}$, if $P_1 = (3,4,7)$ and $P_2 = (4,-1,6)$.
11. Write down the vector represented by the directed line segment OP, if O is the origin and $P(x,y,z)$ is a general point in space.
12. Let $\mathbf{D} = \mathbf{i} + \mathbf{j} + \mathbf{k}$, $\mathbf{E} = \mathbf{i} + \mathbf{j} - \mathbf{k}$, and $\mathbf{F} = \mathbf{i} - \mathbf{j}$. Determine scalars s, t, and r, such that $4\mathbf{i} + 6\mathbf{j} - \mathbf{k} = s\mathbf{D} + t\mathbf{E} + r\mathbf{F}$.
13. What are the direction cosines of the vector $2\mathbf{i} - 2\mathbf{j} + \mathbf{k}$?
14. Derive the identity $\cos^2\alpha + \cos^2\beta + \cos^2\gamma = 1$.
15. Give a geometrical description of the locus of all points P for which OP represents a vector with direction cosine $\cos\alpha = \frac{1}{2}$ (O is the origin).
16. How many unit vectors are there for which $\cos\alpha = \frac{1}{2}$ and also $\cos\beta = \frac{1}{2}$? Illustrate with a diagram.
17. $\mathbf{A}$ is a vector with direction cosines $\cos\alpha$, $\cos\beta$, and $\cos\gamma$, respectively. What are the direction cosines of the reflected image of $\mathbf{A}$ in the yz plane? (Think of the yz plane as a mirror.)
18. Determine all unit vectors for which $\cos\alpha = \cos\beta = \cos\gamma$.
19. Verify the commutative and associative laws of addition for space vectors by expressing them componentwise.

20. If one is given the direction angles α and β of a vector, to what extent can γ be determined?
21. Why is it impossible for a vector to have direction angles $\alpha = 30°$ and $\beta = 30°$? Answer this both geometrically and in terms of the constraint eq. (1.5).
22. Generalize exercise 21: show that neither β nor γ can be less than $90° - \alpha$.

1.6 Types of Vectors

A first step in solving some problems in mechanics is to choose a coordinate system. For instance, if the problem involves a particle sliding down an inclined plane, it may be convenient to take one of the axes, say the x axis, parallel to the plane, and another axis, say the z axis, perpendicular to the plane. After we have chosen a particular coordinate system, we can speak of the *position vector* of the particle. This is the vector represented by the directed line segment extending from the origin (0,0,0) to the point (x,y,z) where the particle is located, and (in terms of **i**, **j**, and **k**) it is the vector $\mathbf{R} = x\mathbf{i} + y\mathbf{j} + z\mathbf{k}$. Strictly speaking, we should not say "position vector of a particle" because this might give the false impression that it is an intrinsic property of the particle, whereas it also depends on the location of the origin of the coordinate system.

If a particle moves from an initial position (x_1,y_1,z_1) to another position (x_2,y_2,z_2), the *displacement* of the particle is the vector represented by the directed line segment extending from its initial position to its final position. This vector is $(x_2 - x_1)\mathbf{i} + (y_2 - y_1)\mathbf{j} + (z_2 - z_1)\mathbf{k}$. Notice that if the initial position vector is $\mathbf{R}_1 = x_1\mathbf{i} + y_1\mathbf{j} + z_1\mathbf{k}$ and the final position vector is $\mathbf{R}_2 = x_2\mathbf{i} + y_2\mathbf{j} + z_2\mathbf{k}$, the displacement is $\mathbf{R}_2 - \mathbf{R}_1$. *The displacement of a particle is the final position vector minus the initial position vector* (fig. 1.15).

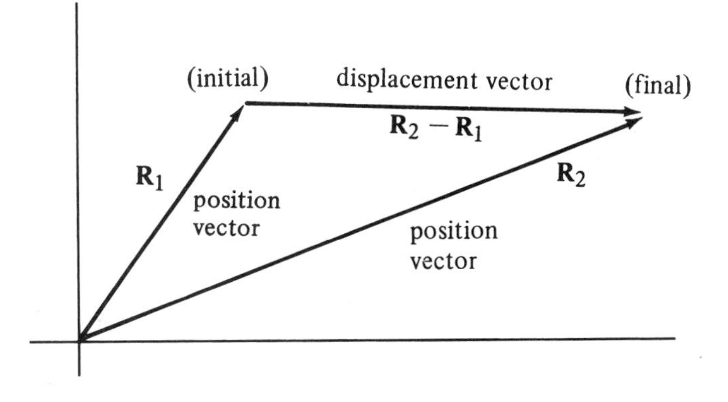

Figure 1.15

The displacement vector, unlike the position vector, *is* an intrinsic property of the particle; it does not depend on the choice of a coordinate system (although its *components* will be different in different coordinate systems). In fact, the displacement of a particle is a perfect model for a vector, at this stage. We have defined the addition of vectors so that they add in the same way that displacements "add." Thus, if a particle undergoes a displacement **A**, and then another displacement **B**, it is clear that the resultant displacement is $\mathbf{A} + \mathbf{B}$. That is, $\mathbf{A} + \mathbf{B}$ is the single displacement that produces the same net effect as the two displacements **A** and **B**. From the physicist's viewpoint, this is the reason for defining vector addition in this way.

Occasionally it is helpful to think of vectors as representing displacements, even when no physics is involved. For example, consider exercise 1 of section 1.2, where we are asked to write **C** in terms of **E**, **D**, and **F**. The answer is $\mathbf{C} = -\mathbf{F} + \mathbf{E} - \mathbf{D}$, which is clear since the net result of the three displacements $-\mathbf{F}$, **E**, and $-\mathbf{D}$ is **C**, as one can see by looking at figure 1.6.

Do not get the mistaken impression that when we represent a displacement by a vector **A**, the path of the particle has necessarily been straight. The directed line segment representing a displacement extends directly from the initial position to the final position, but the particle itself may have gone by way of the North Pole!

Forces are also vector quantities. This may seem obvious since a force is conveniently represented geometrically by a directed line segment. It is *not* so obvious, however. How do we know that forces "add" in the same way as vectors? We shall simply take the word of the physicists that they do, and direct the interested reader to the laboratory. If $\mathbf{F}_1$ and $\mathbf{F}_2$ are forces acting on a particle, their vector sum $\mathbf{F}_1 + \mathbf{F}_2$ is the single force that would produce the same effect, and it is sometimes called the *resultant* of the two forces. In elementary physics the resultant of two or more forces is usually found in the following manner: one draws a diagram showing the forces, then systematically marks out each force, replacing it by its components along the coordinate axes. The forces along each axis are summed algebraically, so that one has a single force remaining along each of the coordinate axes. The magnitude of the resultant force **F** can then be found by the pythagorean theorem, since the axes are perpendicular. This is discussed in every introductory physics book. Obviously, the process is equivalent to writing each force in terms of **i**, **j**, and **k**, and adding them in the manner of the preceding section.

It is rather surprising that *rotations* in space are *not* vector quantities. Clearly, a rotation can be represented by a directed line segment; the *direction* would be the axis of rotation, and the *length* would be the angle through which the body is rotated. But the result of two successive rotations is not represented by the vector sum of these line segments. In fact, the "sum" of two rotations is not even commutative! A body rotated through 90° about, first, the x direction, then the y direction, will achieve a final position quite different from the one resulting from the rotations performed in the other order (try this with the textbook). In this light, it is even more remarkable that *angular velocity* is, nonetheless, a vector quantity. This matter is discussed in Appendix C.

EXERCISES

1. A particle moves from (3,7,8) to (5,2,0). Write its displacement in terms of **i**, **j**, and **k**.
2. Write down the position vector of a particle located at the point (1,2,9).
3. The position vector of a moving particle at time t is $\mathbf{R} = 3\mathbf{i} + 4t^2\mathbf{j} - t^3\mathbf{k}$. Find its displacement during the time interval from $t = 1$ to $t = 3$.
4. What is the *magnitude* of the resultant of the following two displacements: 6 miles east, 8 miles north?
5. Strings are tied to a small metal ring and, by an arrangement of pulleys and weights, four forces are exerted on the ring. One force is directed upward with magnitude 3 lb, another is directed east with magnitude 6 lb, and a third is directed north with magnitude 2 lb. The ring is in equilibrium (i.e., it is not moving). What is the magnitude of the fourth force that is counterbalancing the other three?
6. The *center of mass* of a system of n particles is defined by the position vector

$$\mathbf{R}_{\text{cm}} = \frac{m_1\mathbf{R}_1 + m_2\mathbf{R}_2 + \cdots + m_n\mathbf{R}_n}{m_1 + m_2 + \cdots + m_n}$$

where the ith particle is located at $\mathbf{R}_i$ and has mass m_i. The *mass unbalance* of the system, measured at the position $\mathbf{R}$, is defined to be

$$m_1(\mathbf{R}_1 - \mathbf{R}) + m_2(\mathbf{R}_2 - \mathbf{R}) + \cdots + m_n(\mathbf{R}_n - \mathbf{R})$$

Show that the mass unbalance, measured at the center of mass, is zero. (*Hint:* Try it first for $n = 1$ and $n = 2$.)

7. Suppose a particle of electrical charge q_1 is located at $\mathbf{R}_1$, and q_2 is located at $\mathbf{R}_2$. The *Coulomb force* on particle 1 due to particle 2 is proportional to q_1 and q_2, and inversely proportional to the square of the distance between them; it is directed along the line from q_2 to q_1. Write down a vector formula for this force.

1.7 Some Problems in Geometry

To avoid circumlocution, practically everybody who works with vectors makes no distinction between vectors and directed line segments. It is easier to say "the vector **A**" than to say "the vector represented by the directed line segment **A**." When we do this, it is still important to recognize that the concept of a vector is an *abstraction* from the concept of a directed line segment, in which we ignore the actual location of the directed line segment: we say "**A** equals **B**" when we really mean "the directed line segments **A** and **B** are equivalent and therefore represent the same vector." If **A** extends from (2,3,4) to (2,3,5), and if **B** extends from (3,−2,8) to (3,−2,9), then we have as vectors **A** = **B** even though they extend from different points.

What we are saying is that two things are equal when they are really not identical but are only "equivalent" according to some definition. We are already familiar with this in elementary arithmetic. We say the fractions $\frac{2}{3}$ and $\frac{4}{6}$ are "equal" when in fact they are not identical but are only "equivalent" in a certain way. Strictly speaking, we should say that $\frac{2}{3}$ and $\frac{4}{6}$ are fractions that represent the same rational number: as fractions, they are not equal, but they represent the same rational number.

Similarly, if we have two directed line segments **A** and **B**, we may write **A** = **B** even when the directed line segments are not equal (because they extend from different points) but are equivalent according to the definition given in section 1.1.

With this in mind, we now turn to the practical utility of vector algebra. The simplest applications are in geometry and will be considered first.

Example 1.2 If the midpoints of the consecutive sides of a quadrilateral are joined by line segments, is the resulting quadrilateral a parallelogram?

Let *PQRS* be the quadrilateral and *T, U, V,* and *W* the midpoints of its sides. In the case shown in figure 1.16, it certainly appears that *TUVW* is a parallelogram. Keep in mind, however, that *PQRS* need not be a plane figure; perhaps *S* is a point several inches above the plane containing *P, Q,* and *R*. In view of this possibility, is *TUVW* a parallelogram?

Solution Let the sides be made into directed line segments **A, B, C,** and **D,** as shown in figure 1.16. Then one very obvious relationship is

$$\mathbf{A} + \mathbf{B} + \mathbf{C} + \mathbf{D} = \mathbf{0}$$

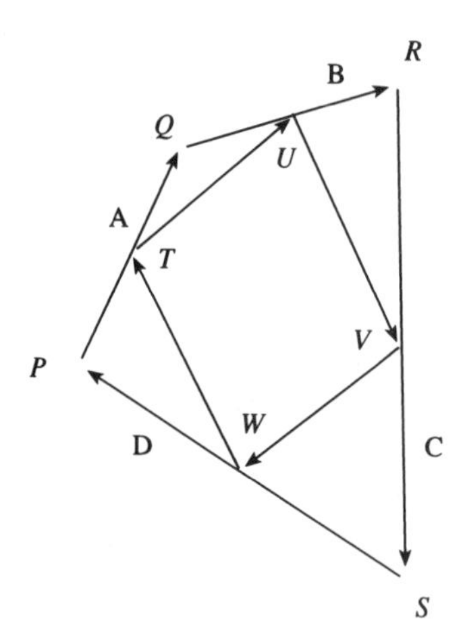

Figure 1.16

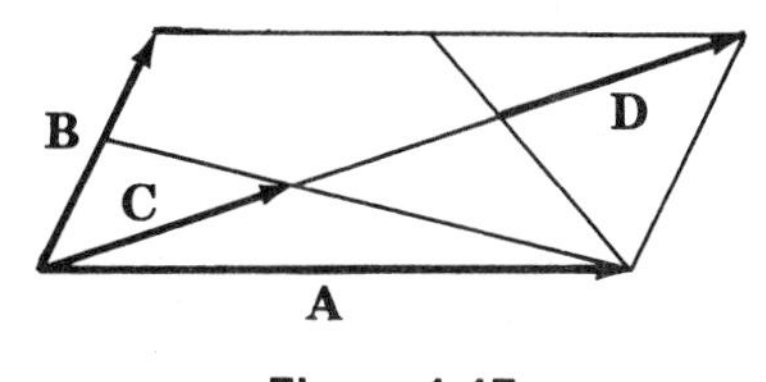

Figure 1.17

To conclude that $TUVW$ is a parallelogram, we need to show that $TU = -VW$. From the figure, TU may be expressible in terms of **A** and **B**; in fact, TU equals the "tip half" of **A** plus the "tail half" of **B.** Thus

$$TU = \tfrac{1}{2}\mathbf{A} + \tfrac{1}{2}\mathbf{B} = \tfrac{1}{2}(\mathbf{A} + \mathbf{B})$$

Similarly,

$$VW = \tfrac{1}{2}(\mathbf{C} + \mathbf{D})$$

But our basic relationship shows that $\mathbf{A} + \mathbf{B} = -(\mathbf{C} + \mathbf{D})$. Thus $TU = -VW$.

Example 1.3 Line segments are drawn from a vertex of a parallelogram to the midpoints of the opposite sides. Show that they trisect a diagonal.

Solution We have diagrammed the situation in figure 1.17, labeling certain vectors for convenience. Since the diagonal is $\mathbf{A} + \mathbf{B}$, the problem reduces to showing $\mathbf{C} = \mathbf{D} = \tfrac{1}{3}(\mathbf{A} + \mathbf{B})$. Let us try to express **C** in terms of **A** and **B.** First of all, certainly $\mathbf{C} = s(\mathbf{A} + \mathbf{B})$ for some scalar s. Also, since the tip of **C** lies on the line connecting the tip of **A** to the tip of $\tfrac{1}{2}\mathbf{B}$, we have $\mathbf{C} - \mathbf{A} = t(\tfrac{1}{2}\mathbf{B} - \mathbf{A})$ for some scalar t. If we equate the two expressions for **C,**

$$s(\mathbf{A} + \mathbf{B}) = \mathbf{A} + t(\tfrac{1}{2}\mathbf{B} - \mathbf{A})$$

we derive

$$(s - \tfrac{1}{2}t)\mathbf{B} = (1 - s - t)\mathbf{A}$$

Since **A** and **B** are not parallel, this equation can be true only if the scalars are zero:

$$s - \tfrac{1}{2}t = 0$$
$$1 - s - t = 0$$

Solving, we obtain $s = \tfrac{1}{3}$, so $\mathbf{C} = \tfrac{1}{3}(\mathbf{A} + \mathbf{B})$.

The reader should try to complete the solution as an exercise, manipulating **D** in an analogous manner.

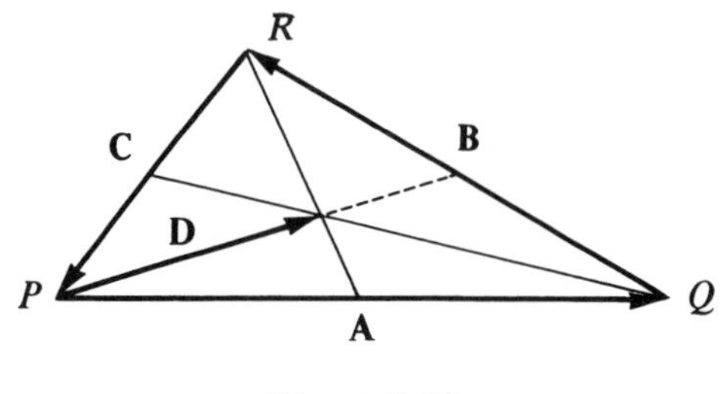

Figure 1.18

Example 1.4 Prove that the medians of a triangle intersect at a single point.

Solution In figure 1.18, **D** is the vector from the corner P to the point of intersection of the medians from Q and R. We must show that **D** lies along the median from P (i.e., that it is a multiple of $\mathbf{A} + \frac{1}{2}\mathbf{B}$). The condition that **D** lies along the median from R is expressed as

$$\mathbf{C} + \mathbf{D} = s(\mathbf{C} + \tfrac{1}{2}\mathbf{A})$$

for some number s, while the fact that **D** lies on the median from Q implies that, for some number t,

$$\mathbf{A} - \mathbf{D} = t(\tfrac{1}{2}\mathbf{C} + \mathbf{A})$$

Solving for **D** and equating the expressions, we derive

$$(s + \tfrac{1}{2}t - 1)\mathbf{C} = (1 - t - \tfrac{1}{2}s)\mathbf{A}$$

As in example 1.3, we conclude that both coefficients must vanish; thus

$$s = t = \tfrac{2}{3}$$

Using this in either equation for **D** and writing **C** in terms of **A** and **B**, we find

$$\mathbf{D} = \tfrac{2}{3}(\mathbf{A} + \tfrac{1}{2}\mathbf{B})$$

which is the form that we sought. Exercise 11 provides a delightful generalization of this example.

Example 1.5 Let θ denote the angle between two nonzero vectors **A** and **B**. Show that

$$\cos\theta = \frac{A_1B_1 + A_2B_2 + A_3B_3}{|\mathbf{A}||\mathbf{B}|} \tag{1.6}$$

(*Note: This is one of the most important identities in vector algebra.* Its significance is revealed in section 1.9.)

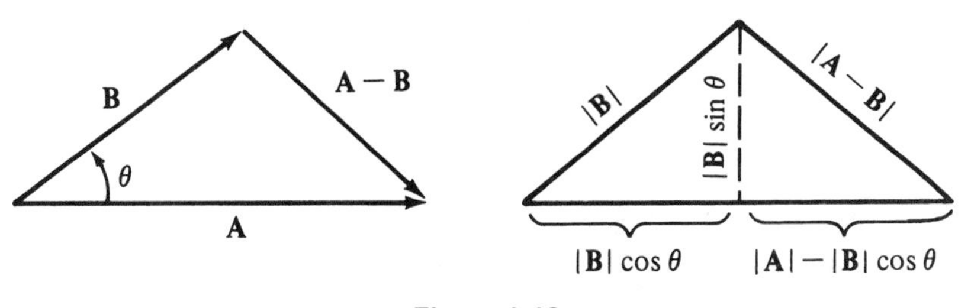

Figure 1.19

Solution This formula will "pop out" if we compare two expressions for $|\mathbf{A} - \mathbf{B}|^2$, one derived componentwise and one derived geometrically. Using components, we know that

$$|\mathbf{A} - \mathbf{B}|^2 = (A_1 - B_1)^2 + (A_2 - B_2)^2 + (A_3 - B_3)^2$$

Expanding powers and regrouping terms, we can write this as

$$|\mathbf{A} - \mathbf{B}|^2 = |\mathbf{A}|^2 + |\mathbf{B}|^2 - 2(A_1B_1 + A_2B_2 + A_3B_3)$$

Now for the geometric formula. **A, B,** and θ are depicted in figure 1.19; also, the perpendicular from the tip of **B** to **A** is drawn, with the lengths of the appropriate segments indicated. We can visualize **A** − **B** as the hypotenuse of a right triangle and, according to Pythagoras,

$$\begin{aligned} |\mathbf{A} - \mathbf{B}|^2 &= (|\mathbf{B}| \sin \theta)^2 + (|\mathbf{A}| - |\mathbf{B}| \cos \theta)^2 \\ &= |\mathbf{B}|^2 (\sin^2 \theta + \cos^2 \theta) + |\mathbf{A}|^2 - 2|\mathbf{A}||\mathbf{B}| \cos \theta \\ |\mathbf{A} - \mathbf{B}|^2 &= |\mathbf{A}|^2 + |\mathbf{B}|^2 - 2|\mathbf{A}||\mathbf{B}| \cos \theta \end{aligned} \tag{1.7}$$

Comparing this with the componentwise expression, we conclude

$$|\mathbf{A}||\mathbf{B}| \cos \theta = A_1B_1 + A_2B_2 + A_3B_3 \tag{1.8}$$

which is equivalent to the desired identity.

Incidentally, by referring to figure 1.19, the alert reader will recognize eq. (1.7) as the *law of cosines* from trigonometry.

As applications of this formula, consider examples 1.6 and 1.7.

Example 1.6 Show that the vectors $\mathbf{A} = 2\mathbf{i} - \mathbf{j} + 5\mathbf{k}$ and $\mathbf{B} = \mathbf{i} + 7\mathbf{j} + \mathbf{k}$ are perpendicular.

Solution

$$\cos \theta = \frac{2 - 7 + 5}{\sqrt{30}\,\sqrt{51}} = 0$$

Hence $\theta = 90°$.

Example 1.7 Find the angle between
(a) $\mathbf{i} + 2\mathbf{j} + 2\mathbf{k}$ and $\mathbf{i} - \mathbf{j}$
(b) $\mathbf{i}$ and $\mathbf{i} + \mathbf{j} + \mathbf{k}$

Solutions

(a) $\cos\theta = \dfrac{1-2}{\sqrt{9}\sqrt{2}} = -0.2357;\ \theta = \cos^{-1}(-0.2357) = 103.63°$

(b) $\cos\theta = \dfrac{1}{\sqrt{1}\sqrt{3}} = 0.57735;\ \theta = \cos^{-1}(0.57735) = 54.74°$

Notice that the natural interpretation of the angle between two vectors always lies between 0° and 180°. Thus the principal value of the arccosine is perfectly appropriate for these calculations.

Summary: Geometrical and analytical descriptions

Now is a good time to catch our breath and get an overview of what we have learned. There are two ways of looking at vectors—geometrically and analytically. Geometric descriptions are more physical; a vector has magnitude and direction, and relationships are described in terms of lengths and angles. But it is often difficult to compute with these quantities, especially if the problem is three-dimensional and hard to sketch. Thus, to solve such problems as finding the resultant of several forces, we introduce a cartesian coordinate system and represent all vectors by their components. Then a vector becomes an ordered triple of numbers. (Another reason for using this rather unphysical component description is communication. How, for instance, does an astronaut on the moon convey information to his earthbound colleagues about a quantity with magnitude and direction? The astronaut must describe its components in some coordinate system common to both, as determined by, for instance, the fixed stars.)

Let us summarize the equations we have derived; they tell us how to relate one description to the other. The geometrical concept of length of a vector is computed in terms of components by

$$|\mathbf{A}| = (A_1^2 + A_2^2 + A_3^2)^{1/2}$$

The angle θ between two vectors $\mathbf{A}$ and $\mathbf{B}$ is computed from components using eq. (1.6):

$$\cos\theta = \frac{A_1B_1 + A_2B_2 + A_3B_3}{|\mathbf{A}||\mathbf{B}|}$$

In particular, the direction cosines of **A**, which are the cosines of the angles between **A** and the positive coordinate axes, can be computed by substituting **i**, **j**, or **k** for **B** in the above; thus

$$\cos\alpha = \frac{A_1}{|\mathbf{A}|} \qquad \cos\beta = \frac{A_2}{|\mathbf{A}|} \qquad \cos\gamma = \frac{A_3}{|\mathbf{A}|}$$

Viewed another way, these equations can be used to compute the component description of a vector from its geometric characteristics; we have

$$A_1 = |\mathbf{A}|\cos\alpha \qquad A_2 = |\mathbf{A}|\cos\beta \qquad A_3 = |\mathbf{A}|\cos\gamma$$

Table 1.1 Geometrical and analytical descriptions of vectors

Geometrical Quantities: Lengths, Angles, Cosines	Analytical Quantities: Cartesian Coordinates
Length of $\mathbf{A} = \lvert\mathbf{A}\rvert$	$\sqrt{(A_1^2 + A_2^2 + A_3^2)}$
Angle between **A** and **B**	$\cos^{-1}[(A_1B_1 + A_2B_2 + A_3B_3)/\lvert\mathbf{A}\rvert\lvert\mathbf{B}\rvert]$
Direction cosines	$A_1/\lvert\mathbf{A}\rvert, A_2/\lvert\mathbf{A}\rvert, A_3/\lvert\mathbf{A}\rvert$

Hence the cycle is complete, and we are free to exploit whichever description, geometrical or analytical, is more convenient. Exercises 1 through 5 illustrate these ideas; see table 1.1.

EXERCISES

1. Find the angle between $2\mathbf{i} + \mathbf{j} + 2\mathbf{k}$ and $3\mathbf{i} - 4\mathbf{k}$.
2. Find the angle between the x axis and $\mathbf{i} + \mathbf{j} + \mathbf{k}$.
3. Find the three angles of the triangle with vertices $(2,-1,1)$, $(1,-3,-5)$, $(3,-4,-4)$.
4. Find the angle between the xy plane and $2\mathbf{i} + 2\mathbf{j} - \mathbf{k}$. (Note that $\mathbf{k}$ is perpendicular to the xy plane. You will have to decide what is meant by the angle between a vector and a plane.)
5. Show that $\mathbf{i} + \mathbf{j} + \mathbf{k}$ is perpendicular to the plane $x + y + z = 0$. (*Hint:* This plane passes through the origin. Show that $\mathbf{i} + \mathbf{j} + \mathbf{k}$ is perpendicular to every vector extending from the origin to a point in the plane.)
6. Imitate the solution of example 1.2, but instead of proving that $TU = -VW$, prove that $UV = -WT$.
7. Using vector methods, prove directly that if two sides of a quadrilateral are parallel and equal in magnitude, the other two sides are also.
8. By vector methods, show that the line segment joining the midpoints of two sides of a triangle is parallel to the third side, and has length equal to one half the length of the third side.

9. Show that the diagonals of a parallelogram bisect each other.

10. Construct another proof of the fact that the medians of a triangle intersect at a point, based on the following observation: if **D**, **E**, and **F** are vectors drawn from some fixed point to the corners of the triangle, then

$$\mathbf{D} + \tfrac{3}{2}[\tfrac{1}{3}(\mathbf{D} + \mathbf{E} + \mathbf{F}) - \mathbf{D}] = \tfrac{1}{2}(\mathbf{E} + \mathbf{F})$$

Verify this algebraically and then interpret it geometrically. [*Hint:* The tip of the vector $\tfrac{1}{3}(\mathbf{D} + \mathbf{E} + \mathbf{F})$ *is* this point of intersection.]

The following simple exercises are inserted here to help you recall some of the basic ideas of analytic geometry.

11. A treasure map has n villages marked on it, and it contains the following instructions. Start at village A, go $\frac{1}{2}$ of the way to village B, $\frac{1}{3}$ of the way to village C, $\frac{1}{4}$ of the way to village D, and so forth. The treasure is buried at the last stop. *Problem:* You lose the instructions, and don't know in what order to select the villages. *Show that it doesn't matter!* Then relate this to example 1.4 for $n = 3$.

12. Let PQR be a triangle. By vector methods, show there exists a triangle whose sides are parallel and equal in length to the medians of PQR.

13. True or false: $3x - 4y + 5z = 0$ represents a plane passing through the origin.

14. True or false: The yz plane is represented by the equation $x = 0$.

15. True or false: The locus of points for which $x = 3$ and $y = 4$ is a line parallel to the z axis whose distance from the z axis is 5.

16. True or false: $x^2 + y^2 + z^2 = 9$ is the equation of a sphere centered at the origin having radius 9.

17. Write down the equation of a sphere centered at the point (2,3,4) having radius 3.

18. Write down an equation for the cylinder concentric with the z axis having radius 2.

19. Do the equations $x = y = z$ represent a *line* or a *plane?*

20. What is the locus of points for which $x^2 + z^2 = 0$?

21. What is the locus of points for which $(x - 2)^2 + (y + 3)^2 + (z - 4)^2 = 0$?

22. What geometrical figure is represented by the equation $xyz = 0$? (Keep in mind that a product of numbers is zero if and only if at least one of the numbers is zero.)

23. What is the distance between the points (2,3,4) and (5,3,8)?

24. What is the distance between the point (3,8,9) and the xz plane? (Distance in such cases always means *shortest distance* or *perpendicular distance.*)

25. What is the distance between the point (0,3,0) and the cylinder $x^2 + y^2 = 4$? (You probably won't find a formula for this in any of your books. Just use some common sense.)

26. The expression $x^2 + y^2$ gives the square of the distance between (x,y,z) and the z axis. In view of this, what figure is represented by $x^2 + y^2 = z^2$?

27. Do you know what figure is represented by the equation $(x/2)^2 + (y/3)^2 + (z/4)^2 = 1$? (If so, you know more analytic geometry than is required to read this book.)

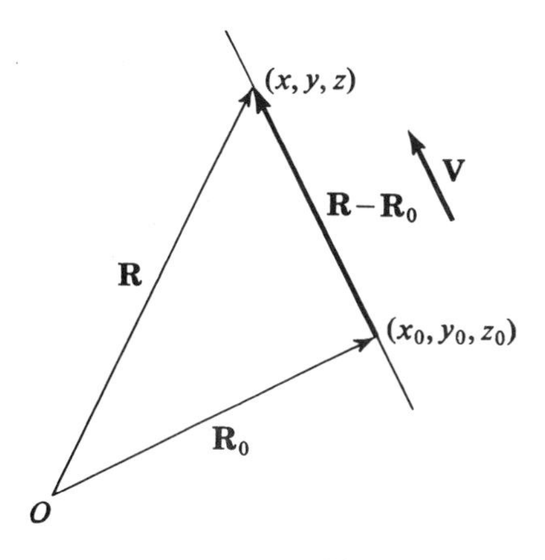

Figure 1.20

1.8 Equations of a Line

Recall that the *position vector of a point* is the vector extending from the origin to the point; the position vector of the point (x,y,z) is the vector $x\mathbf{i} + y\mathbf{j} + z\mathbf{k}$. This correspondence between points and vectors is the fundamental means whereby problems in analytic geometry can be studied by vector methods.

As an elementary example, let us derive the equations of a line passing through a given point (x_0,y_0,z_0) and parallel to a given nonzero vector $\mathbf{V} = a\mathbf{i} + b\mathbf{j} + c\mathbf{k}$ (see fig. 1.20).

Let $\mathbf{R}_0$ be the position vector of (x_0,y_0,z_0) and let $\mathbf{R}$ be the position vector of a point (x,y,z). It is not immediately obvious what conditions on the vector $\mathbf{R}$ itself will make the point (x,y,z) lie on the desired line, but the vector from (the tip of) $\mathbf{R}_0$ to (the tip of) $\mathbf{R}$ *must be parallel to* $\mathbf{V}$. This vector, describing $\mathbf{R}$ "relative to $\mathbf{R}_0$," is, of course, $\mathbf{R} - \mathbf{R}_0$. It will be parallel to $\mathbf{V}$ if and only if it equals some scalar multiple of $\mathbf{V}$, so the condition that (x,y,z) be on the line is that $\mathbf{R} - \mathbf{R}_0 = t\mathbf{V}$ for some number t. Rewriting this as $\mathbf{R} = \mathbf{R}_0 + t\mathbf{V}$ and expressing it in terms of the components of the vectors, we obtain

$$\mathbf{R} = \mathbf{R}_0 + t\mathbf{V} \qquad \text{or} \qquad \begin{aligned} x &= x_0 + at \\ y &= y_0 + bt \\ z &= z_0 + ct \end{aligned} \tag{1.9}$$

A point (x,y,z) is on the line passing through (x_0,y_0,z_0) and parallel to $\mathbf{V} = a\mathbf{i} + b\mathbf{j} + c\mathbf{k}$ if and only if its coordinates satisfy all three of the eqs. (1.9) for some value of the scalar t between $-\infty$ and $+\infty$.

Let us dwell for a moment on the significance of the scalar t. It seems to be somewhat artificial in the description. After all, physically speaking we can draw the line once we know $\mathbf{R}_0$ and $\mathbf{V}$; no other data are needed. The introduction of this

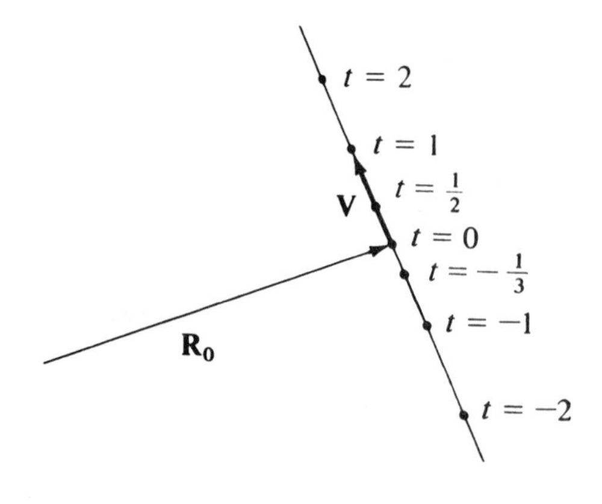

Figure 1.21

element t is just a mathematical device to help us say, *with equations,* that $\mathbf{R} - \mathbf{R}_0$ is parallel to $\mathbf{V}$. Equations (1.9) are called the *parametric form* of the equations of the line, and the "dummy" variable t is called the *parameter.*

An interpretation of the role of t can be gleaned from observing points $\mathbf{R}$ on the line for various values of t. Thus, for instance, if $t = 0$, $\mathbf{R} = \mathbf{R}_0$; if $t = 1$, $\mathbf{R} = \mathbf{R}_0 + \mathbf{V}$; if $t = -1$, $\mathbf{R} = \mathbf{R}_0 - \mathbf{V}$; other points, in between and beyond, are indicated in figure 1.21. If we think of the parameter t as representing time, we can think of eqs. (1.9) as giving the position of a moving particle at time t. This particle traverses a line parallel to $\mathbf{V}$ and passes through the point (x_0,y_0,z_0) at time $t = 0$.

As far as the line itself is concerned, the scalar multiple t in eqs. (1.9) could be replaced by any scalar function of t, such as $t/2$, $-t$, or t^3, as long as the function takes all values between $-\infty$ and $+\infty$. However, if we wrote

$$\mathbf{R} = \mathbf{R}_0 + t^2\mathbf{V}$$

we would be adding only *positive* multiples of $\mathbf{V}$ to $\mathbf{R}_0$, so we would be generating only "half" of the line (i.e., a *ray;* see fig. 1.22). If we wrote

$$\mathbf{R} = \mathbf{R}_0 + (\sin t)\mathbf{V}$$

we would generate just the *segment* of the line between $\mathbf{R}_0 - \mathbf{V}$ and $\mathbf{R}_0 + \mathbf{V}$ (since $-1 \leq \sin t \leq 1$), and we would be covering this segment infinitely often; interpreting t as time, the particle would oscillate forever from one end of the segment to the other. This same segment could be generated by the original parametric equations, eqs. (1.9), if we restrict t to the interval $-1 \leq t \leq 1$.

Clearly, the parametric form is not unique. Most people, however, would agree that eqs. (1.9) are the simplest form. Even so, notice that $\mathbf{R}_0$ could be replaced by the position vector of any other point on the line, and $\mathbf{V}$ could be replaced by any other vector having the same direction.

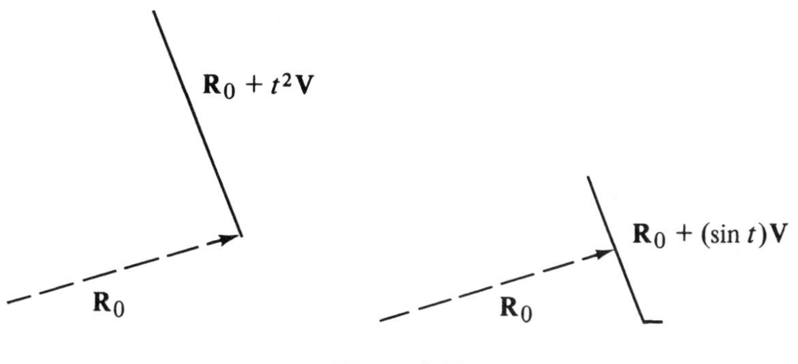

Figure 1.22

The parameter t can be eliminated by manipulating eqs. (1.9). The reader can easily verify that if none of the components of $\mathbf{V}$ are zero, one can derive

$$\frac{x - x_0}{a} = \frac{y - y_0}{b} = \frac{z - z_0}{c} \tag{1.10}$$

This is a nonparametric form, and from it one can immediately read off the components of $\mathbf{V}$ and of $\mathbf{R}_0$. [In using eq. (1.10), keep in mind the essential feature that the coefficients of x, y, and z are 1's! Also observe that eq. (1.10) represents two equations.]

Example 1.8 Find equations of the line passing through (2,0,4) and parallel to $2\mathbf{i} + \mathbf{j} + 3\mathbf{k}$, both in parametric and nonparametric form.

Solution The condition that $\mathbf{R} - \mathbf{R}_0$ is parallel to $\mathbf{V}$ becomes

$$x - 2 = 2t \qquad y - 0 = 1t \qquad z - 4 = 3t$$

Thus

$$x = 2 + 2t \qquad y = t \qquad z = 4 + 3t$$

Nonparametrically,

$$\frac{x - 2}{2} = y = \frac{z - 4}{3}$$

Example 1.9 Find equations of the line passing through $(0,3,-1)$ parallel to $3\mathbf{i} + 4\mathbf{k}$.

Solution In parametric form, we have

$$x = 3t \qquad y = 3 \qquad z = -1 + 4t$$

For the nonparametric form, $b = 0$, so eq. (1.10) does not make sense. If we eliminate t from the first and third equations above, we find

$$\frac{x}{3} = \frac{z+1}{4}$$

To this equation we append $y = 3$, which is already nonparametric.

Example 1.10 Find a unit vector parallel to the line

$$x - 2 = 2y - 3 = \frac{-2z+1}{2}$$

Solution By comparison with eq. (1.10), we have $a = 1, b = \frac{1}{2}$, and $c = -1$, so a vector parallel to the line is $\mathbf{i} + \frac{1}{2}\mathbf{j} - \mathbf{k}$. Dividing this vector by its own length, we obtain a unit vector $\frac{2}{3}\mathbf{i} + \frac{1}{3}\mathbf{j} - \frac{2}{3}\mathbf{k}$. The negative of this vector is also a correct solution. The point $(2, \frac{3}{2}, \frac{1}{2})$ lies on the line.

Example 1.11 Find the point of intersection of the two straight lines

$$\mathbf{R} = 3\mathbf{i} + 2\mathbf{j} + (2\mathbf{i} + \mathbf{j} + \mathbf{k})t$$
$$\mathbf{R} = \mathbf{i} - 2\mathbf{k} + (\mathbf{j} + \mathbf{k})t$$

Solution This is a little deceptive. Although we have used the same letter, t, for the parameter on both lines, we do not imply that at the point of intersection t takes the same values for each of the two lines; in terms of the particle-motion interpretation we are saying that the two paths may intersect, but the individual particles can go through the point of intersection at different times. The solution is more straightforward if we go to a nonparametric description.

In order that the point (x,y,z) lie on the first line, we must have

$$\frac{x-3}{2} = y - 2 = z$$

The condition for the second line reads

$$x = 1 \qquad y = z + 2$$

These constitute four equations that the three unknowns (x,y,z) must satisfy. If we consider just the first three equations

$$\frac{x-3}{2} = y - 2 \qquad y - 2 = z \qquad x = 1$$

we find that they have a solution $(x,y,z) = (1,1,-1)$. We must still check that the fourth equation,

$$y = z + 2$$

is satisfied; otherwise, there is no point of intersection (which is quite possible in space!). In this case it checks, so the point of intersection has $\mathbf{i} + \mathbf{j} - \mathbf{k}$ as its position vector.

Example 1.12 Find the angle between the lines in example 1.11.

Solution We have already verified that the lines do intersect, so the problem makes sense. The first line is parallel to $2\mathbf{i} + \mathbf{j} + \mathbf{k}$, the second to $\mathbf{j} + \mathbf{k}$. The angle between these vectors satisfies

$$\cos\theta = \frac{(2)(0) + (1)(1) + (1)(1)}{(2^2 + 1^2 + 1^2)^{1/2}(0^2 + 1^2 + 1^2)^{1/2}} = \frac{2}{12^{1/2}} = \frac{1}{3^{1/2}}$$

Therefore

$$\theta = \cos^{-1}\left(\frac{\sqrt{3}}{3}\right) = 54.74°$$

EXERCISES

1. Find parametric equations of the line passing through the origin parallel to $3\mathbf{i} - 2\mathbf{j} + 7\mathbf{k}$.
2. Find the equations of the line parallel to the z axis passing through the point (1,2,3).
3. Find equations of the line perpendicular to the yz plane, passing through (1,2,3).
4. Find the two unit vectors parallel to the line

$$\frac{x-1}{3} = \frac{y+2}{4} \qquad z = 9$$

5. Find two unit vectors parallel to the line $x = 2y = 3z + 3$. These equations can be written in form (1.10) as follows:

$$x = \frac{y}{\frac{1}{2}} = \frac{z+1}{\frac{1}{3}}$$

6. Find two unit vectors parallel to the line represented by the equations $x + y = 1$, $x - 3z = 5$. [*Hint:* Rewrite in form (1.10).]
7. Find equations of the line passing through the origin and parallel to the line

$$x - 3 = \frac{y+2}{4} = 1 - z$$

8. Find equations of the line passing through the points (3,4,5) and (3,4,7).
9. Find equations of the line passing through the points (1,4,−1) and (2,2,7).
10. By vector methods, find the cosine of the angle between the lines

$$\frac{x-1}{3} = \frac{y-0.5}{2} = z \qquad \text{and} \qquad x = y = z$$

11. Find the angle between the two intersecting lines

$$\frac{x-1}{3} = \frac{y-3}{4} = \frac{z}{5} \qquad \text{and} \qquad \frac{x-1}{2} = 3 - y = 2z$$

12. Let A and B be two points with position vectors $\mathbf{A}$ and $\mathbf{B}$, respectively. Show that the line passing through these points may be represented by the vector equation

$$\mathbf{R} = s\mathbf{A} + t\mathbf{B} \quad (s + t = 1) \tag{1.11}$$

13. Solve exercise 9 by making use of eq. (1.11).

14. (*Points of Division*) If the points A, B, and P are collinear, P is said to divide the segment AB in the ratio λ when the segments AP and PB are related by

$$AP = \lambda(PB) \tag{1.12}$$

(a) For what values of λ does P lie between A and B? to the left of A? to the right of B?

(b) Show that, relative to an origin O, eq. (1.12) can be written

$$OP = \frac{OA + \lambda(OB)}{1 + \lambda}$$

Relate this to eq. (1.11).

(c) If P and P' divide AB internally and externally in the same numerical ratios $\pm\lambda$, show that A and B divide PP' internally and externally in the ratios $\pm(1 - \lambda)/(1 + \lambda)$.

15. Find the point(s) of intersection of the following pairs of straight lines:

(a) $\mathbf{R} = (5\mathbf{i} + 4\mathbf{j} + 5\mathbf{k})t + 7\mathbf{i} + 6\mathbf{j} + 8\mathbf{k}$ and
$\mathbf{R} = (6\mathbf{i} + 4\mathbf{j} + 6\mathbf{k})t + 8\mathbf{i} + 6\mathbf{j} + 9\mathbf{k}$

(b) $\mathbf{R} = (3\mathbf{i} + 2\mathbf{j} + \mathbf{k})t + 2\mathbf{k}$ and
$\mathbf{R} = (6\mathbf{i} + 4\mathbf{j} + 2\mathbf{k})t + 3\mathbf{i} + 2\mathbf{j} + 3\mathbf{k}$

(c) $\mathbf{R} = (3\mathbf{i} - \mathbf{j} + \mathbf{k})t$ and
$\mathbf{R} = (-6\mathbf{i} + 2\mathbf{j} - 2\mathbf{k})t + 2\mathbf{i}$

(d) $\mathbf{R} = (\mathbf{i} + \mathbf{j} + \mathbf{k})t$ and
$\mathbf{R} = (\mathbf{i} + \mathbf{j} - 3\mathbf{k})t - \mathbf{i} + \mathbf{j}$

16. Redo example 1.11 using the parametric description.

17. Equations (1.9) seem to indicate that six numbers (x_0, y_0, z_0, a, b, c) are needed to specify a straight line. Is this correct?

18. Find the points of intersection of the lines

$$l_1\colon \mathbf{R} = 2\mathbf{i} + 3\mathbf{j} + 3\mathbf{k} + t(\mathbf{i} - 2\mathbf{j} + 5\mathbf{k})$$

$$l_2\colon \frac{x + 3}{2} = \frac{y + 1}{2} = -z$$

1.9 Scalar Products

The *scalar product* of two vectors is the number

$$\mathbf{A} \cdot \mathbf{B} = |\mathbf{A}||\mathbf{B}| \cos\theta \tag{1.13}$$

where θ denotes the angle between the vectors. Although $\mathbf{A}$ and $\mathbf{B}$ are vectors, $\mathbf{A} \cdot \mathbf{B}$ is a scalar. The scalar product is also called the *dot product*, or the *inner*

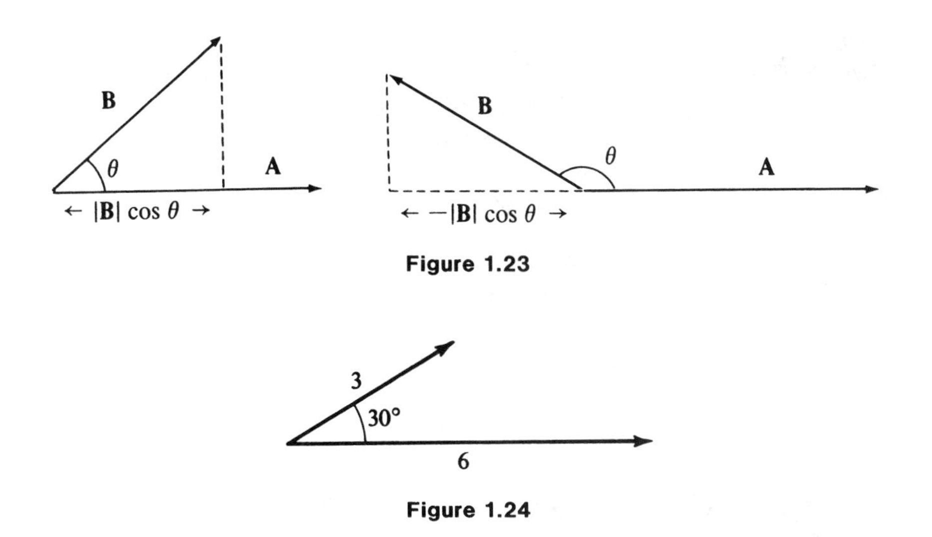

Figure 1.23

Figure 1.24

product. From figure 1.23, we identify $|\mathbf{B}| \cos \theta$ as the component of **B** parallel to **A**, that is, the length of the orthogonal projection of **B** in the direction of **A**, with the appropriate sign. Thus we can interpret $\mathbf{A} \cdot \mathbf{B}$ as

$$(\textit{length of } \mathbf{A})(\textit{signed component of } \mathbf{B} \textit{ along } \mathbf{A})$$

Since the definition is symmetric in **A** and **B**, it can equally well be interpreted as

$$(\textit{length of } \mathbf{B})(\textit{signed component of } \mathbf{A} \textit{ along } \mathbf{B})$$

In a few simple cases the scalar product of two vectors is easily computed directly from this definition. For example, the scalar product of the vectors shown in figure 1.24 is $9\sqrt{3}$.

If either **A** or **B** is the zero vector, we have $|\mathbf{A}| = 0$ or $|\mathbf{B}| = 0$, so by (1.13) it follows that $\mathbf{A} \cdot \mathbf{B} = 0$. (We ignore the fact that θ is not defined in this case.)

On the other hand, it is possible to have $\mathbf{A} \cdot \mathbf{B} = 0$ even though both **A** and **B** are nonzero vectors. For example, if **A** and **B** are perpendicular, then $\cos \theta = \cos 90° = 0$ and hence $\mathbf{A} \cdot \mathbf{B} = 0$.

Recall that in section 1.7 we derived a component expression for the right-hand side of eq. (1.13), namely eq. (1.8):

$$|\mathbf{A}||\mathbf{B}| \cos \theta = A_1B_1 + A_2B_2 + A_3B_3 \tag{1.8}$$

Combining this with eq. (1.13) gives us

$$\mathbf{A} \cdot \mathbf{B} = A_1B_1 + A_2B_2 + A_3B_3 \tag{1.14}$$

Thus we have two important formulas for the scalar product. Equation (1.13) describes $\mathbf{A} \cdot \mathbf{B}$ in terms of geometric concepts and provides a visualization, while eq. (1.14) gives the componentwise description and is useful for computations. Memorize both formulas now. They are important.

Example 1.13 Find the scalar product of $4\mathbf{i} - 5\mathbf{j} - \mathbf{k}$ and $\mathbf{i} + 2\mathbf{j} + 3\mathbf{k}$.

Solution $(4)(1) + (-5)(2) + (-1)(3) = -9$. (The negative sign indicates that the angle between the vectors must be greater than $90°$.)

Example 1.14 Find the angle between the vectors $\mathbf{A} = 2\mathbf{i} + 2\mathbf{j} - \mathbf{k}$ and $\mathbf{B} = 3\mathbf{i} + 4\mathbf{j}$.

Solution We have $|\mathbf{A}| = 3$ and $|\mathbf{B}| = 5$. Using (1.14), we see that $\mathbf{A} \cdot \mathbf{B} = 14$. Substituting these values in eq. (1.13), we solve to get $\theta = \cos^{-1} 14/15$.

Example 1.15 If $\mathbf{F}$ is a constant force acting through a displacement $\mathbf{D}$, the work done by $\mathbf{F}$ is defined to be the product of the magnitude of the displacement with the component of the force in the direction of the displacement. In vector notation,

$$\textit{Work} = \mathbf{F} \cdot \mathbf{D}$$

The following properties of the scalar product are easily verified from eq. (1.14):

$$\begin{aligned} \mathbf{A} \cdot \mathbf{B} &= \mathbf{B} \cdot \mathbf{A} \\ (s\mathbf{A} + \mathbf{B}) \cdot \mathbf{C} &= s\mathbf{A} \cdot \mathbf{C} + \mathbf{B} \cdot \mathbf{C} \\ \mathbf{A} \cdot (s\mathbf{B} + \mathbf{C}) &= s\mathbf{A} \cdot \mathbf{B} + \mathbf{A} \cdot \mathbf{C} \\ |\mathbf{A}|^2 &= \mathbf{A} \cdot \mathbf{A} \end{aligned}$$

Thus the magnitude of $\mathbf{A}$ can be expressed as $(\mathbf{A} \cdot \mathbf{A})^{1/2}$.

Example 1.16 (*A Maximum Principle*) Let there be given a nonzero vector $\mathbf{D}$, and let $\mathbf{n}$ denote a unit vector. Then $|\mathbf{n}| = 1$ and $\mathbf{D} \cdot \mathbf{n} = |\mathbf{D}||\mathbf{n}| \cos \theta = |\mathbf{D}| \cos \theta$. This will be a maximum when $\cos \theta = 1$ (i.e., when $\theta = 0$). Thus we have derived the following maximum principle, which will be useful to us in later sections:

The unit vector $\mathbf{n}$ *making* $\mathbf{D} \cdot \mathbf{n}$ *a maximum is the unit vector pointing in the same direction as* $\mathbf{D}$.

Example 1.17 The scalar product can be used to express components along the axes, of course; thus the component of $\mathbf{D}$ in the x direction is $\mathbf{D} \cdot \mathbf{i}$, and so forth. In fact, for any vector $\mathbf{D}$, we can write

$$\mathbf{D} = (\mathbf{D} \cdot \mathbf{i})\,\mathbf{i} + (\mathbf{D} \cdot \mathbf{j})\,\mathbf{j} + (\mathbf{D} \cdot \mathbf{k})\,\mathbf{k} \qquad (1.15)$$

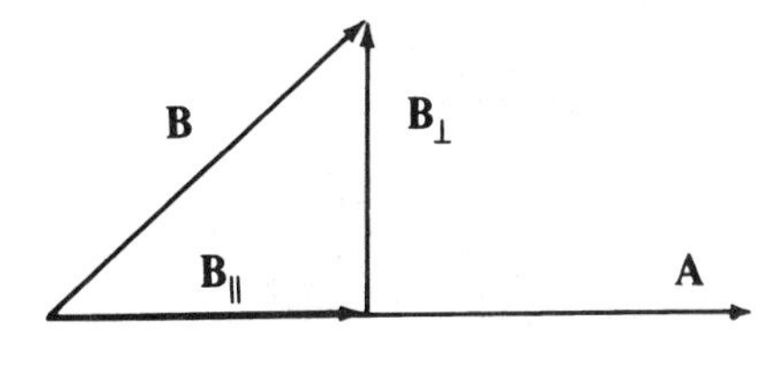

Figure 1.25

As an important example of the use of the scalar product, consider the following problem. One is given two vectors, **A** and **B**, and one wishes to decompose **B** into a vector parallel to **A** plus a vector perpendicular to **A**. In other words, one wishes to find expressions for the vectors $\mathbf{B}_\parallel$ and $\mathbf{B}_\perp$ in figure 1.25. Clearly, the (signed) length of $\mathbf{B}_\parallel$ is $\mathbf{B} \cdot \mathbf{A}/|\mathbf{A}|$. To construct a vector of this length in the direction of **A**, we take the unit vector along **A** and multiply by this scalar. Since $\mathbf{A}/|\mathbf{A}|$ is the unit vector, we have the following simple formula:

$$\mathbf{B}_\parallel = \frac{\mathbf{B} \cdot \mathbf{A}}{|\mathbf{A}|}\frac{\mathbf{A}}{|\mathbf{A}|} = \frac{\mathbf{B} \cdot \mathbf{A}}{\mathbf{A} \cdot \mathbf{A}}\mathbf{A}$$

Having computed $\mathbf{B}_\parallel$, we see that $\mathbf{B}_\perp$ is just the rest of **B**:

$$\mathbf{B}_\perp = \mathrm{B} - \mathrm{B}_\parallel = \mathbf{B} - \frac{\mathbf{B} \cdot \mathbf{A}}{\mathbf{A} \cdot \mathbf{A}}\mathbf{A}$$

Example 1.18 Resolve the vector $6\mathbf{i} + 2\mathbf{j} - 2\mathbf{k}$ into vectors parallel and perpendicular to $\mathbf{i} + \mathbf{j} + \mathbf{k}$.

Solution The parallel vector is

$$\frac{6 + 2 - 2}{1 + 1 + 1}(\mathbf{i} + \mathbf{j} + \mathbf{k}) = 2(\mathbf{i} + \mathbf{j} + \mathbf{k})$$

The perpendicular vector is

$$6\mathbf{i} + 2\mathbf{j} - 2\mathbf{k} - 2(\mathbf{i} + \mathbf{j} + \mathbf{k}) = 4\mathbf{i} - 4\mathbf{k}$$

Example 1.19 Find a formula for $\mathbf{V}'$, the mirror image of a vector **V**, reflected in a plane mirror with unit normal **n**. (See fig. 1.26*a*).

Solution In figure 1.26*b*, we have drawn representatives of **n**, **V**, and $\mathbf{V}'$ with a common tail. The dotted lines illustrate the fact that **V** and $\mathbf{V}'$ have the same component perpendicular to **n**, but the parallel components are opposite. To obtain $\mathbf{V}'$ from **V**, we must subtract this parallel component *twice*. Hence, keeping in mind that **n** was given as a *unit* normal, we see that

$$\mathbf{V}' = \mathbf{V} - 2\mathbf{V}_\parallel = \mathbf{V} - 2(\mathbf{V} \cdot \mathbf{n})\mathbf{n}$$

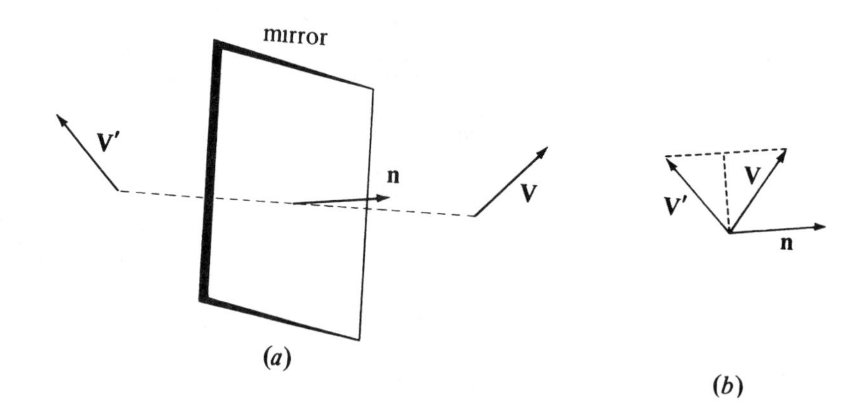

Figure 1.26

EXERCISES

1. Find the scalar product of $3\mathbf{i} + 8\mathbf{j} - 2\mathbf{k}$ with $5\mathbf{i} + \mathbf{j} + 2\mathbf{k}$.
2. Find the scalar product of $2\mathbf{i} + 3\mathbf{j} + 4\mathbf{k}$ with $4\mathbf{i} + 9\mathbf{j} - 3\mathbf{k}$.
3. Find the scalar product of $3\mathbf{i} + 4\mathbf{j}$ with $5\mathbf{j} - 10\mathbf{k}$.
4. Determine the angle between $2\mathbf{i} + \mathbf{j} - 2\mathbf{k}$ and $3\mathbf{i} - 4\mathbf{j}$.
5. Find the angle between $2\mathbf{i}$ and $3\mathbf{i} + 4\mathbf{j}$.
6. A force $\mathbf{F} = 2\mathbf{i} + 3\mathbf{j} + \mathbf{k}$ acts through a displacement $\mathbf{D} = -2\mathbf{i} + \mathbf{j} - \mathbf{k}$. Find the work done.
7. Find the component of $8\mathbf{i} + \mathbf{j}$ in the direction of $\mathbf{i} + 2\mathbf{j} - 2\mathbf{k}$.
8. Find the component of $\mathbf{i} + \mathbf{j} + \mathbf{k}$ in the direction of $\mathbf{i} + \mathbf{j}$.
9. Find the component of the force $5\mathbf{i} + 7\mathbf{j} - \mathbf{k}$ in the direction of the displacement PQ, where $P(3,0,1)$ and $Q(4,4,4)$ are points in space.
10. Find the vector in the same direction as $\mathbf{i} + \mathbf{j}$ whose component in the direction of $2\mathbf{i} - 4\mathbf{k}$ is unity.
11. If $\mathbf{A} \cdot \mathbf{A} = 0$ and $\mathbf{A} \cdot \mathbf{B} = 0$, what can you conclude about the vector $\mathbf{B}$?
12. Decompose $6\mathbf{i} - 3\mathbf{j} - 6\mathbf{k}$ into vectors parallel and perpendicular to
    (a) the vector $\mathbf{i} + \mathbf{j} + \mathbf{k}$.
    (b) the vector $2\mathbf{i} - \mathbf{j} - 2\mathbf{k}$.
    (c) the vector $2\mathbf{j} - \mathbf{k}$.
13. Determine s and t so that $\mathbf{C} - s\mathbf{A} - t\mathbf{B}$ is perpendicular to both $\mathbf{A}$ and $\mathbf{B}$, given that

$$\mathbf{A} = \mathbf{i} + \mathbf{j} + 2\mathbf{k}$$
$$\mathbf{B} = 2\mathbf{i} - \mathbf{j} + \mathbf{k}$$
$$\mathbf{C} = 2\mathbf{i} - \mathbf{j} + 4\mathbf{k}$$

14. The vector $\mathbf{n} = (3\mathbf{i} + 2\mathbf{j} + 6\mathbf{k})/7$ is perpendicular to a plane. A line segment representing the vector $\mathbf{A} = 2\mathbf{i} + 5\mathbf{j} + 6\mathbf{k}$ lies on one side of this plane. Regarding the plane as a mirror, write down the vector represented by the mirror image of $\mathbf{A}$.

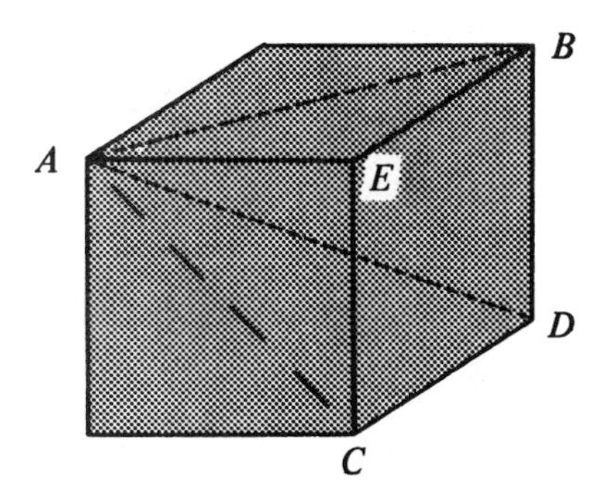

Figure 1.27

15. By interpreting $2x + 3y + 4z$ as a scalar product, show that $2\mathbf{i} + 3\mathbf{j} + 4\mathbf{k}$ is perpendicular to the plane $2x + 3y + 4z = 0$.

16. If $\mathbf{A}$ is a fixed nonzero vector, interpret geometrically $(\mathbf{R} - \mathbf{A}) \cdot \mathbf{R} = 0$,
(a) in the plane, $\mathbf{R} = x\mathbf{i} + y\mathbf{j}$.
(b) in space, $\mathbf{R} = x\mathbf{i} + y\mathbf{j} + z\mathbf{k}$.

17. If $\mathbf{u}$ and $\mathbf{v}$ are unit vectors, and θ is the angle between them, find $\frac{1}{2}|\mathbf{u} - \mathbf{v}|$ in terms of θ.

18. Let $\mathbf{A} = (\cos \phi)\mathbf{i} + (\sin \phi)\mathbf{j}$ and $\mathbf{B} = (\cos \theta)\mathbf{i} + (\sin \theta)\mathbf{j}$. Draw these vectors in the xy plane. By interpreting the scalar product $\mathbf{A} \cdot \mathbf{B}$ geometrically, prove that $\cos(\phi - \theta) = \cos \phi \cos \theta + \sin \phi \sin \theta$.

19. Prove, by vector methods, that the median from the vertex angle of an isosceles triangle is perpendicular to the base.

20. Prove the parallelogram equality, that is, the sum of the squares of the diagonals of a parallelogram equals the sum of the squares of its sides.

21. Prove the triangle inequality of section 1.3, $|\mathbf{A} + \mathbf{B}| \leq |\mathbf{A}| + |\mathbf{B}|$. (*Hint:* Square both sides, and use the scalar product.)

22. Prove that $|\mathbf{A}|\mathbf{B} + |\mathbf{B}|\mathbf{A}$ is orthogonal to $|\mathbf{A}|\mathbf{B} - |\mathbf{B}|\mathbf{A}$, for any vectors $\mathbf{A}$ and $\mathbf{B}$.

23. Consider the cube in figure 1.27. Find the angles between
(a) the face diagonals AB and AC.
(b) the principal diagonal AD and the face diagonal AB.
(c) the principal diagonal AD and the edge AE.

24. Suppose the line l_1 passes through the points $(5,1,-2)$ and $(2,-3,1)$, and the line l_2 passes through $(3,8,1)$ and $(-3,0,7)$. Are these lines perpendicular, parallel, coincident, or none of these?

25. Prove: the diagonals of a rectangle are perpendicular if and only if the rectangle is a square.

26. Describe the set of points located by $\mathbf{R}$ such that $(\mathbf{R} - \mathbf{a}) \cdot (\mathbf{R} + \mathbf{a}) = 0$, where $\mathbf{a}$ is fixed. (*Hint:* Draw a diagram.)

27. Prove: the sum of the squares of the sides of any quadrilateral, minus the sum of the squares of the two diagonals, equals four times the square of the distance between the midpoints of the diagonals.

28. By vector methods, prove that the angle subtended at the circumference by a diameter of a circle is a right angle.

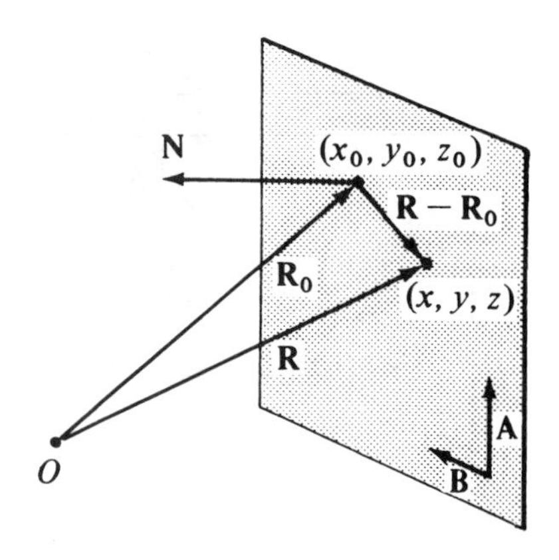

Figure 1.28

1.10 Equations of a Plane

Recall that in section 1.8 we specified a straight line by giving a point on the line and a vector parallel to the line. By analogy, then, we might specify a plane by giving a point (x_0,y_0,z_0) in the plane, and *two* vectors **A** and **B** parallel to the plane. Of course, **A** and **B** must not be parallel to each other. Introducing the position vectors $\mathbf{R}_0 = x_0\mathbf{i} + y_0\mathbf{j} + z_0\mathbf{k}$ and $\mathbf{R} = x\mathbf{i} + y\mathbf{j} + z\mathbf{k}$, we seek the condition on **R** guaranteeing that (x,y,z) lies in the plane. It is not immediately obvious how **R** depends on $\mathbf{R}_0$, **A**, and **B**, but, clearly, the "relative vector" $\mathbf{R} - \mathbf{R}_0$ must lie in the plane (more precisely, it has a representative that lies in the plane; fig. 1.28); hence it can be expressed as a combination of **A** and **B.** Thus we have

$$\mathbf{R} - \mathbf{R}_0 = s\mathbf{A} + t\mathbf{B}$$

for some scalars s and t, each taking values between $-\infty$ and $+\infty$.

These scalars s and t play a similar role to the single parameter t in the equation [eq. (1.9)] for a straight line. The need for *two* parameters to locate a point in a plane is indicative of the fact that a plane is a *two*-dimensional object. The vectors **A** and **B** are said to *span the plane.*

The experience we have gained in deriving this parametric equation for a plane will be helpful in chapter 4, when we analyze other two-dimensional surfaces. The fact of the matter, however, is that we can derive a *nonparametric* equation that is much simpler, and the above parametric form is almost never used. So let us start afresh and try a different tack.

The key to the nonparametric description is the observation that, instead of specifying *two* vectors **A** and **B** lying in the plane, it suffices to give *one* vector **N** *that is perpendicular, or normal, to the plane.* Given a point $(\mathbf{R}_0)$ in the plane and a direction (**N**) normal to the plane, one can reconstruct the plane unambiguously.

The condition that $\mathbf{R}$ is the position vector of a point in the plane can be expressed by saying that the "relative vector" $\mathbf{R} - \mathbf{R}_0$, which lies in the plane as before, is perpendicular to $\mathbf{N}$ (see fig. 1.28 again). According to the previous section, this condition can be written

$$(\mathbf{R} - \mathbf{R}_0) \cdot \mathbf{N} = 0 \tag{1.16}$$

Conversely, if eq. (1.16) is satisfied, then $\mathbf{R} - \mathbf{R}_0$ is perpendicular to $\mathbf{N}$. This ensures that $\mathbf{R}$ is the position vector of a point in the plane.

Hence eq. (1.16) is a vector equation describing the plane. In terms of the components of $\mathbf{N} = a\mathbf{i} + b\mathbf{j} + c\mathbf{k}$, it becomes

$$a(x - x_0) + b(y - y_0) + c(z - z_0) = 0 \tag{1.17}$$

By lumping the constant terms, this can be written

$$ax + by + cz = d \tag{1.18}$$

where $d = ax_0 + by_0 + cz_0$.

Example 1.20 Find an equation of the plane passing through $(1,3,-6)$ perpendicular to the vector $3\mathbf{i} - 2\mathbf{j} + 7\mathbf{k}$.

Solution By eq. (1.17) we can write the equation at once: $3(x - 1) - 2(y - 3) + 7(z + 6) = 0$. This can be simplified to $3x - 2y + 7z = -45$.

Example 1.21 Find an equation of the plane passing through $(1,2,3)$ perpendicular to the line

$$\frac{x - 1}{4} = \frac{y}{5} = \frac{z + 5}{6}$$

Solution We recall from section 1.8 that we can find a *vector* parallel to the given line by reading off the coefficients in the denominators: $4\mathbf{i} + 5\mathbf{j} + 6\mathbf{k}$. This vector is perpendicular to the desired plane, and therefore the equation of the plane is $4(x - 1) + 5(y - 2) + 6(z - 3) = 0$.

To be logically complete we should show that any equation of the form (1.18) does represent a plane, with normal $\mathbf{N} = a\mathbf{i} + b\mathbf{j} + c\mathbf{k}$ (assumed nonzero). This is straightforward: let $\mathbf{R}_0$ be the position vector of some point satisfying eq. (1.18), as, for instance, $(d/c)\mathbf{k}$. Then if $\mathbf{R}$ also satisfies eq. (1.18), we have $\mathbf{R} \cdot \mathbf{N} = d = \mathbf{R}_0 \cdot \mathbf{N}$, so $(\mathbf{R} - \mathbf{R}_0) \cdot \mathbf{N} = 0$, and we have recovered the form (1.16).

Example 1.22 Find a unit vector perpendicular to the plane $2x + y - 2z = 7$.

Solution Reading off the coefficients, we see that $2\mathbf{i} + \mathbf{j} - 2\mathbf{k}$ is perpendicular to the plane. Its magnitude is 3, so the desired unit vector is $\frac{2}{3}\mathbf{i} + \frac{1}{3}\mathbf{j} - \frac{2}{3}\mathbf{k}$. The negative of this vector is also a correct answer.

Example 1.23 Find the angle between the two planes $3x + 4y = 0$ and $2x + y - 2z = 5$.

Solution The desired angle equals the angle between the normals $\mathbf{N}_1 = 3\mathbf{i} + 4\mathbf{j}$ and $\mathbf{N}_2 = 2\mathbf{i} + \mathbf{j} - 2\mathbf{k}$. By the methods of section 1.9

$$\cos\theta = \frac{\mathbf{N}_1 \cdot \mathbf{N}_2}{|\mathbf{N}_1||\mathbf{N}_2|} = \frac{6 + 4}{(5)(3)} = \frac{2}{3}$$

The desired angle is approximately 48°.

Example 1.24 In books on analytic geometry it is shown that the distance between an arbitrary point (x_1,y_1,z_1) and the plane $ax + by + cz = d$ is given by the expression

$$\frac{|ax_1 + by_1 + cz_1 - d|}{(a^2 + b^2 + c^2)^{1/2}}$$

Derive this expression by vector methods.

Solution Let $\mathbf{R}_0$ be the position vector of a point in the plane, and let $\mathbf{R}_1 = x_1\mathbf{i} + y_1\mathbf{j} + z_1\mathbf{k}$ and $\mathbf{N} = a\mathbf{i} + b\mathbf{j} + c\mathbf{k}$. The desired distance is the absolute value (distance is never negative!) of the component of $\mathbf{R}_1 - \mathbf{R}_0$ in the direction of $\mathbf{N}$. Hence this distance is

$$\frac{|(\mathbf{R}_1 - \mathbf{R}_0) \cdot \mathbf{N}|}{|\mathbf{N}|} = \frac{|\mathbf{R}_1 \cdot \mathbf{N} - d|}{|\mathbf{N}|}$$

which, written out in terms of components, is the expression given above.

Example 1.25 Find the distance between the parallel planes $x + y + z = 5$ and $x + y + z = 10$.

Solution Take an arbitrary point in the first plane, say (1,1,3), and find its distance to the second plane by the expression derived in example 1.24. We obtain

$$\frac{|5 - 10|}{\sqrt{3}} = \frac{5\sqrt{3}}{3}$$

EXERCISES

1. Find unit vectors normal to the planes
   (a) $2x + y + 2z = 8$
   (b) $4x - 4z = 0$
   (c) $-y + 6z = 0$
   (d) $x = 5$
   (e) $y = z + 2$
   (f) $x = y$
2. Find an equation of the plane through the origin perpendicular to $2\mathbf{i} - 8\mathbf{j} + 2\mathbf{k}$.

3. Find an equation of the plane perpendicular to **D** and through P, where

$$\mathbf{D} = 10\mathbf{i} - 10\mathbf{j} + 5\mathbf{k}$$

and P is $(1,1,-3)$.

4. Find a plane crossing through $(1,3,3)$, parallel to the plane $3x + y - z = 8$.
5. Is it possible to find a plane perpendicular to both **i** and **j**?
6. By vector methods find the distance from the point $(3,4,7)$ to the plane $2x - y - 2z = 4$.
7. Find the distances between the pairs of planes
   (a) $x + 2y + 3z = 5$ and $x + 2y + 3z = 19$
   (b) $x + y = 4$ and $x + y = 10$
   (c) $x = 5$ and $x = 7$ (no calculations needed here!)
8. Determine $\cos\theta$, where θ is the angle between the planes $x + y + z = 0$ and $x = 0$.
9. By vector methods show that the line $x = y = \frac{1}{3}(z + 2)$ is parallel to the plane $2x - 8y + 2z = 5$.
10. The vertices of a regular tetrahedron are $OABC$. Prove that the vector $OA + OB + OC$ is perpendicular to the plane ABC.
11. Find the angle that the plane OAB makes with the z axis, if A is the point $(1,3,2)$ and B is $(2,1,1)$.
12. Given the points $O(0,0,0)$, $A(1,2,3)$, $B(0,-1,1)$, $C(2,0,2)$
    (a) Find a vector perpendicular to the plane OAB.
    (b) Find the distance from C to the plane OAB.
13. By vector methods find the angle between the line $x = y = 2z$ and the plane $x + y + z = 0$.
14. Find the angle between the plane $x + y + z = 21$ and the line $x - 1 = y + 2 = 2z + 3$.
15. Find the equation of a line in the xy plane perpendicular to the vector $3\mathbf{i} - \mathbf{j}$.
16. Find the distance between the lines $x + y = 0$ and $x + y = 5$ in the xy plane.
17. Find a line in the xy plane parallel to $3x + 2y = 4$ passing through the point $(3,1)$.
18. Write the equation of the plane containing the lines

$$x = y = \frac{4 - z}{4} \qquad 2x = 2 - y = z$$

19. We are given two distinct parallel planes and are told the distance between the planes is d. A vector **v** is perpendicular to the planes, and its magnitude is $1/d$. The planes intersect the y axis in the points $(0,1,0)$ and $(0,4,0)$ respectively. What is the y component of **v**? (There are two possible answers, depending on the two possible directions of **v**.)
20. Find the distance from the point $A(3,7,2)$ to the plane passing through $B(5,10,8)$ that is perpendicular to the line AB.
21. Find the distance from the origin to the plane through $(3,2,6)$ that is perpendicular to the z axis.

22. Find the distance from the origin to the plane passing through (3,4,2) that is perpendicular to the line joining (1,2,3) and (3,5,9).

23. A plane has intercepts (4,0,0), (0,6,0), and (0,0,12). Find the equations of another plane through (6,−2,4) that is parallel to this plane.

24. Find the intersection of the following geometric objects:
   (a) the plane $3x + 2y - z = -9$ and the line $\frac{1}{2}x = y - 2 = -\frac{1}{4}(z - 1)$
   (b) the plane $x + y + 2z = 6$ and the line $-x = 2y = 4z + 1$
   (c) the plane $3x - y + z = 3$ and the plane $2x + z = 0$
   (d) the plane $x - y + 2z = 4$ and the plane $-2x + 2y - 4z = 1$

25. What is the distance from the origin to the plane intersecting the x, y, and z axes at $x = a$, $y = b$, and $z = c$ respectively?

26. Find the distance between (1,2,3) and the plane $2x - 2y + z = 4$.

27. Find the distance between the planes $2x + y + z = 2$ and $2x + y + z = 4$.

28. In the study of crystals with cubic symmetry the *Miller indices* of a plane are numbers (h,k,l) proportional to the reciprocals of the intercepts of the plane with the x, y, and z axes. Write down a normal vector to a plane with Miller indices (h,k,l).

1.11 Orientation

In working in the xy plane, it is conventional to take the positive x direction to the right and the positive y direction upward. Angles are then taken to be *positive* in the *counterclockwise* direction.

When working with planes in space, there is no generally accepted convention for determining the positive sense for angles. The choice is quite arbitrary. Given any plane in space, we may arbitrarily decree in which direction we shall consider angles to be positive. The plane is then said to be oriented.

One way of orienting a plane is as follows. Let **A** and **B** be nonzero vectors, not parallel, represented by arrows in the given plane. Let these arrows extend from the same point. Let **A** be rotated through the smallest angle possible to coincide in direction with **B.** The sense of this rotation is then said to be "positive" and the plane is thereby oriented. *The plane is oriented by giving the vectors* **A, B,** *in that order.*

For example, the usual orientation of the xy plane is obtained by giving the vectors **i, j,** in that order. By a 90° rotation the direction of **i** can be made to coincide with that of **j,** and this rotation has the conventional "positive" sense. We obtain the same orientation by giving the vectors **i** + **j** and **j,** in that order (fig. 1.29). On the other hand, if we specified the orientation by giving **j, i,** in that order, we would obtain the opposite orientation, whereby angles would be measured positive in the clockwise sense (which is not conventional, but is perfectly satisfactory).

Another way of orienting a plane is as follows. Let there be given a single vector that is not parallel to the plane. Let this vector be represented by an arrow that has

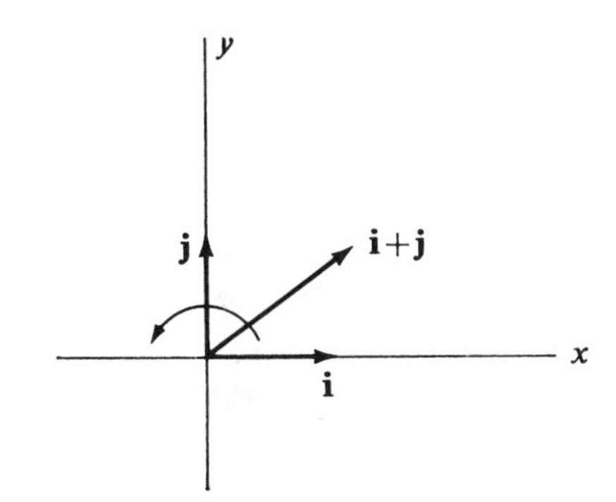

Figure 1.29

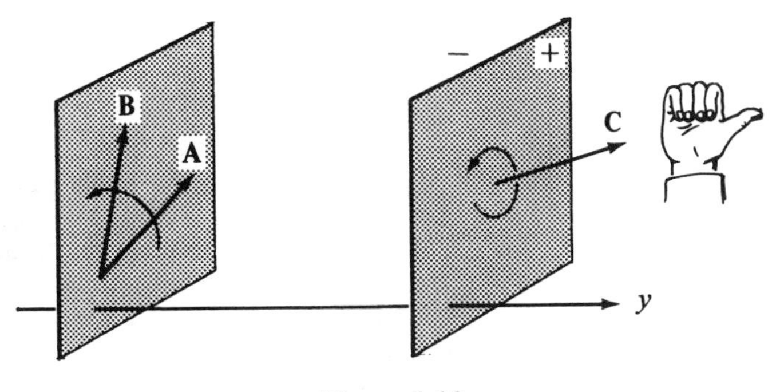

Figure 1.30

its initial point in the plane. Then the terminal point of the arrow will be on one side of the plane, which we call (arbitrarily) the *positive* side. If we imagine the right hand grasping the given vector, with the thumb pointing in the direction of the arrowhead, then the fingers will curl around the shank of the arrow in the positive sense of rotation in the plane.

In figure 1.30 both methods of orienting a plane are illustrated for planes perpendicular to the y axis. At the left, the plane is oriented by prescribing two vectors in the plane, **A** and **B**, in that order. On the right, the same orientation is achieved by prescribing a vector **C** extending from a point in the plane.

Now let **A**, **B**, and **C** be nonzero vectors, not all parallel to the same plane, represented by arrows with initial points at the origin (fig. 1.31). The vectors **A** and **B** determine a plane passing through the origin. If the orientation of this plane, as determined by **A**, **B**, in that order, is identical to its orientation as determined by **C**, we say that **A**, **B**, and **C**, in that order, form a *right-handed system*. One reason for this terminology is that if the thumb and first two fingers of the right hand are held so that they are mutually perpendicular, the thumb, forefinger, and second finger

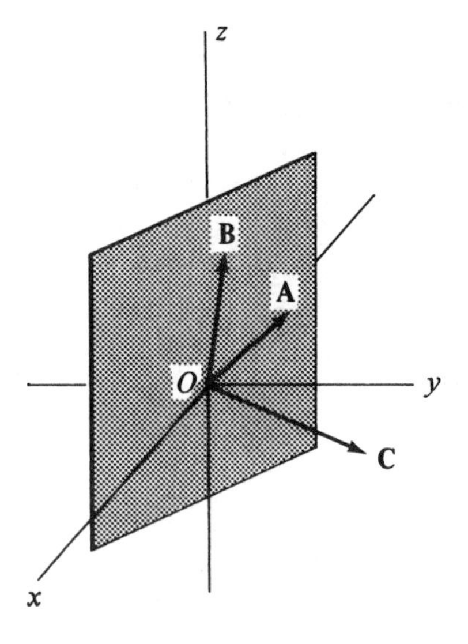

Figure 1.31

form such a system. Another reason is that if **A, B,** and **C,** in that order, form a *right-handed* system, the rotation of **A** into **B** (through an angle less than 180°) will advance a right-handed screw into the general direction of **C** (regardless of which way the screw is pointing!). The vectors **A, B,** and **C** of figure 1.30 form a right-handed system, as do the vectors **i, j,** and **k** for the coordinate axes depicted in figure 1.31; this is a right-handed coordinate system.

EXERCISE

If an oriented plane area is represented by a vector perpendicular to the area, with magnitude numerically equal to the area, what is the geometrical significance of the components of the vector?

1.12 Vector Products

We have seen that the scalar product of two vectors **A** and **B** can be interpreted as the length of **A** times the component of **B** parallel to **A;** in mechanics, it expresses the work done by a force **B** exerted through a displacement **A,** and it is also a very useful tool in analytic geometry. So we are naturally led to explore the possible

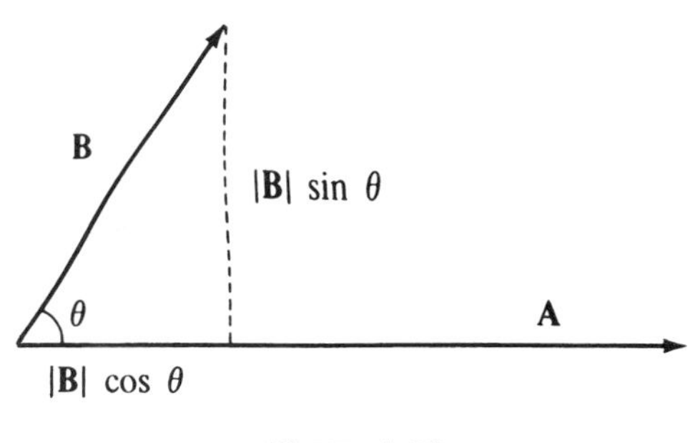

Figure 1.32

advantages of defining another kind of product, given by the length of **A** times the component of **B** *perpendicular* to **A** (i.e., $|\mathbf{B}| \sin \theta$ in fig. 1.32). Mechanics again lends a provocative interpretation to this operation.

Let us suppose we have a rigid body and, for purposes of reference, we define a right-handed coordinate system fixed in this body. We interpret **B** as a force applied to the body at the point located by the vector **A** (relative to the origin, held stationary). Observe that the component of this force *perpendicular* to **A** tends to *rotate* the body about the axis normal to the plane of **A** and **B.** The rotational effect of this force is enhanced if the point of application is moved further from the origin, increasing the "leverage" of the force. In fact, the overall effect is measured by the vector product we just proposed. Consequently, in physics the *torque* due to the force **B** applied at the point **A** is defined to be a vector whose magnitude is this product ("lever arm times perpendicular force"), and *whose direction is perpendicular to the plane of* **A** *and* **B,** so that **A, B,** and the torque vector form a right-handed system (i.e., if the fingers of the right hand rotate **A** into **B** as in fig. 1.30, the extended thumb gives the direction of the torque).

Motivated by these considerations, we now define the *vector product* of **A** and **B** to be the vector

$$\mathbf{A} \times \mathbf{B} = |\mathbf{A}||\mathbf{B}| \sin \theta \, \mathbf{n}$$

where θ is the angle between the vectors, and the unit vector **n** is perpendicular to both **A** and **B,** with **A, B,** and **n** forming a right-handed system (see fig. 1.33). Sometimes $\mathbf{A} \times \mathbf{B}$ is called the *cross product.*

Notice that $|\mathbf{A} \times \mathbf{B}|$ is the area of the parallelogram determined by **A** and **B** (computed as base times height). Observe further that because of the rule determining the direction of **n,** we have

$$\mathbf{A} \times \mathbf{B} = -\mathbf{B} \times \mathbf{A}$$

From these geometric considerations we see that *if two vectors are parallel, their vector product is zero.* Of course, $\mathbf{A} \times \mathbf{B}$ is also zero if either **A** or **B** is zero.

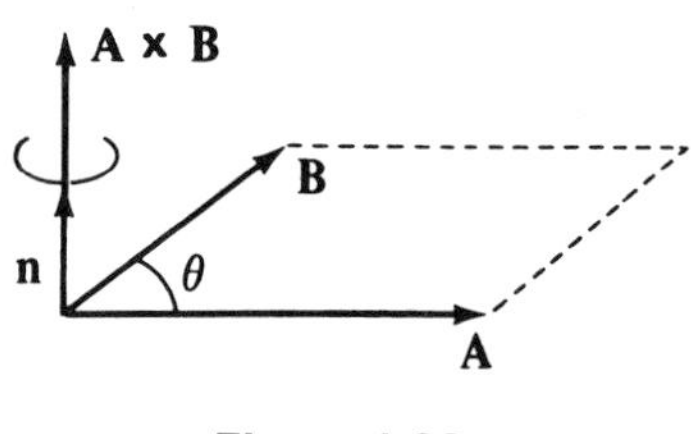

Figure 1.33

As particular instances of the definition, note that

$$\mathbf{i} \times \mathbf{i} = \mathbf{j} \times \mathbf{j} = \mathbf{k} \times \mathbf{k} = \mathbf{0}$$

$$\mathbf{i} \times \mathbf{j} = \mathbf{k} \qquad \mathbf{j} \times \mathbf{k} = \mathbf{i} \qquad \mathbf{k} \times \mathbf{i} = \mathbf{j}$$

$$\mathbf{i} \times \mathbf{k} = -\mathbf{j} \qquad \mathbf{j} \times \mathbf{i} = -\mathbf{k} \qquad \mathbf{k} \times \mathbf{j} = -\mathbf{i}$$

As with the scalar product, it is convenient to have a representation of $\mathbf{A} \times \mathbf{B}$ in terms of the components of **A** and **B.** The derivation of such a formula hinges on the validity of the distributive laws for the vector product, that is,

$$\mathbf{A} \times (\mathbf{B} + \mathbf{C}) = \mathbf{A} \times \mathbf{B} + \mathbf{A} \times \mathbf{C} \tag{1.19}$$

$$(\mathbf{A} + \mathbf{B}) \times \mathbf{C} = \mathbf{A} \times \mathbf{C} + \mathbf{B} \times \mathbf{C} \tag{1.20}$$

The proof of the distributive laws appears at the end of this section as optional reading. If we accept the laws for now, we can compute the componentwise expression for $\mathbf{A} \times \mathbf{B}$ easily:

$$\begin{aligned}\mathbf{A} \times \mathbf{B} &= (A_1\mathbf{i} + A_2\mathbf{j} + A_3\mathbf{k}) \times (B_1\mathbf{i} + B_2\mathbf{j} + B_3\mathbf{k}) \\ &= A_1\mathbf{i} \times B_1\mathbf{i} + A_2\mathbf{j} \times B_1\mathbf{i} + A_3\mathbf{k} \times B_1\mathbf{i} \\ &\quad + A_1\mathbf{i} \times B_2\mathbf{j} + A_2\mathbf{j} \times B_2\mathbf{j} + A_3\mathbf{k} \times B_2\mathbf{j} \\ &\quad + A_1\mathbf{i} \times B_3\mathbf{k} + A_2\mathbf{j} \times B_3\mathbf{k} + A_3\mathbf{k} \times B_3\mathbf{k}\end{aligned}$$

The vector products in this expression are easy to evaluate from the definition: $A_1\mathbf{i} \times B_1\mathbf{i} = \mathbf{0}$, $A_2\mathbf{j} \times B_1\mathbf{i} = -A_2B_1\mathbf{k}$, etc. Thus we finally arrive at the componentwise expression for the vector product:

$$\mathbf{A} \times \mathbf{B} = (A_2B_3 - A_3B_2)\mathbf{i} + (A_3B_1 - A_1B_3)\mathbf{j} + (A_1B_2 - A_2B_1)\mathbf{k} \tag{1.21}$$

This formula may be conveniently memorized in determinant form:

$$\mathbf{A} \times \mathbf{B} = \begin{vmatrix} \mathbf{i} & \mathbf{j} & \mathbf{k} \\ A_1 & A_2 & A_3 \\ B_1 & B_2 & B_3 \end{vmatrix} \tag{1.21$'$}$$

This symbolic determinant is interpreted to be the vector whose x, y, and z components are the cofactors, respectively, of the first, second, and third entries in the first row.

Example 1.26 Find the vector product $\mathbf{A} \times \mathbf{B}$ if $\mathbf{A} = 3\mathbf{i} + 4\mathbf{j}$ and $\mathbf{B} = \mathbf{i} - 2\mathbf{j} + 5\mathbf{k}$.

Solution

$$\mathbf{A} \times \mathbf{B} = \begin{vmatrix} \mathbf{i} & \mathbf{j} & \mathbf{k} \\ 3 & 4 & 0 \\ 1 & -2 & 5 \end{vmatrix} = 20\mathbf{i} - 15\mathbf{j} - 10\mathbf{k}$$

For convenience, we list the algebraic properties of the vector product here:

$$\mathbf{A} \times \mathbf{B} = -(\mathbf{B} \times \mathbf{A})$$
$$(s\mathbf{A} + \mathbf{B}) \times \mathbf{C} = s(\mathbf{A} \times \mathbf{C}) + (\mathbf{B} \times \mathbf{C})$$
$$\mathbf{A} \times (s\mathbf{B} + \mathbf{C}) = s(\mathbf{A} \times \mathbf{B}) + (\mathbf{A} \times \mathbf{C})$$

Example 1.27 Find two unit vectors perpendicular to both $\mathbf{A} = 2\mathbf{i} + 2\mathbf{j} - 3\mathbf{k}$ and $\mathbf{B} = \mathbf{i} + 3\mathbf{j} + \mathbf{k}$.

Solution We have seen that $\mathbf{A} \times \mathbf{B}$ is perpendicular to both $\mathbf{A}$ and $\mathbf{B}$. We have

$$\mathbf{A} \times \mathbf{B} = \begin{vmatrix} \mathbf{i} & \mathbf{j} & \mathbf{k} \\ 2 & 2 & -3 \\ 1 & 3 & 1 \end{vmatrix} = 11\mathbf{i} - 5\mathbf{j} + 4\mathbf{k}$$

The length of this vector is $9\sqrt{2}$. The desired *unit* vector is therefore

$$\mathbf{n} = \frac{11}{9\sqrt{2}}\mathbf{i} - \frac{5}{9\sqrt{2}}\mathbf{j} + \frac{4}{9\sqrt{2}}\mathbf{k}$$

If we had taken $\mathbf{B} \times \mathbf{A}$ instead, we would have obtained the negative of this vector. The two answers are

$$\pm\left(\frac{11\sqrt{2}}{18}\mathbf{i} - \frac{5\sqrt{2}}{18}\mathbf{j} + \frac{2\sqrt{2}}{9}\mathbf{k}\right)$$

Example 1.28 Find the area of the parallelogram determined by $\mathbf{A} = \mathbf{i} + \mathbf{j} - 3\mathbf{k}$ and $\mathbf{B} = -6\mathbf{j} + 5\mathbf{k}$.

Solution

$$\mathbf{A} \times \mathbf{B} = \begin{vmatrix} \mathbf{i} & \mathbf{j} & \mathbf{k} \\ 1 & 1 & -3 \\ 0 & -6 & 5 \end{vmatrix} = -13\mathbf{i} - 5\mathbf{j} - 6\mathbf{k}$$

$$|\mathbf{A} \times \mathbf{B}| = \sqrt{13^2 + 5^2 + 6^2} = \sqrt{230}$$

which is the desired area.

Example 1.29 Find the equations of the line passing through $(3,2,-4)$ parallel to the line of intersection of the two planes $x + 3y - 2z = 8$, $x - 3y + z = 0$.

Solution Observe that $\mathbf{A} = \mathbf{i} + 3\mathbf{j} - 2\mathbf{k}$ and $\mathbf{B} = \mathbf{i} - 3\mathbf{j} + \mathbf{k}$ are the normals to the planes, and $\mathbf{A} \times \mathbf{B}$ is perpendicular to both $\mathbf{A}$ and $\mathbf{B}$. It follows that $\mathbf{A} \times \mathbf{B}$ is parallel to both planes. Hence $\mathbf{A} \times \mathbf{B}$ is parallel to the line of intersection. We have

$$\mathbf{A} \times \mathbf{B} = \begin{vmatrix} \mathbf{i} & \mathbf{j} & \mathbf{k} \\ 1 & 3 & -2 \\ 1 & -3 & 1 \end{vmatrix} = -3\mathbf{i} - 3\mathbf{j} - 6\mathbf{k}$$

Equations of the desired line are

$$\frac{x-3}{-3} = \frac{y-2}{-3} = \frac{z+4}{-6}$$

or, equivalently,

$$x - 3 = y - 2 = \frac{z+4}{2}$$

Now consider a rigid body rotating about a fixed axis with constant angular speed ω. The *angular velocity* is represented by a vector $\boldsymbol{\omega}$ of magnitude ω extending along the axis of rotation with sense determined by the right-hand rule: if the fingers of the right hand are wrapped about the axis in the direction of rotation, the thumb points in the direction of $\boldsymbol{\omega}$ (fig. 1.34).

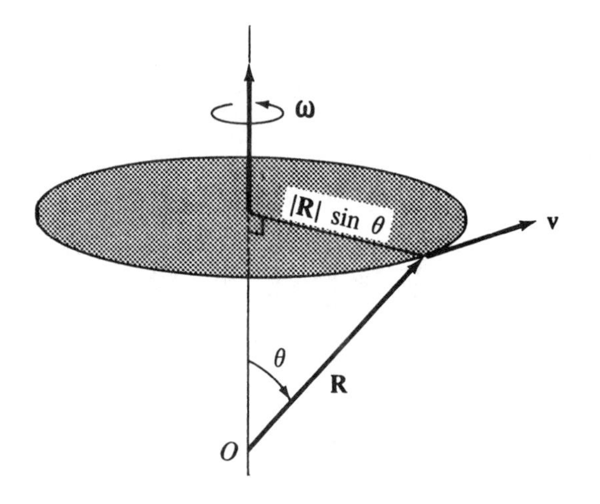

Figure 1.34

Let us assume that the origin O is on the axis of rotation, and let $\mathbf{R}$ denote the position vector of a particle in the body. Then the velocity $\mathbf{v}$ of the particle is given by

$$\mathbf{v} = \boldsymbol{\omega} \times \mathbf{R} \tag{1.22}$$

To see this, we first note that $|\mathbf{R}| \sin \theta$ is the distance of the particle from the axis of rotation, so $\mathbf{v}$ has magnitude $\omega|\mathbf{R}| \sin \theta$. Moreover, the velocity $\mathbf{v}$ is necessarily perpendicular to both $\mathbf{R}$ and $\boldsymbol{\omega}$, and the sense of $\boldsymbol{\omega}$ is such that $\mathbf{v}$ equals $\boldsymbol{\omega} \times \mathbf{R}$ rather than $\mathbf{R} \times \boldsymbol{\omega}$, as we see from figure 1.34.

Example 1.30 A rigid body rotates with constant angular velocity $\boldsymbol{\omega}$ about the line $x = y/2 = z/2$. Find the speed of a particle at the instant it passes through the point (2,3,5).

Solution The vector $\mathbf{i} + 2\mathbf{j} + 2\mathbf{k}$ is parallel to the axis. A unit vector parallel to the axis is $\frac{1}{3}\mathbf{i} + \frac{2}{3}\mathbf{j} + \frac{2}{3}\mathbf{k}$. Therefore

$$\boldsymbol{\omega} = \pm\omega(\tfrac{1}{3}\mathbf{i} + \tfrac{2}{3}\mathbf{j} + \tfrac{2}{3}\mathbf{k})$$

(The statement of the problem leaves the sign ambiguous.) The velocity is

$$\mathbf{v} = \boldsymbol{\omega} \times \mathbf{R} = \pm\omega \begin{vmatrix} \mathbf{i} & \mathbf{j} & \mathbf{k} \\ \frac{1}{3} & \frac{2}{3} & \frac{2}{3} \\ 2 & 3 & 5 \end{vmatrix} = \pm\omega(\tfrac{4}{3}\mathbf{i} - \tfrac{1}{3}\mathbf{j} - \tfrac{1}{3}\mathbf{k})$$

The speed is

$$|\mathbf{v}| = \omega(\tfrac{16}{9} + \tfrac{1}{9} + \tfrac{1}{9})^{1/2} = \sqrt{2}\omega$$

Optional Reading: The Proof of the Distributive Laws

Observe that we have to prove only eq. (1.19); eq. (1.20) will then follow since

$$\begin{aligned}(\mathbf{A} + \mathbf{B}) \times \mathbf{C} &= -\mathbf{C} \times (\mathbf{A} + \mathbf{B}) \\ &= -(\mathbf{C} \times \mathbf{A} + \mathbf{C} \times \mathbf{B}) \\ &= \mathbf{A} \times \mathbf{C} + \mathbf{B} \times \mathbf{C}\end{aligned}$$

We begin by proving eq. (1.19) in the special case where $\mathbf{B}$ and $\mathbf{C}$ are both perpendicular to $\mathbf{A}$; then, of course, $(\mathbf{B} + \mathbf{C})$ is also. In this case it follows from the definition of the vector product that $\mathbf{A} \times \mathbf{B}$ is a vector that can be formed from the vector $\mathbf{B}$ by multiplying its length by the factor $|\mathbf{A}|$, and rotating it counterclockwise through 90° about $\mathbf{A}$ as an axis. In figure 1.35 think of the vector $\mathbf{A}$ as perpendicular

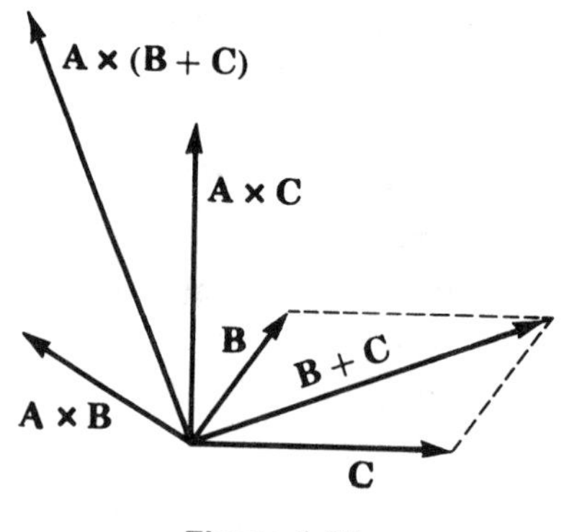

Figure 1.35

to the page, pointing to the reader. Then **B**, **C**, and **B** + **C** all lie in the plane of the page, as do the "rescaled and rotated" vectors **A** × **B**, **A** × **C**, and **A** × (**B** + **C**).

Now eq. (1.19) makes a claim about the sum of vectors; geometrically, it can be interpreted as saying that **A** × (**B** + **C**) is the diagonal of the parallelogram whose sides are **A** × **B** and **A** × **C**. This can be seen by considering the similar triangles resulting from the equal angles and proportional sides in figure 1.35.

To prove eq. (1.19) in the general case, with no assumptions about the directions of the vectors, we resolve **B** and **C** into their vector components parallel and perpendicular to **A**, as in section 1.10 (recall fig. 1.25):

$$\mathbf{B} = \mathbf{B}_{\parallel} + \mathbf{B}_{\perp} \qquad \mathbf{C} = \mathbf{C}_{\parallel} + \mathbf{C}_{\perp}$$

Then it follows from the definition of vector product that

$$\mathbf{A} \times \mathbf{B} = \mathbf{A} \times \mathbf{B}_{\perp} \qquad \mathbf{A} \times \mathbf{C} = \mathbf{A} \times \mathbf{C}_{\perp}$$

(Think this over: neither the direction nor magnitude of **A** × **B** is changed if we replace **B** by $\mathbf{B}_{\perp}$.) Furthermore, it is easy to see that the identity

$$\mathbf{B} + \mathbf{C} = (\mathbf{B}_{\parallel} + \mathbf{C}_{\parallel}) + (\mathbf{B}_{\perp} + \mathbf{C}_{\perp})$$

resolves the sum **B** + **C** into vector components parallel and perpendicular to **A**, and therefore

$$\mathbf{A} \times (\mathbf{B} + \mathbf{C}) = \mathbf{A} \times (\mathbf{B}_{\perp} + \mathbf{C}_{\perp})$$

Since we have proved the validity of eq. (1.19) for vectors perpendicular to **A**, the general validity is seen as follows:

$$\mathbf{A} \times (\mathbf{B} + \mathbf{C}) = \mathbf{A} \times (\mathbf{B}_{\perp} + \mathbf{C}_{\perp}) = \mathbf{A} \times \mathbf{B}_{\perp} + \mathbf{A} \times \mathbf{C}_{\perp} = \mathbf{A} \times \mathbf{B} + \mathbf{A} \times \mathbf{C}$$

(Another proof is outlined in exercise 19 of section 1.14.)

Summary: Multiplying vectors

Now we have defined all the essential elements of vector algebra; let us review their interpretations and applications.

We started by learning how to add two vectors; the sum has the usual algebraic properties of commutativity and associativity, and it is compatible with scalar multiplication.

The effect of multiplying two vectors is rather more involved. We have defined two kinds of multiplication, and they have quite different properties. If we multiply two vectors by the scalar product, the result is not a vector—it is a scalar. If we multiply by the vector product, the result is a vector, but its direction is quite distinct from the directions of the original vectors—perpendicular to both, in fact. Furthermore, it depends on the order of the original vectors, changing sign when we switch the order.

The geometric formula for the scalar product of **A** and **B** is

$$\mathbf{A} \cdot \mathbf{B} = |\mathbf{A}||\mathbf{B}| \cos \theta$$

where θ is the angle between the vectors, while the componentwise expression is

$$\mathbf{A} \cdot \mathbf{B} = A_1B_1 + A_2B_2 + A_3B_3$$

For the vector product, the geometric formula is

$$\mathbf{A} \times \mathbf{B} = |\mathbf{A}||\mathbf{B}| \sin \theta \, \mathbf{n}$$

where **n** is the unit vector perpendicular to **A** and **B** so that **A**, **B**, and **n** form a right-handed system; and the componentwise expression is most conveniently represented by

$$\mathbf{A} \times \mathbf{B} = \begin{vmatrix} \mathbf{i} & \mathbf{j} & \mathbf{k} \\ A_1 & A_2 & A_3 \\ B_1 & B_2 & B_3 \end{vmatrix}$$

From the geometric formulas, we saw that a zero scalar product is a test for orthogonality, while a zero vector product is an indication of parallelism.

Example 1.31 Derive the nonparametric equations for the straight line passing through $\mathbf{R}_0 = x_0\mathbf{i} + y_0\mathbf{j} + z_0\mathbf{k}$, and parallel to $\mathbf{V} = a\mathbf{i} + b\mathbf{j} + c\mathbf{k}$, using the vector product.

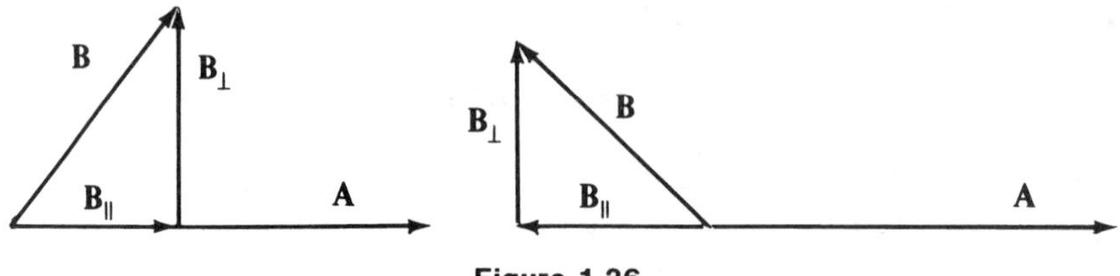

Figure 1.36

Solution Recall that in section 1.8 we observed that $\mathbf{R} = x\mathbf{i} + y\mathbf{j} + z\mathbf{k}$ would be the position vector of a point on the line if $\mathbf{R} - \mathbf{R}_0$ was parallel to $\mathbf{V}$. Setting the vector product equal to zero,

$$(\mathbf{R} - \mathbf{R}_0) \times \mathbf{V} = \begin{vmatrix} \mathbf{i} & \mathbf{j} & \mathbf{k} \\ x - x_0 & y - y_0 & z - z_0 \\ a & b & c \end{vmatrix} = \mathbf{0}$$

we derive the equations

$$\begin{aligned} (x - x_0)b &= (y - y_0)a \\ (y - y_0)c &= (z - z_0)b \\ (x - x_0)c &= (z - z_0)a \end{aligned}$$

which are equivalent to eq. (1.10).

The geometric interpretations of the scalar and vector products can be visualized with the aid of the by-now familiar figure 1.36. We see that the length of the component of **B** parallel to **A** can be computed from the scalar product:

$$|\mathbf{B}_{\|}| = \frac{\mathbf{A} \cdot \mathbf{B}}{|\mathbf{A}|}$$

while the length of the perpendicular component is computed from the vector product:

$$|\mathbf{B}_{\perp}| = \frac{|\mathbf{A} \times \mathbf{B}|}{|\mathbf{A}|}$$

To express the *vector* $\mathbf{B}_{\|}$, we use a unit vector in the direction of **A**:

$$\mathbf{B}_{\|} = \frac{\mathbf{A} \cdot \mathbf{B}}{|\mathbf{A}|} \frac{\mathbf{A}}{|\mathbf{A}|} = \frac{\mathbf{A} \cdot \mathbf{B}}{\mathbf{A} \cdot \mathbf{A}} \mathbf{A}$$

while $\mathbf{B}_{\perp}$ can be computed as the difference

$$\mathbf{B}_{\perp} = \mathbf{B} - \mathbf{B}_{\|}$$

Example 1.32 Derive an expression for $\mathbf{B}_\perp$ directly in terms of **A** and **B**.

Solution Clearly we need an expression for a vector in the direction of $\mathbf{B}_\perp$. The key here is to analyze the vector $(\mathbf{A} \times \mathbf{B}) \times \mathbf{A}$.

Referring to figure 1.36, we see that $\mathbf{A} \times \mathbf{B}$ points toward the reader, perpendicular to the page. Now taking the vector product of this with **A**, we find that the resulting vector falls back in the plane of **A** and **B**, and in the direction of $\mathbf{B}_\perp$! Keeping in mind that the angle between $\mathbf{A} \times \mathbf{B}$ and **A** is 90°, we compute the length:

$$\begin{aligned} |(\mathbf{A} \times \mathbf{B}) \times \mathbf{A}| &= |\mathbf{A} \times \mathbf{B}||\mathbf{A}| \sin 90^\circ \\ &= (|\mathbf{A}||\mathbf{B}| \sin \theta)|\mathbf{A}|(1) \\ &= |\mathbf{A}|^2|\mathbf{B}| \sin \theta \end{aligned}$$

Since $|\mathbf{B}_\perp| = |\mathbf{B}| \sin \theta$, we have

$$\mathbf{B}_\perp = \frac{(\mathbf{A} \times \mathbf{B}) \times \mathbf{A}}{|\mathbf{A}|^2} = \frac{(\mathbf{A} \times \mathbf{B}) \times \mathbf{A}}{\mathbf{A} \cdot \mathbf{A}}$$

The parallel–perpendicular decomposition of **B** can thus be expressed

$$\mathbf{B} = \frac{\mathbf{A} \cdot \mathbf{B}}{\mathbf{A} \cdot \mathbf{A}} \mathbf{A} + \frac{(\mathbf{A} \times \mathbf{B}) \times \mathbf{A}}{\mathbf{A} \cdot \mathbf{A}}$$

EXERCISES

1. Find $\mathbf{A} \times \mathbf{B}$, where
   (a) $\mathbf{A} = 3\mathbf{i} - \mathbf{j} + 2\mathbf{k}$, $\mathbf{B} = \mathbf{i} + \mathbf{j} - 4\mathbf{k}$
   (b) $\mathbf{A} = 2\mathbf{i} + \mathbf{j} + 7\mathbf{k}$, $\mathbf{B} = 3\mathbf{i} + \mathbf{j} - \mathbf{k}$
   (c) $\mathbf{A} = \mathbf{j} + 6\mathbf{k}$, $\mathbf{B} = \mathbf{k} + 2\mathbf{j} - \mathbf{i}$
   (d) $\mathbf{A} = \mathbf{i}$, $\mathbf{B} = \mathbf{j}$
   (e) $\mathbf{B} \times \mathbf{A}$ is known to be $\mathbf{i} - \mathbf{j}$
2. Find the area of the parallelogram determined by $3\mathbf{i} + 4\mathbf{j}$ and $\mathbf{i} + \mathbf{j} + \mathbf{k}$.
3. Find the area of the triangle with vertices (1,1,2), (2,3,5), and (1,5,5).
4. Find $\mathbf{A} \times \mathbf{B}$ if $\mathbf{A} = \mathbf{i} - \mathbf{j} + \mathbf{k}$ and $\mathbf{B} = 3\mathbf{i} - 3\mathbf{j} + 3\mathbf{k}$. What is the geometrical significance of this answer?
5. Find a unit vector perpendicular to both $3\mathbf{i} + \mathbf{j}$ and $2\mathbf{i} - \mathbf{j} - 5\mathbf{k}$.
6. By vector methods, find the equations of the line through (2,3,7) parallel to the line of intersection of the planes $2x + y + z = 0$ and $x - y + 7z = 0$.
7. Find equations of a line perpendicular to the lines $x = y = z$ and $x = 2y = 3z$, passing through the origin.
8. Compute $(\mathbf{A} \times \mathbf{B}) \times \mathbf{C}$ and also $\mathbf{A} \times (\mathbf{B} \times \mathbf{C})$, given that $\mathbf{A} = 2\mathbf{i} + 2\mathbf{j}$, $\mathbf{B} = 3\mathbf{i} - \mathbf{j} + \mathbf{k}$, and $\mathbf{C} = 8\mathbf{i}$. Does the associative law hold for vector products?

9. By vector methods, determine the equation of the plane determined by the points (2,0,1), (1,1,3), and (4,7,−2).

10. Find a unit vector in the plane of the vectors $\mathbf{A} = \mathbf{i} + 2\mathbf{j}$ and $\mathbf{B} = \mathbf{j} + 2\mathbf{k}$, perpendicular to the vector $\mathbf{C} = 2\mathbf{i} + \mathbf{j} + 2\mathbf{k}$.

11. By taking the vector cross product of $(\cos\theta)\mathbf{i} + (\sin\theta)\mathbf{j}$ and $(\cos\psi)\mathbf{i} + (\sin\psi)\mathbf{j}$ and interpreting geometrically, derive a well-known trigonometric identity.

12. If **A**, **B**, and **C** are vectors from the origin to points A, B, and C respectively, show that $(\mathbf{A} \times \mathbf{B}) + (\mathbf{B} \times \mathbf{C}) + (\mathbf{C} \times \mathbf{A})$ is perpendicular to the plane ABC. [*Hint:* Consider $(\mathbf{B} - \mathbf{A}) \times (\mathbf{C} - \mathbf{A})$.]

13. Find the distance from the point (5,7,14) to the line passing through (2,3,8) and (3,6,12). (*Hint:* Use the parallel–perpendicular decomposition.)

14. Determine the shortest distance from the point (3,4,5) to the line through the origin parallel to the vector $2\mathbf{i} - \mathbf{j} + 2\mathbf{k}$.

15. Write the scalar equations of the line parallel to the intersection of the planes $3x + y + z = 5$, $x - 2y + 3z = 1$, and passing through the point (4,2,1).

16. Vectors from the origin O to four points A, B, C, D are given as follows:

$$\mathbf{A} = 2\mathbf{i} \qquad \mathbf{B} = 3\mathbf{j} \qquad \mathbf{C} = 4\mathbf{k} \qquad \mathbf{D} = \mathbf{i} + \mathbf{j} + 2\mathbf{k}$$

    (a) Find the length of the perpendicular drawn from A to the plane BCD.
    (b) Find the length of the common perpendicular to the lines AB and CD.
    (c) Find a vector parallel to this perpendicular.

17. Write an expression for a vector five units long, parallel to the plane $3x + 4y + 5z = 10$ and perpendicular to the vector $\mathbf{i} + 2\mathbf{j} + 2\mathbf{k}$.

18. Find r and s if $(2\mathbf{i} + 6\mathbf{j} - 27\mathbf{k}) \times (\mathbf{i} + r\mathbf{j} + s\mathbf{k}) = \mathbf{0}$.

19. Given that $\mathbf{A} \cdot \mathbf{B} = 0$ and $\mathbf{A} \times \mathbf{B} = \mathbf{0}$, what can you conclude about the vectors **A** and **B**?

20. If $\mathbf{A} \neq \mathbf{0}$, do $\mathbf{A} \cdot \mathbf{B} = \mathbf{A} \cdot \mathbf{C}$ and $\mathbf{A} \times \mathbf{B} = \mathbf{A} \times \mathbf{C}$ together imply $\mathbf{B} = \mathbf{C}$?

21. Express $2\mathbf{i} - \mathbf{j} + 3\mathbf{k}$ as the sum of a vector parallel, plus a vector perpendicular, to $2\mathbf{i} + 4\mathbf{j} - 2\mathbf{k}$.

22. Given that **A** and **B** are parallel to the yz plane, that $|\mathbf{A}| = 2$, $|\mathbf{B}| = 4$, and $\mathbf{A} \cdot \mathbf{B} = 0$, what can you say about $\mathbf{A} \times \mathbf{B}$?

23. (a) Do the lines $x/3 = y/2 = z/2$ and $x/5 = y/3 = (z - 4)/2$ intersect?
    (b) Find equations for a line perpendicular to both of these lines.
    (c) What is the distance between these lines?

24. If ω points in the direction of $\mathbf{i} + \mathbf{j} + \mathbf{k}$ and the body rotates about an axis through the origin with angular velocity $10\sqrt{3}$ rad/sec, find the locus of points having speed 20 ft/sec. What does this locus represent?

25. Supply the missing details of the proof of the distributive law for vector products. (Use similar triangles.)

26. If **u**, **v**, and **w** are mutually perpendicular unit vectors and $\mathbf{u} \times \mathbf{v} = \mathbf{w}$, show that $\mathbf{v} = \mathbf{w} \times \mathbf{u}$ and $\mathbf{u} = \mathbf{v} \times \mathbf{w}$.

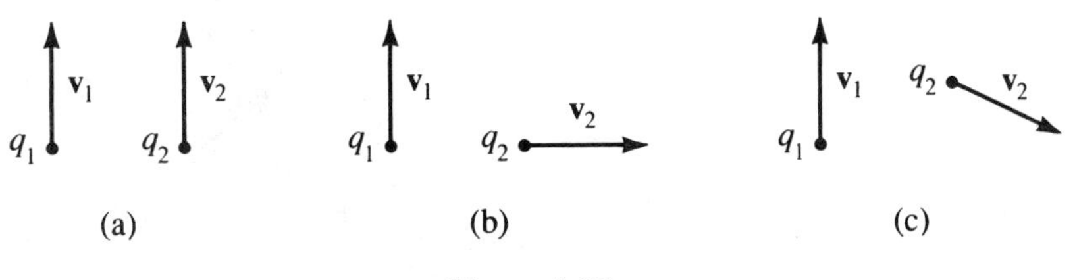

Figure 1.37

27. Given two nonintersecting lines $x/2 = y = (z - 1)/3$ and $x/3 = y = z$, find points P and Q, one on each line, such that PQ is perpendicular to both lines.
28. Starting with arbitrary $\mathbf{A}_1$ and $\mathbf{B}$, define the sequence of vectors $\mathbf{A}_n$ by $\mathbf{A}_{n+1} = \mathbf{B} \times \mathbf{A}_n$. What is the ultimate behavior of the sequence?
29. *Ampere's force* is the magnetic force that one moving charged particle exerts on another moving charged particle. Suppose particle 1 has charge q_1 and velocity $\mathbf{v}_1$, and is located at $\mathbf{R}_1$; q_2, $\mathbf{v}_2$, and $\mathbf{R}_2$ are the corresponding parameters for particle 2. Particle 2 produces a "magnetic flux density" vector $\mathbf{B}$ at $\mathbf{R}_1$ that is proportional to q_2 and $|\mathbf{v}_2|$ and inversely proportional to the square of the distance between q_1 and q_2; its direction is perpendicular to $\mathbf{v}_2$ and to the line from q_2 to q_1. The force on particle 1 is proportional to q_1, $|\mathbf{B}|$, and $|\mathbf{v}_1|$, and is directed perpendicular to $\mathbf{v}_1$ and to $\mathbf{B}$. Show that the formula

$$\text{Force on particle 1} = kq_1q_2\mathbf{v}_1 \times \left\{\mathbf{v}_2 \times \frac{\mathbf{R}_1 - \mathbf{R}_2}{|\mathbf{R}_1 - \mathbf{R}_2|^3}\right\}$$

has all these properties.
30. What is the direction of the force on particle 1 for the situations depicted in figure 1.37? Assume $q_2 = q_1$.
31. What is the force on particle 2 in the situations depicted in figure 1.37? Is it equal and opposite to the force on particle 1?

1.13 Triple Scalar Products

The *triple scalar product* of three vectors **A**, **B**, and **C** is defined to be the scalar

$$[\mathbf{A},\mathbf{B},\mathbf{C}] = \mathbf{A} \cdot (\mathbf{B} \times \mathbf{C}) \tag{1.23}$$

Notice that the parentheses can be omitted because there is no other sensible way of interpreting $\mathbf{A} \cdot \mathbf{B} \times \mathbf{C}$. Using the componentwise expression derived in the previous section for the cross product, we have

$$\begin{aligned}[\mathbf{A},\mathbf{B},\mathbf{C}] = {} & A_1B_2C_3 - A_1B_3C_2 + A_2B_3C_1 \\ & -A_2B_1C_3 + A_3B_1C_2 - A_3B_2C_1\end{aligned} \tag{1.24}$$

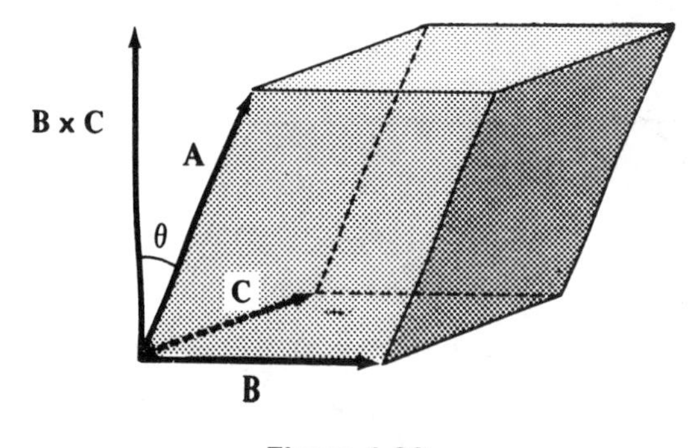

Figure 1.38

Alternatively, from the determinant expression for the cross product we can express [**A**,**B**,**C**] in the form

$$[\mathbf{A},\mathbf{B},\mathbf{C}] = \mathbf{A} \cdot \begin{vmatrix} \mathbf{i} & \mathbf{j} & \mathbf{k} \\ B_1 & B_2 & B_3 \\ C_1 & C_2 & C_3 \end{vmatrix} = \begin{vmatrix} A_1 & A_2 & A_3 \\ B_1 & B_2 & B_3 \\ C_1 & C_2 & C_3 \end{vmatrix} \tag{1.24'}$$

The triple scalar product has a geometric interpretation. Consider the parallelepiped with **A, B,** and **C** as coterminal edges, as in figure 1.38. The base of this solid is a parallelogram whose area is given, as we saw previously, by $|\mathbf{B} \times \mathbf{C}|$. Its height is the length of the component of **A** perpendicular to the base, which can be regarded as the component of **A** *parallel* to **B** × **C**, or $|\mathbf{A}| \cos \theta$, as shown in figure 1.38. To be precise, we should say that this height is the *magnitude* of $|\mathbf{A}| \cos \theta$, because $\cos \theta$ would be negative if **A** pointed to the opposite side of the plane of **B** and **C**, that is, if **A, B,** and **C** formed a left-handed system. Thus we see that the volume of the parallelepiped, computed as base area times height, equals the magnitude of $|\mathbf{B} \times \mathbf{C}| \cos \theta |\mathbf{A}|$. But this is precisely $\mathbf{A} \cdot \mathbf{B} \times \mathbf{C}$, the triple scalar product! Summarizing, we can state that *the volume of the parallelepiped with coterminal edges* **A, B,** *and* **C** *is given, up to sign, by* [**A**,**B**,**C**]. Furthermore, [**A**,**B**,**C**] *is positive if and only if* **A, B,** *and* **C** *form a right-handed system.*

Example 1.33 Compute [**A**,**B**,**C**] if $\mathbf{A} = 2\mathbf{i} + \mathbf{k}$, $\mathbf{B} = 3\mathbf{i} + \mathbf{j} + \mathbf{k}$, and $\mathbf{C} = \mathbf{i} + \mathbf{j} + 4\mathbf{k}$.

Solution

$$[\mathbf{A},\mathbf{B},\mathbf{C}] = [2\mathbf{i} + \mathbf{k}, 3\mathbf{i} + \mathbf{j} + \mathbf{k}, \mathbf{i} + \mathbf{j} + 4\mathbf{k}]$$

$$= \begin{vmatrix} 2 & 0 & 1 \\ 3 & 1 & 1 \\ 1 & 1 & 4 \end{vmatrix} = 8 + 3 - 1 - 2 = 8$$

Example 1.34 Compute $[\mathbf{i}, \mathbf{j}, \mathbf{i} + 2\mathbf{j}]$.

Solution

$$[\mathbf{i}, \mathbf{j}, \mathbf{i} + 2\mathbf{j}] = \begin{vmatrix} 1 & 0 & 0 \\ 0 & 1 & 0 \\ 1 & 2 & 0 \end{vmatrix} = 0$$

(The vectors are coplanar, so the parallelepiped has zero volume.)

We now list some properties of the triple scalar product that can be verified from eq. (1.24). They will also be familiar to students who have studied determinants.

First, notice that the absolute value of the triple scalar product does not depend on the order of the vectors, but the sign changes whenever two of the vectors are switched:

$$[\mathbf{A},\mathbf{B},\mathbf{C}] = -[\mathbf{B},\mathbf{A},\mathbf{C}] = [\mathbf{B},\mathbf{C},\mathbf{A}] \tag{1.25}$$

This shows that the position of the dot and cross can be changed freely, because

$$\mathbf{A} \cdot \mathbf{B} \times \mathbf{C} = [\mathbf{A},\mathbf{B},\mathbf{C}] = [\mathbf{C},\mathbf{A},\mathbf{B}] = \mathbf{C} \cdot \mathbf{A} \times \mathbf{B} = \mathbf{A} \times \mathbf{B} \cdot \mathbf{C} \tag{1.26}$$

Second, the triple scalar product is linear in each of its factors:

$$\begin{aligned} [s\mathbf{A} + \mathbf{B}, \mathbf{C}, \mathbf{D}] &= s[\mathbf{A},\mathbf{C},\mathbf{D}] + [\mathbf{B},\mathbf{C},\mathbf{D}] \\ [\mathbf{A}, s\mathbf{B} + \mathbf{C}, \mathbf{D}] &= s[\mathbf{A},\mathbf{B},\mathbf{D}] + [\mathbf{A},\mathbf{C},\mathbf{D}] \\ [\mathbf{A}, \mathbf{B}, s\mathbf{C} + \mathbf{D}] &= s[\mathbf{A},\mathbf{B},\mathbf{C}] + [\mathbf{A},\mathbf{B},\mathbf{D}] \end{aligned} \tag{1.27}$$

Third, we have the obvious identity

$$[\mathbf{i},\mathbf{j},\mathbf{k}] = 1 \tag{1.28}$$

Clearly, if any two of the vectors **A, B,** or **C** are equal, the triple scalar product will be zero (the parallelepiped will have zero volume). Furthermore, if any one of the three vectors is replaced by the sum of that one vector with a linear combination of the other two, the triple scalar product is unchanged. For example, if we replace **A** by $\mathbf{A} + s\mathbf{B} + t\mathbf{C}$, where s and t are any numbers, then $[\mathbf{A} + s\mathbf{B} + t\mathbf{C}, \mathbf{B}, \mathbf{C}] = [\mathbf{A},\mathbf{B},\mathbf{C}]$. The proof is easy:

$$[\mathbf{A} + s\mathbf{B} + t\mathbf{C}, \mathbf{B}, \mathbf{C}] = [\mathbf{A},\mathbf{B},\mathbf{C}] + s[\mathbf{B},\mathbf{B},\mathbf{C}] + t[\mathbf{C},\mathbf{B},\mathbf{C}]$$

and the last two terms are zero.

It is amusing to notice that these properties make it possible to evaluate any triple scalar product without using eqs. (1.24) or (1.24′). For example, let $\mathbf{A} = \mathbf{i} + 3\mathbf{j}$, $\mathbf{B} = \mathbf{i} + \mathbf{k}$, and $\mathbf{C} = -\mathbf{k}$; then

$$\begin{aligned}[\mathbf{A},\mathbf{B},\mathbf{C}] &= [\mathbf{i} + 3\mathbf{j}, \mathbf{i} + \mathbf{k}, -\mathbf{k}] \\ &= [\mathbf{i}, \mathbf{i} + \mathbf{k}, -\mathbf{k}] + [3\mathbf{j}, \mathbf{i} + \mathbf{k}, -\mathbf{k}] \\ &= [\mathbf{i},\mathbf{i},-\mathbf{k}] + [\mathbf{i},\mathbf{k},-\mathbf{k}] + [3\mathbf{j},\mathbf{i},-\mathbf{k}] + [3\mathbf{j},\mathbf{k},-\mathbf{k}] \\ &= -[\mathbf{i},\mathbf{i},\mathbf{k}] - [\mathbf{i},\mathbf{k},\mathbf{k}] - 3[\mathbf{j},\mathbf{i},\mathbf{k}] - 3[\mathbf{j},\mathbf{k},\mathbf{k}] \\ &= -3[\mathbf{j},\mathbf{i},\mathbf{k}] = 3[\mathbf{i},\mathbf{j},\mathbf{k}] = 3\end{aligned}$$

As a final note, let us show how the scalar triple product can be used to relate the parametric and nonparametric equations of a plane derived in section 1.10. The parametric equation was based on specifying a point in the plane with position vector $\mathbf{R}_0 = x_0\mathbf{i} + y_0\mathbf{j} + z_0\mathbf{k}$, and two vectors $\mathbf{A}$ and $\mathbf{B}$ parallel to the plane. Clearly $\mathbf{R} =$

Table 1.2 Vector products

Concept	Geometrical Formula	Analytical Formula
Scalar product, dot product $\mathbf{A} \cdot \mathbf{B}$	$\lvert\mathbf{A}\rvert\lvert\mathbf{B}\rvert \cos\theta$	$A_1B_1 + A_2B_2 + A_3B_3$
Vector product, cross product $\mathbf{A} \times \mathbf{B}$	$\lvert\mathbf{A}\rvert\lvert\mathbf{B}\rvert \sin\theta\, \mathbf{n}$	$\begin{vmatrix} \mathbf{i} & \mathbf{j} & \mathbf{k} \\ A_1 & A_2 & A_3 \\ B_1 & B_2 & B_3 \end{vmatrix}$
Triple scalar product, volume $[\mathbf{A},\mathbf{B},\mathbf{C}]$	$\mathbf{A} \cdot \mathbf{B} \times \mathbf{C} = \mathbf{A} \times \mathbf{B} \cdot \mathbf{C}$	$\begin{vmatrix} A_1 & A_2 & A_3 \\ B_1 & B_2 & B_3 \\ C_1 & C_2 & C_3 \end{vmatrix}$
Parallel–perpendicular decomposition	$\mathbf{B}_\parallel$	$\dfrac{\mathbf{A} \cdot \mathbf{B}}{\mathbf{A} \cdot \mathbf{A}}\mathbf{A}$
	$\mathbf{B}_\perp$	$\dfrac{(\mathbf{A} \times \mathbf{B}) \times \mathbf{A}}{\mathbf{A} \cdot \mathbf{A}}$ or $\mathbf{B} - \mathbf{B}_\parallel$

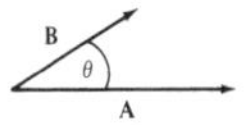

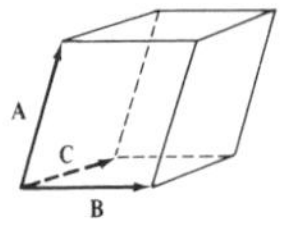

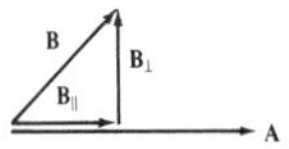

$x\mathbf{i} + y\mathbf{j} + z\mathbf{k}$ will be the position vector of a point in the plane if the parallelepiped formed by $\mathbf{R} - \mathbf{R}_0$, $\mathbf{A}$, and $\mathbf{B}$ is flat (i.e., has zero volume). Hence the equation for this plane can be expressed

$$[\mathbf{R} - \mathbf{R}_0, \mathbf{A}, \mathbf{B}] = 0$$

Inserting the definition (1.23) for the triple scalar product, we identify $\mathbf{A} \times \mathbf{B}$ as being a vector $\mathbf{N}$ normal to the plane, and we have

$$(\mathbf{R} - \mathbf{R}_0) \cdot \mathbf{N} = 0$$

This agrees with eq. (1.16), the nonparametric equation for the plane.

Table 1.2 summarizes the properties of the various vector products.

EXERCISES

1. Find the triple scalar product $[\mathbf{A},\mathbf{B},\mathbf{C}]$ given that
   (a) $\mathbf{A} = 2\mathbf{i}, \mathbf{B} = 3\mathbf{j}, \mathbf{C} = 5\mathbf{k}$
   (b) $\mathbf{A} = \mathbf{i} + \mathbf{j} + \mathbf{k}, \mathbf{B} = 3\mathbf{i} + \mathbf{j}, \mathbf{C} = 5\mathbf{k} - \mathbf{j}$
   (c) $\mathbf{A} = 2\mathbf{i} - \mathbf{j} + \mathbf{k}, \mathbf{B} = \mathbf{i} + \mathbf{j} + \mathbf{k}, \mathbf{C} = 2\mathbf{i} + 3\mathbf{k}$
   (d) $\mathbf{A} = \mathbf{k}, \mathbf{B} = \mathbf{i}, \mathbf{C} = \mathbf{j}$
2. Find the volume of the parallelepiped whose coterminal edges are arrows representing the vectors $3\mathbf{i} + 4\mathbf{j}$, $2\mathbf{i} + 3\mathbf{j} + 4\mathbf{k}$, $5\mathbf{k}$.
3. Find the volume of the parallelepiped with coterminal edges AB, AC, and AD, where $A = (3,2,1)$, $B = (4,2,1)$, $C = (0,1,4)$, and $D = (0,0,7)$.
4. Find the volume of the tetrahedron with coterminal edges representing the vectors $\mathbf{i} + \mathbf{j}$, $\mathbf{i} - \mathbf{j}$, $2\mathbf{k}$. Illustrate with a sketch. (*Note:* The volume of the tetrahedron is one sixth the volume of the parallelepiped having the same coterminal edges.)
5. Find the area of the parallelogram in the plane with vertices at (0,0), (1,1), (3,4), (4,5). (*Hint:* Convert this to a three-dimensional problem, finding the volume of the parallelepiped with this parallelogram as base, taking the third edge to be of unit length along the z axis.)
6. Find the equation of the plane passing through the origin parallel to the vectors $\mathbf{A} = 3\mathbf{i} + \mathbf{j} - 2\mathbf{k}$ and $\mathbf{B} = \mathbf{i} - \mathbf{j} + 5\mathbf{k}$.
7. Find the equation of the plane passing through $(3,4,-1)$ parallel to the vectors $\mathbf{A} = 2\mathbf{i} + \mathbf{j} + \mathbf{k}$ and $\mathbf{B} = \mathbf{i} - 3\mathbf{k}$.
8. (a) Show that the vectors $\mathbf{i} - \mathbf{j}$, $\mathbf{j} - \mathbf{k}$, $\mathbf{k} - \mathbf{i}$ are parallel to a plane.
   (b) Find an equation of the plane passing through the origin that is parallel to these three vectors.
9. Given the points $P_1(2,-1,4)$, $P_2(-1,0,3)$, $P_3(4,3,1)$, and $P_4(3,-5,0)$, determine
   (a) the volume of the tetrahedron $P_1P_2P_3P_4$.
   (b) the equation of the plane containing the points P_1, P_2, and P_3.
   (c) the cosine of the angle between the line segments P_1P_2 and P_1P_3.

10. Consider

$$\mathbf{A} = \mathbf{i} + \mathbf{j} + \mathbf{k}$$
$$\mathbf{B} = \mathbf{i}$$
$$\mathbf{C} = C_1\mathbf{i} + C_2\mathbf{j} + C_3\mathbf{k}$$

(a) If $C_1 = 1$ and $C_2 = 2$, find C_3 to make the three vectors coplanar.
(b) If $C_2 = -1$ and $C_3 = 1$, show that no value of C_1 can be found to make the three vectors coplanar.
(c) Discuss the geometrical reason for the result in part (b).

11. Find the altitude of a parallelepiped determined by **A**, **B**, and **C**, if the base is taken to be the parallelogram determined by **A** and **B**, and if

$$\mathbf{A} = \mathbf{i} + \mathbf{j} + \mathbf{k}$$
$$\mathbf{B} = 2\mathbf{i} + 4\mathbf{j} - \mathbf{k}$$
$$\mathbf{C} = \mathbf{i} + \mathbf{j} + 3\mathbf{k}$$

(*Hint:* Think of the geometrical interpretation of $[\mathbf{A},\mathbf{B},\mathbf{C}]/|\mathbf{A} \times \mathbf{B}|$.)

12. Let **A**, **B**, **C**, and **D** be position vectors of the points $A(1,3,-2)$, $B(3,5,-3)$, $C(-5,9,-5)$, and $D(4,-1,10)$ respectively. Find
(a) $|\mathbf{A} - \mathbf{D}|$
(b) $\mathbf{A} \times \mathbf{B}$
(c) $(\mathbf{A} - \mathbf{C}) \cdot (\mathbf{A} - \mathbf{B})$
(d) $\mathbf{A} \cdot \mathbf{B} \times \mathbf{C}$

13. Given the four points specified in exercise 12, determine
(a) the area of the triangle *OAB*.
(b) the volume of the tetrahedron *OABC*.
(c) the angle *CAB*.

14. Sketch the vectors $\mathbf{A} = \mathbf{i} + \mathbf{j}$, $\mathbf{B} = \mathbf{i} + 2\mathbf{j} + 2\mathbf{k}$, and $\mathbf{C} = \mathbf{i} + 3\mathbf{k}$. Determine from your sketch whether or not **A**, **B**, and **C**, in that order, form a right-handed system. Check by computing the sign of $[\mathbf{A},\mathbf{B},\mathbf{C}]$.

15. What can you conclude about nonzero vectors **A**, **B**, **C**, and **D**, given that $|(\mathbf{A} \times \mathbf{B}) \cdot \mathbf{C}| + |(\mathbf{B} \times \mathbf{C}) \cdot \mathbf{D}| = 0$?

16. Let **u**, **v**, and **w** be mutually perpendicular unit vectors, forming a right-handed system.
(a) Show that the vector $\mathbf{A} = \mathbf{i} \times \mathbf{u} + \mathbf{j} \times \mathbf{v} + \mathbf{k} \times \mathbf{w}$ makes the same angle with **i** that it does with **u**.
(b) Find a vector extending along the axis of the rotation that carries **i**, **j**, and **k** into **u**, **v**, and **w**, respectively.

17. Show that an arbitrary vector **V** can be expressed in terms of any three noncoplanar vectors **A**, **B**, and **C**, according to

$$\mathbf{V} = \frac{[\mathbf{V},\mathbf{B},\mathbf{C}]}{[\mathbf{A},\mathbf{B},\mathbf{C}]}\mathbf{A} + \frac{[\mathbf{V},\mathbf{C},\mathbf{A}]}{[\mathbf{A},\mathbf{B},\mathbf{C}]}\mathbf{B} + \frac{[\mathbf{V},\mathbf{A},\mathbf{B}]}{[\mathbf{A},\mathbf{B},\mathbf{C}]}\mathbf{C} \quad (1.29)$$

(*Hint:* We know that **V** can be expressed as $a\mathbf{A} + b\mathbf{B} + c\mathbf{C}$; to find a, take the scalar product of **V** with $\mathbf{B} \times \mathbf{C}$.)

18. Devise a geometric proof of eq. (1.26) based on the interpretation of the triple scalar product as a volume.

19. Construct another proof of the distributive law for the vector product, based on the interchange of $\times$ and $\cdot$ (see exercise 18) and the distributivity of the scalar product. (*Hint:* Derive the identity

$$\mathbf{D} \cdot \mathbf{A} \times (\mathbf{B} + \mathbf{C}) = \mathbf{D} \cdot \mathbf{A} \times \mathbf{B} + \mathbf{D} \cdot \mathbf{A} \times \mathbf{C}$$

and then let $\mathbf{D}$ be $\mathbf{i}$, $\mathbf{j}$, and $\mathbf{k}$, in turn.)

1.14 Vector Identities

Of the following identities, the first is the most important because the other three can be derived from it fairly easily:

$$\mathbf{A} \times (\mathbf{B} \times \mathbf{C}) = (\mathbf{A} \cdot \mathbf{C})\mathbf{B} - (\mathbf{A} \cdot \mathbf{B})\mathbf{C} \tag{1.30}$$

$$(\mathbf{A} \times \mathbf{B}) \times \mathbf{C} = (\mathbf{A} \cdot \mathbf{C})\mathbf{B} - (\mathbf{B} \cdot \mathbf{C})\mathbf{A} \tag{1.31}$$

$$(\mathbf{A} \times \mathbf{B}) \times (\mathbf{C} \times \mathbf{D}) = [\mathbf{A},\mathbf{C},\mathbf{D}]\mathbf{B} - [\mathbf{B},\mathbf{C},\mathbf{D}]\mathbf{A} \tag{1.32}$$

$$(\mathbf{A} \times \mathbf{B}) \cdot (\mathbf{C} \times \mathbf{D}) = (\mathbf{A} \cdot \mathbf{C})(\mathbf{B} \cdot \mathbf{D}) - (\mathbf{A} \cdot \mathbf{D})(\mathbf{B} \cdot \mathbf{C}) \tag{1.33}$$

In formula (1.30), if $\mathbf{V} = \mathbf{A} \times (\mathbf{B} \times \mathbf{C})$ is not the zero vector, then it must be perpendicular to $\mathbf{B} \times \mathbf{C}$. Since $\mathbf{B} \times \mathbf{C}$ is itself perpendicular to both $\mathbf{B}$ and $\mathbf{C}$, it follows that $\mathbf{V}$ must be in the plane of $\mathbf{B}$ and $\mathbf{C}$, and since they are nonzero vectors that are not parallel (otherwise $\mathbf{V}$ would be the zero vector), $\mathbf{V}$ must be a linear combination of $\mathbf{B}$ and $\mathbf{C}$. Thus $\mathbf{V} = m\mathbf{B} + n\mathbf{C}$ for suitable scalars m and n. The fact that $m = \mathbf{A} \cdot \mathbf{C}$ and $n = -\mathbf{A} \cdot \mathbf{B}$ is not obvious, of course. The actual verification of eq. (1.30) can be accomplished by working out the componentwise expression for each side of the equality. We leave this laborious computation to energetic readers. (Or they can read section 1.16.)

We suggest the following device for memorizing eq. (1.30). As we observed, $\mathbf{A} \times (\mathbf{B} \times \mathbf{C})$ must be expressible as a linear combination of $\mathbf{B}$ and $\mathbf{C}$. If you can remember only that the coefficients in this expression are scalar products of the other two vectors, and that the terms have opposite signs, you will be able to write

$$\mathbf{A} \times (\mathbf{B} \times \mathbf{C}) = \pm[(\mathbf{A} \cdot \mathbf{C})\mathbf{B} - (\mathbf{A} \cdot \mathbf{B})\mathbf{C}]$$

To get the proper sign, use the familiar vectors $\mathbf{i}$, $\mathbf{j}$, and $\mathbf{k}$; thus

$$\mathbf{i} \times (\mathbf{i} \times \mathbf{j}) = \mathbf{i} \times \mathbf{k} = -\mathbf{j} = \pm[(\mathbf{i} \cdot \mathbf{j})\mathbf{i} - (\mathbf{i} \cdot \mathbf{i})\mathbf{j}]$$

so the plus sign is correct. [This also works for formula (1.31), of course.]

Formula (1.31) is easily proved by observing that

$$(\mathbf{A} \times \mathbf{B}) \times \mathbf{C} = -\mathbf{C} \times (\mathbf{A} \times \mathbf{B})$$

and using eq. (1.30) for the right-hand side.

To derive eq. (1.32), let $\mathbf{U} = \mathbf{C} \times \mathbf{D}$, whence

$$(\mathbf{A} \times \mathbf{B}) \times \mathbf{U} = (\mathbf{A} \cdot \mathbf{U})\mathbf{B} - (\mathbf{B} \cdot \mathbf{U})\mathbf{A} = [\mathbf{A},\mathbf{C},\mathbf{D}]\mathbf{B} - [\mathbf{B},\mathbf{C},\mathbf{D}]\mathbf{A}$$

To derive eq. (1.33),

$$\begin{aligned}(\mathbf{A} \times \mathbf{B}) \cdot \mathbf{U} &= [\mathbf{A},\mathbf{B},\mathbf{U}] = \mathbf{A} \cdot (\mathbf{B} \times \mathbf{U}) = \mathbf{A} \cdot [\mathbf{B} \times (\mathbf{C} \times \mathbf{D})] \\ &= \mathbf{A} \cdot [(\mathbf{B} \cdot \mathbf{D})\mathbf{C} - (\mathbf{B} \cdot \mathbf{C})\mathbf{D}] \\ &= (\mathbf{B} \cdot \mathbf{D})(\mathbf{A} \cdot \mathbf{C}) - (\mathbf{B} \cdot \mathbf{C})(\mathbf{A} \cdot \mathbf{D})\end{aligned}$$

The reader is advised to attach a permanent bookmark in this section, as an aid in referring to the identities in the future.

EXERCISES

1. Derive the identity

$$(\mathbf{A} \times \mathbf{B}) \times (\mathbf{C} \times \mathbf{D}) = [\mathbf{A},\mathbf{B},\mathbf{D}]\mathbf{C} - [\mathbf{A},\mathbf{B},\mathbf{C}]\mathbf{D}$$

2. Derive the identity

$$(\mathbf{A} \times \mathbf{B}) \cdot (\mathbf{B} \times \mathbf{C}) \times (\mathbf{C} \times \mathbf{A}) = [\mathbf{A},\mathbf{B},\mathbf{C}]^2$$

3. Derive the identity

$$\mathbf{A} \times (\mathbf{B} \times \mathbf{C}) + \mathbf{B} \times (\mathbf{C} \times \mathbf{A}) + \mathbf{C} \times (\mathbf{A} \times \mathbf{B}) = \mathbf{0}$$

4. Verify formula (1.30) by working out the componentwise expression.
5. If the vector ω in figure 1.34 is constant, then the acceleration of a particle with position vector $\mathbf{R}$ is $\mathbf{a} = \omega \times (\omega \times \mathbf{R})$. Simplify this expression.
6. Are any of the following identities generally valid for vectors?
   (a) $\mathbf{A} \times \mathbf{B} = \mathbf{B} \times \mathbf{A}$
   (b) $(\mathbf{A} \times \mathbf{B}) \times \mathbf{C} = \mathbf{A} \times (\mathbf{B} \times \mathbf{C})$
   (c) $\mathbf{A} \times \mathbf{B} = \mathbf{A} \times \mathbf{C}$ if and only if $\mathbf{B} = \mathbf{C}$
   (d) $\mathbf{A} \times \mathbf{B} = \mathbf{0}$ if and only if $\mathbf{A} = \mathbf{0}$ or $\mathbf{B} = \mathbf{0}$
7. Simplify $|\mathbf{A} \times \mathbf{B}|^2 + (\mathbf{A} \cdot \mathbf{B})^2 - |\mathbf{A}|^2|\mathbf{B}|^2$.
8. Let

$$\begin{aligned}\mathbf{A} &= 3\mathbf{i} + \mathbf{j} + 2\mathbf{k} \\ \mathbf{B} &= 4\mathbf{i} + \mathbf{j} + 5\mathbf{k} \\ \mathbf{C} &= \mathbf{i} - \mathbf{j} + \mathbf{k}\end{aligned}$$

   Find $\mathbf{A} \times \mathbf{B}$, $[\mathbf{A},\mathbf{B},\mathbf{C}]$, $|\mathbf{A} \times \mathbf{B}|$, and the distance from the tip of $\mathbf{C}$ to the plane through the origin spanned by $\mathbf{A}$ and $\mathbf{B}$.
9. Prove, for any vector $\mathbf{A}$, that

$$\mathbf{i} \times (\mathbf{i} \times \mathbf{A}) + \mathbf{j} \times (\mathbf{j} \times \mathbf{A}) + \mathbf{k} \times (\mathbf{k} \times \mathbf{A}) = -2\mathbf{A}$$

10. Prove: if $\mathbf{A} + \mathbf{B} + \mathbf{C} = \mathbf{0}$, then $\mathbf{A} \times \mathbf{B} = \mathbf{B} \times \mathbf{C} = \mathbf{C} \times \mathbf{A}$. Interpret geometrically.
11. Simplify $[\mathbf{A} \times (\mathbf{A} \times \mathbf{B})] \times \mathbf{A} \cdot \mathbf{C}$.

12. Given that **u**, **v**, and **w** are nonzero vectors having the same magnitude and $(\mathbf{u} \times \mathbf{v}) \times \mathbf{w} = \mathbf{u} \times (\mathbf{v} \times \mathbf{w})$, what can you say about **u**, **v**, and **w**?

Exercises 13 and 14 are quite advanced.

13. Prove the following theorem of Desargues. Given two (nondegenerate) triangles *ABC* and *DEF* with the property that the line through *AD*, the line through *BE*, and the line through *CF* have a point in common; moreover, let the lines through *AB* and *DE* intersect at *P*, the lines through *BC* and *EF* intersect at *Q*, and the lines through *AC* and *DF* intersect at *R*. Then *P*, *Q*, and *R* are collinear.
14. Prove the converse of Desargues' theorem in exercise 13.
15. Write $(\mathbf{u} \times \mathbf{v}) \cdot (\mathbf{u} \times \mathbf{v})$ as a determinant involving only scalar products.

1.15 Optional Reading: Tensor Notation

The use of distinguished symbols such as **A**, $\mathbf{A} \times \mathbf{B}$, and the like, to denote vectors and vector operations provides an excellent and often suggestive shorthand for expressing laws in geometry and physics. However, when we ultimately come down to the actual computations of a concrete problem, these expressions must be dealt with componentwise. Furthermore, the verification (and discovery!) of some of the more complicated vector identities such as those appearing in the previous section is often accomplished most efficiently by dealing with the components. In this section we shall introduce some notation that often facilitates this process; it is widely known as *tensor notation.* Although we do not intend to discuss tensors themselves here, we see no reason to designate the notational system by anything other than its proper name.

The boldface vector symbol **A** suggests, as we have said, a quantity with magnitude and direction; this quantity is equally well represented by three numbers, A_1, A_2, and A_3, the components of the vector. Every statement about the vector is actually a statement about its components. Thus $\mathbf{A} = \mathbf{B}$ means $A_1 = B_1$, $A_2 = B_2$, and $A_3 = B_3$; briefly

$$A_i = B_i \qquad (i = 1, 2, 3) \tag{1.34}$$

Expressed simply, the basic idea in tensor notation is to try to write all vector equations in component form, but using dummy subscripts such as i in eq. (1.34) rather than explicitly writing out the equation for the first component, then the second, then the third. (Whether or not this is always possible will not be discussed here. For now, we will be satisfied with using tensor notation *when we can.*) We will indicate the components of a vector **A** by A_i, or $(\mathbf{A})_i$ if it is more convenient; we shall regard the parenthetical phrase "$(i = 1, 2, 3)$" as understood, and delete it. Let's try some examples.

The expression of the fact that vectors add componentwise becomes

$$(\mathbf{A} + \mathbf{B})_i = A_i + B_i$$

That is, the ith component of $\mathbf{A} + \mathbf{B}$ is the sum of the ith components of $\mathbf{A}$ and of $\mathbf{B}$. Scalar multiplication is expressed

$$(s\mathbf{A})_i = sA_i$$

The associative law for vectors, expressed componentwise, merely reduces to the associative law for *numbers:*

$$\begin{aligned}[(\mathbf{A} + \mathbf{B}) + \mathbf{C}]_i &= (\mathbf{A} + \mathbf{B})_i + C_i \\ &= (A_i + B_i) + C_i \\ &= A_i + (B_i + C_i) \\ &= [\mathbf{A} + (\mathbf{B} + \mathbf{C})]_i\end{aligned}$$

The condition that $\mathbf{R}$ lies on the line through the tip of $\mathbf{V}$ and parallel to $\mathbf{W}$ is expressed

$$R_i = V_i + tW_i$$

For the scalar product, we have

$$\mathbf{A} \cdot \mathbf{B} = A_1B_1 + A_2B_2 + A_3B_3$$

This can be compacted by using the Greek letter Σ to denote summation. For any set of n numbers $\{a_l\}$ $(l = 1, 2, \ldots, n)$, we abbreviate

$$a_1 + a_2 + \cdots + a_n$$

by the expression

$$\sum_{l=1}^{n} a_l$$

Thus our scalar product becomes

$$\mathbf{A} \cdot \mathbf{B} = \sum_{i=1}^{3} A_iB_i \tag{1.35}$$

The componentwise expression of the cross product is a bit complicated. Observe that each component of $\mathbf{A} \times \mathbf{B}$ is a sum of products of components of $\mathbf{A}$ times components of $\mathbf{B}$. If we (conceptually) form all the products $\{A_jB_k\}$, we can say that the ith component of $\mathbf{A} \times \mathbf{B}$ is a linear combination of these with coefficients $+1$, -1, or 0 (if the term doesn't actually appear). So by defining ϵ_{ijk} appropriately, we can write

$$(\mathbf{A} \times \mathbf{B})_i = \sum_{j=1}^{3} \sum_{k=1}^{3} \epsilon_{ijk}A_jB_k \tag{1.36}$$

ϵ_{ijk} is the coefficient of A_jB_k in the ith component of $\mathbf{A} \times \mathbf{B}$. Comparison of this with expression (1.21) in section 1.12 shows

$$\epsilon_{ijk} = \begin{cases} +1 & \text{if } (ijk) \text{ is either } (123), (231), \text{ or } (312) \\ -1 & \text{if } (ijk) \text{ is either } (321), (213), \text{ or } (132) \\ 0 & \text{otherwise} \end{cases} \qquad (1.37)$$

In fact, ϵ_{ijk} is the coefficient of $\lambda_i\mu_j\eta_k$ in the determinant

$$\begin{vmatrix} \lambda_1 & \lambda_2 & \lambda_3 \\ \mu_1 & \mu_2 & \mu_3 \\ \eta_1 & \eta_2 & \eta_3 \end{vmatrix}$$

—a fact that we could have anticipated by comparing the expression (1.36) with the determinant formula for the cross product in section 1.12.

A few observations about the symbol ϵ_{ijk} are in order:

(*i*) $\epsilon_{ijk} = 0$ if any of the subscripts are equal.
(*ii*) $\epsilon_{ijk} = \epsilon_{jki} = \epsilon_{kij}$; that is, the subscripts can be permuted cyclically.
(*iii*) $\epsilon_{ijk} = -\epsilon_{jik}$; that is, the sign changes if two subscripts are switched.

Of course, the scalar product $\mathbf{A} \cdot \mathbf{B}$ is also composed of products of $\mathbf{A}$'s components with $\mathbf{B}$'s components, and if the expression (1.35) weren't so simple already, we would write it

$$\mathbf{A} \cdot \mathbf{B} = \sum_{i=1}^{3} \sum_{j=1}^{3} \delta_{ij} A_i B_j$$

where

$$\delta_{ij} = \begin{cases} 1 & \text{if } i = j \\ 0 & \text{otherwise} \end{cases} \qquad (1.38)$$

The effect of δ_{ij} in an expression is simple if the subscripts are summed; since $\delta_{ij} = 0$ unless $i = j$, one can merely drop the δ and substitute i for j. Thus

$$\sum_{i=1}^{3} \sum_{j=1}^{3} \delta_{ij} A_i B_j = \sum_{j=1}^{3} A_j B_j \qquad (= \mathbf{A} \cdot \mathbf{B})$$

For this reason, δ_{ij} is sometimes called the *substitution tensor*. It's also known as the *Kronecker delta*.

Example 1.35 Show that the triple scalar product can be computed as a determinant.

Solution In the expression $\mathbf{A} \cdot \mathbf{B} \times \mathbf{C}$, we first use tensor notation for the scalar product:

$$\mathbf{A} \cdot \mathbf{B} \times \mathbf{C} = \sum_{i=1}^{3} A_i(\mathbf{B} \times \mathbf{C})_i$$

Now using (1.36) for the vector product,

$$\mathbf{A} \cdot \mathbf{B} \times \mathbf{C} = \sum_{i=1}^{3} A_i \sum_{j=1}^{3} \sum_{k=1}^{3} \epsilon_{ijk} B_j C_k$$

$$= \sum_{i=1}^{3} \sum_{j=1}^{3} \sum_{k=1}^{3} \epsilon_{ijk} A_i B_j C_k \tag{1.39}$$

As we observed earlier, this is the expansion of the determinant

$$\begin{vmatrix} A_1 & A_2 & A_3 \\ B_1 & B_2 & B_3 \\ C_1 & C_2 & C_3 \end{vmatrix}$$

Notice that every time we have used the summation symbol Σ, the subscript over which we were summing occurred *twice* in the term expressing the addend; i is repeated in eq. (1.35), j and k are repeated in eq. (1.36), and so on. This happens so often that the "Einstein summation convention" is used in tensor notation: *whenever a subscript appears more than once in a single term, it is understood that this particular term is to be summed over all values* (1, 2, and 3) *of the repeated subscript.* Scalar products are thus written A_iB_i, and the ith component of $\mathbf{A} \times \mathbf{B}$ is $\epsilon_{ijk}A_jB_k$. In fact, we have $\delta_{ii} = 3$. *Exceptions to this rule must be explicitly indicated.*

The manipulation of expressions involving more than one cross product is aided by the following identity:

$$\epsilon_{ikm}\epsilon_{psm} = \delta_{ip}\,\delta_{ks} - \delta_{is}\,\delta_{kp} \tag{1.40}$$

(Observe that m is summed over.) To prove this, we notice that the right-hand side is zero unless it has the form $1 - 0$ or $0 - 1$. Thus

$$\delta_{ip}\,\delta_{ks} - \delta_{is}\,\delta_{kp} = \begin{cases} 1 & \text{if } i = p \text{ and } k = s \text{ but } i \neq s \text{ (or } k \neq p) \\ -1 & \text{if } i = s \text{ and } k = p \text{ but } i \neq p \text{ (or } k \neq s) \\ 0 & \text{otherwise} \end{cases}$$

On the left-hand side of eq. (1.40), ϵ is zero unless all subscripts are different; in which case, $i \neq k$ and $p \neq s$ and m must be different from i, p, k, or s. So there is actually only one (at most) nonzero term in the sum over values of m! The product is $+1$ if (ikm) is a cyclic permutation of (psm), which can happen only if $i = p$ and $k = s$; and -1 results if (ikm) is in the opposite order as (psm), which requires $i = s$ and $k = p$. Comparing these conditions, we see that the left- and right-hand sides of eq. (1.40) are equal.

Example 1.36 Simplify $\mathbf{A} \times (\mathbf{B} \times \mathbf{C})$.

Solution The ith component is

$$\begin{aligned} \epsilon_{ijk}A_j(\mathbf{B} \times \mathbf{C})_k &= \epsilon_{ijk}A_j\epsilon_{klm}B_lC_m \\ &= \epsilon_{ijk}\epsilon_{klm}A_jB_lC_m \end{aligned}$$

Remember that we are summing over repeated subscripts. First we sum over k. Since the other terms do not depend on k, we can compute $\epsilon_{ijk}\epsilon_{klm}$; this is the same as $\epsilon_{ijk}\epsilon_{lmk}$, which, by eq. (1.40), is $\delta_{il}\,\delta_{jm} - \delta_{im}\,\delta_{jl}$. So the above expression equals

$$(\delta_{il}\,\delta_{jm} - \delta_{im}\,\delta_{jl})A_jB_lC_m$$

Now sum out the substitution tensors one at a time. Summing over m, we get

$$\delta_{il}A_jB_lC_j - \delta_{jl}A_jB_lC_i$$

and, summing over l,

$$A_jC_jB_i - A_jB_jC_i$$

Identifying the scalar products, we recognize that this is the ith component of $(\mathbf{A} \cdot \mathbf{C})\mathbf{B} - (\mathbf{A} \cdot \mathbf{B})\mathbf{C}$. We have proved formula (1.30)!

Example 1.37 Simplify $(\mathbf{A} \times \mathbf{B}) \times (\mathbf{C} \times \mathbf{D})$.

Solution The ith component is

$$\epsilon_{ijk}(\mathbf{A} \times \mathbf{B})_j(\mathbf{C} \times \mathbf{D})_k = \epsilon_{ijk}\epsilon_{jmn}A_mB_n\epsilon_{kpq}C_pD_q$$

If we sum over j first, only the two factors are involved. Rewriting them as $\epsilon_{kij}\,\epsilon_{mnj}$, we use eq. (1.40) to transform the expression to

$$(\delta_{km}\,\delta_{in} - \delta_{kn}\,\delta_{im})A_mB_n\epsilon_{kpq}C_pD_q$$

Summing over the subscripts of the substitution tensors is easy, yielding

$$A_kB_i\epsilon_{kpq}C_pD_q - A_iB_k\epsilon_{kpq}C_pD_q$$

Now we identify $\epsilon_{kpq}A_kC_pD_q$ as the triple scalar product, recalling eq. (1.39); similarly for $\epsilon_{kpq}B_kC_pD_q$. So we are left with

$$[\mathbf{A},\mathbf{C},\mathbf{D}]B_i - [\mathbf{B},\mathbf{C},\mathbf{D}]A_i$$

which is the ith component of $[\mathbf{A},\mathbf{C},\mathbf{D}]\mathbf{B} - [\mathbf{B},\mathbf{C},\mathbf{D}]\mathbf{A}$. We have "discovered" formula (1.32).

EXERCISES

1. Simplify $(\mathbf{A} \times \mathbf{B}) \cdot \mathbf{C}$.
2. Simplify $(\mathbf{A} \times \mathbf{B}) \cdot (\mathbf{C} \times \mathbf{D})$.
3. Simplify $(\mathbf{A} \times \mathbf{B}) \cdot (\mathbf{B} \times \mathbf{C}) \times (\mathbf{C} \times \mathbf{A})$.

2 Vector Functions of a Single Variable

2.1 Differentiation

The theory of vector functions parallels that of real-valued functions. A vector-valued function $\mathbf{F}(t)$ is a rule that associates a vector $\mathbf{F}$ with each real number t in some set, usually an interval $(t_1 \leq t \leq t_2)$ or a collection of intervals. For example, $\mathbf{F}(t) = (1/t)\mathbf{i}$ is defined for $-\infty < t < 0$ and $0 < t < \infty$.

The concept of a limit can be applied to vector functions. The expression

$$\lim_{t \to t_0} \mathbf{F}(t) = \mathbf{A} \tag{2.1}$$

means that, given any positive number ϵ, no matter how small, one can find a positive number δ such that $|\mathbf{F}(t) - \mathbf{A}| < \epsilon$ whenever $0 < |t - t_0| < \delta$.

This has a simple intuitive meaning. It means that the magnitude of $\mathbf{F}(t)$ is approaching the magnitude of $\mathbf{A}$, and that (if $\mathbf{A}$ is nonzero) the angle between them is approaching zero (see fig. 2.1). Equivalently, the *components* of $\mathbf{F}(t)$ are approaching the *components* of $\mathbf{A}$.

The definition just given is identical to that provided in calculus books for real-valued functions, except that the expression $|\mathbf{F}(t) - \mathbf{A}|$ now refers to the magnitude of the vector rather than to the absolute value of a number.

A vector function $\mathbf{F}$ is said to be *continuous* at t_0 if

$$\lim_{t \to t_0} \mathbf{F}(t) = \mathbf{F}(t_0) \tag{2.2}$$

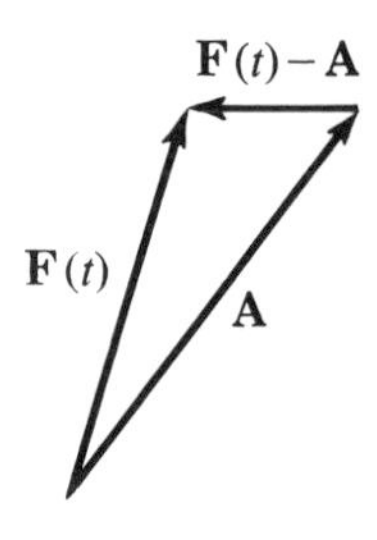

Figure 2.1

It is said to be *differentiable* at t_0 if the limit

$$\lim_{\Delta t \to 0} \frac{\mathbf{F}(t_0 + \Delta t) - \mathbf{F}(t_0)}{\Delta t} \tag{2.3}$$

exists; this is then called the *derivative* of $\mathbf{F}(t)$ at t_0 and is written $\mathbf{F}'(t_0)$ or $\dfrac{d\mathbf{F}}{dt}(t_0)$. The derivative is also a vector function.

If $\mathbf{F}(t)$ is continuous (or differentiable) at *every* point t for which it is defined, we shall simply say that $\mathbf{F}(t)$ is *continuous* (or *differentiable*).

Example 2.1 Use eq. (2.3) to compute the derivative of $\mathbf{F}(t) = t^2\mathbf{i} + t^3\mathbf{j}$.

Solution We are inclined to write $\mathbf{F}'(t) = 2t\mathbf{i} + 3t^2\mathbf{j}$. Is this correct? By eq. (2.3) we have

$$\begin{aligned}\mathbf{F}'(t) &= \lim_{\Delta t \to 0} \frac{(t + \Delta t)^2\mathbf{i} + (t + \Delta t)^3\mathbf{j} - t^2\mathbf{i} - t^3\mathbf{j}}{\Delta t} \\ &= \lim_{\Delta t \to 0} \frac{(2t\Delta t + \Delta t^2)\mathbf{i} + (3t^2\Delta t + 3t\Delta t^2 + \Delta t^3)\mathbf{j}}{\Delta t} \\ &= 2t\mathbf{i} + 3t^2\mathbf{j}\end{aligned}$$

In general, to differentiate a vector expressed in terms of **i, j,** and **k,** simply differentiate the coefficients.

The fundamental theorems concerning differentiation of vector-valued functions are similar to those for real-valued functions, except that when differentiating the vector product of two vector functions, one must be careful to preserve the order of factors, since the vector product is not a commutative operation.

THEOREM 2.1 *If* **F** *and* **G** *are differentiable vector functions, then so also is their sum* **F** + **G**, *and the derivative of the function* **F** + **G** *is the sum of the derivatives of* **F** *and* **G** *respectively:*

$$\frac{d}{dt}(\mathbf{F} + \mathbf{G}) = \frac{d\mathbf{F}}{dt} + \frac{d\mathbf{G}}{dt} \tag{2.4}$$

THEOREM 2.2 *If* **F** *is a differentiable vector function, and s is a differentiable scalar function, then the product s***F** *is a differentiable vector function, and*

$$\frac{d}{dt}(s\mathbf{F}) = \frac{ds}{dt}\mathbf{F} + s\frac{d\mathbf{F}}{dt} \tag{2.5}$$

THEOREM 2.3 *If* **F** *and* **G** *are differentiable vector functions, then* **F** · **G** *is a differentiable scalar function, and*

$$\frac{d}{dt}(\mathbf{F} \cdot \mathbf{G}) = \frac{d\mathbf{F}}{dt} \cdot \mathbf{G} + \mathbf{F} \cdot \frac{d\mathbf{G}}{dt} \tag{2.6}$$

THEOREM 2.4 *If* **F** *and* **G** *are differentiable vector functions, then* **F** × **G** *is also a differentiable vector function, and*

$$\frac{d}{dt}(\mathbf{F} \times \mathbf{G}) = \frac{d\mathbf{F}}{dt} \times \mathbf{G} + \mathbf{F} \times \frac{d\mathbf{G}}{dt} \tag{2.7}$$

The reader who is familiar with the proofs of the sum and product formulas of elementary calculus will have no difficulty filling in the proofs of these theorems.

Example 2.2 Prove Theorem 2.4.

Solution With the definition of the derivative in mind, we write

$$\frac{\mathbf{F}(t + \Delta t) \times \mathbf{G}(t + \Delta t) - \mathbf{F}(t) \times \mathbf{G}(t)}{\Delta t}$$
$$= \frac{[\mathbf{F}(t + \Delta t) - \mathbf{F}(t)] \times \mathbf{G}(t + \Delta t)}{\Delta t} + \frac{\mathbf{F}(t) \times [\mathbf{G}(t + \Delta t) - \mathbf{G}(t)]}{\Delta t}$$

As Δt goes to zero, the right-hand side approaches a limiting value given by

$$\frac{d\mathbf{F}}{dt} \times \mathbf{G} + \mathbf{F} \times \frac{d\mathbf{G}}{dt}$$

Since the limit of the left-hand side is $(d/dt)(\mathbf{F} \times \mathbf{G})$, we have proved the theorem.

It follows from (2.4) and (2.5) that if

$$\mathbf{F}(t) = P(t)\mathbf{i} + Q(t)\mathbf{j} + R(t)\mathbf{k}$$

then

$$\mathbf{F}'(t) = P'(t)\mathbf{i} + Q'(t)\mathbf{j} + R'(t)\mathbf{k} \tag{2.8}$$

Thus vector differentiation is like scalar differentiation, treating **i**, **j**, and **k** as constants.

Example 2.3 Let $\mathbf{F}(t) = \mathbf{i} + 2\mathbf{j} - \mathbf{k}$. Then $\mathbf{F}$ is a constant vector-valued function, and its derivative with respect to t is identically equal to the zero vector for all t.

Example 2.4 If $\mathbf{F}(t) = \sin t\mathbf{i} + \cos t\mathbf{j} + t\mathbf{k}$, then $\mathbf{F}'(t) = \cos t\mathbf{i} - \sin t\mathbf{j} + \mathbf{k}$.

Example 2.5 If $\mathbf{F}(t) = t^3\mathbf{j} - \mathbf{k}$, then $\mathbf{F}'(t) = 3t^2\mathbf{j}$.

Example 2.6 If $\mathbf{F}'(t) = \mathbf{0}$, then $\mathbf{F}(t) = \mathbf{C}$, where the constant $\mathbf{C}$ is a *vector.*

Example 2.7 Prove that, if $\mathbf{F}(t)$ has constant nonzero magnitude (varies only in direction), then $\mathbf{F}'(t)$ is either the zero vector or a nonzero vector perpendicular to $\mathbf{F}(t)$.

Solution If $|\mathbf{F}(t)| =$ constant, then we must have

$$\mathbf{F} \cdot \mathbf{F} = \text{constant}$$

and differentiating with respect to t, using (2.6), we have

$$\frac{d\mathbf{F}}{dt} \cdot \mathbf{F} + \mathbf{F} \cdot \frac{d\mathbf{F}}{dt} = 0$$

$$2\mathbf{F} \cdot \frac{d\mathbf{F}}{dt} = 0$$

Hence the scalar product of $\mathbf{F}$ with $d\mathbf{F}/dt$ is identically zero. This can happen only if the vectors $\mathbf{F}$ and $d\mathbf{F}/dt$ are perpendicular, or if one of them is the zero vector. This fact is well worth remembering: *the derivative of a vector of constant length is perpendicular to the vector, or is zero.*

EXERCISES

1. Let $\mathbf{F}(t) = \sin t\mathbf{i} + \cos t\mathbf{j} + \mathbf{k}$.
   (a) Find $\mathbf{F}'(t)$.
   (b) Show that $\mathbf{F}'(t)$ is always parallel to the xy plane.
   (c) For what values of t is $\mathbf{F}'(t)$ parallel to the xz plane?
   (d) Does $\mathbf{F}(t)$ have constant magnitude?
   (e) Does $\mathbf{F}'(t)$ have constant magnitude?
   (f) Compute $\mathbf{F}''(t)$.
2. Find $\mathbf{F}'(t)$ in each of the following cases:
   (a) $\mathbf{F}(t) = 3t\mathbf{i} + t^3\mathbf{j}$
   (b) $\mathbf{F}(t) = \sin t\mathbf{i} + e^{-t}\mathbf{j} + 3\mathbf{k}$
   (c) $\mathbf{F}(t) = (e^t\mathbf{i} + \mathbf{j} + t^2\mathbf{k}) \times (t^3\mathbf{i} + \mathbf{j} - \mathbf{k})$
   (d) $\mathbf{F}(t) = (\sin t + t^3)(\mathbf{i} + \mathbf{j} + 2\mathbf{k})$
   (e) $\mathbf{F}(t) = 3\mathbf{i} + \mathbf{k}$

3. Find $f'(t)$ in each of the following cases:
   (a) $f(t) = (3t\mathbf{i} + 5t^2\mathbf{j}) \cdot (t\mathbf{i} - \sin t\mathbf{j})$
   (b) $f(t) = |2t\mathbf{i} + 2t\mathbf{j} - \mathbf{k}|$
   (c) $f(t) = [(\mathbf{i} + \mathbf{j} - 2\mathbf{k}) \times (3t^4\mathbf{i} + t\mathbf{j})] \cdot \mathbf{k}$
4. Show that

$$\frac{d}{dt}\left(\mathbf{R} \times \frac{d\mathbf{R}}{dt}\right) = \mathbf{R} \times \frac{d^2\mathbf{R}}{dt^2}$$

5. Given the three vectors $\mathbf{A} = 3\mathbf{i} + 2\mathbf{j} + 6\mathbf{k}$, $\mathbf{B} = 3\mathbf{i} + 4\mathbf{k}$, and $\mathbf{C} = 2\mathbf{i} - 2\mathbf{j} + \mathbf{k}$, evaluate
   (a) $|\mathbf{A}|$
   (b) $\mathbf{A} \cdot \mathbf{B}$
   (c) $\mathbf{B} \times \mathbf{C}$
   (d) $\mathbf{B} \cdot \mathbf{B} \times \mathbf{C}$
   (e) $[\mathbf{A},\mathbf{B},\mathbf{C}]$
   (f) $\mathbf{A}/|\mathbf{B}|$
   (g) $\mathbf{A} \times (\mathbf{B} \times \mathbf{C})$
   (h) $\dfrac{d}{dt}(\mathbf{A} + \mathbf{B}t)$
   (i) $\dfrac{d}{dt}(\mathbf{B} \times t\mathbf{C})$

2.2 Space Curves, Velocities, and Tangents

In the first chapter we showed that the parametric equations of a line can be written in vector form:

$$\mathbf{R}(t) = \mathbf{R}_0 + t\mathbf{V} \tag{2.9}$$

Here $\mathbf{R}_0$ is the position vector of a fixed point on the line, $\mathbf{V}$ is parallel to the line, and as t assumes values from $-\infty$ to $+\infty$, the tip of the vector $\mathbf{R}$ traces out the line in (x,y,z) space. We can also regard eq. (2.9) as defining $\mathbf{R}$ as a vector function of t (whose derivative is, of course, $\mathbf{V}$).

In this section we shall consider equations of the form $\mathbf{R} = \mathbf{R}(t)$ where the function $\mathbf{R}(t)$ is more complicated than eq. (2.9). Of course, the equation $\mathbf{R} = \mathbf{R}(t)$ can be written out in terms of its components, giving the system of equations

$$\begin{aligned} x &= x(t) \\ y &= y(t) \\ z &= z(t) \end{aligned} \tag{2.10}$$

where x, y, and z are simply real-valued functions of t.

As t increases from its initial value t_1 to the value t_2, the point (x,y,z) [i.e., the tip of the position vector $\mathbf{R}(t)$] traces out some geometric object in space. In the case of eq. (2.9), the object is a segment of a straight line. For more complicated (continuous) vector functions, this locus of points will be some more general kind of one-dimensional object, which we can call a *space curve* or an *arc*. [We say it's *one*-dimensional because any point on it can be located, via the continuous function $\mathbf{R}(t)$, by specifying the *single* number t.] We use the term "curve" even if the trace of $\mathbf{R}(t)$ is a straight line.

Thus we have associated with every continuous vector function $\mathbf{R}(t)$ a curve in space, which is the set of values assumed by $\mathbf{R}(t)$ as t varies over an interval. Notice that this is quite different from *graphing* x, y, and/or z as functions of t; the *curve* traced by $\mathbf{R}(t)$ is a threadlike collection of points in (x,y,z) space. One cannot read, from the curve alone, the value of t corresponding to a given point. So the curve itself contains much less information than the *function* $\mathbf{R}(t)$. Remember that as far as the curve is concerned, t is a sort of invisible dummy variable, which we have glamorized by awarding it the officious title "parameter."

Remember also that there are any number of different *parametrizations* for a given curve. For example, if $\mathbf{W} = \frac{1}{3}\mathbf{V}$, the function

$$\mathbf{R}_1(t) = \mathbf{R}_0 + t\,\mathbf{W}$$

traces out exactly the same straight line as (2.9) for $-\infty < t < \infty$, as does the function

$$\mathbf{R}_2(t) = \mathbf{R}_0 + \tan t\,\mathbf{W}$$

for $-\pi/2 < t < \pi/2$. Thus we must keep in mind that many different *functions* may parametrize the same *curve*.

There are a few "standard" parametrizations with which the reader should become familiar, in order to have a supply of examples for applying the theorems in this chapter. The first is the *straight line* described by eq. (2.9). The demarcations in figure 2.2 suggest that if the parameter t were to be interpreted as *time*, then $\mathbf{V}$ would correspond to the *velocity* of a particle tracing out the line in accordance with eq. (2.9).

To parametrize the *circle* $x^2 + y^2 = 1$ in the $z = 0$ plane, it is most convenient to use the trigonometric functions, with t interpreted as the angle depicted in figure 2.3. Then we have $x = \cos t$, $y = \sin t$, or

$$\mathbf{R}(t) = \cos t\,\mathbf{i} + \sin t\,\mathbf{j} \tag{2.11}$$

If the circle is stretched to one of radius ρ, and its center shifted to $\mathbf{R}_0 = x_0\mathbf{i} + y_0\mathbf{j} + z_0\mathbf{k}$, the parametrization becomes

$$x = \rho\cos t + x_0 \qquad y = \rho\sin t + y_0 \qquad z = z_0$$

$$\mathbf{R}(t) = \mathbf{R}_0 + \rho\cos t\,\mathbf{i} + \rho\sin t\,\mathbf{j} \tag{2.12}$$

Finally, it is obvious that $\mathbf{i}$ and $\mathbf{j}$ can be replaced by any pair of perpendicular unit vectors $(\mathbf{e}_1, \mathbf{e}_2)$ to depict *a circle of radius* ρ, *centered at* $\mathbf{R}_0$, *lying in the* $\mathbf{e}_1, \mathbf{e}_2$ *plane:*

$$\mathbf{R}(t) = \mathbf{R}_0 + \rho\cos t\,\mathbf{e}_1 + \rho\sin t\,\mathbf{e}_2 \tag{2.13}$$

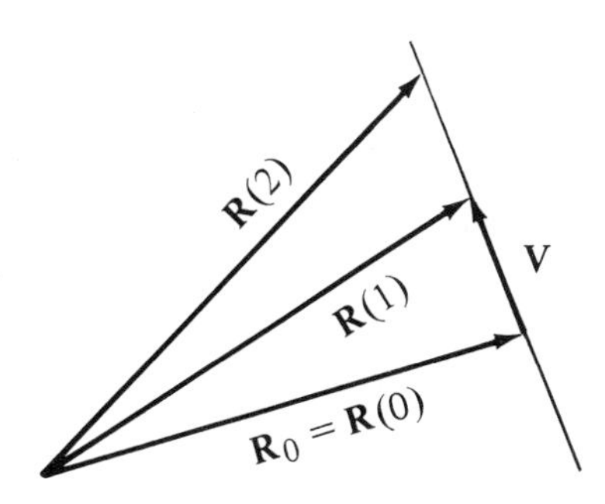

Figure 2.2

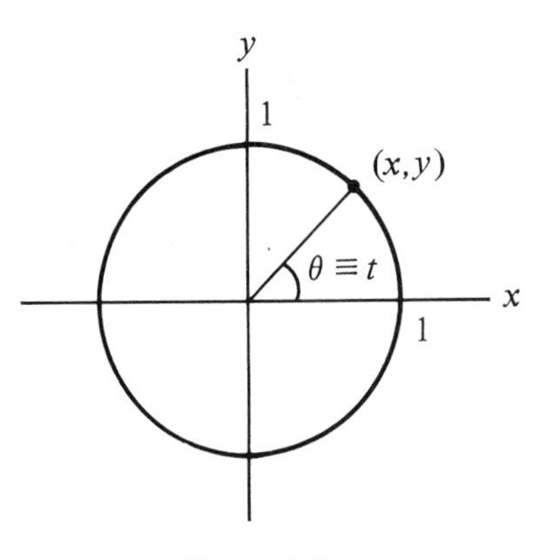

Figure 2.3

Example 2.8 Parametrize the circle of radius 2, parallel to the xz plane, centered at (0,1,0).

Solution We use $\mathbf{R_0} = \mathbf{j}$, $\mathbf{e_1} = \mathbf{i}$, and $\mathbf{e_2} = \mathbf{k}$ in (2.13) to derive

$$\mathbf{R}(t) = \mathbf{j} + 2 \cos t\, \mathbf{i} + 2 \sin t\, \mathbf{k}$$

Example 2.9 Parametrize the circle cutting the axes at the points $x = 1$, $y = 1$, and $z = 1$ respectively (fig. 2.4).

Solution Clearly the plane of the circle is normal to the direction $\mathbf{i} + \mathbf{j} + \mathbf{k}$. Thus, for example, the mutually perpendicular unit vectors

$$\mathbf{e_1} = \frac{\mathbf{i} - \mathbf{j}}{\sqrt{2}} \qquad \mathbf{e_2} = \frac{\mathbf{i} + \mathbf{j} - 2\mathbf{k}}{\sqrt{6}}$$

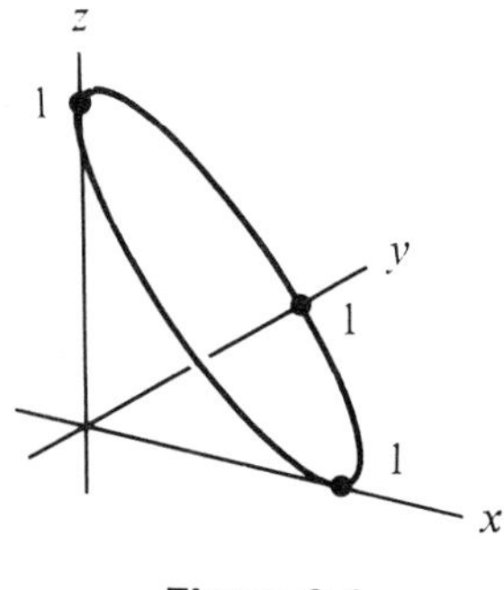

Figure 2.4

lie in this plane. The midpoint of the circle is the center of the equilateral triangle, which is given by $\frac{1}{3}(\mathbf{i} + \mathbf{j} + \mathbf{k})$ (if this is not obvious, please reread exercise 5, section 1.7). The radius is the distance from the midpoint to, say, **i**: $\rho = |\mathbf{i} - \frac{1}{3}(\mathbf{i} + \mathbf{j} + \mathbf{k})| = \sqrt{6}/3$. Therefore

$$\begin{aligned}\mathbf{R}(t) &= \frac{\mathbf{i} + \mathbf{j} + \mathbf{k}}{3} + \frac{\sqrt{6}}{3} \cos t \, \frac{(\mathbf{i} - \mathbf{j})}{\sqrt{2}} + \frac{\sqrt{6}}{3} \sin t \, \frac{(\mathbf{i} + \mathbf{j} - 2\mathbf{k})}{\sqrt{6}} \\ &= \frac{\mathbf{i} + \mathbf{j} + \mathbf{k}}{3} + \frac{\sqrt{3}}{3} \cos t \, (\mathbf{i} - \mathbf{j}) + \frac{1}{3} \sin t \, (\mathbf{i} + \mathbf{j} - 2\mathbf{k})\end{aligned}$$

The simplest nonplanar curve is the *helix,* depicted in figure 2.5. Here the x and y coordinates describe a circle as before, but the z coordinate increases (or decreases) in direct proportion to the angle $\theta = t$; thus

$$x = \rho \cos t \qquad y = \rho \sin t \qquad z = at \tag{2.14}$$

The vertical distance between the "coils" equals the increase in z as t jumps by 2π. Thus the *pitch* (as it is called) is given by $2\pi a$.

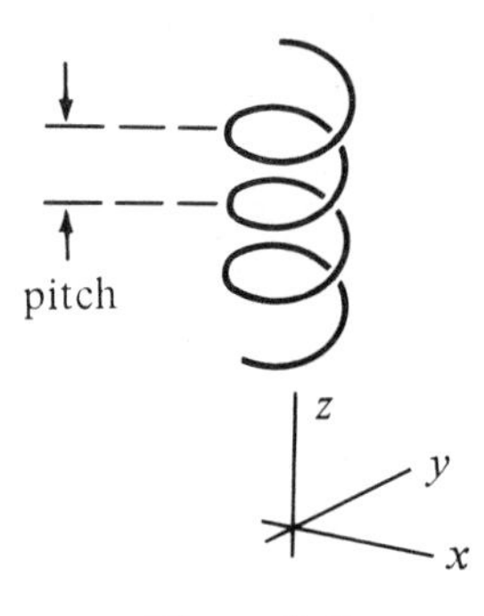

Figure 2.5

More generally, *if* $\mathbf{e}_1$, $\mathbf{e}_2$, *and* $\mathbf{e}_3$ *form a right-handed system of mutually perpendicular unit vectors, then*

$$\mathbf{R}(t) = \mathbf{R}_0 + \rho \cos t\, \mathbf{e}_1 + \rho \sin t\, \mathbf{e}_2 + at\, \mathbf{e}_3 \tag{2.15}$$

parametrizes a helix of radius ρ, *with axis passing through* $\mathbf{R}_0$ *and parallel to* $\mathbf{e}_3$. *Its pitch is* $2\pi|a|$, *and the helix is said to be right-handed if* $a > 0$, *left-handed if* $a < 0$.

Example 2.10 Describe the helix generated by $x = \cos t$, $y = 2t$, $z = \sin t$.

Solution This has the form of eq. (2.15), with $\mathbf{R}_0 = \mathbf{0}$, radius $\rho = 1$, $\mathbf{e}_1 = \mathbf{i}$, and $\mathbf{e}_2 = \mathbf{k}$. Note, however, that $\mathbf{i}, \mathbf{k}, \mathbf{j}$ is *not* a right-handed system, so we must take $\mathbf{e}_3 = -\mathbf{j}$. We rewrite the equations as

$$\mathbf{R} = \cos t\, \mathbf{e}_1 + \sin t\, \mathbf{e}_2 - 2t\mathbf{e}_3$$

and see that this helix is left-handed and has pitch 4π.

Example 2.11 Describe the helix of example 2.10 using nonparametric equations.

Solution We can eliminate t in favor of y; $t = y/2$. This results in

$$x = \cos y/2 \qquad z = \sin y/2$$

as a nonparametric form.

The interpretation of the parameter t as denoting time, and the equation $\mathbf{R} = \mathbf{R}(t)$ as giving the trajectory of a particle, is useful for curves in general. During a time interval of duration Δt, the position vector of the particle changes from the value $\mathbf{R}(t)$ to a new value, $\mathbf{R}(t + \Delta t)$. The *displacement* of the particle during this interval of time is

$$\Delta\mathbf{R} = \mathbf{R}(t + \Delta t) - \mathbf{R}(t) = \Delta x\, \mathbf{i} + \Delta y\, \mathbf{j} + \Delta z\, \mathbf{k} \tag{2.16}$$

If the displacement is divided by the scalar Δt, we obtain the *average velocity* of the particle during the time interval:

$$\frac{\Delta\mathbf{R}}{\Delta t} = \frac{\Delta x}{\Delta t}\mathbf{i} + \frac{\Delta y}{\Delta t}\mathbf{j} + \frac{\Delta z}{\Delta t}\mathbf{k} \tag{2.17}$$

(In fig. 2.6, we take Δt less than unity; hence the vector $\Delta\mathbf{R}/\Delta t$ is greater in magnitude than $\Delta\mathbf{R}$.)

If $\mathbf{R}$ is differentiable, the average velocity $\Delta\mathbf{R}/\Delta t$ tends to a limit as Δt tends to zero. This limit is, by definition, the (instantaneous) *velocity* $\mathbf{v}$:

$$\mathbf{v}(t) = \mathbf{R}'(t) = \frac{d\mathbf{R}}{dt} = \frac{dx}{dt}\mathbf{i} + \frac{dy}{dt}\mathbf{j} + \frac{dz}{dt}\mathbf{k} \tag{2.18}$$

The magnitude of $\mathbf{v}$ is called the *speed;* it may be denoted v.

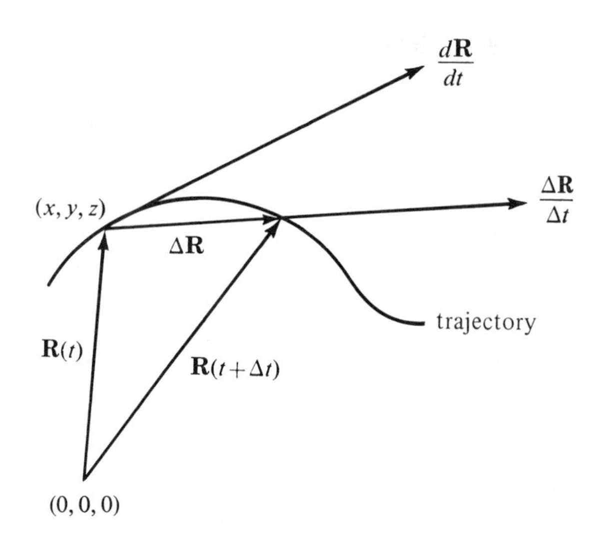

Figure 2.6

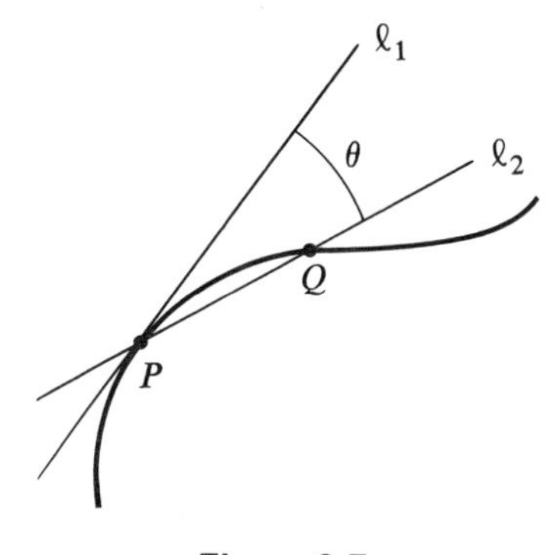

Figure 2.7

Figure 2.6 seems to indicate that the velocity vector $d\mathbf{R}/dt$ is tangent to the curve. Let us explore this further. Referring to figure 2.7, we say informally that the line l_1 is tangent to the curve at P if the angle θ, between l_1 and the secant line l_2 determined by P and Q, goes to zero as Q approaches P along the curve; that is, the direction of the secant line l_2 approaches that of l_1 as Q approaches P.

If we try to apply this to the situation in figure 2.6, we identify the direction of the secant line as that of $\Delta\mathbf{R}/\Delta t$. Thus, as Δt goes to zero, the secant line must have a limiting direction, namely that of $d\mathbf{R}/dt$, unless, of course, the latter is the zero vector, which has no direction. We have shown the following: if the vector function $\mathbf{R}(t)$ has a nonzero derivative at t_0, then the curve parametrized by $\mathbf{R} = \mathbf{R}(t)$ has a tangent at $\mathbf{R}(t_0)$ whose direction coincides with that of $d\mathbf{R}/dt$. In short, *$d\mathbf{R}/dt$ is tangent to the curve.*

It is conventional to denote by the letter $\mathbf{T}$ a *unit vector* tangent to a curve. Such a $\mathbf{T}$ is defined by the expression

$$\mathbf{T} = \frac{(dx/dt)\mathbf{i} + (dy/dt)\mathbf{j} + (dz/dt)\mathbf{k}}{\sqrt{(dx/dt)^2 + (dy/dt)^2 + (dz/dt)^2}} \tag{2.19}$$

obtained by dividing the above vector by its own magnitude.

Example 2.12 Determine the unit vector tangent to the circle $x = \cos t$, $y = \sin t$, $z = 0$, at
(a) $t = 0$
(b) $t = \pi/2$

Solutions The answers are obviously (a) $\mathbf{j}$, (b) $-\mathbf{i}$, as can be seen from figure 2.8. These answers can be obtained also by use of eq. (2.19), which gives

$$\mathbf{T} = \frac{-\sin t\,\mathbf{i} + \cos t\,\mathbf{j}}{\sqrt{\sin^2 t + \cos^2 t}} = -\sin t\,\mathbf{i} + \cos t\,\mathbf{j}$$

At $t = 0$, we have $\mathbf{T} = -\sin 0\,\mathbf{i} + \cos 0\,\mathbf{j} = \mathbf{j}$, and at $t = \pi/2$, $\mathbf{T} = -\sin(\pi/2)\,\mathbf{i} + \cos(\pi/2)\,\mathbf{j} = -\mathbf{i}$.

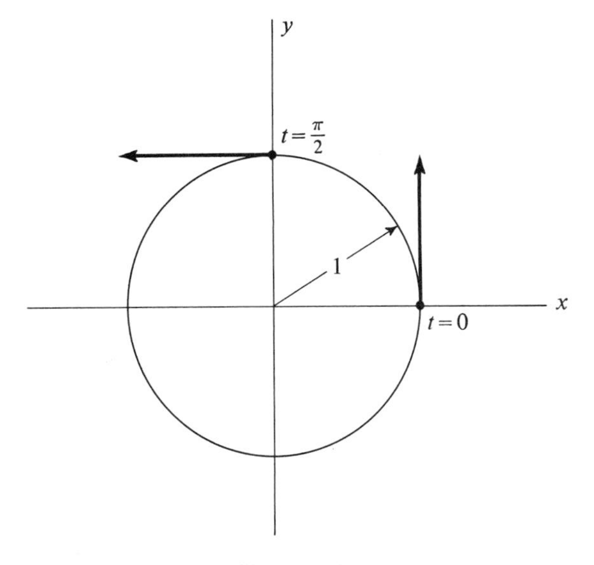

Figure 2.8

Example 2.13 Find the unit vector tangent to the curve $x = t$, $y = t^2$, $z = t^3$, at the point (2,4,8).

Solution By eq. (2.19) we have

$$\mathbf{T} = \frac{\mathbf{i} + 2t\,\mathbf{j} + 3t^2\,\mathbf{k}}{\sqrt{1 + 4t^2 + 9t^4}}$$

When $t = 2$ we have $(x,y,z) = (2,4,8)$ and $\mathbf{T} = (1/\sqrt{161})(\mathbf{i} + 4\mathbf{j} + 12\mathbf{k})$.

We now wish to introduce some nomenclature for describing the curves pictured in figures 2.9–2.12. The simplest of these is the one in figure 2.9, which has a continuously turning tangent at every point and no self-intersections. From the above discussion we can see how to guarantee these properties. Accordingly, we say that an arc is *smooth* if it has a parametrization $\mathbf{R} = \mathbf{R}(t)$, $t_1 \leq t \leq t_2$, satisfying the following conditions:

(*i*) $d\mathbf{R}/dt$ exists and is a continuous function of t for all values of t in the interval $t_1 \leq t \leq t_2$.
(*ii*) To distinct values of t in the interval $t_1 < t < t_2$ there correspond distinct points.
(*iii*) There is no value of t in the interval $t_1 \leq t \leq t_2$ for which $d\mathbf{R}/dt$ is the zero vector.

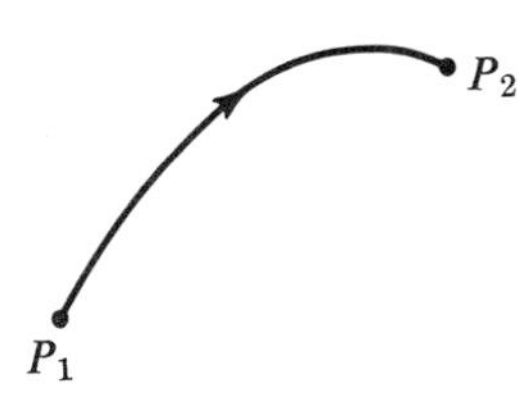

Figure 2.9

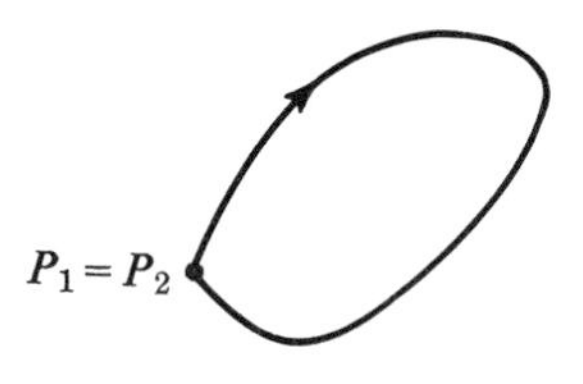

Figure 2.10

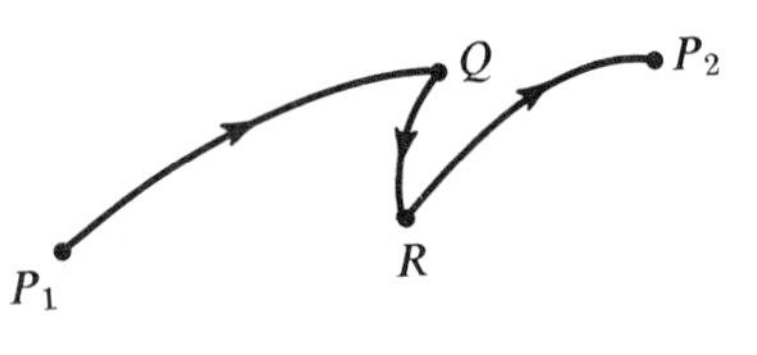

Figure 2.11

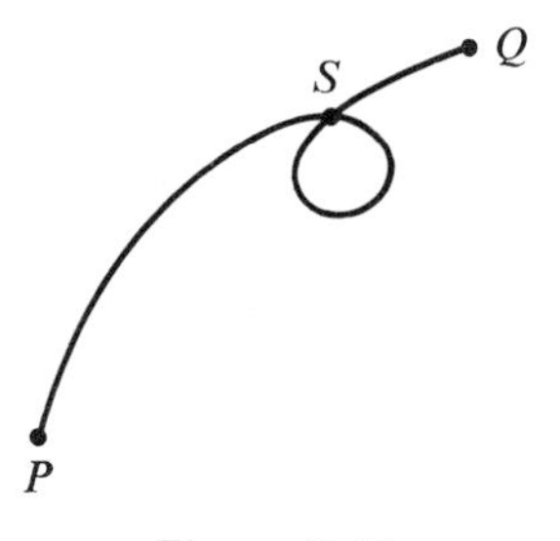

Figure 2.12

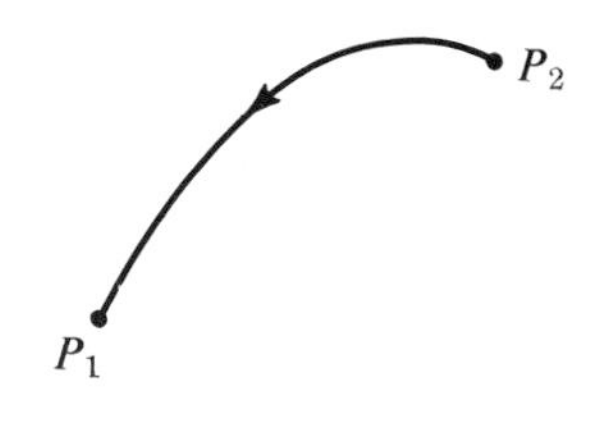

Figure 2.13

We allow the possibility that a smooth arc can be *closed,* as in figure 2.10, if $\mathbf{R}(t_1) = \mathbf{R}(t_2)$.

Notice that to show an arc is smooth we need produce only *one* such parametrization; there may be others, which violate the three conditions. For example, a straight line segment can be parametrized by $\mathbf{R}(t) = \mathbf{R}_0 + t^3\mathbf{V}$, $-1 \leq t \leq 1$, in violation of (iii); yet it is a smooth arc.

The arc in figure 2.11 is not smooth since it fails to have a tangent at Q and R. However, it consists of a finite number of smooth arcs joined together, and it does not cross itself; such a curve is called *regular.* The curve in figure 2.12 is not regular, because of the crossing at S.

In figures 2.9 through 2.11, we have indicated the direction in which the particle is traversing the curve by a small arrow. Strictly speaking, any curve is nothing more than a collection of points in space. When, however, we indicate a direction along a smooth arc, as we have in these diagrams, then we say that the arc has been *oriented.* Obviously, a smooth arc can be oriented in only two ways. The arc in figure 2.13 is a replica of that in figure 2.9, but is oriented in the opposite way.

When an arc is described by equations such as (2.10), in terms of a parameter t, the orientation is usually understood to be determined by that parameter: the direction is that of increasing t. For example, the closed arc

$$\begin{aligned} x &= \cos t \\ y &= \sin t \\ z &= 0 \end{aligned} \tag{2.20}$$

is simply a circle of unit radius in the xy plane. As t increases from 0 to 2π the point moves counterclockwise around the circle, as shown in figure 2.14. The same arc with opposite orientation can be given parametrically by

$$\begin{aligned} x &= \cos t \\ y &= -\sin t \\ z &= 0 \end{aligned} \tag{2.21}$$

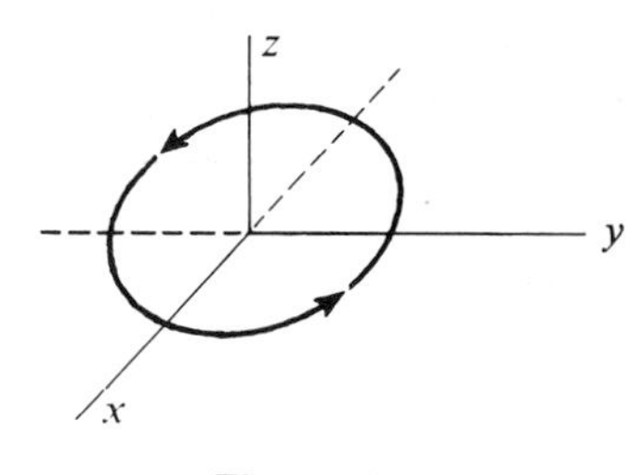

Figure 2.14

The eqs. (2.21) specify the same arc as eq. (2.20), but with opposite orientation, since as t increases from 0 to 2π, the point (x,y,z) traverses the circle in the opposite direction.

The same circle can be represented *nonparametrically* (i.e., without a dummy variable) by the equations

$$x^2 + y^2 = 1$$
$$z = 0 \tag{2.22}$$

There is no way of knowing from eqs. (2.22) which orientation is intended. Note that eqs. (2.22) represent the arc as the intersection of two surfaces (a cylinder and a plane). When one specifies an oriented arc as the intersection of two surfaces, by giving two equations, it is necessary to specify the orientation separately, either verbally or by drawing a diagram.

Undoubtedly the reader is familiar with the notion of arc length, which is discussed in calculus books (at least for plane curves). This notion generalizes easily to space curves.

Suppose C is a smooth space curve. Let us subdivide C into smaller arcs, and approximate it by a polygonal path consisting of n straight-line segments joining the endpoints of the arcs (fig. 2.15). That is, we select points $Q_0, Q_1, \ldots, Q_n$ along C, in that order, with Q_0 and Q_n the endpoints of C. For each $k = 0, 1, \ldots, n$, let $\mathbf{R}_k$ be the position vector to the point Q_k, and let $\Delta\mathbf{R}_k = \mathbf{R}_k - \mathbf{R}_{k-1}$, for $k = 1, 2, \ldots, n$. The total length of the polygonal path is then $\Sigma_{k=1}^{n}|\Delta\mathbf{R}_k|$. The length of the space curve C is defined to be the limit of sums of this form, where the approximating polygonal paths are obtained by taking increasingly smaller subdivisions while n increases without bound.

We can compute this limit when the curve is parametrized by $\mathbf{R}(t)$ for, say, $a \leq t \leq b$ as follows. The length of $\Delta\mathbf{R}_k$ is

$$|\Delta\mathbf{R}_k| = \sqrt{(\Delta x_k)^2 + (\Delta y_k)^2 + (\Delta z_k)^2}$$

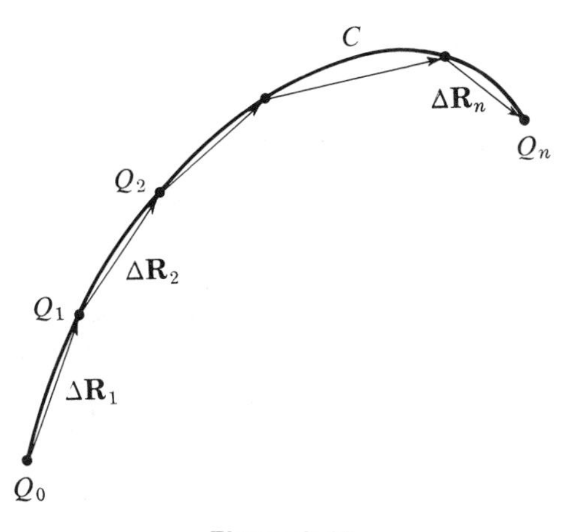

Figure 2.15

Let t_k be the values of t that correspond to the points $\mathbf{R}_k$; that is, $\mathbf{R}_k = \mathbf{R}(t_k)$. Then, because $d\mathbf{R}/dt$ is continuous, the mean value theorem of calculus ensures that for some number τ_k between t_{k-1} and t_k,

$$\Delta x_k = x_k - x_{k-1} = (t_k - t_{k-1}) \frac{dx}{dt}(\tau_k)$$

Similarly, there are numbers τ_k' and τ_k'' in the same interval such that

$$\Delta y_k = (t_k - t_{k-1}) \frac{dy}{dt}(\tau_k')$$

$$\Delta z_k = (t_k - t_{k-1}) \frac{dz}{dt}(\tau_k'')$$

So for the length of the polygonal path we have

$$\sum_{k=1}^{n} |\Delta \mathbf{R}_k| = \sum_{k=1}^{n} \left[\left(\frac{dx}{dt}(\tau_k)\right)^2 + \left(\frac{dy}{dt}(\tau_k')\right)^2 + \left(\frac{dz}{dt}(\tau_k'')\right)^2 \right]^{1/2} (t_k - t_{k-1})$$

Now if the polygonal subdivision is made finer, the differences $t_k - t_{k-1}$ become smaller and this sum approaches the integral

$$\int_a^b \left[\left(\frac{dx}{dt}\right)^2 + \left(\frac{dy}{dt}\right)^2 + \left(\frac{dz}{dt}\right)^2 \right]^{1/2} dt \qquad (2.23)$$

as a limit. Recognizing the integrand as $|d\mathbf{R}/dt|$, we see that *the length of the curve C is given by*

$$\int_a^b \left|\frac{d\mathbf{R}}{dt}\right| dt \tag{2.24}$$

The arc length of a regular curve is defined to be the sum of the lengths of the various smooth curves that constitute it.

Sometimes it is possible to write two of the variables, say y and z, in terms of the other, say x. In that case dy and dz may be expressed in terms of x and dx, and the integral is taken with respect to x, the limits of integration being the values of x corresponding to t_1 and t_2.

If the arc P_1P_2 lies entirely in the xy plane, which is the simplest case treated in calculus books, then z is identically equal to zero and so $dz/dt = 0$ and, by eliminating the parameter t, eq. (2.23) may be written in the familiar alternative form

$$\int_{x_1}^{x_2} \left[1 + \left(\frac{dy}{dx}\right)^2\right]^{1/2} dx \tag{2.25}$$

provided the integral exists, or

$$\int_{y_1}^{y_2} \left[\left(\frac{dx}{dy}\right)^2 + 1\right]^{1/2} dy \tag{2.26}$$

provided this integral exists. It is possible that these integrals may not exist. For example, if the arc P_1P_2 contains a segment that is parallel to the y axis, then dy/dx will not exist along this segment (i.e., dy/dx is "infinite") and (2.25) will not make sense.

Example 2.14 Find the arc length between (0,0,1) and (1,0,1) of the helix $y = \sin 2\pi x$, $z = \cos 2\pi x$.

Solution

$$dx^2 + dy^2 + dz^2 = dx^2 + 4\pi^2 \cos^2 2\pi x\, dx^2 + 4\pi^2 \sin^2 2\pi x\, dx^2$$

Hence the integral is

$$\int_0^1 (1 + 4\pi^2)^{1/2}\, dx = (1 + 4\pi^2)^{1/2}$$

The expression (2.23) is sometimes written

$$\int_C [(dx)^2 + (dy)^2 + (dz)^2]^{1/2} \qquad \text{or} \qquad \int_C |d\mathbf{R}| \tag{2.27}$$

where it is understood that dx, dy, and dz are expressed in terms of the parameter t and the differential dt (so that t is the variable over which the integration is performed). The form (2.27) emphasizes that the arc length is a property of the curve alone and does not depend on the particular parametrization.

Returning to eq. (2.24), we see that the arc length measured along the curve from some arbitrary initial position $\mathbf{R}(t_1)$ to a *variable* position $\mathbf{R}(t)$ is given by

$$s = s(t) = \int_{t_1}^{t} \left| \frac{d\mathbf{R}}{dt} \right| dt \qquad (t \geq t_1)$$

This suggests the possibility of using s itself as the parameter. In principle, at least, we may invert the above equation to get t in terms of s; substituting into the function $\mathbf{R}(t)$ gives $\mathbf{R}$ as a function of s.

In practice, the direct computation of $\mathbf{R}(s)$ is prohibitively difficult except for some standard, contrived examples, to wit:

Example 2.15 Reparametrize the curves

$$(i)\ \mathbf{R}(t) = \frac{t^2}{2}\mathbf{i} + \frac{t^3}{3}\mathbf{k} \qquad (0 \leq t \leq 2)$$

$$(ii)\ \mathbf{R}(t) = (2\cos t)\mathbf{i} + (2\sin t)\mathbf{j} \qquad (0 \leq t \leq 2\pi)$$

in terms of arc length.

Solution (i) Choosing $t_1 = 0$, we have

$$s = \int_0^t \left| \frac{d\mathbf{R}}{dt} \right| dt = \int_0^t (t^2 + t^4)^{1/2}\, dt = \frac{(t^2 + 1)^{3/2} - 1}{3}$$

Inverting this produces

$$t = [(3s + 1)^{2/3} - 1]^{1/2}$$

and the new parametrization is

$$\mathbf{R}(s) = \frac{(3s + 1)^{2/3} - 1}{2}\mathbf{i} + \frac{[(3s + 1)^{2/3} - 1]^{3/2}}{3}\mathbf{k}$$

(ii) Again with $t_1 = 0$, we find

$$s = \int_0^t \left|\frac{d\mathbf{R}}{dt}\right| dt = \int_0^t (4 \sin^2 t + 4 \cos^2 t)^{1/2}\, dt = 2t$$

Hence $t = s/2$ and the arc length parametrization reads

$$\mathbf{R}(s) = 2 \cos \frac{s}{2}\, \mathbf{i} + 2 \sin \frac{s}{2}\, \mathbf{j}$$

The arc length parametrization possesses some advantages. By the fundamental theorem of calculus, we have

$$\frac{ds}{dt} = \left|\frac{d\mathbf{R}}{dt}\right| \qquad (= |\mathbf{v}|) \tag{2.28}$$

(This identifies the *speed* with the rate of change of arc length, a reassuring fact.) In coordinate form, this becomes

$$\frac{ds}{dt} = \left[\left(\frac{dx}{dt}\right)^2 + \left(\frac{dy}{dt}\right)^2 + \left(\frac{dz}{dt}\right)^2\right]^{1/2}$$

Because our assumptions guarantee that $ds/dt \neq 0$, it follows from the chain rule that

$$\frac{d\mathbf{R}}{ds} = \frac{d\mathbf{R}/dt}{ds/dt}$$

Since $d\mathbf{R}/dt$ is tangent to the curve, this shows $d\mathbf{R}/ds$ is also. (This reflects the obvious fact that the tangent *direction* is independent of the parametrization used to describe the curve.) Moreover, $d\mathbf{R}/ds$ is a *unit* tangent vector; so by eq. (2.19)

$$\mathbf{T} = \frac{d\mathbf{R}}{ds}$$

In example 2.15 these vectors are

(i) $\dfrac{d\mathbf{R}}{ds} = (3s + 1)^{-1/3}\mathbf{i} + [(3s + 1)^{2/3} - 1]^{1/2}(3s + 1)^{-1/3}\mathbf{k}$

(ii) $\dfrac{d\mathbf{R}}{ds} = -\sin \dfrac{s}{2}\, \mathbf{i} + \cos \dfrac{s}{2}\, \mathbf{j}$

and the reader can verify that both are unit vectors.

EXERCISES

1. Find the unit vector tangent to the oriented closed curve

$$x = a\cos t \qquad y = b\sin t \qquad z = 0$$

at $t = \frac{3}{2}\pi$.

2. For the curve

$$x = \sin t - t\cos t \qquad y = \cos t + t\sin t \qquad z = t^2$$

find
(a) the arc length between $(0,1,0)$ and $(-2\pi,1,4\pi^2)$
(b) $\mathbf{T}(t)$
(c) $\mathbf{T}(\pi)$

3. Observe that

$$x = \frac{t}{2\pi} \qquad y = \sin t \qquad z = \cos t$$

is a parametrization of the helix in example 2.14. Compute the arc length between the same two endpoints using formula (2.24). What is the unit tangent vector at $(0,0,1)$?

4. If $\mathbf{T}$ denotes the unit tangent to the curve

$$x = t \qquad y = 2t + 5 \qquad z = 3t$$

show that $d\mathbf{T}/dt = 0$. Interpret this.

5. (a) Determine the arc length of the curve

$$x = e^t\cos t \qquad y = e^t\sin t \qquad z = 0$$

between $t = 0$ and $t = 1$.
(b) Reparametrize the curve in terms of arc length.
(c) This curve is a spiral. Sketch it to see why.

6. Find the arc length of the curve described in exercise 4, between $(0,5,0)$ and $(1,7,3)$, (a) by using (2.24), and (b) by using common sense.

7. By using identities concerning hyperbolic functions, eliminate the parameter t from the equations

$$x = \cosh t \qquad y = \sinh t \qquad z = 0$$

8. Show that the curve $x = t$, $y = 2t^2$, $z = t^3$ intersects the plane $x + 8y + 12z = 162$ at right angles.

9. As t varies from -1 to 1, the point (x,y,z) where

$$x = t \qquad y = |t| \qquad z = 0$$

traces a regular curve. At what point on this curve is there no tangent?

10. The helix $\mathbf{R} = \cos t\,\mathbf{i} + \sin t\,\mathbf{j} + t\,\mathbf{k}$ is right-handed, according to the characterization of this section. Show that if the orientation is reversed by changing t to $-t$, the helix remains right-handed.
11. Is the helix of example 2.14 right- or left-handed?
12. Parametrize a right-handed helix with unit pitch that is wrapped around the cylinder described by $y^2 + z^2 = 1$.
13. Suppose that P_1P_2 is a smooth arc in the xy plane. Is it necessarily true that dy/dx exists at every point on this arc?
14. Study the consequences of dropping condition (*iii*) in the definition of a smooth arc. (*Hint:* Sketch the arc $\mathbf{R} = t^2\mathbf{i} + t^3\mathbf{j}$.)
15. Show that the graph of any continuously differentiable function $y = f(x)$ is a smooth curve. [*Hint:* Check the parametrization $x = t$, $y = f(t)$, $z = 0$.]

2.3 Acceleration and Curvature

The *acceleration* of a particle is defined to be the time rate of change of its velocity. Since velocity is a vector quantity, this acceleration may be associated with a change in either the magnitude or the direction of the velocity, or both.

Suppose first that the direction of the velocity is constant. Then the motion of the particle takes place along a straight line and the magnitude of the acceleration is the rate of change of speed:

$$|\mathbf{a}| = \frac{d}{dt}|\mathbf{v}| = \frac{d^2s}{dt^2}$$

where s is arc length along the trajectory [recall eq. (2.28)]. The acceleration is directed along the straight line. On the other hand, if the particle moves at constant speed around a circle of radius ρ, it is well known that it undergoes a "centripetal" acceleration of magnitude

$$|\mathbf{a}| = \frac{|\mathbf{v}|^2}{\rho} = \frac{1}{\rho}\left(\frac{ds}{dt}\right)^2$$

directed toward the center of the circle. This is due solely to the change of direction.

One of the aims of this section is to show that for motion along a *general* curve with $\mathbf{v}$ changing direction and magnitude, the acceleration vector can be expressed as the sum of two orthogonal vectors, one giving the rate of change of speed and the other giving the instantaneous centripetal acceleration corresponding to a related circular trajectory.

If motion along a curve is to be related to motion on a circle, we clearly need to select the circle that "best" approximates the curve at a given point. In figure 2.16 we indicate the circle approximating the curve at P_1. Two properties that the

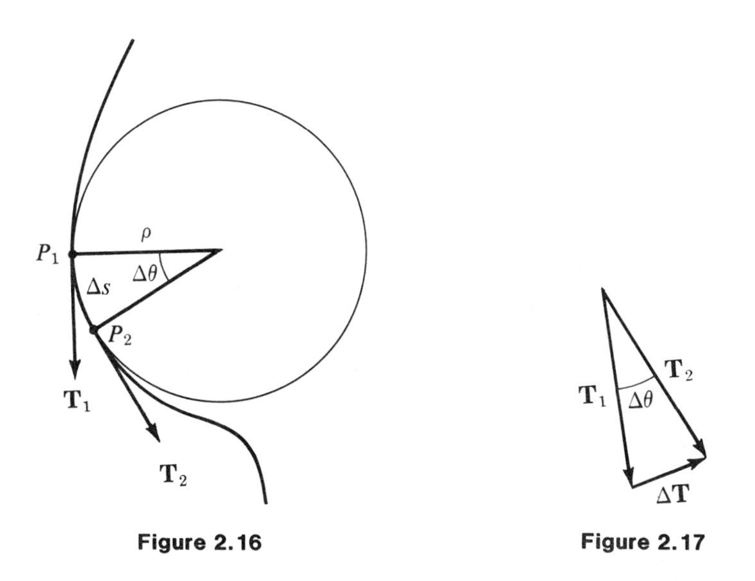

Figure 2.16 **Figure 2.17**

circle must have are clear: it should pass through the point P_1, and its tangent must coincide with the tangent to the curve at P_1. It remains for us to decide what radius ρ the circle should have in order that it fit the curve as well as possible.

Observe that circles with small radii are more sharply curved than circles with large radii. Thus, by choosing ρ appropriately, we should be able to select a circle with the same *curvature* as the given curve, at P_1. But how do we measure this curvature? Intuitively, curvature arises as a result of the tangent direction changing as we move along the curve; a straight line has no curvature and an arc is more sharply curved if the tangent turns faster along the length of the curve. Let us therefore *define* the curvature k as the rate at which the *unit* tangent vector turns, with respect to arc length along the curve:

$$k = \left|\frac{d\mathbf{T}}{ds}\right| = \frac{|d\mathbf{T}/dt|}{ds/dt} \tag{2.29}$$

What does this give for the curvature of a circle of radius ρ? In figure 2.16 the arc length between P_1 and P_2 on the circle is $\Delta s = \rho\Delta\theta$. The unit tangent vectors $\mathbf{T}_1$ and $\mathbf{T}_2$ also make an angle $\Delta\theta$, and the change in the unit tangent as we proceed from P_1 to P_2 is

$$\Delta\mathbf{T} = \mathbf{T}_2 - \mathbf{T}_1$$

For small $\Delta\theta$, the magnitude of $\Delta\mathbf{T}$ is approximately $\Delta\theta$, as we see from figure 2.17 (keep in mind that the magnitudes of $\mathbf{T}_1$ and $\mathbf{T}_2$ are unity). Thus

$$\left|\frac{\Delta\mathbf{T}}{\Delta s}\right| \approx \frac{\Delta\theta}{\Delta s} = \frac{1}{\rho}$$

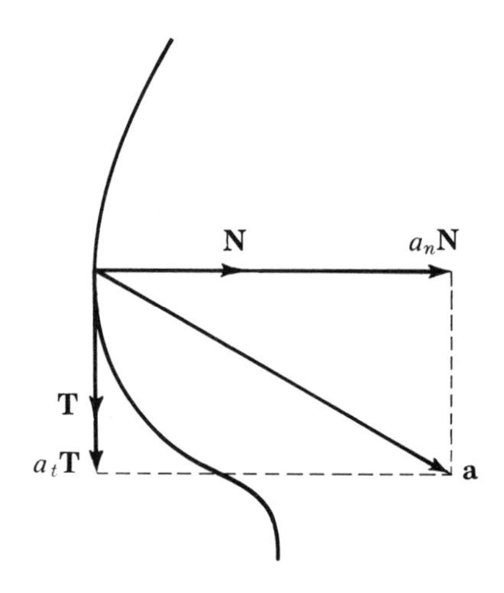

Figure 2.18

This approximation becomes exact as Δs approaches zero, so we can write

$$\left|\frac{d\mathbf{T}}{ds}\right| = \frac{1}{\rho}$$

Thus the curvature of a circle as we have defined it is the reciprocal of its radius. This is in harmony with our intuition, and so we shall feel confident in adopting the definition (2.29).

Consequently, the radius ρ of our approximating circle (called the "osculating circle") is given by

$$\rho = \frac{1}{k} = \frac{1}{|d\mathbf{T}/ds|}$$

Now we let **N** denote a *unit* vector pointing toward the center of the approximating circle, and as usual let **T** denote the unit tangent vector. The directions of both **T** and **N** may vary at different points along the curve, but they are always at right angles with each other, as shown in figure 2.18.

If our earlier considerations about circular motion can be generalized to motion along a curve, then we are led to anticipate that the acceleration **a** can be expressed as a sum of two components:

$$\mathbf{a} = a_t\mathbf{T} + a_n\mathbf{N} \tag{2.30}$$

where $a_t = d^2s/dt^2$ is the rate of change of speed, and $a_n = |\mathbf{v}|^2/\rho$ results from the change in the direction of the velocity. To see that this is, in fact, the case, we start over with a more careful analysis.

The position vector of the particle, as usual, is taken to be

$$\mathbf{R}(t) = x(t)\mathbf{i} + y(t)\mathbf{j} + z(t)\mathbf{k}$$

which we visualize as the directed line segment extending from the origin to the point where the particle is located. We restrict our attention to a portion of the trajectory where $\mathbf{R}(t)$ defines a smooth arc and is twice differentiable. Its derivatives, which are the velocity $\mathbf{v}$ and the acceleration $\mathbf{a}$ respectively, are computed as in section 2.1:

$$\mathbf{v} = \frac{d\mathbf{R}}{dt} = \frac{dx}{dt}\mathbf{i} + \frac{dy}{dt}\mathbf{j} + \frac{dz}{dt}\mathbf{k} \tag{2.31}$$

$$\mathbf{a} = \frac{d^2\mathbf{R}}{dt^2} = \frac{d^2x}{dt^2}\mathbf{i} + \frac{d^2y}{dt^2}\mathbf{j} + \frac{d^2z}{dt^2}\mathbf{k} \tag{2.32}$$

It is convenient to visualize $\mathbf{v}(t)$ as a directed line segment with its tail at the point where the particle is located. As t varies, the corresponding vector $\mathbf{v}(t)$ may vary either in direction or magnitude, or both (fig. 2.19). The speed of the particle is the magnitude of the velocity ds/dt, where the arc length s is measured along the curve from some arbitrary initial point:

$$v = |\mathbf{v}(t)| = \left[\left(\frac{dx}{dt}\right)^2 + \left(\frac{dy}{dt}\right)^2 + \left(\frac{dz}{dt}\right)^2\right]^{1/2} = \frac{ds}{dt} \tag{2.33}$$

The unit tangent vector $\mathbf{T}$ may be obtained by dividing the velocity $\mathbf{v}(t)$ by the speed $|\mathbf{v}(t)|$, since our assumptions guarantee that $|\mathbf{v}(t)|$ is never zero:

$$\mathbf{T} = \frac{\mathbf{v}(t)}{|\mathbf{v}(t)|} \tag{2.34}$$

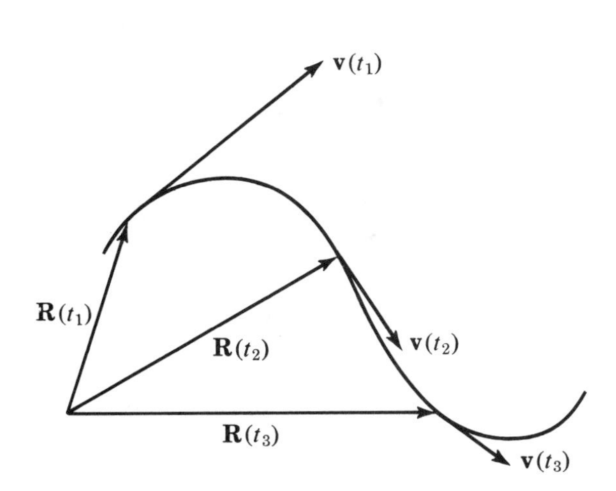

Figure 2.19

We note that $\mathbf{T}$ is also given by the expression

$$\mathbf{T} = \frac{d\mathbf{R}}{ds} \tag{2.35}$$

The *curvature* k of the curve at any point is defined to be the *magnitude* of the vector $d\mathbf{T}/ds$ at that point [recall eq. (2.29)]:

$$k = \left|\frac{d\mathbf{T}}{ds}\right| \tag{2.36}$$

If $k \neq 0$, the *radius of curvature* ρ is defined to be the reciprocal of the curvature:

$$\rho = \frac{1}{k} \tag{2.37}$$

The motivation for this definition of ρ was given above. By introducing k we will be able to avoid using the term "infinite radius of curvature." Thus the curvature of a straight line is $k = 0$.

Since $\mathbf{T}$ has constant *magnitude,* the derivative of $\mathbf{T}$ with respect to t is either the zero vector or a nonzero vector perpendicular to $\mathbf{T}$. This was proved in example 2.7; moreover, it is clear geometrically from figure 2.17, where we see that $\Delta\mathbf{T}$ is approximately perpendicular to $\mathbf{T}$ if $\Delta\mathbf{T}$ is small.

If $d\mathbf{T}/dt$ is not the zero vector, we define the unit vector $\mathbf{N}$ to be $d\mathbf{T}/dt$ divided by its own magnitude:

$$\mathbf{N} = \frac{d\mathbf{T}/dt}{|d\mathbf{T}/dt|} \tag{2.38}$$

This vector is called the *principal normal.* If we apply the chain rule $d\mathbf{T}/dt = (d\mathbf{T}/ds)(ds/dt)$ to both numerator and denominator of this fraction, we can cancel ds/dt and obtain the alternative expression

$$\mathbf{N} = \frac{d\mathbf{T}/ds}{|d\mathbf{T}/ds|}$$

and since $k = |d\mathbf{T}/ds|$ we can write

$$\frac{d\mathbf{T}}{ds} = k\mathbf{N}$$

In words, we may say that $\mathbf{T}$ turns in the direction $\mathbf{N}$, at a rate k (with respect to arc length).

Now we are ready to derive the representation for the acceleration of the particle. This has been defined as the time rate of change of the velocity:

$$\mathbf{a}(t) = \mathbf{v}'(t) = \frac{d\mathbf{v}}{dt} = \frac{d^2x}{dt^2}\mathbf{i} + \frac{d^2y}{dt^2}\mathbf{j} + \frac{d^2z}{dt^2}\mathbf{k} \tag{2.39}$$

Since $|\mathbf{v}(t)| = ds/dt$, we can write

$$\mathbf{v}(t) = \frac{ds}{dt}\mathbf{T} \tag{2.40}$$

So by the product rule for derivatives (section 2.1),

$$\begin{aligned}\mathbf{a}(t) = \mathbf{v}'(t) &= \frac{d^2s}{dt^2}\mathbf{T} + \frac{ds}{dt}\frac{d\mathbf{T}}{dt} = \frac{d^2s}{dt^2}\mathbf{T} + \frac{ds}{dt}\frac{d\mathbf{T}}{ds}\frac{ds}{dt} \\ &= \frac{d^2s}{dt^2}\mathbf{T} + \left(\frac{ds}{dt}\right)^2 k\mathbf{N}\end{aligned}$$

In other words, we have

$$\mathbf{a} = a_t\mathbf{T} + a_n\mathbf{N} \tag{2.41}$$

where $a_t = d^2s/dt^2$ and $a_n = kv^2$. This is exactly what we anticipated in eq. (2.30).

We note that at any point on the curve where $k = 0$, the normal vector $\mathbf{N}$ is not defined. This does not matter, since we have $a_n = 0$ in that case and hence have no need for $\mathbf{N}$ in eq. (2.41). In case $k \neq 0$, we can write $a_n = v^2/\rho$, the way we did in the previous heuristic discussion.

Since $\mathbf{T}$ and $\mathbf{N}$ are mutually perpendicular vectors at any point where they are defined, we have, by the pythagorean theorem,

$$a^2 = a_t^2 + a_n^2 \tag{2.42}$$

To compute a, we need only find $d^2\mathbf{R}/dt^2$ by differentiation, and calculate the magnitude of this vector. To compute a_t we need only find $\mathbf{v} = d\mathbf{R}/dt$, calculate its magnitude $|d\mathbf{R}/dt| = ds/dt$, and differentiate this with respect to t. Having computed a and a_t, it is then easy to obtain a_n by using eq. (2.42). In some problems, this is more convenient than using the expression kv^2.

Example 2.16 The position of a particle moving around the circle $x^2 + y^2 = r^2$ in the xy plane, with angular velocity ω, is

$$x = r\cos\omega t \qquad y = r\sin\omega t \qquad z = 0$$

Find the normal and tangential components of acceleration of the particle, and determine the curvature of the circle.

Solution We have

$$\begin{aligned}\mathbf{R} &= r\cos\omega t\,\mathbf{i} + r\sin\omega t\,\mathbf{j} \\ \frac{d\mathbf{R}}{dt} &= -r\omega\sin\omega t\,\mathbf{i} + r\omega\cos\omega t\,\mathbf{j} \\ \frac{d^2\mathbf{R}}{dt^2} &= -r\omega^2\cos\omega t\,\mathbf{i} - r\omega^2\sin\omega t\,\mathbf{j}\end{aligned}$$

The magnitudes of these vectors are

$$v = \frac{ds}{dt} = \left|\frac{d\mathbf{R}}{dt}\right| = (r^2\omega^2 \sin^2 \omega t + r^2\omega^2 \cos^2 \omega t)^{1/2} = \omega r$$

$$a = \left|\frac{d^2\mathbf{R}}{dt^2}\right| = \omega^2 r$$

Since ds/dt is a constant, $a_t = d^2s/dt^2 = 0$ and $a = a_n$. Therefore $kv^2 = \omega^2 r$; and since $v = \omega r$, we have $k = \omega^2 r/\omega^2 r^2 = 1/r$. This verifies that the curvature of a circle is the reciprocal of its radius. The answers are: $a_n = \omega^2 r$, $a_t = 0$, $k = 1/r$.

Example 2.17 The coordinates of a particle at time t are

$$x = \sin t - t \cos t \qquad y = \cos t + t \sin t \qquad z = t^2$$

Find the speed, the normal and tangential components of acceleration, and the curvature of the path, in terms of t.

Solution

$$\mathbf{R} = (\sin t - t \cos t)\mathbf{i} + (\cos t + t \sin t)\mathbf{j} + t^2\mathbf{k}$$

$$\frac{d\mathbf{R}}{dt} = (t \sin t)\mathbf{i} + (t \cos t)\mathbf{j} + 2t\mathbf{k}$$

$$\frac{d^2\mathbf{R}}{dt^2} = (t \cos t + \sin t)\mathbf{i} + (-t \sin t + \cos t)\mathbf{j} + 2\mathbf{k}$$

The speed is $ds/dt = |d\mathbf{R}/dt| = (t^2 \sin^2 t + t^2 \cos^2 t + 4t^2)^{1/2} = \sqrt{5}t$. The tangential component of acceleration is $a_t = d^2s/dt^2 = \sqrt{5}$.

From eq. (2.42),

$$\begin{aligned} a_n &= (a^2 - a_t^2)^{1/2} \\ &= [(t \cos t + \sin t)^2 + (-t \sin t + \cos t)^2 + 2^2 - 5]^{1/2} = t \end{aligned}$$

Since $a_n = kv^2$, we have $k = a_n/v^2 = t/5t^2 = 1/5t$.

One can derive a fairly simple expression for the curvature k by taking the vector cross product of

$$\mathbf{R}'(t) = |\mathbf{v}|\mathbf{T} \qquad \text{and} \qquad \mathbf{R}''(t) = \frac{d^2s}{dt^2}\mathbf{T} + k|\mathbf{v}|^2\mathbf{N}$$

which, since $\mathbf{T} \times \mathbf{T} = \mathbf{0}$, gives

$$\mathbf{R}' \times \mathbf{R}'' = k|\mathbf{v}|^3 (\mathbf{T} \times \mathbf{N})$$

Since $\mathbf{T}$ and $\mathbf{N}$ are mutually perpendicular unit vectors, their cross product $\mathbf{B} = \mathbf{T} \times \mathbf{N}$ is a unit vector; this vector is called the *binormal*. We have $\mathbf{R}' \times \mathbf{R}'' = k|\mathbf{v}|^3\mathbf{B}$ and

$$|\mathbf{R}' \times \mathbf{R}''| = k|\mathbf{v}|^3$$

and hence

$$k = \frac{|\mathbf{R}' \times \mathbf{R}''|}{(\mathbf{R}' \cdot \mathbf{R}')^{3/2}} = \frac{|\mathbf{R}' \times \mathbf{R}''|}{|\mathbf{R}'|^3} \tag{2.43}$$

However, in most cases it is easier to use eq. (2.29).

Once again we point out that although many of these formulas involve derivatives with respect to arc length s, one never actually needs to compute the reparameterization $\mathbf{R}(s)$, because of the chain rule.

Optional Reading: The Frenet Formulas

Because of its importance in geometry, it may be well to say more about the vector **B**, which is a unit vector mutually perpendicular to both **T** and **N**. The vectors **T**, **N**, **B**, in that order, form a right-handed system. It is useful to think of these three vectors as attached to a particle moving along the curve: as the particle moves, its associated triad of mutually perpendicular unit vectors moves and rotates (see fig. 2.20).* For a *plane curve,* **T** and **N** lie in the plane of the curve, so that **B** is a constant unit vector always perpendicular to the plane.

Let us try to describe how the triad rotates as a particle proceeds along a space curve. As we have seen, the vector **T** turns toward the vector **N** at a rate k, measured with respect to arc length:

$$\frac{d\mathbf{T}}{ds} = k\mathbf{N} \tag{2.44}$$

But since **N** is always perpendicular to **T**, these vectors will turn *together* like a rigid body. **N** must therefore turn toward the direction $-\mathbf{T}$ at the same rate, k. In addition, it is also possible for **N** to *rotate about* **T** *as an axis;* this would happen if the instantaneous plane of the curve were to "tilt."** In such a case, $d\mathbf{N}/ds$ would have a component perpendicular to both **T** and **N** (i.e., along **B**). Thus we would have

$$\frac{d\mathbf{N}}{ds} = -k\mathbf{T} + \tau\mathbf{B} \tag{2.45}$$

where τ measures the rate at which the curve *twists;* accordingly, it is known as the *torsion.*

* The triad can be visualized as an aircraft, whose nose points in the direction **T** (parallel to the velocity), with wings aligned along **N** and tail fin along **B**.

** Or if the aircraft were to "bank."

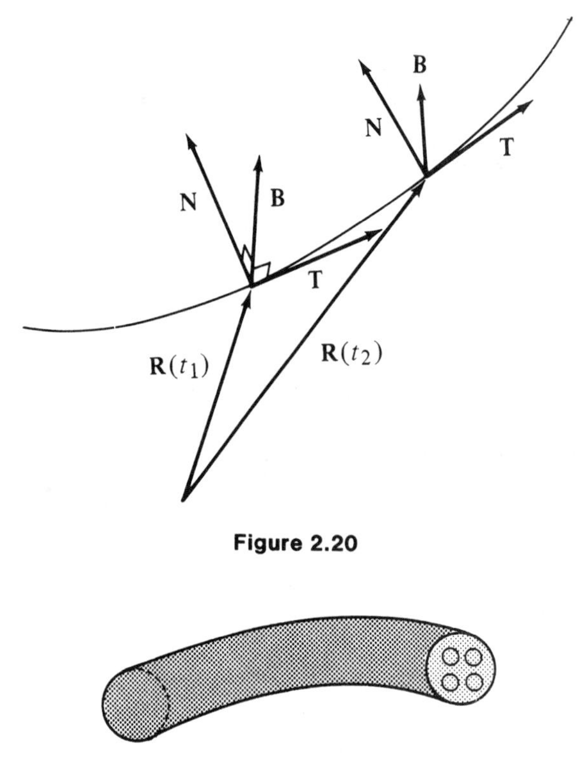

Figure 2.20

Figure 2.21

The torsion can be visualized by observing the cross section of a piece of solder wire bent into the shape of the curve, as in figure 2.21. The torsion, or twisting, of the wire will affect a *rotation* of the cross-sectional pattern. One can shape the wire into any plane curve without introducing torsion, but if the curve is nonplanar, the wire must twist.

Once again, the fact that **N** turns toward **B** at a rate τ, and the fact that **N** and **B** are rigidly fixed at right angles, imply that **B** turns toward $-\mathbf{N}$ at the same rate. At first glance it seems conceivable that **B** might also rotate about **N** as an axis, thus turning in the direction of **T**. However, if this happened, **T** would be forced by rigidity into turning in the direction $-\mathbf{B}$; but **T** turns only in the direction **N**, *by definition!* Thus we have

$$\frac{d\mathbf{B}}{ds} = -\tau\mathbf{N} \tag{2.46}$$

Equations (2.44) through (2.46) are called the *Frenet formulas*. They are important in differential geometry, where it is shown that any two curves with identical corresponding values of curvature and torsion are congruent (as usual, subject to certain restrictions).

The curve formulas we have derived are summarized in table 2.1.

Table 2.1 Formulas for Curves

Parametrization	$\mathbf{R} = \mathbf{R}(t)$
Velocity	$\mathbf{v} = \dfrac{d\mathbf{R}}{dt}$
Arc length	$ds = \lvert d\mathbf{R}\rvert = \lvert\mathbf{v}\rvert dt$
Acceleration	$\mathbf{a} = \dfrac{d^2\mathbf{R}}{dt^2} = \dfrac{d^2s}{dt^2}\mathbf{T} + k\left(\dfrac{ds}{dt}\right)^2\mathbf{N}$
	$= \dfrac{d\lvert\mathbf{v}\rvert}{dt}\mathbf{T} + k\lvert\mathbf{v}\rvert^2\mathbf{N}$
Curvature	$k = \left\lvert\dfrac{d\mathbf{T}}{ds}\right\rvert$
Radius of curvature	$\rho = \dfrac{1}{k}$
Tangent	$\mathbf{T} = \dfrac{\mathbf{v}}{\lvert\mathbf{v}\rvert}$
Normal	$\mathbf{N} = \dfrac{1}{k}\dfrac{d\mathbf{T}}{ds}$
Binormal	$\mathbf{B} = \mathbf{T} \times \mathbf{N}$
Torsion	$\tau = -\mathbf{N} \cdot \dfrac{d\mathbf{B}}{ds} = (\pm)\left\lvert\dfrac{d\mathbf{B}}{ds}\right\rvert$
Frenet formulas	$\dfrac{d\mathbf{T}}{ds} = k\mathbf{N}$
	$\dfrac{d\mathbf{N}}{ds} = -k\mathbf{T} + \tau\mathbf{B}$
	$\dfrac{d\mathbf{B}}{ds} = -\tau\mathbf{N}$
(Chain rule)	$\left(\dfrac{d}{ds} = \dfrac{1}{\lvert\mathbf{v}\rvert}\dfrac{d}{dt}\right)$

EXERCISES

In the first four problems below, the coordinates of a moving particle are given as a function of the time t. Find (a) the speed, (b) the tangential and normal components of acceleration, (c) the unit tangent vector **T**, and (d) the curvature of the curve, as functions of time.

1. $x = e^t \cos t,\ y = e^t \sin t,\ z = 0$
2. $x = 3t \cos t,\ y = 3t \sin t,\ z = 4t$
3. $x = e^t \cos t,\ y = e^t \sin t,\ z = e^t$

4. $x = 5 \sin 4t$, $y = 5 \cos 4t$, $z = 10t$
5. The position vector of a moving particle is

$$\mathbf{R} = \cos t\,(\mathbf{i} - \mathbf{j}) + \sin t\,(\mathbf{i} + \mathbf{j}) + \tfrac{1}{2}t\,\mathbf{k}$$

(a) Determine the velocity and the speed of the particle.
(b) Determine the acceleration of the particle.
(c) Find a unit tangent to the path of the particle, in the direction of motion.
(d) Show that the curve traversed by the particle has constant curvature k, and find its value.
(e) Show that the curve is a helix.

6. Find the curvature of the space curve

$$x = 3t^2 - t^3 \qquad y = 3t^2 \qquad z = 3t + t^3$$

7. Let C be the curve given by the equation

$$\mathbf{R}(t) = \sin t\,\mathbf{i} + \cos t\,\mathbf{j} + \log \sec t\,\mathbf{k} \qquad (0 \leq t < \pi/2)$$

Find
(a) the element of arc length, ds, along C, in terms of t.
(b) the unit tangent $\mathbf{T}$.
(c) the unit normal $\mathbf{N}$.
(d) the curvature k.

8. A particle moves so that its coordinates at time t are given by

$$x(t) = e^{-t} \cos t \qquad y(t) = e^{-t} \sin t \qquad z(t) = e^{-t}$$

Find its velocity, speed, and acceleration, and the curvature of its path at time t.

9. A particle moves so that its position $\mathbf{R}$ at time t is given by

$$\mathbf{R}(t) = \log(t^2 + 1)\mathbf{i} + (t - 2 \arctan t)\mathbf{j} + 2\sqrt{2}t\mathbf{k}$$

(a) Show that this particle moves with constant speed $v = 3$.
(b) Find the curvature of the path of this particle.

10. A point moves along a curve so that its position $\mathbf{R}$ is given by

$$\mathbf{R} = t^2\mathbf{i} + t^2\mathbf{j} + t\mathbf{k}$$

Find
(a) its speed v.
(b) the unit tangent $\mathbf{T}$ to its path.
(c) the vector $k\mathbf{N}$.

11. Graph the planar curve $y = \sin x$. Without writing equations, demonstrate on the graph how the normal $\mathbf{N}$ jumps discontinuously each time the curve crosses the axis. What is the curvature at these points?
12. If $\mathbf{F}$ is a function of t possessing derivatives of all orders, find the derivative of

$$\mathbf{F} \times \frac{d\mathbf{F}}{dt} \cdot \frac{d^2\mathbf{F}}{dt^2}$$

13. Find the curvature and torsion for the helix

$$x = t \qquad y = \sin t \qquad z = \cos t$$

14. The position vector of a particle is given by

$$\mathbf{R}(t) = \sqrt{2}\cos 3t\,\mathbf{i} + \sqrt{2}\cos 3t\mathbf{j} + 2\sin 3t\,\mathbf{k}$$

Find its speed, the curvature and torsion of its path, and describe the path geometrically.

15. By inspection, write down the values of each of the following:

(a) $\dfrac{d\mathbf{R}}{ds}\cdot\mathbf{T}$ (d) $\mathbf{T}\cdot\mathbf{N}$ (g) $[\mathbf{T},\mathbf{N},\mathbf{B}]$

(b) $\dfrac{d}{ds}(\mathbf{T}\cdot\mathbf{T})$ (e) $\dfrac{d\mathbf{R}}{dt}\cdot\mathbf{T}$ (h) $\left|\dfrac{d^2\mathbf{R}}{ds^2}\right|$

(c) $\dfrac{d^2\mathbf{R}}{dt^2}\cdot\mathbf{T}$ (f) $\dfrac{d\mathbf{N}}{ds}\cdot\mathbf{B}$ (i) $\dfrac{d\mathbf{B}}{ds}$

16. If C is the curve given parametrically by

$$\mathbf{R}(t) = \cos t\,\mathbf{i} + \sin t\,\mathbf{j} + 2t\,\mathbf{k}$$

find

(a) the normal $\mathbf{N}$ and the binormal $\mathbf{B}$ for this curve at $t = 0$.

(b) the equation of the plane passing through the point $\mathbf{R}(0)$ and parallel to both vectors $\mathbf{N}$ and $\mathbf{B}$ of part (a).

17. Find the unit tangent **T**, the principal normal **N**, the binormal **B**, the curvature, and the torsion for

$$x = \cos^3 t \qquad y = \sin^3 t \qquad z = 2\sin^2 t \qquad (0 < t \leq \pi/2)$$

18. True or false:

(a) If $\mathbf{R}$ is the position vector of a particle, t denotes time, and s denotes arc length, $d^2\mathbf{R}/dt^2$ is a scalar multiple of $d^2\mathbf{R}/ds^2$.

(b) A moving particle achieves its maximum speed at the instant $t = 3$. (Before and after that instant, its speed is less than its speed at $t = 3$.) It follows from this that its acceleration is zero at the instant $t = 3$.

(c) The acceleration of a particle moving along a curve with binormal $\mathbf{B}$ is always perpendicular to $\mathbf{B}$. [More precisely, $\mathbf{a}(t)$ and $\mathbf{B}(t)$ are orthogonal for each fixed value of t.]

19. Express the curvature and torsion of a helix (eq. 2.15) in terms of its radius and pitch.

20. The *evolute* of a curve $\mathbf{R}(t)$ is the locus of the centers of curvature of the curve. Using the parametric formulas, show that the tangent to the evolute is normal to the original curve.

21. If the curve $\mathbf{R}(t)$ lies on a sphere $|\mathbf{R}(t)| = \textit{constant}$, prove that

$$\mathbf{R} = -\rho\mathbf{N} - \frac{1}{\tau}\frac{d\rho}{ds}\mathbf{B}$$

(*Hint:* Keep differentiating $\mathbf{R}\cdot\mathbf{R} = \textit{constant}$, using the Frenet formulas.)

22. A rigorous derivation of the Frenet formulas proceeds as follows:

(a) Regard eq. (2.44) as the defining equation for k and **N**.

(b) Show that $d\mathbf{N}/ds + k\mathbf{T}$ is perpendicular to both **T** and **N**. (Here it is helpful to differentiate the relation $\mathbf{T} \cdot \mathbf{N} = 0$.) Thus eq. (2.45) can be regarded as the defining equation for τ.

(c) Prove eq. (2.46) from eqs. (2.44) and (2.45) by differentiating the relation $\mathbf{B} = \mathbf{T} \times \mathbf{N}$.

Carry out the details of this program.

23. *The Darboux vector* is defined to be

$$\omega = \tau\mathbf{T} + k\mathbf{B}$$

Show that the equation

$$\frac{d\mathbf{U}}{ds} = \omega \times \mathbf{U}$$

is satisfied for $\mathbf{U} = \mathbf{T}$, **N**, and **B**. Note the resemblance of this equation to the angular velocity equation (1.22).

2.4 Planar Motion in Polar Coordinates

In this section we consider the motion of a particle in the xy plane where the position of the particle is given in polar coordinates, r and θ. We remind the reader that r and θ provide alternative descriptions of points in the plane, and they are sometimes more convenient when circular symmetries are present. They are depicted in figure 2.22 and are related to (x,y) coordinates through the equations

$$x = r\cos\theta \qquad r = (x^2 + y^2)^{1/2}$$

$$y = r\sin\theta \qquad \theta = \sin^{-1}\frac{y}{(x^2 + y^2)^{1/2}} = \cos^{-1}\frac{x}{(x^2 + y^2)^{1/2}}$$

The extra equations for θ are necessary to avoid quadrant ambiguities; customarily, one takes $-\pi < \theta \leq \pi$.*

We assume that the particle's trajectory is specified by giving r and θ as functions of the time t, and that these functions possess second derivatives.

In order to work directly with polar coordinates, it is convenient to introduce unit vectors $\mathbf{u}_r$ and $\mathbf{u}_\theta$, which point respectively along the position vector and at right angles to it (in the direction of increasing θ), as shown in figure 2.22.

* Thus, for example, if $x < 0$ and $y < 0$, θ would actually lie in the third quadrant. The principal value of arcsine would output an angle in the fourth quadrant, and arccosine would yield a second-quadrant angle.

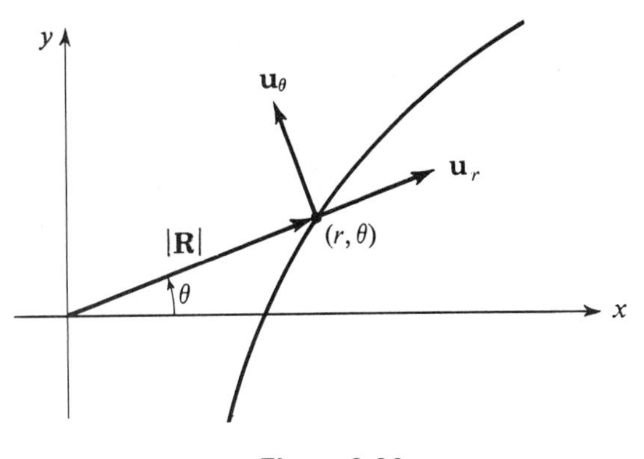

Figure 2.22

It is easy to see that $\mathbf{u}_r$ and $\mathbf{u}_\theta$ can be written in terms of $\mathbf{i}$ and $\mathbf{j}$ as follows:

$$\mathbf{u}_r = \cos\theta\,\mathbf{i} + \sin\theta\,\mathbf{j}$$
$$\mathbf{u}_\theta = -\sin\theta\,\mathbf{i} + \cos\theta\,\mathbf{j} \tag{2.47}$$

Note that $\mathbf{u}_r$ and $\mathbf{u}_\theta$ are functions of θ and are defined at every point in space except the origin. Unlike $\mathbf{i}$ and $\mathbf{j}$, $\mathbf{u}_r$ and $\mathbf{u}_\theta$ are not constants. For example, along the positive x axis, $\mathbf{u}_r = \mathbf{i}$, but along the positive y axis, $\mathbf{u}_r = \mathbf{j}$. It follows that we must be careful in differentiating vectors written in terms of $\mathbf{u}_r$ and $\mathbf{u}_\theta$.

Directly from eq. (2.47) we see that

$$\frac{d\mathbf{u}_r}{d\theta} = \mathbf{u}_\theta$$
$$\frac{d\mathbf{u}_\theta}{d\theta} = -\mathbf{u}_r \tag{2.48}$$

(Notice that these important formulas reinforce the observations we made in the previous section about the derivatives of unit vectors rigidly attached to each other.)

The position vector of a particle located at a point (r,θ) is

$$\mathbf{R} = r\mathbf{u}_r \tag{2.49}$$

We obtain the velocity by differentiating eq. (2.49) and using the chain rule:

$$\frac{d\mathbf{R}}{dt} = \frac{dr}{dt}\mathbf{u}_r + r\frac{d\mathbf{u}_r}{dt}$$
$$= \frac{dr}{dt}\mathbf{u}_r + r\frac{d\mathbf{u}_r}{d\theta}\frac{d\theta}{dt}$$

Hence by eq. (2.48) the velocity is given by

$$\mathbf{v} = \frac{d\mathbf{R}}{dt} = \frac{dr}{dt}\mathbf{u}_r + r\frac{d\theta}{dt}\mathbf{u}_\theta \tag{2.50}$$

This expresses the velocity as the sum of a radial component, directed away from or toward the origin with magnitude $|dr/dt|$, and a transverse component with magnitude $|r\, d\theta/dt|$.

Example 2.18 A particle moves around the circle $r = 2$ with angular velocity $d\theta/dt = 5$ rad/sec. Find its speed.

Solution Since r is a constant, $dr/dt = 0$. Hence

$$\mathbf{v} = r\frac{d\theta}{dt}\mathbf{u}_\theta = 10\,\mathbf{u}_\theta$$

Therefore $|\mathbf{v}| = 10$.

Example 2.19 A circular disc rotates with constant angular velocity 3 rad/sec. A fly walks from the center of the disc outward to the rim at a rate of 2 cm/sec (relative to the disc). Find the speed of the fly 4 seconds after he starts at the center.

Solution Since $dr/dt = 2$, we have $r = r_0 + 2t$. Because the fly starts at the center, $r_0 = 0$. Hence by eq. (2.50)

$$\mathbf{v} = 2\mathbf{u}_r + 3r\mathbf{u}_\theta$$

At time $t = 4$, $r = 2t = 8$, so $\mathbf{v} = 2\mathbf{u}_r + 24\mathbf{u}_\theta$. The speed is then $(2^2 + 24^2)^{1/2} = (580)^{1/2}$ cm/sec.

Returning to eq. (2.50), we differentiate again to obtain the acceleration:

$$\begin{aligned}
\mathbf{a} = \frac{d\mathbf{v}}{dt} &= \frac{d^2r}{dt^2}\mathbf{u}_r + \frac{dr}{dt}\frac{d\mathbf{u}_r}{dt} + \frac{dr}{dt}\frac{d\theta}{dt}\mathbf{u}_\theta + r\frac{d^2\theta}{dt^2}\mathbf{u}_\theta + r\frac{d\theta}{dt}\frac{d\mathbf{u}_\theta}{dt} \\
&= \frac{d^2r}{dt^2}\mathbf{u}_r + \frac{dr}{dt}\frac{d\mathbf{u}_r}{d\theta}\frac{d\theta}{dt} + \frac{dr}{dt}\frac{d\theta}{dt}\mathbf{u}_\theta + r\frac{d^2\theta}{dt^2}\mathbf{u}_\theta + r\frac{d\theta}{dt}\frac{d\theta}{dt}\frac{d\mathbf{u}_\theta}{d\theta} \\
&= \frac{d^2r}{dt^2}\mathbf{u}_r + 2\frac{dr}{dt}\frac{d\theta}{dt}\mathbf{u}_\theta + r\frac{d^2\theta}{dt^2}\mathbf{u}_\theta - r\left(\frac{d\theta}{dt}\right)^2\mathbf{u}_r
\end{aligned}$$

Combining terms, we find

$$\mathbf{a} = \left[\frac{d^2r}{dt^2} - r\left(\frac{d\theta}{dt}\right)^2\right]\mathbf{u}_r + \left[r\frac{d^2\theta}{dt^2} + 2\frac{dr}{dt}\frac{d\theta}{dt}\right]\mathbf{u}_\theta \tag{2.51}$$

The first term in eq. (2.51), $(d^2r/dt^2)\mathbf{u}_r$, gives the acceleration for pure radial motion; and the third term, $r(d^2\theta/dt^2)\mathbf{u}_\theta$, measures the effect of angular acceleration. In the special case that r is a constant, we have motion in a circle with center at the origin; then $\mathbf{u}_\theta$ and $\mathbf{u}_r$ are, respectively, the vectors $\mathbf{T}$ and $-\mathbf{N}$ of the preceding section. In this special case the second term is the centripetal acceleration term.

The fourth term,

$$2\frac{dr}{dt}\frac{d\theta}{dt}\mathbf{u}_\theta$$

is more complicated and is usually not discussed in elementary physics textbooks. Under certain circumstances it is known as the *Coriolis acceleration.* As a careful examination of the above derivation will show, this term is due partly to the change in *direction* of the radial component of velocity, and partly to the fact that, as r changes, the transverse component of velocity changes, even if the angular velocity $d\theta/dt$ is constant.

According to Newton's second law, $\mathbf{F} = m\mathbf{a}$, where $\mathbf{F}$ is the total force acting on the particle. This force $\mathbf{F}$ may be written as the sum of two components:

$$\mathbf{F} = F_r\mathbf{u}_r + F_\theta\mathbf{u}_\theta$$

The motion of the particle is then governed by the two differential equations

$$F_r = m\frac{d^2r}{dt^2} - mr\left(\frac{d\theta}{dt}\right)^2 \tag{2.52}$$

$$F_\theta = mr\frac{d^2\theta}{dt^2} + 2m\frac{dr}{dt}\frac{d\theta}{dt} \tag{2.53}$$

If both sides of eq. (2.53) are multiplied by r, eq. (2.53) can be written in the form

$$rF_\theta = \frac{d}{dt}\left(mr^2\frac{d\theta}{dt}\right) \tag{2.54}$$

which may be interpreted as stating that the torque applied to the particle equals the time rate of change of its angular momentum.

If $F_\theta = 0$, (2.54) may be integrated to yield $mr^2\,d\theta/dt = C$. In other words, if the force is always directed radially toward or away from the origin (a "central force field"), then the angular momentum of the particle will be constant. This immediately implies Kepler's second law of planetary motion, which states that the radius vector in a central force field sweeps over area at a constant rate, since the rate at which the vector $\mathbf{R}$ sweeps out area is

$$\frac{dA}{dt} = \frac{1}{2}r^2\frac{d\theta}{dt}$$

EXERCISES

1. Find $\mathbf{v}$ and $\mathbf{a}$ if a particle moves such that

$$r = b(1 - \cos\theta) \qquad \frac{d\theta}{dt} = 4$$

2. Find $\mathbf{v}$ and $\mathbf{a}$ if

$$r = b(1 + \sin t) \qquad \theta = e^{-t} - 1$$

3. A particle moves so that its position (r,θ) in polar coordinates is given by

$$r = 2(1 + \sin\theta) \qquad \theta = e^{-t}$$

Find its velocity $\mathbf{v}$ in terms of the vectors $\mathbf{u}_r$ and $\mathbf{u}_\theta$.

4. A particle, starting at $t = 0$ from the point $r = 2$, $\theta = 0$ in polar coordinates, moves such that

$$r = 2 + \sin t \qquad v = \sqrt{2}\cos t$$

Find a formula for the angle θ in terms of t, and determine the position of this particle at time $t = \pi/2$. (Assume that $\theta \geq 0$ for all t.)

5. A particle moves along a straight line not passing through the origin. Is $r(d\theta/dt)^2$ nonzero?

6. Which terms in eq. (2.51) will be nonzero in each of the following cases?
   (a) A particle moves around a circle with center at the origin with constant nonzero angular velocity.
   (b) A particle moves around a circle with center at the origin with constant nonzero angular acceleration.
   (c) A particle moves along a straight line not passing through the origin, with constant speed.
   (d) A person is walking from the center of a merry-go-round toward its outer edge (discuss various possibilities).

7. A particle moves along the curve

$$r = \frac{1}{1 + 2\cos\theta} \qquad \frac{d\theta}{dt} = \frac{1}{r^2}$$

   (a) By differentiating the equation $\mathbf{R} = r\mathbf{u}_r$, show that

$$\frac{d\mathbf{R}}{dt} = 2\sin\theta\,\mathbf{u}_r + \frac{1}{r}\,\mathbf{u}_\theta$$

   (b) Find $d^2\mathbf{R}/dt^2$ and simplify.

8. A disc rotates back and forth with angular velocity $\cos t$ rad/sec. An insect starting 1 cm from the center of the disc at time $t = 0$ crawls outward at a rate of $2t$ cm/sec. Find the position, velocity, and speed of the insect after 2π seconds.

9. Find the magnitude of the Coriolis acceleration of a particle moving in the xy plane with position given by

$$x = 3t\cos 4\pi t \qquad y = 3t\sin 4\pi t$$

10. A particle moves with constant radial speed 2 cm/sec away from the center of a platform rotating with uniform angular velocity 30 rev/min.
(a) What is its radial acceleration?
(b) What is its Coriolis acceleration?

11. The force **F** exerted by a magnetic field **B** on a particle carrying a charge q is given by $\mathbf{F} = q(\mathbf{v} \times \mathbf{B})$, where **v** is the velocity of the particle. Draw a diagram showing the relative directions of **v**, **B**, and **F**, in some special cases. Under what circumstances will the field exert no force on the particle?

12. A particle of mass m and charge q moves in a constant magnetic field **B** directed parallel to the z axis. If the resulting trajectory is a circle of radius r in the xy plane, express q/m in terms v, r, and **B**.

13. An experiment is being designed in which a particle of mass 1 is to exhibit the following planar motion in polar coordinates:

$$\left.\begin{aligned} r(t) &= 1 + t \\ \theta(t) &= \frac{\pi}{1+t} \end{aligned}\right\} \quad (t \geq 0)$$

(a) Determine the position and velocity of this particle at time $t = 1$, illustrating your answer in a diagram.
(b) Find the radial and transverse forces $F_r(t)$ and $F_\theta(t)$ needed on the particle to attain the desired motion.
(c) If the forces acting on this particle are removed at $t = 1$, find its position at $t = 5$.

14. Find $d^3\mathbf{R}/dt^3$ in terms of $\mathbf{u}_r$ and $\mathbf{u}_\theta$.

15. A particle moves in a plane with constant angular velocity ω about the origin, but r varies such that the rate of increase of its acceleration is parallel to the position vector **R**. Show that $d^2r/dt^2 = r\omega^2/3$.

16. The equations of a curve are given in polar coordinates as $r = r(t)$, $\theta = \theta(t)$. Identify the Frenet parameters k, τ, **T**, **N**, and **B** in terms of $\mathbf{u}_r$, $\mathbf{u}_\theta$, and the derivatives of $r(t)$ and $\theta(t)$.

2.5 Optional Reading: Tensor Notation

As shown in section 2.1, differentiation of a vector function proceeds componentwise. That is, the ith component of $d\mathbf{F}/dt$ is the derivative of the ith component of **F**:

$$\left(\frac{d\mathbf{F}}{dt}\right)_i = \frac{dF_i}{dt}$$

This happy circumstance makes the tensor notation for the rules in theorems 2.1 through 2.4, and their proofs, quite apparent. Thus for the cross product we have

$$\frac{d}{dt}\epsilon_{ijk}F_jG_k = \epsilon_{ijk}\frac{dF_j}{dt}G_k + \epsilon_{ijk}F_j\frac{dG_k}{dt}$$

by the rules of ordinary calculus. Interpreted in vector notation, this says

$$\frac{d}{dt}(\mathbf{F} \times \mathbf{G}) = \frac{d\mathbf{F}}{dt} \times \mathbf{G} + \mathbf{F} \times \frac{d\mathbf{G}}{dt}$$

which is eq. (2.7).

The other theorems are equally straightforward.

EXERCISE

Derive the rule for the derivative of the dot product.

3 *Scalar and Vector Fields*

3.1 Scalar Fields: Isotimic Surfaces: Gradients

If to each point (x,y,z) of a region in space there is made to correspond a number $f(x,y,z)$, we say that f is a *scalar field.* In other words, a scalar field is simply a scalar-valued function of three variables.

Here are some physical examples of scalar fields: the mass density of the atmosphere, the temperature at each point in an insulated wall, the water pressure at each point in the ocean, the gravitational potential of points in astronomical space, the electrostatic potential of the region between two condenser plates. Such scalar fields as density and pressure are only approximate idealizations of a complicated physical situation, since they take no account of the atomic properties of matter.

For the sake of fixing ideas, the following scalar fields are given as examples that will be referred to repeatedly:

1. $f(x,y,z) = x + 2y - 3z$
2. $f(x,y,z) = x^2 + y^2 + z^2$
3. $f(x,y,z) = x^2 + y^2$
4. $f(x,y,z) = \dfrac{x^2}{4} + \dfrac{y^2}{9} + z^2$
5. $f(x,y,z) = \sqrt{x^2 + y^2} - z$
6. $f(x,y,z) = \dfrac{1}{x^2 + y^2}$

The fields in examples 1 through 5 are defined at every point in space. The field in example 6 is defined at every point (x,y,z) except where $x^2 + y^2 = 0$, that is, everywhere except on the z axis.

If f is a scalar field, any surface defined by $f(x,y,z) = C$, where C is a constant, is called an *isotimic surface* (from the Greek *isotimos,* meaning "of equal value"). Sometimes, in physics, more specialized terms are used. For instance, if f denotes either electric or gravitational field potential, such surfaces are called *equipotential surfaces.* If f denotes temperature, they are called *isothermal surfaces.* If f denotes pressure, they are called *isobaric surfaces.*

In the above examples, the isotimic surfaces are (see fig. 3.1):

1. all planes perpendicular to the vector $\mathbf{i} + 2\mathbf{j} - 3\mathbf{k}$
2. all spheres with center at the origin
3. all right circular cylinders with the z axis as axis of symmetry
4. a family of ellipsoids
5. a family of cones
6. the same as in example 3

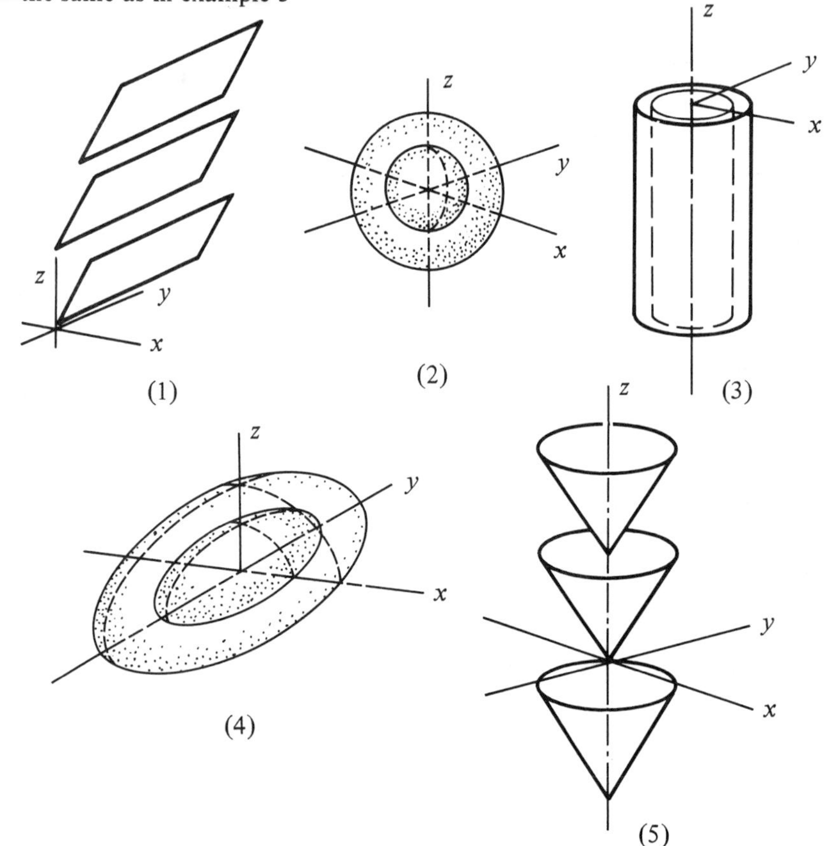

Figure 3.1

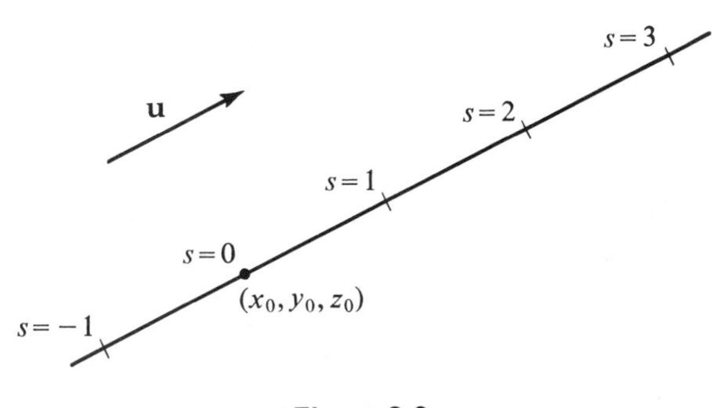

Figure 3.2

It is impossible for distinct isotimic surfaces of the same scalar field to intersect, since only one number $f(x,y,z)$ is associated with any one point (x,y,z).

Let us consider the behavior of a scalar field in the neighborhood of a point (x_0,y_0,z_0) within its region of definition. We imagine a line segment passing through (x_0,y_0,z_0) parallel to a given vector **u**. Let s denote the displacement measured along the line segment in the direction of **u** (fig. 3.2), with $s = 0$ corresponding to (x_0,y_0,z_0). To each value of the parameter s there corresponds a point (x,y,z) on the line segment, and hence a corresponding scalar $f(x,y,z)$. The derivative df/ds at $s = 0$, if this derivative exists, is called the ***directional derivative*** of f at (x_0,y_0,z_0), in the direction of the vector **u.**

In other words, the directional derivative of f is simply the rate of change of f, per unit distance, in some prescribed direction. The directional derivative df/ds will generally depend on the location of the point (x_0,y_0,z_0) and also on the direction prescribed.

The directional derivative of a scalar field f in a direction parallel to the x axis, with s measured as increasing in the positive x direction, is conventionally denoted $\partial f/\partial x$, and is called the partial derivative of f with respect to x. Similarly, the directional derivative of f in the positive y direction is called $\partial f/\partial y$, and that in the positive z direction, $\partial f/\partial z$. We assume that the reader has had some experience with partial derivatives.

The directional derivative of a scalar field f in a direction that is not parallel to any of the coordinate axes is conventionally denoted df/ds, but of course this symbol is ambiguous; it would not make sense to ask "what is df/ds" without specifying the direction in which s is to be measured.

A convenient way of specifying the desired direction is by prescribing a vector **u** pointing in that direction. Although the magnitude of **u** is immaterial, it is conventional to take **u** to be a unit vector. We have already seen (section 2.2) that a unit vector in a desired direction can be obtained by computing $d\mathbf{R}/ds$ in that direction, where $\mathbf{R} = x\mathbf{i} + y\mathbf{j} + z\mathbf{k}$. That is,

$$\mathbf{u} = \frac{dx}{ds}\mathbf{i} + \frac{dy}{ds}\mathbf{j} + \frac{dz}{ds}\mathbf{k} \qquad (3.1)$$

is a unit vector pointing in the direction in which s is measured. Here we are thinking of x, y, and z as functions of the parameter s, for points (x,y,z) on the line segment; s is, of course, arc length along the segment.

If the partial derivatives $\partial f/\partial x$, $\partial f/\partial y$, and $\partial f/\partial z$ exist and are continuous throughout a region, then it is well known (see Appendix B for a proof) that the following chain rule is valid:

$$\frac{df}{ds} = \frac{\partial f}{\partial x}\frac{dx}{ds} + \frac{\partial f}{\partial y}\frac{dy}{ds} + \frac{\partial f}{\partial z}\frac{dz}{ds} \tag{3.2}$$

If we define the *gradient* of f to be the vector

$$\mathbf{grad}\, f = \frac{\partial f}{\partial x}\mathbf{i} + \frac{\partial f}{\partial y}\mathbf{j} + \frac{\partial f}{\partial z}\mathbf{k} \tag{3.3}$$

we see that the right side of (3.2) is the dot product of $\mathbf{u}$ with $\mathbf{grad}\, f$:

$$\frac{df}{ds} = \mathbf{u} \cdot \mathbf{grad}\, f \tag{3.2$'$}$$

Since $\mathbf{u}$ is a unit vector, $\mathbf{u} \cdot \mathbf{grad}\, f = |\mathbf{u}||\mathbf{grad}\, f| \cos\theta = |\mathbf{grad}\, f| \cos\theta$, where θ is the angle between $\mathbf{grad}\, f$ and $\mathbf{u}$. This gives us the first fundamental property of the gradient.

PROPERTY 3.1 *The component of* **grad** *f in any given direction gives the directional derivative df/ds in that direction.*

By the maximum principle (example 1.16), the largest possible value of $\mathbf{u} \cdot \mathbf{grad}\, f$, for unit vectors $\mathbf{u}$, is obtained when $\mathbf{u}$ is in the same direction as $\mathbf{grad}\, f$ (assuming that $\mathbf{grad}\, f \neq \mathbf{0}$). Since $\mathbf{u} \cdot \mathbf{grad}\, f = df/ds$, it follows that the maximum value of df/ds is obtained in the direction of $\mathbf{grad}\, f$. This is the second fundamental property of the gradient.

PROPERTY 3.2 **grad** *f points in the direction of the maximum rate of increase of the function f.*

If $\mathbf{u}$ points in the direction of $\mathbf{grad}\, f$, then

$$\mathbf{u} \cdot \mathbf{grad}\, f = |\mathbf{u}||\mathbf{grad}\, f| \cos\theta = |\mathbf{grad}\, f|$$

which gives the third fundamental property of the gradient.

PROPERTY 3.3 *The magnitude of* **grad** *f equals the maximum rate of increase of f per unit distance.*

Experience has shown that the wording of these fundamental properties makes them rather easy to memorize [and they *should* be memorized, together with the definition (3.3)].

The fourth fundamental property of the gradient of a function makes it possible to use the gradient concept in solving geometrical problems.

PROPERTY 3.4 *Through any point* (x_0,y_0,z_0) *where* $\mathbf{grad}\,f \neq \mathbf{0}$, *there passes an isotimic surface* $f(x,y,z) = C$; $\mathbf{grad}\,f$ *is normal (i.e., perpendicular) to this surface at the point* (x_0,y_0,z_0).

This property holds only when $\partial f/\partial x$, $\partial f/\partial y$, and $\partial f/\partial z$ exist and are continuous in a neighborhood of the point in question. The constant C is, of course, equal to $f(x_0,y_0,z_0)$. If $\mathbf{grad}\,f = \mathbf{0}$, the locus of points satisfying $f(x,y,z) = C$ might not form a surface. (Consider, for example, this locus if f is a *constant* function.)

We omit a detailed proof of this fourth property, but the following discussion may make it seem reasonable. Let C denote the value of f at (x_0,y_0,z_0). Since $\mathbf{grad}\,f \neq \mathbf{0}$, it follows from the preceding fundamental properties that df/ds will be positive in some directions. If, then, we proceed away from (x_0,y_0,z_0) in one of these directions, the value of $f(x,y,z)$ will increase, and if we proceed in the opposite direction, its value will decrease. Since f and its partial derivatives are continuous, it seems reasonable that there will be a surface passing through (x_0,y_0,z_0) on which f remains at the constant value C; on one side of this surface the values of f will be greater than (and on the other, less than) C. Now suppose we consider any smooth arc passing through (x_0,y_0,z_0) and entirely contained in this surface. Then $f(x,y,z) = C$ for all points on this arc, and so $df/ds = 0$, where s is measured along this arc. Since $df/ds = \mathbf{u} \cdot \mathbf{grad}\,f$, and in this case $\mathbf{u}$ is a unit vector tangent to this arc, we see that $\mathbf{u} \cdot \mathbf{grad}\,f = df/ds = 0$, implying that $\mathbf{grad}\,f$ is perpendicular to $\mathbf{u}$. This reasoning applies to any smooth arc in the surface passing through (x_0,y_0,z_0). Hence $\mathbf{grad}\,f$ is perpendicular to every such arc, at that point, which can be the case only if $\mathbf{grad}\,f$ is perpendicular to the surface (fig. 3.3).

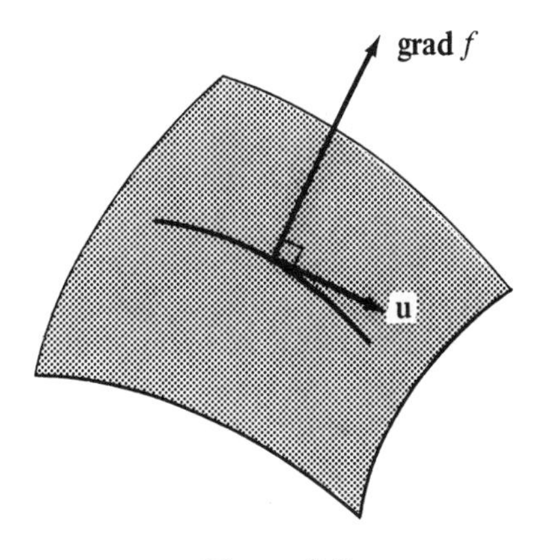

Figure 3.3

We now return to the six examples given previously. In each case the gradient is easily computed using the definition (3.3):

1. $\mathbf{grad}\, f = \mathbf{i} + 2\mathbf{j} - 3\mathbf{k}$
2. $\mathbf{grad}\, f = 2x\mathbf{i} + 2y\mathbf{j} + 2z\mathbf{k}$
3. $\mathbf{grad}\, f = 2x\mathbf{i} + 2y\mathbf{j}$
4. $\mathbf{grad}\, f = \dfrac{x}{2}\mathbf{i} + \dfrac{2y}{9}\mathbf{j} + 2z\mathbf{k}$
5. $\mathbf{grad}\, f = \dfrac{x\mathbf{i} + y\mathbf{i}}{\sqrt{x^2 + y^2}} - \mathbf{k}$
6. $\mathbf{grad}\, f = \dfrac{-2x}{(x^2 + y^2)^2}\mathbf{i} - \dfrac{2y}{(x^2 + y^2)^2}\mathbf{j}$

1. (This is the only one of the six examples for which $\mathbf{grad}\, f$ is a constant.) We already know from section 1.10 that $\mathbf{i} + 2\mathbf{j} - 3\mathbf{k}$ is perpendicular to any plane of the form $x + 2y - 3z = C$. We see that $\mathbf{grad}\, f = \mathbf{i} + 2\mathbf{j} - 3\mathbf{k}$. Thus we have verified the fourth fundamental property in this special case.
2. In this case the isotimic surfaces are spheres centered at the origin, so the normals to these surfaces must be vectors pointing directly away from the origin. Sure enough, we have $\mathbf{grad}\, f = 2x\mathbf{i} + 2y\mathbf{j} + 2z\mathbf{k} = 2\mathbf{R}$, and we know that the vector $2\mathbf{R}$ always points directly away from the origin. To see the significance of the number 2 here, let R denote the distance from the origin to the point (x,y,z). Then we can, in this example, write the function in terms of R: it is simply R^2. Moreover, if we move away from any point in the direction of maximum increase of R^2, which obviously means moving directly away from the origin, then the element of arc length is simply dR. In this direction, the derivative df/ds is df/dR, and $(d/dR)(R^2) = 2R$. Also, $|2\mathbf{R}| = 2R$, so we have verified the third fundamental property in this special case.
3. The reader familiar with cylindrical coordinates can do the same thing here as we just did with example 2. Let $\rho = (x^2 + y^2)^{1/2}$, the distance from the point (x,y,z) to the z axis. The function f in this example is simply ρ^2 and obviously increases most rapidly in a direction perpendicular to the z axis. Its derivative in this direction is 2ρ, which is also the magnitude $|\mathbf{grad}\, f| = (4x^2 + 4y^2)^{1/2}$. The direction is clearly normal to the isotimic surfaces, since the latter are right circular cylinders centered on the z axis. The second, third, and fourth fundamental properties are transparent in this case, as they were in example 2.
5. [We skip example 4.] All we care to note here is the elementary geometrical significance of the $-\mathbf{k}$ term in $\mathbf{grad}\, f$. The isotimic surfaces of this function are conical; each has an apex on the z axis and spreads outward with increasing z. Thus, we see easily that the normal to one such surface will not point directly away from the z axis, as it does in example 3, but will have an additional, constant component in the negative z direction.

The following examples are some sample problems that illustrate the use of the fundamental properties of the gradient of a scalar field.

Example 3.1 Find df/ds in the direction of the vector $4\mathbf{i} + 4\mathbf{j} - 2\mathbf{k}$ at the point (1,1,2) if $f(x,y,z) = x^2 + y^2 - z$.

Solution $\mathbf{grad}\, f = 2x\mathbf{i} + 2y\mathbf{j} - \mathbf{k} = 2\mathbf{i} + 2\mathbf{j} - \mathbf{k}$ at (1,1,2). A unit vector in the desired direction is $\mathbf{u} = \frac{2}{3}\mathbf{i} + \frac{2}{3}\mathbf{j} - \frac{1}{3}\mathbf{k}$ (obtained by dividing $4\mathbf{i} + 4\mathbf{j} - 2\mathbf{k}$ by its own length). Property 3.1 then gives $df/ds = \mathbf{u} \cdot \mathbf{grad}\, f = \frac{4}{3} + \frac{4}{3} + \frac{1}{3} = 3$. This means that the value of the function f is increasing 3 units per unit distance if we proceed from (1,1,2) in the direction stated.

Example 3.2 The temperature of points in space is given by $f(x,y,z) = x^2 + y^2 - z$. A mosquito located at (1,1,2) desires to fly in such a direction that he will get cool as soon as possible. In what direction should he move?

Solution As we saw in example 3.1, $\mathbf{grad}\, f = 2\mathbf{i} + 2\mathbf{j} - \mathbf{k}$ at (1,1,2). The mosquito should move in the direction $-\mathbf{grad}\, f$, since $\mathbf{grad}\, f$ is in the direction of increasing temperature.

Example 3.3 A mosquito is flying at a speed of 5 units of distance per second, in the direction of the vector $4\mathbf{i} + 4\mathbf{j} - 2\mathbf{k}$. The temperature is given by $f(x,y,z) = x^2 + y^2 - z$. What is his rate of increase of temperature, per unit time, at the instant he passes through the point (1,1,2)?

Solution As shown in example 3.1, df/ds in this direction is 3 units per unit distance. The rate of increase of temperature per unit time is thus $df/dt = (df/ds)(ds/dt) = (3)(5) = 15$ degrees per second.

Example 3.4 What is the maximum possible df/ds, if $f(x,y,z) = x^2 + y^2 - z$, at the point (1,4,2)?

Solution $|\mathbf{grad}\, f| = |2\mathbf{i} + 8\mathbf{j} - \mathbf{k}| = \sqrt{69}$. The answer is approximately 8.31 units per unit distance.

Example 3.5 Find a unit vector normal to the surface $x^2 + y^2 - z = 6$ at the point (2,3,7).

Solution This is an isotimic surface for the function $f(x,y,z) = x^2 + y^2 - z$. At (2,3,7) we have $\mathbf{grad}\, f = 2x\mathbf{i} + 2y\mathbf{j} - \mathbf{k} = 4\mathbf{i} + 6\mathbf{j} - \mathbf{k}$. The length of this vector is $\sqrt{53}$. Thus an answer is $(\sqrt{53}/53)(4\mathbf{i} + 6\mathbf{j} - \mathbf{k})$. (The negative of this vector is also a correct answer.)

The reader may have observed that the number 6, the constant on the right-hand side of the equation defining the isotimic surface in example 3.5, appears to have no effect on the normal, $\mathbf{grad}\, f$. This is not quite true. Granted, the formula for **grad** f ignores the 6, but when it is evaluated at (x,y,z), the numbers x, y, and z must satisfy $x^2 + y^2 - z = 6$. Clearly $2^2 + 3^2 - 7 = 6$.

EXERCISES

1. Compute **grad** f if
   (a) $f = \sin x + e^{xy} + z$
   (b) $f = 1/|\mathbf{R}|$
   (c) $f = \mathbf{R} \cdot \mathbf{i} \times \mathbf{j}$
2. If $f(x,y,z) = x^2 + y^2$, what is the locus of points in space for which **grad** f is parallel to the y axis?
3. What can you say about a function whose gradient is everywhere parallel to the y axis?
4. Find all functions $f(x,y,z)$ such that $\mathbf{grad}\, f = 2x\mathbf{i} + z\mathbf{j} + y\mathbf{k}$.
5. Can you find a scalar whose gradient is $y\mathbf{i}$?
6. Describe **grad** f in words, without actually doing any calculating, given that $f(x,y,z)$ is the distance between (x,y,z) and the z axis.
7. Given $f(x,y,z) = x^2 + y^2 + z^2$, find the maximum value of df/ds at the point (3,0,4),
   (a) by using the gradient of f.
   (b) by interpreting f geometrically.
8. A volcano just erupted and lava is streaming down from the mountain top. Suppose that the altitude of the mountain is given by

$$z(x,y) = he^{-(x^2 + 2y^2)}$$

   where h is the maximum height, and suppose also that lava flows in the direction of steepest descent (fastest change in z). Find
   (a) the projection on the xy plane of the direction in which lava flows away from the point $(1,2,he^{-9})$.
   (b) the equation of the projection on the xy plane of the flow line of the lava passing through the point $(1,2,he^{-9})$.
9. Find the derivative of $f(x,y,z) = x + xyz$ at the point $(1,-2,2)$ in the direction of
   (a) $2\mathbf{i} + 2\mathbf{j} - \mathbf{k}$
   (b) $2\mathbf{i} + 2\mathbf{j} + \mathbf{k}$
10. Find the directional derivative df/ds at $(1,3,-2)$ in the direction of $-\mathbf{i} + 2\mathbf{j} + 2\mathbf{k}$ if
    (a) $f(x,y,z) = yz + xy + xz$
    (b) $f(x,y,z) = x^2 + 2y^2 + 3z^2$
    (c) $f(x,y,z) = xy + x^3y^3$
    (d) $f(x,y,z) = \sqrt{x^2 + y^2 + z^2}$
11. Find the magnitude of the greatest rate of change of $f(x,y,z) = (x^2 + z^2)^3$ at $(1,3,-2)$. Interpret geometrically.

12. Find the direction of maximal increase of the function $f(x,y,z) = e^{-xy} \cos z$ at the point (1,1,0).
13. By vector methods, find the point on the curve $x = t$, $y = t^2$, $z = 2$ at which the temperature $\phi(x,y,z) = x^2 - 6x + y^2$ takes its minimum value.
14. Find a vector normal to the surface $x^2 + yz = 5$ at (2,1,1).
15. Find an equation of the plane tangent to the sphere $x^2 + y^2 + z^2 = 21$ at $(2,4,-1)$.
16. Find a vector normal to the cylinder $x^2 + z^2 = 8$ at (2,0,2),
    (a) by inspection (draw a diagram).
    (b) by finding the gradient of the function $f(x,y,z) = x^2 + z^2$ at (2,0,2).
17. Find an equation of the plane tangent to the surface $z^2 - xy = 14$ at (2,1,4).
18. Find equations of the line normal to the sphere $x^2 + y^2 + z^2 = 2$ at (1,1,0),
    (a) by inspection (draw a diagram).
    (b) by computing the gradient of $f(x,y,z) = x^2 + y^2 + z^2$ at (1,1,0), and using this to find the normal.
19. Find a unit vector normal to the plane $3x - y + 2z = 3$,
    (a) by the methods of section 1.10.
    (b) by the methods of the preceding section.
20. Find an equation of the plane tangent to the surface $z = x^2 + y^2$ at (2,3,13). [*Hint:* Consider the function $f(x,y,z) = x^2 + y^2 - z$.]
21. Let $T(x,y,z) = x^2 + 2y^2 + 3z^2$, and let S be the isotimic surface: $T = 1$. Find all points (x,y,z) on S that have tangent planes with normals (1,1,1).
22. If $\phi(x,y,z) = x^2y + zy + z^3$, find
    (a) the gradient of ϕ.
    (b) the equation of the plane passing through the point $(1,-1,1)$ and tangent to the level surface of ϕ at that point.
23. Find a unit vector tangent to the curve of intersection of the cylinder $x^2 + y^2 = 4$ and the sphere $x^2 + y^2 + z^2 = 9$ at the point $(\sqrt{2},\sqrt{2},\sqrt{5})$,
    (a) by drawing a diagram, obtaining the answer by inspection.
    (b) by finding the vector product of the normals to the two surfaces at that point.
    (c) by writing the equation of the curve in parametric form. (*Hint:* Let $x = 2 \sin t$ and $y = 2 \cos t$.)
24. Determine the angle between the normals of the intersecting spheres $x^2 + y^2 + z^2 = 16$ and $(x - 1)^2 + y^2 + z^2 = 16$ at the point $(1/2, 3/2, 3\sqrt{6}/2)$.
25. At what angle does the line $2x = y = 2z$ intersect the ellipsoid $2x^2 + y^2 + 2z^2 = 8$?
26. What is the angle between the tangent to the curve

$$\mathbf{R}(t) = t\mathbf{i} + t^2\mathbf{j} + 2t^2\mathbf{k} \qquad (0 \leq t \leq 3)$$

    and the normal to the surface $z = 16 - x^2 - y$ at their point of intersection?
27. At what angle does the curve $x = t$, $y = 2t - t^2$, $z = 2t^4$ intersect the surface $x^2 + y^3 + 3z^2 = 14$ at the point (1,1,2)?
28. Find the angle between the surfaces $z = x^2 + y^2$ and $x^2 + y^2 + (z - 3)^2 = 9$ at the point $(2,-1,5)$.

29. Let S_1 and S_2 be the surfaces with equations

$$\frac{x^2}{a^2} + \frac{y^2}{b^2} + \frac{z^2}{c^2} = 1 \qquad \frac{x^2}{A^2} + \frac{y^2}{B^2} + \frac{z^2}{C^2} = 1$$

Show that, if $a^2B^2 - b^2A^2 = 0$, then the curve of intersection of S_1 and S_2 must be parallel to the xy plane.

30. If $\mathbf{R}_1$ denotes the position vector of a point P relative to an origin O_1 in the xy plane, and $\mathbf{R}_2$ denotes the position vector of the same point relative to another origin O_2, then $|\mathbf{R}_1| + |\mathbf{R}_2| =$ constant is the equation of an ellipse with foci O_1 and O_2. Use this observation to prove that lines O_1P and O_2P make equal angles with the tangent to the ellipse at P. [*Hint:* **grad** $(|\mathbf{R}_1| + |\mathbf{R}_2|)$ is normal to the ellipse.]

31. Find the point on the sphere $x^2 + y^2 + z^2 = 84$ that is nearest the plane $x + 2y + 4z = 77$.

32. Find the point on the ellipsoid $x^2 + 2y^2 + 3z^2 = 6$ that is nearest the plane $x + 2y + 3z = 8$.

33. What point on the curve $x = t$, $y = t^2$, $z = 2$ is closest to the surface $x^2 - 6x + y^2 + 7 = 0$?

34. Show that any level curve $\mathbf{R}(t)$ for the function $f(x,y,z)$ satisfies

$$\frac{d\mathbf{R}}{dt} \cdot \mathbf{grad}\, f = 0$$

35. Given $\phi = \tan^{-1} x + \tan^{-1} y$ and $\psi = (x + y)/(1 - xy)$, show that $\nabla\phi \times \nabla\psi = 0$. [*Hint:* It is easy if you recognize the formula for $\tan (A + B)$.]

36. Given $w = uv$, where u and v are scalar fields, show that $\nabla w \cdot \nabla u \times \nabla v = 0$,
(a) by direct calculation.
(b) without calculation.

37. Generalize the result of the preceding exercise.

3.2 Vector Fields and Flow Lines

A *vector field* $\mathbf{F}$ is a rule associating with each point (x,y,z) in a region a vector $\mathbf{F}(x,y,z)$. In other words, a vector field is a vector-valued function of three variables.

In visualizing a vector field, we imagine that from each point in the region there extends a vector. Both direction and magnitude may vary with position (fig. 3.4). A good visualization of a vector field is a sandstorm, with $\mathbf{F}(\mathbf{R})$ giving the velocity of the sand particle at $\mathbf{R}$.

Some vector fields are not defined for all points in space. For example, the vector field

$$\mathbf{F}(x,y,z) = \frac{x\mathbf{i} + y\mathbf{j}}{x^2 + y^2}$$

is not defined along the z axis, since $x^2 + y^2 = 0$ for points on the z axis.

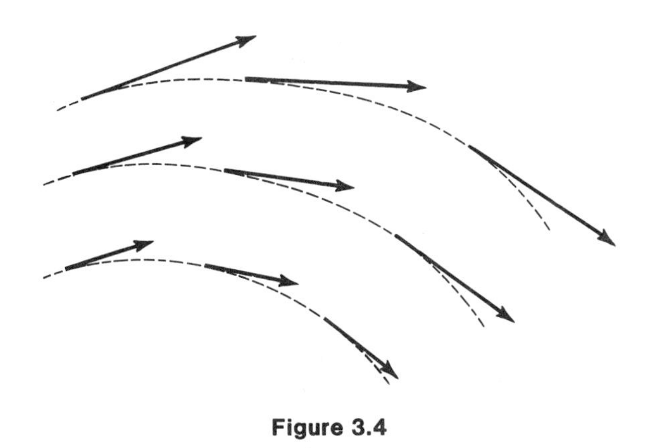

Figure 3.4

A vector field may be written in terms of its components:

$$\mathbf{F}(x,y,z) = F_1(x,y,z)\mathbf{i} + F_2(x,y,z)\mathbf{j} + F_3(x,y,z)\mathbf{k}$$

Example 3.6 If $f(x,y,z)$ is a *scalar* field, **grad** f is a *vector* field.

Example 3.7 Each of the "vectors" $\mathbf{u}_r$ and $\mathbf{u}_\theta$ (section 2.5) is a vector field defined in the plane.

Example 3.8 In hydrodynamics, one associates with each point of a region the velocity of the fluid passing that point. In this manner one obtains, at any instant of time, a vector field describing the instantaneous velocity of the fluid at every point.

Example 3.9 In theoretical physics, there is associated with each point in space an electric field intensity vector, representing the force that would be exerted, per unit charge, on a charged particle, if it were located at that point. This electric field, at any instant of time, constitutes a vector field. (Magnetic fields and gravitational fields also provide examples of vector fields defined in space.)

Let us consider a vector field **F** that is defined and nonzero at every point of a region in space. Any curve passing through the region is called a *flow line* of **F** provided that, at every point on the curve, **F** is tangent to the curve. (Flow lines are also called *stream lines* or *characteristic curves* of **F.** If **F** is a force field, the flow lines are commonly called *lines of force.*) In figure 3.4, three flow lines are indicated as dotted curves.

This may be visualized another way. The vector field **F** determines, at each point in the region, a direction. If a particle moves in such a manner that the direction of its velocity at any point coincides with the direction of the vector field **F** at that point, the space curve traced out is a flow line.

If the vector field $\mathbf{F}(x,y,z)$ describes the velocity at each point in a hydrodynamic system, the flow lines are the paths traversed by the component particles of the fluid, assuming that $\mathbf{F}$ is not a function of time. (The situation is more complicated for time-varying flows.)

Note that if $g(x,y,z)$ is a scalar field that is not zero at any point, the flow lines of the vector field $g(x,y,z)\mathbf{F}(x,y,z)$ will be the same as those of $\mathbf{F}(x,y,z)$, since only the *direction* of $\mathbf{F}$ at any point is relevant in determining the flow lines.

Since the direction of a flow line is uniquely determined by the field $\mathbf{F}$, it is impossible to have two different flow directions at the same point, and therefore it is impossible for two flow lines to cross. If the magnitude of $\mathbf{F}$ is zero at some point in space, then no direction is defined at that point and no flow line passes through that point. Now let's see how to calculate flow lines.

If $\mathbf{R}$ is the position vector to an arbitrary point of a flow line, and if s represents arc length measured along the flow line, then the unit vector tangent to the curve at that point is given by

$$\mathbf{T} = \frac{d\mathbf{R}}{ds} = \frac{dx}{ds}\mathbf{i} + \frac{dy}{ds}\mathbf{j} + \frac{dz}{\mathrm{ds}}\mathbf{k} \tag{3.4}$$

The requirement that $\mathbf{T}$ have the same direction as $\mathbf{F}$ can be written

$$\mathbf{T} = \beta\mathbf{F} \tag{3.5}$$

where β is a scalar-valued function of x, y, and z. This can be written in terms of components:

$$\beta F_1 = \frac{dx}{ds} \qquad \beta F_2 = \frac{dy}{ds} \qquad \beta F_3 = \frac{dz}{ds} \tag{3.6}$$

If F_1, F_2, and F_3 are all nonzero, we may eliminate β and write eq. (3.6) in differential form:

$$\frac{dx}{F_1} = \frac{dy}{F_2} = \frac{dz}{F_3} \tag{3.7}$$

If one of these functions (say F_3) is identically zero in a region, then we obtain directly from eq. (3.6) that the curve lies in a plane (say, z = constant) parallel to one of the coordinate planes.

Example 3.10 If $\mathbf{F} = x\mathbf{i} + y\mathbf{j} + \mathbf{k}$, then $F_1 = x$, $F_2 = y$, and $F_3 = 1$, giving $dx/x = dy/y = dz$. Solving the differential equations $dx/x = dz$ and $dy/y = dz$, we obtain $x = C_1e^z$, $y = C_2e^z$. Thus the equations of the flow line passing through the point (3,4,7) are $x = 3e^{z-7}$, $y = 4e^{z-7}$. The equations of the flow line passing through the origin are $x = 0$, $y = 0$—i.e., the z axis.

Example 3.11 If $\mathbf{F} = x\mathbf{i} + y\mathbf{j}$, then $F_1 = x$, $F_2 = y$, and $F_3 = 0$. In this case eq. (3.6) becomes $\beta x = dx/ds$, $\beta y = dy/ds$, and $0 = dz/ds$. Eliminating β from the first two equations, we obtain $dx/x = dy/y$, and, solving, we obtain $y = Cx$. From the third equation we obtain $z =$ constant. The field is zero when both x and y equal zero, and so the flow lines are not defined along the z axis. The flow lines are straight half-lines parallel to the xy plane, extending outward from the z axis.

Example 3.12 If $\mathbf{F} = -y\mathbf{i} + x\mathbf{j}$, then $-\beta y = dx/ds$, $\beta x = dy/ds$, and $0 = dz/ds$. Thus $-dx/y = dy/x$, and hence $x^2 + y^2 =$ constant. Also, we have $z =$ constant. The flow lines are circles surrounding the z axis and are parallel to the xy plane. As in example 3.11, no flow lines pass through points on the z axis.

Flow lines may be infinite in extent, as in examples 3.10 and 3.11, or they may close upon themselves, as in example 3.12.

EXERCISES

1. For the vector field $\mathbf{F}$ of example 3.12, draw a diagram similar to figure 3.4 showing the values of F at the points $(1,0)$, $(0,1)$, $(-1,0)$, $(0,-1)$, $(1,1)$, $(-1,1)$, $(-1,-1)$, $(1,-1)$, and a scattering of other points. Indicate flow lines.
2. Let $\mathbf{F} = x^2\mathbf{i} + y^2\mathbf{j} + \mathbf{k}$.
   (a) Find the general equation of a flow line.
   (b) Find the flow line through the point $(1,1,2)$.
3. Without doing any calculating, describe the flow lines of the vector field $\mathbf{R} = x\mathbf{i} + y\mathbf{j} + z\mathbf{k}$. [*Hint:* If a particle located at (x,y,z) has velocity $\mathbf{R}$, in what direction is it moving relative to the origin?]
4. The flow lines of the gradient of a scalar field cross the isotimic surfaces orthogonally. Explain.

3.3 Divergence

The concept of *gradient*, as we have presented it, describes the rate of change of a scalar field. We now consider the more complicated problem of describing the rate of change of a *vector* field. There are two fundamental measures of this rate of change: the *divergence* and the *curl*.

Roughly speaking, the divergence of a vector field is a scalar field that tells us, at each point, the extent to which the field *explodes*, or diverges, from that point. The curl of a vector field is a vector field that gives us, at each point, an indication of how the field swirls in the vicinity of that point (fig. 3.5). However, to describe divergence and curl in such a brief manner is both useless and a bit dangerous, since

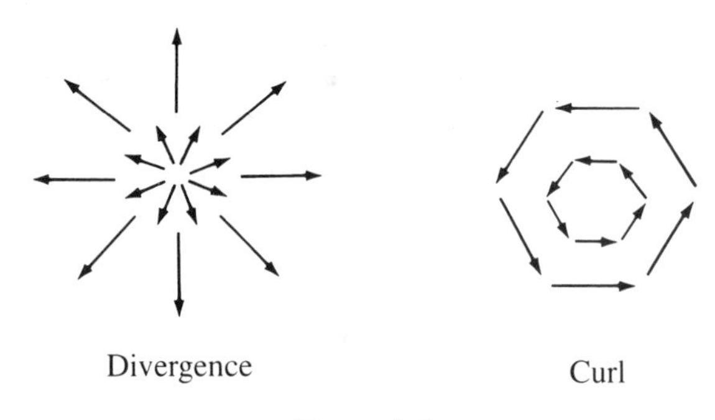

Figure 3.5

(if taken literally) both of these preceding sentences are not only vague but technically incorrect. As we shall see, it is possible for a field to have a positive divergence without appearing to "diverge" at all, and it is possible for a field to have a nontrivial curl and yet have flow lines that do not bend at all.

In this section we consider only the divergence. We begin by presenting a heuristic derivation, which will serve to motivate the formal definition. As usual the vector field will be denoted by

$$\mathbf{F} = F_1\mathbf{i} + F_2\mathbf{j} + F_3\mathbf{k}$$

Let us once again picture the velocity field for a stream of flowing particles, such as sand particles in a sandstorm, electrons in a wire or plasma, or fluid particles in a jet emerging from a nozzle. Denote the number of particles per unit volume—the *particle density*—by ν. Now if each particle weighs m grams (say), then the *mass density* μ equals $m\nu$ grams per unit volume.

Next let $\mathbf{v}(x,y,z)$ be the velocity of the fluid particle located at (x,y,z); $\mathbf{v}$ is the velocity field of the fluid. The vector field

$$\mathbf{F}(x,y,z) = \mu(x,y,z)\mathbf{v}(x,y,z)$$

is called the *mass flow rate density* of the fluid. We shall use F to calculate the mass flow rate, or number of grams per unit time, that crosses any hypothetical "window" in the flow pattern.

Thus consider a small planar patch of surface area δS inside the fluid, as in figure 3.6; the arrows depict the velocity field $\mathbf{v}(x,y,z)$. If we start counting particles crossing δS for the next Δt seconds, which will be the last particles to make it through? Clearly they are the ones that are $-(\mathbf{v}\Delta t)$ away from the patch at the start—and as figure 3.6(b) illustrates, all the particles in the cylinder with base ΔS and slant height $|\mathbf{v}|\Delta t$ cross ΔS in that time.

How many particles are in this cylinder? Its *volume* is given by

$$(\text{base area}) \text{ times } (\text{height}) = \Delta S\, \mathbf{n} \cdot \mathbf{v}\Delta t$$

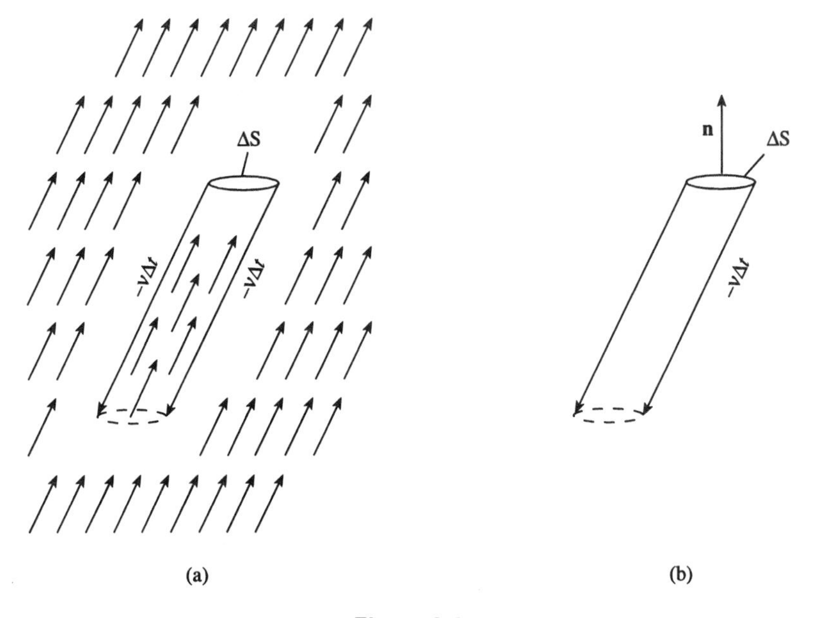

Figure 3.6

where $\mathbf{n}$ is the unit normal to the patch as shown; therefore it contains $\nu(\Delta S)\mathbf{n} \cdot \mathbf{v}\Delta t$ particles. As a result the mass flow rate through the small area ΔS *per unit time* is given by

$$[m\nu\mathbf{v}] \cdot \mathbf{n}\Delta S = \mu\mathbf{v} \cdot \mathbf{n}\Delta S = \mathbf{F} \cdot \mathbf{n}\Delta S$$

This is called the *flux of the vector field* **F** *through* ΔS.

[If we had used the charge q per particle instead of the mass m and defined the vector field $\mathbf{j}(x,y,z)$ in terms of the charge density $\rho = q\nu$

$$\mathbf{j}(x,y,z) = \rho(x,y,z)\mathbf{v}(x,y,z)$$

then the flux of the "current density" $\mathbf{j}$ through ΔS would give the charge flow rate, or *current*, through the patch.]

To define the divergence of the field **F**, we imagine an infinitesimal rectangular parallelepiped having corners at (x,y,z), $(x + \Delta x, y, z)$, $(x, y + \Delta y, z)$, $(x, y, z + \Delta z)$, and so on (fig. 3.7). We shall compute the total flux of the field **F** through the six sides of this box in the outward direction (i.e., on each side we choose $\mathbf{n}$ to be the outward normal). We then divide this flux by the volume of the box and take the limit as the dimensions of the box go to zero. *This limit is called the divergence of* **F** *at the point* (x,y,z). In other words, the divergence is the net outflux per unit volume.

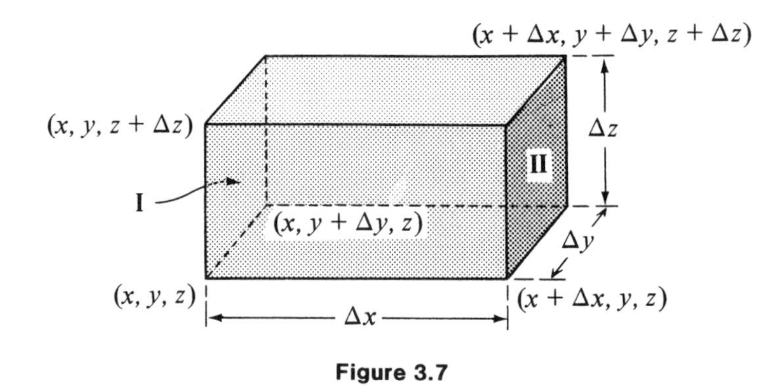

Figure 3.7

The computation of this limit proceeds as follows. On face I in figure 3.7 the outward normal is $-\mathbf{i}$. Thus, according to the above analysis, the flux out of this face is approximately $-F_1\Delta y\Delta z$. The flux out of face II, whose outward normal is $\mathbf{i}$, is $F_1\Delta y\Delta z$. These need not cancel, however, since F_1 may grow or diminish along the length Δx of the box. To account for this we estimate the total flux out of faces I and II as

$$[F_1(x + \Delta x, y, z) - F_1(x,y,z)]\Delta y\Delta z$$

The difference in these values of F_1 is given, to the same order of accuracy, by

$$\frac{\partial F_1}{\partial x}\Delta x$$

Thus the contribution to the net outward flux from faces I and II is

$$\frac{\partial F_1}{\partial x}\Delta x\Delta y\Delta z$$

Similarly, the two faces in the y direction contribute

$$\frac{\partial F_2}{\partial y}\Delta y\Delta x\Delta z$$

and, adding the contribution of the two remaining faces, we see that the net outward flux is approximately

$$\left(\frac{\partial F_1}{\partial x} + \frac{\partial F_2}{\partial y} + \frac{\partial F_3}{\partial z}\right)\Delta x\Delta y\Delta z$$

After we divide by the volume $\Delta x\Delta y\Delta z$, our approximations become accurate as we take the limit, and we are led to the following statement, which we take as our formal *definition* of divergence:

The divergence of a vector field

$$\mathbf{F} = F_1\mathbf{i} + F_2\mathbf{j} + F_3\mathbf{k}$$

is a scalar field, denoted div **F**, *defined by*

$$\text{div } \mathbf{F} = \frac{\partial F_1}{\partial x} + \frac{\partial F_2}{\partial y} + \frac{\partial F_3}{\partial z} \tag{3.8}$$

It is easy to compute the divergence of a vector field, as we demonstrate with examples. Keep in mind that div **F** *is defined* by eq. (3.8), and that our heuristic discussion gives us the interpretation of div **F** as net outflux per unit volume.

Example 3.13 Find div **F** if $\mathbf{F} = x\,\mathbf{i} + y^2z\,\mathbf{j} + xz^3\,\mathbf{k}$.

Solution

$$\begin{aligned} \text{div } \mathbf{F} &= \frac{\partial}{\partial x}(x) + \frac{\partial}{\partial y}(y^2z) + \frac{\partial}{\partial z}(xz^3) \\ &= 1 + 2yz + 3xz^2 \end{aligned}$$

Example 3.14 Find div **F** if $\mathbf{F} = xe^y\,\mathbf{i} + e^{xy}\,\mathbf{j} + \sin yz\,\mathbf{k}$.

Solution

$$\begin{aligned} \text{div } \mathbf{F} &= \frac{\partial}{\partial x}(xe^y) + \frac{\partial}{\partial y}(e^{xy}) + \frac{\partial}{\partial z}(\sin yz) \\ &= e^y + xe^{xy} + y\cos yz \end{aligned}$$

Example 3.15 Give an example of a vector field **F** that has divergence equal to 3 at every point in space.

Solution Many solutions can be given, for instance $\mathbf{F} = 3x\mathbf{i}$ or $\mathbf{F} = x\mathbf{i} + y\mathbf{j} + z\mathbf{k}$.

Example 3.16 In figure 3.8, is the divergence of **F** at point P positive or negative? Assume no variation of **F** in the z direction and that F_3 is identically zero.

Solution Heuristically, we can see from the diagram that the flux through the x faces of a parallelepiped at P will cancel, while there is definitely flux *out* of both y faces. Since there is no flux in the z direction, we expect that the divergence is positive.

More rigorously, we observe that F_1 is approximately constant, so $\partial F_1/\partial x = 0$. Below P, F_2 is negative, and above P, F_2 is positive, so $\partial F_2/\partial y$ is positive. Since $F_3 = 0$, we have $\partial F_3/\partial z = 0$. It follows that div **F** is positive at point P.

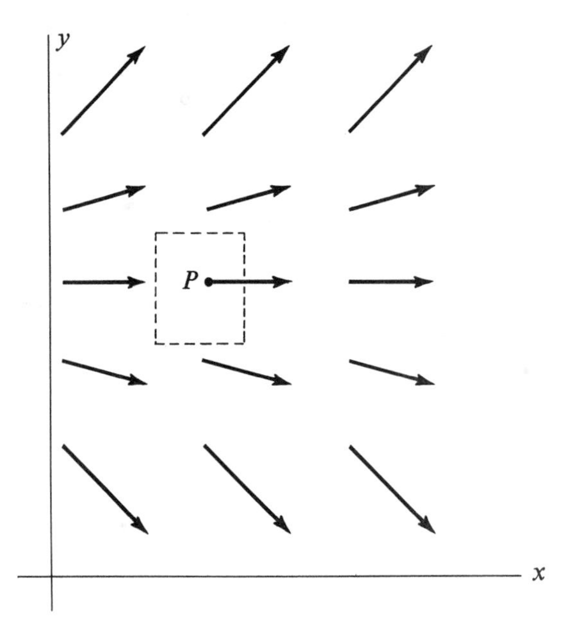

Figure 3.8

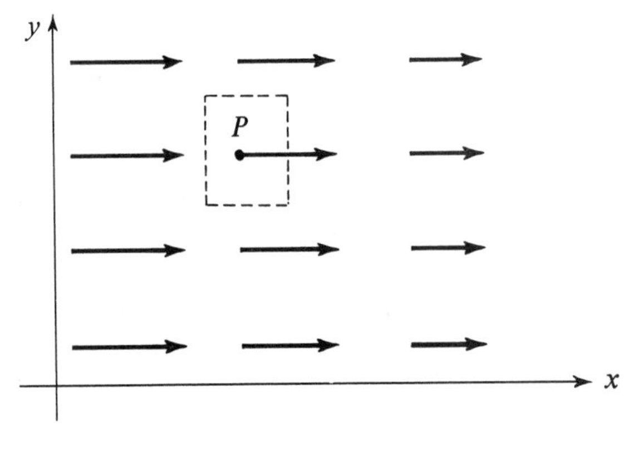

Figure 3.9

Example 3.17 In figure 3.9, is the divergence of **F** at point P positive or negative? Assume no variation of **F** in the z direction and that F_3 is identically zero.

Solution Again heuristically, there is no flux in the y or z direction, and the flux in the x direction *decreases* as we move to the right. So the net flux through the sides of a box at P is inward, and the divergence must be negative.

More precisely, we note that F_1 is decreasing with increasing x; hence $\partial F_1/\partial x$ is negative. F_2 and F_3 are zero at every point. It follows that the divergence of **F** is negative at every point.

In figure 3.8 where the divergence is positive, the lines of flux do, in a sense, diverge in a neighborhood of P. This is the picture that motivates the common (incorrect) statement that "positive divergence means the field is diverging, negative divergence means the field is converging." Note that in figure 3.9 the divergence is negative, but the flow lines are not converging. The divergence is negative because more fluid enters a given region from the left than leaves it to the right.

Returning to our mental image of **F** as a mass flow rate density $\mu\mathbf{v}$, we can derive an important relationship by considering the consequences of the principle of conservation of mass. The quantity

$$\operatorname{div}(\mu\mathbf{v})\ \Delta x\Delta y\Delta z$$

measures the flux of $(\mu\mathbf{v})$ out of a box with dimensions Δx, Δy, and Δz. But we have seen that the flux measures the mass of fluid crossing the faces of the box. Therefore this outflux must result in a decrease in the amount of fluid in the box

$$\mu\Delta x\Delta y\Delta z$$

and hence a decrease in the density. Thus we can write

$$\operatorname{div}(\mu\mathbf{v}) = -\frac{\partial\mu}{\partial t} \tag{3.9}$$

This is called the *equation of continuity* in fluid mechanics. The corresponding equation for charge density

$$\operatorname{div}(\rho\mathbf{v}) = -\frac{\partial\rho}{\partial t}$$

expresses the conservation of charge.

The heuristic reasoning employed in this section is, of course, subject to criticism, as are most arguments involving "infinitesimals." Its rigorous justification rests on a result known, appropriately enough, as the *divergence theorem,* and we will study it in the next chapter. For the present, we are satisfied with having a formal, precise definition of div **F** in eq. (3.8), and an intuitive picture of what it represents.

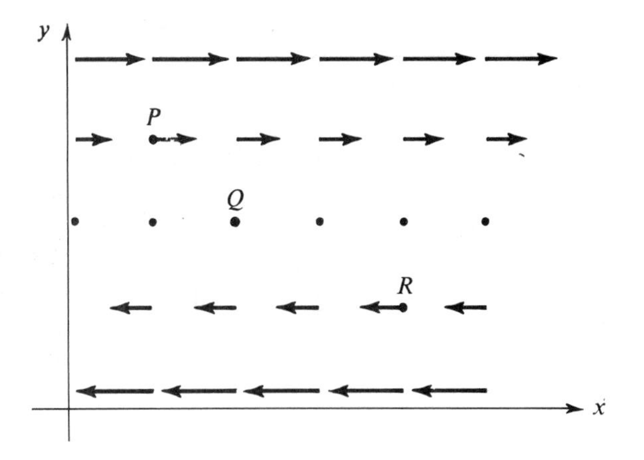

Figure 3.10

EXERCISES

1. Find div **F**, given that $\mathbf{F} = e^{xy}\,\mathbf{i} + \sin xy\,\mathbf{j} + \cos^2 zx\,\mathbf{k}$.
2. Find div **F**, given that $\mathbf{F} = x\mathbf{i} + y\mathbf{j} + z\mathbf{k}$.
3. Find div **F**, given that $\mathbf{F} = \mathbf{grad}\,\phi$, where $\phi = 3x^2y^3z$.
4. Find the divergence of the field

$$\frac{x\mathbf{i} + y\mathbf{j} + z\mathbf{k}}{(x^2 + y^2 + z^2)^{3/2}}$$

   Is the divergence of this field defined at every point in space?
5. Show in detail that $\text{div}(\phi\mathbf{F}) = \phi\,\text{div}\,\mathbf{F} + \mathbf{F}\cdot\mathbf{grad}\,\phi$.
6. Construct an example of a scalar field ϕ and a vector field **F**, neither of which is constant, for which $\text{div}(\phi\mathbf{F})$ is identically equal to $\phi\,\text{div}\,\mathbf{F}$.
7. Give an example of a nonconstant field with zero divergence.
8. Give an example of a field with a constant negative divergence.
9. Give an example of a field whose divergence depends only on x, is always positive, and increases with increasing x. (*Hint:* The function e^x is positive for every x.)
10. What can you say about the divergence of the vector field in figure 3.10 at points P, Q, and R? Assume no variation of **F** in the z direction and that F_3 is identically zero.
11. What can you say about the divergence of the vector field in figure 3.11 at points P, Q, and R? Assume no variation of **F** in the z direction and that F_3 is identically zero.
12. Another hydrodynamic interpretation of divergence is as follows. Let **F** be the velocity field of a fluid. Consider a small rectangular parallelepiped of fluid located at (x,y,z). Then the divergence of **F** is the time rate of change of volume of this body of fluid, per

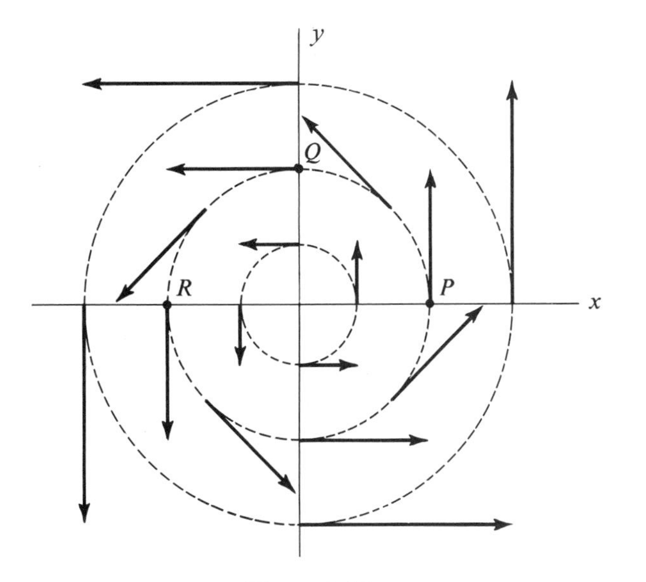

Figure 3.11

unit volume, as the size of the box goes to zero. Show this. [*Hint:* With $\mathbf{R} = x\mathbf{i} + y\mathbf{j} + z\mathbf{k}$, the box initially has corners at $\mathbf{R}$, $\mathbf{R} + \Delta x\mathbf{i}$, $\mathbf{R} + \Delta y\mathbf{j}$, $\mathbf{R} + \Delta z\mathbf{k}$, etc. After time Δt these corners have moved to the new positions $\mathbf{R} + \mathbf{F}(x,y,z)\Delta t$, $\mathbf{R} + \Delta x\mathbf{i} + \mathbf{F}(x + \Delta x, y, z)\Delta t$, $\mathbf{R} + \Delta y\mathbf{j} + \mathbf{F}(x, y + \Delta y, z)\Delta t$, $\mathbf{R} + \Delta z\mathbf{k} + \mathbf{F}(x, y, z + \Delta z)\Delta t$, etc. Calculate the new volume using the triple scalar product, and compute the limit described above.]

3.4 Curl

As in the previous section, we shall preface our formal definition of the curl of a vector field with some heuristic considerations. Once again imagine a flowing liquid with velocity field $\mathbf{v}(x,y,z)$ and *unit* density $\mu = 1$; thus the mass flow rate density is $\mathbf{F} = \mu\mathbf{v}$ is simply $\mathbf{v}$ itself. Consider a small paddle wheel, like that shown in figure 3.12, that is free to rotate about its axis AA'.

Imagine that we immerse this paddle wheel in the liquid. Because of the flow of the liquid, it will tend to rotate with some angular velocity. This angular velocity will vary, depending on where we locate the paddle wheel and on the positioning of its axis. For definiteness we shall compute the angular velocity with the paddle wheel lined up along the z axis.

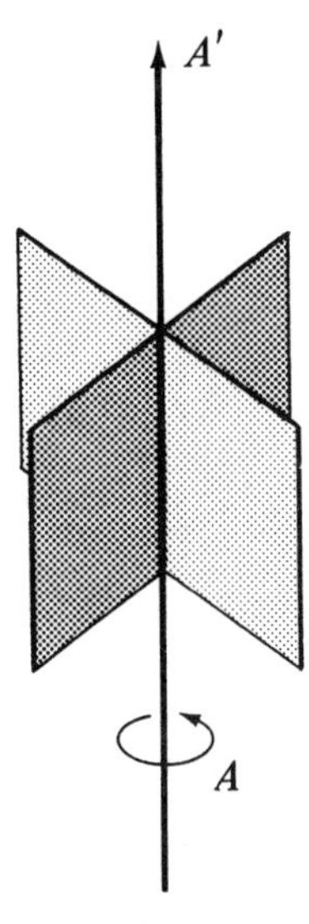

Figure 3.12

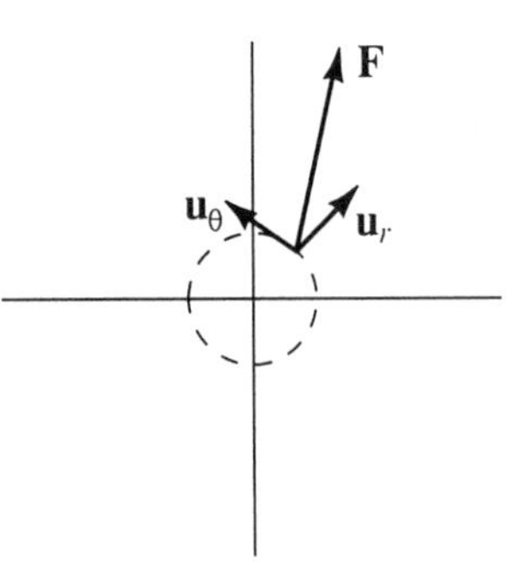

Figure 3.13

The mechanism that rotates the wheel is provided by the tendency of the fluid to *swirl* around the z axis; this motion is due to the counterclockwise components of the velocity near the axis. If we impose a polar coordinate system centered around the paddle wheel axis, as in figure 3.13, then the counterclockwise component of $\mathbf{F}$ at the point (r,θ) is given by $\mathbf{F} \cdot \mathbf{u}_\theta$. This component of the velocity would turn a blade of the paddle wheel at an *angular* rate of $\mathbf{F} \cdot \mathbf{u}_\theta / r$, in radians per second. Of course, this rate will differ from point to point near the axis, so that the different blades of the wheel are "pushed" at different speeds. But it seems plausible to expect that if we take the *average* counterclockwise velocity component over a small circle around the axis, and then divide by the radius of the circle, the quotient would give the angular velocity of the paddle wheel (whose blades we regard as rigidly fixed to each other). [Recall eq. (1.22).]

Let us perform this computation. In figure 3.13, (x,y,z) are the coordinates of the center of the circle; the z axis comes out of the page toward the reader. At the point on the circle with coordinates $(x + \Delta x, y + \Delta y, z)$, the unit vector $\mathbf{u}_\theta$ is given in terms of the angle θ by

$$\mathbf{u}_\theta = -\sin\theta\,\mathbf{i} + \cos\theta\,\mathbf{j}$$

The components of the velocity $\mathbf{F}(x + \Delta x, y + \Delta y, z)$ at this point can be expressed to first order of accuracy by

$$F_1(x + \Delta x, y + \Delta y, z) \cong F_1(x,y,z) + \frac{\partial F_1}{\partial x}\Delta x + \frac{\partial F_1}{\partial y}\Delta y$$

$$F_2(x + \Delta x, y + \Delta y, z) \cong F_2(x,y,z) + \frac{\partial F_2}{\partial x}\Delta x + \frac{\partial F_2}{\partial y}\Delta y$$

F_3 does not concern us, since we are interested in only the counterclockwise component $\mathbf{F} \cdot \mathbf{u}_\theta$. Expressing Δx and Δy in terms of r and θ,

$$\Delta x = r\cos\theta$$

$$\Delta y = r\sin\theta$$

we have, for the counterclockwise component of velocity at (r,θ) on the circle,

$$\mathbf{F} \cdot \mathbf{u}_\theta \cong -\left(F_1 + \frac{\partial F_1}{\partial x} r\cos\theta + \frac{\partial F_1}{\partial y} r\sin\theta\right)\sin\theta + \left(F_2 + \frac{\partial F_2}{\partial x} r\cos\theta + \frac{\partial F_2}{\partial y} r\sin\theta\right)\cos\theta$$

The average counterclockwise component around the circle will be

$$\frac{1}{2\pi}\int_0^{2\pi} \mathbf{F} \cdot \mathbf{u}_\theta\, d\theta$$

Since the integrals, over one period, of $\cos\theta$, $\sin\theta$, and $\sin\theta\cos\theta$ are zero, and since $\int_0^{2\pi} \cos^2\theta\, d\theta = \int_0^{2\pi} \sin^2\theta\, d\theta = \pi$, this average is seen to be

$$\frac{1}{2}r\left(\frac{\partial F_2}{\partial x} - \frac{\partial F_1}{\partial y}\right)$$

and dividing by r we conclude that *the angular velocity of the fluid about the z axis is*

$$\frac{1}{2}\left(\frac{\partial F_2}{\partial x} - \frac{\partial F_1}{\partial y}\right)$$

The computation of the angular velocity about the x axis yields

$$\frac{1}{2}\left(\frac{\partial F_3}{\partial y} - \frac{\partial F_2}{\partial z}\right)$$

and, for the y axis,

$$\frac{1}{2}\left(\frac{\partial F_1}{\partial z} - \frac{\partial F_3}{\partial x}\right)$$

We want the curl of a vector field to express its tendency to swirl; so, dropping the factor $\frac{1}{2}$ for convenience, we formulate the following definition: ***the curl of a vector field*** $\mathbf{F} = F_1\mathbf{i} + F_2\mathbf{j} + F_3\mathbf{k}$ ***is the vector field***

$$\left(\frac{\partial F_3}{\partial y} - \frac{\partial F_2}{\partial z}\right)\mathbf{i} + \left(\frac{\partial F_1}{\partial z} - \frac{\partial F_3}{\partial x}\right)\mathbf{j} + \left(\frac{\partial F_2}{\partial x} - \frac{\partial F_1}{\partial y}\right)\mathbf{k} \tag{3.10}$$

Rather than memorize eq. (3.10), the student is advised to write the curl in the form of a symbolic determinant:

$$\mathbf{curl\,F} = \begin{vmatrix} \mathbf{i} & \mathbf{j} & \mathbf{k} \\ \dfrac{\partial}{\partial x} & \dfrac{\partial}{\partial y} & \dfrac{\partial}{\partial z} \\ F_1 & F_2 & F_3 \end{vmatrix} \tag{3.11}$$

Example 3.18 Find **curl F** if $\mathbf{F} = xyz\mathbf{i} + x^2y^2z^2\mathbf{j} + y^2z^3\mathbf{k}$.

Solution

$$\mathbf{curl\,F} = \begin{vmatrix} \mathbf{i} & \mathbf{j} & \mathbf{k} \\ \dfrac{\partial}{\partial x} & \dfrac{\partial}{\partial y} & \dfrac{\partial}{\partial z} \\ xyz & x^2y^2z^2 & y^2z^3 \end{vmatrix} = (2yz^3 - 2x^2y^2z)\mathbf{i} + (xy)\mathbf{j} + (2xy^2z^2 - xz)\mathbf{k}$$

Example 3.19 Find **curl F** if $\mathbf{F} = x\mathbf{i} + y\mathbf{j} + z\mathbf{k}$.

Solution

$$\mathbf{curl\,F} = \begin{vmatrix} \mathbf{i} & \mathbf{j} & \mathbf{k} \\ \dfrac{\partial}{\partial x} & \dfrac{\partial}{\partial y} & \dfrac{\partial}{\partial z} \\ x & y & z \end{vmatrix} = \mathbf{0}$$

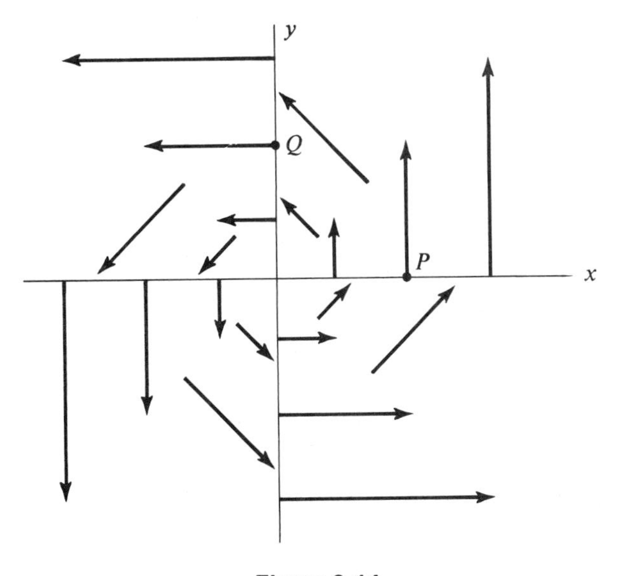

Figure 3.14

Example 3.20 In what direction is **curl F** at points P and Q in figure 3.14? Assume that F_3 is identically zero and that there is no variation of **F** in the z direction. (This field is the same as that shown in figure 3.11.)

Solution It should be clear from our discussion that **curl F** points in the positive z direction. Using the formal definition, observe that at the point P, F_2 is increasing in the x direction, so $\partial F_2/\partial x$ is positive. Although F_1 is zero at P, it is positive below P and negative above, so F_1 is decreasing as we move through P in the y direction; that is, $\partial F_1/\partial y$ is negative. Since we assume that F_3 is identically zero, the derivatives $\partial F_3/\partial y$ and $\partial F_3/\partial x$ are also zero, and since we assume no variation in the z direction, $\partial F_2/\partial z$ and $\partial F_1/\partial z$ are zero. It follows that the only term in eq. (3.10) that does not vanish is the last term, and that the last term is positive.

At point Q, F_2 is zero, but it is negative to the left of Q and positive to the right; hence $\partial F_2/\partial x$ is positive. F_1 is negative at Q and is becoming even more negative with increasing y, and so $\partial F_1/\partial y$ is negative. The term $(\partial F_2/\partial x - \partial F_1/\partial y)$ is therefore positive. The other derivatives in eq. (3.10) equal zero. It follows that **curl F** at point Q is also perpendicular to the xy plane, directed toward the reader. (In fact, a little reflection will convince the reader that **curl F** at point P is equal to **curl F** at point Q.)

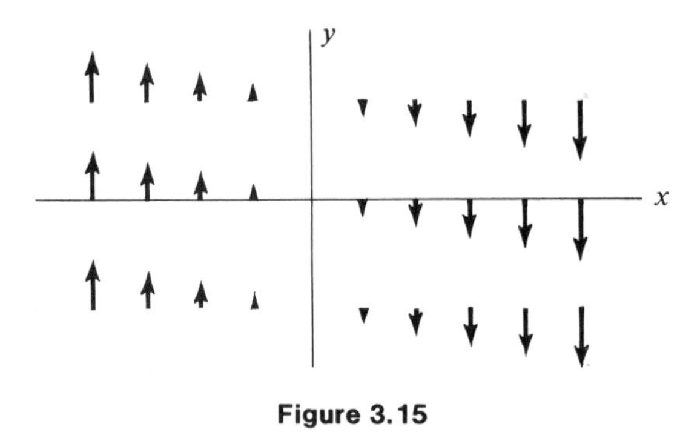

Figure 3.15

Example 3.21 In what direction is **curl F**, if **F** is as shown in figure 3.15?

Solution Since **F** is directed parallel to the y axis and appears to have magnitude proportional to x, we can guess that $\mathbf{F} = Cx\mathbf{j}$, where C is a negative constant. Hence

$$\mathbf{curl\ F} = \begin{vmatrix} \mathbf{i} & \mathbf{j} & \mathbf{k} \\ \dfrac{\partial}{\partial x} & \dfrac{\partial}{\partial y} & \dfrac{\partial}{\partial z} \\ 0 & Cx & 0 \end{vmatrix} = C\mathbf{k}$$

Since C is negative, the curl is directed into the page (negative z direction).

In figure 3.15 the paddle wheel would tend to rotate most rapidly with its axis perpendicular to the page; it will rotate because the velocity of the fluid is greater on one side than on the other. The direction of the curl is into the page, because the paddle wheel will tend to rotate clockwise. Note, however, that the flow lines for this pattern are straight! *Thus it is possible for a vector field to have nonzero curl even when the flow lines are straight lines.*

Example 3.22 Let us imagine that **F** represents the velocity field of a fluid of constant mass density rotating with uniform angular velocity ω about the z axis. (Assume the angular velocity vector $\boldsymbol{\omega}$ to point in the positive z direction.) Verify directly from eq. (3.11) that **curl F** equals twice the angular velocity.

Solution Since $\boldsymbol{\omega} = \omega\mathbf{k}$, we have [eq. (1.22)] $\mathbf{F} = \omega\mathbf{k} \times \mathbf{R}$, where $\mathbf{R} = x\mathbf{i} + y\mathbf{j} + z\mathbf{k}$. Hence $\mathbf{F} = -\omega y\mathbf{i} + \omega x\mathbf{j}$. Using eq. (3.11) we find that $\mathbf{curl\ F} = 2\omega\mathbf{k}$. As we expected, the curl of **F** is twice the angular velocity vector; in this situation, it is the same at every point in space. The reader should convince himself/herself that the field in this example is portrayed in figure 3.14.

We obtained our expression for the curl of **F** by averaging the clockwise component of **F** around the circle depicted in figure 3.13, and dividing by the radius of the circle. We also dropped a bothersome factor of ½ in arriving at eq. (3.10). It will prove convenient to have an alternative description of the curl, visualized by a limiting process similar to that for the divergence.

Thus consider the small rectangle anchored at the point (x, y) in figure 3.16. We will take the "swirl" of **F** around this rectangle to be the sum of the lengths of the four edges, each weighted by the counterclockwise component of **F** along that edge: thus starting from the right we have

$$\begin{aligned}\text{"swirl"} &= F_2\,(\text{edge 1})\,\Delta y + [-F_1\,(\text{edge 2})]\Delta x \\ &\quad + [-F_2\,(\text{edge 3})]\,\Delta y + F_1\,(\text{edge 4})\,\Delta x \\ &= [F_2\,(\text{edge 1}) - F_2\,(\text{edge 3})]\,\Delta y \\ &\quad - [F_1\,(\text{edge 2}) - F_1\,(\text{edge 4})]\,\Delta x\end{aligned}$$

As in the previous section we estimate the differences in the components of **F** by derivatives:

$$\text{"swirl"} = \frac{\partial F_2}{\partial x}\Delta x \Delta y - \frac{\partial F_1}{\partial y}\Delta y \Delta x.$$

Comparing this with eq. (3.10), we see that if we divide the swirl by the area and take the limit as the dimensions approach zero, we obtain the z component of **curl F**. In other words, *the curl can be interpreted as the swirl per unit area.*

This description of curl as "swirl per unit area" is valid for the x and y components of **curl** F too; one simply chooses the plane of the rectangle to be normal to the corresponding component (see exercise 13). Although our paddle wheel velocity interpretation is physically appealing, we shall see that the swirl characterization of the curl is easier to adapt to curvilinear coordinates and transformations of integrals.

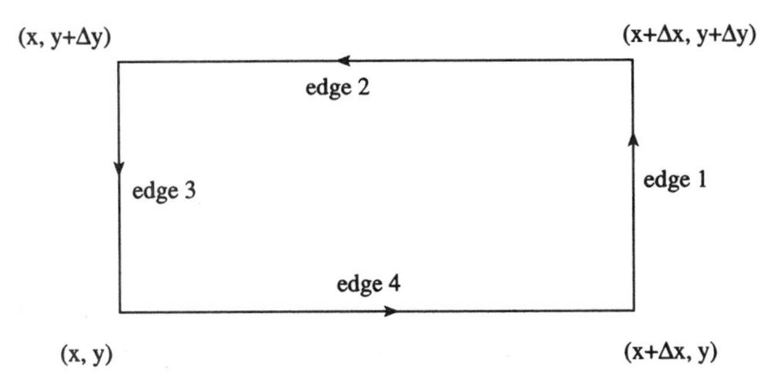

Figure 3.16

EXERCISES

In exercises 1 through 3, find **curl F**.

1. $\mathbf{F} = xy^2\mathbf{i} + xy\mathbf{j} + xy\mathbf{k}$
2. $\mathbf{F} = e^{xy}\,\mathbf{i} + \sin xy\,\mathbf{j} + \cos yz^2\,\mathbf{k}$
3. $\mathbf{F} = z^2x\mathbf{i} + y^2z\mathbf{j} - z^2y\mathbf{k}$
4. Given the vector field $\mathbf{F} = (x + xz^2)\mathbf{i} + xy\mathbf{j} + yz\mathbf{k}$, evaluate
   (a) div **F**.
   (b) **curl F**.
5. Draw a rough picture of the vector field $\mathbf{F} = x\mathbf{i} + y\mathbf{j} + z\mathbf{k}$ and, thinking of the paddle wheel interpretation of **curl F**, explain why **curl F** is identically zero in this case.
6. Give an example of a vector field with curl identically equal to $2\mathbf{i}$.
7. The flow lines of a velocity field **F** are straight lines. Does this imply that $\mathbf{curl\,F} = \mathbf{0}$?
8. Is it possible to tell anything about **curl F**, given only a description of the flow lines of **F**?
9. Can you find a vector field whose curl is $y\mathbf{i}$? $x\mathbf{i}$?
10. Given $\mathbf{F} = y^2\mathbf{i} + z^2\mathbf{j} + x\mathbf{k}$, find
    (a) the curl of **F**.
    (b) the component of **curl F** along the tangent to the curve

$$x = \cos \pi t \qquad y = \sin \pi t \qquad z = t^2 \qquad (t = 1)$$

11. Let $\mathbf{F}(x,y,z)$ be a vector field defined in all space, and consider an intelligent ant living on the xy plane. Suppose all the ant knows about **F** is its values on the xy plane.
    (a) Can this ant compute **curl F**? Explain briefly.
    (b) Can this ant compute $(\mathbf{curl\,F}) \cdot \mathbf{k}$? Explain briefly.
12. Find $\mathbf{curl}[f(R)\mathbf{R}]$, where $\mathbf{R} = x\mathbf{i} + y\mathbf{j} + z\mathbf{k}$, $R = |\mathbf{R}|$, and f is a differentiable function,
    (a) by direct calculation.
    (b) by geometrical interpretation.
13. Show that the x component of **curl F** can be characterized as swirl per unit area for a Δy by Δz rectangle in the y,z plane.

3.5 Del Notation

To understand properly the notion of an "operator," it is necessary to take a broader look at the concept of a "function." For this reason we digress momentarily to consider what is meant by a function.

In elementary calculus, the functions considered are usually "real-valued functions of a real variable." That is, a function f is a rule that associates with every real number x in its domain of definition a single real number $f(x)$. For example, the exponential function is defined for all x, and to each real number x associates a single real number e^x. We define this function by writing $f(x) = e^x$. Other functions are defined by writing, say, $f(x) = x^2$ or $f(x) = \sin x$. Most mathematicians nowadays distinguish rather carefully between the symbol f and the notation $f(x)$. The former denotes the function and the latter denotes the value of the function (which is a number and not a function). Thus, if $f(x) = x^2$, the function f is the *rule* "square the given number," but $f(3)$ is the *number* 9.

In more advanced courses we encounter functions of two or three variables. In this book, "scalar fields" are simply real-valued functions of (usually) three real variables. Thus, the function f defined by $f(x,y,z) = x^2y^2z^2$ says "multiply together the squares of the given numbers." When used alone, in this context, the letter f denotes this *rule,* but if we write, say, $f(2,1,3)$, we mean the *value* of the function at the point (2,1,3), which in this case is the number 36.

Most of the functions we have been considering in this chapter are described by expressions involving x, y, and z. In studying vector analysis it is useful to "visualize" such functions in geometrical or physical terms. Thus the engineering student may think of "an arbitrary function f" as meaning "an arbitrary electric potential," or "an arbitrary temperature distribution," and the mathematics major may think of this as meaning "a rule whereby we tag each point in space with a number." The student who thinks of a function as a jumble of x's, y's, and z's doesn't have much fun, and misses much of the point.

The *vector* fields we discuss are *vector-valued* functions of three real variables. But the idea is still much the same. In this context, a function $\mathbf{F}$ is a *rule* that associates with each point (x,y,z) a single vector $\mathbf{F}(x,y,z)$.

But now we come to what is a big hurdle for some students: passing to the *general* notion of a function. Much of the mystery of modern mathematics vanishes when we realize that a mathematician uses the word "function" in a much more general way to denote any rule that associates an object with each one of a class of objects. Thus we have not only functions that associate numbers with numbers (the functions of elementary calculus), numbers with points in space (scalar fields), and vectors with points in space (vector fields), but also those that associate *functions* with *functions.*

Partly for reasons of convenience, but mainly (we suspect) because so many people have old-fashioned ideas of what the word "function" means, the latter types of functions are usually called "operators." An operator is simply a rule that associates a new function with each member of a particular class of functions.

To take an example from elementary calculus, the process of differentiation defines what is called the derivative operator. This is the operator that associates with every differentiable function f its derivative df/dx. The operator is sometimes denoted d/dx or sometimes, even more simply, D, and it converts each differentiable function f into its derivative. We may formally write this

$$D(f) = \frac{df}{dx}$$

Some textbooks say that D is simply an abbreviation for d/dx, and that the symbol d/dx means nothing by itself, having meaning only when it is applied to some function f. Then we may write df/dx, which of course we all understand. Other differential operators, such as $L = (d^2/dx^2) + 2(d/dx) + 4$, are similarly interpreted as symbols that are meaningless unless followed by a function. In this case we have

$$L(f) = \frac{d^2f}{dx^2} + 2\frac{df}{dx} + 4f$$

This is to miss the whole point of the operator concept, however. It would be much better to visualize this operator as a sort of meat grinder, into which we drop the function f, turn the handle, and out drops the function $(d^2f/dx^2) + 2(df/dx) + 4f$. There is really no insurmountable difficulty in understanding that an operator T is a *rule* that associates with a function f some other function (or possibly even the same function) $T(f)$. It is misleading to say that the symbol d/dx means nothing by itself. It means a great deal: it represents the rule whereby we associate with a differentiable function its derivative. There is no point in recounting here the basic definition of "derivative" or the innumerable techniques involved in actually computing a derivative. The point is that a differentiable function has a derivative and the derivative operator pairs the derivative with the function. (The excellent concept of "pairing" is used in many modern books in discussing the function concept. The derivative of a function is just another function, and the derivative operator is the mathematical twine that binds the two together.)

Another example of an operator is the *gradient*. We recall that the gradient of a scalar field f is a vector field **grad** f. *Divergence* is also an operator. It is an operator that converts a vector field into a scalar field. Similarly, *curl* is an operator, but it is an operator that changes a vector field into another vector field.

The three operators that concern us most are gradient, divergence, and curl. Although they may be written **grad,** div, and **curl,** there is a suggestive and convenient symbolic way of writing them that is commonly used. For this purpose, we introduce the symbol ∇, called "del" (sometimes "nabla"), which is an abbreviation for $\mathbf{i}(\partial/\partial x) + \mathbf{j}(\partial/\partial y) + \mathbf{k}(\partial/\partial z)$. In terms of this symbol, we can write **grad** f as ∇f. Working with ∇ purely formally, pretending for the moment it is a vector, we see that if we form the scalar product of ∇ with a vector field **F,** we obtain

$$\begin{aligned}\nabla \cdot \mathbf{F} &= \left(\mathbf{i}\frac{\partial}{\partial x} + \mathbf{j}\frac{\partial}{\partial y} + \mathbf{k}\frac{\partial}{\partial z}\right) \cdot (\mathbf{i}F_1 + \mathbf{j}F_2 + \mathbf{k}F_3) \\ &= \frac{\partial F_1}{\partial x} + \frac{\partial F_2}{\partial y} + \frac{\partial F_3}{\partial z}\end{aligned}$$

which is the divergence of **F**. Similarly, if we imagine ∇ to be a vector and form the vector cross product of ∇ with **F**, we obtain the curl of **F**:

$$\nabla \times \mathbf{F} = \left(\mathbf{i}\frac{\partial}{\partial x} + \mathbf{j}\frac{\partial}{\partial y} + \mathbf{k}\frac{\partial}{\partial z}\right) \times (\mathbf{i}F_1 + \mathbf{j}F_2 + \mathbf{k}F_3)$$

$$= \begin{vmatrix} \mathbf{i} & \mathbf{j} & \mathbf{k} \\ \dfrac{\partial}{\partial x} & \dfrac{\partial}{\partial y} & \dfrac{\partial}{\partial z} \\ F_1 & F_2 & F_3 \end{vmatrix} = \mathbf{curl}\ \mathbf{F}$$

To recapitulate, ∇ is an abbreviation:

$$\nabla = \mathbf{i}\frac{\partial}{\partial x} + \mathbf{j}\frac{\partial}{\partial y} + \mathbf{k}\frac{\partial}{\partial z} \tag{3.12}$$

The symbols ∇f, $\nabla \cdot \mathbf{F}$, and $\nabla \times \mathbf{F}$ are defined by

$$\nabla f = \mathbf{grad}\, f \tag{3.13}$$

$$\nabla \cdot \mathbf{F} = \text{div}\ \mathbf{F} \tag{3.14}$$

$$\nabla \times \mathbf{F} = \mathbf{curl}\ \mathbf{F} \tag{3.15}$$

After eq. (3.12) is memorized, formulas (3.13) through (3.15) provide very convenient ways of remembering the expressions for gradient, divergence, and curl. We just operate with ∇ as though it were a vector. Henceforth, we will use these abbreviations frequently.

EXERCISES

1. If $f(x,y,z) = x^2y + z$, what is $f(2,3,4)$?

2. If $f(x,y,z) = x^2y + z$, what is the value of ∇f at $(2,3,4)$?

3. If $g(t) = t^3$ and $f(x,y,z) = x^2 + y^2z$, what is $g[f(1,1,3)]$?

4. Given $\mathbf{F}(x,y,z) = x^2y\mathbf{i} + z\mathbf{j} - (x + y - z)\mathbf{k}$, find
(a) $\nabla \cdot \mathbf{F}$
(b) $\nabla \times \mathbf{F}$
(c) $\nabla(\nabla \cdot \mathbf{F})$

5. If **F** is a vector field, is $\nabla \cdot (\nabla \times \mathbf{F})$ a scalar field or a vector field?

6. If **F** is a vector field, is $\nabla \times (\nabla \times \mathbf{F})$ a scalar field or a vector field?

7. Find $\nabla \cdot \mathbf{R}$ and $\nabla \times \mathbf{R}$ where $\mathbf{R} = x\mathbf{i} + y\mathbf{j} + z\mathbf{k}$.

8. If $f(x,y,z) = xyz + e^{xz}$, find $\nabla \cdot (\nabla f)$.

9. (a) Compute $\nabla \times (\nabla f)$ for the scalar field f defined in exercise 8.
(b) Now do the same thing for another scalar field f (use any of the scalar fields defined in preceding problems, or make one up yourself).
(c) What can you conjecture from this?

10. (a) Compute $\nabla \cdot (\nabla \times \mathbf{F})$ for the vector field $\mathbf{F}$ defined in exercise 4.
(b) Do the same for a vector field $\mathbf{F}$ that you have made up yourself.
(c) What can you conjecture from this?

3.6 The Laplacian

In electrostatics, the gradient of the electric potential is a scalar multiple of the electric field intensity, and the divergence of the electric field intensity is related to the charge density. For this and other reasons it is convenient to introduce a single operator that is the composite of the two operators **grad** and div. This operator is called the *laplacian.*

The laplacian of a scalar field f is defined to be div (**grad** f). Note that **grad** f is a vector field and the divergence of **grad** f is a scalar field; hence the laplacian of a scalar field f is a scalar field. In del notation it is $\nabla \cdot (\nabla f)$ and for simplicity is frequently written $\nabla^2 f$, or Δf.

We have

$$\text{laplacian}\,(f) = \nabla^2 f = \nabla \cdot (\nabla f) = \frac{\partial^2 f}{\partial x^2} + \frac{\partial^2 f}{\partial y^2} + \frac{\partial^2 f}{\partial z^2} = \Delta f \tag{3.16}$$

since

$$\nabla \cdot (\nabla f) = \nabla \cdot \left(\frac{\partial f}{\partial x}\mathbf{i} + \frac{\partial f}{\partial y}\mathbf{j} + \frac{\partial f}{\partial z}\mathbf{k}\right) = \frac{\partial^2 f}{\partial x^2} + \frac{\partial^2 f}{\partial y^2} + \frac{\partial^2 f}{\partial z^2}$$

The symbol ∇^2 or Δ may be considered to be simply an abbreviation for

$$\frac{\partial^2}{\partial x^2} + \frac{\partial^2}{\partial y^2} + \frac{\partial^2}{\partial z^2}$$

The laplacian operator is by far the most important operator in mathematical physics. It is a natural tool for use in the analysis of *diffusive processes.* Recall that in section 3.3 (fig. 3.6) we called $\mathbf{F} \cdot \mathbf{n}\,\Delta S$ the flux of the vector field $\mathbf{F}(x,y,z)$ through the surface element ΔS, and when $\mathbf{F}$ was interpreted as fluid velocity times mass density, then $\mathbf{F} \cdot \mathbf{n}\,dS$ measured the amount of fluid crossing dS per unit time. In this context we speak of $\mathbf{F}$ as a *flux density vector.* Flux densities are also used to describe the kinetics of any process wherein one species of matter diffuses through another.

For example, if some chemical is dissolved in a liquid, the concentration of the solute may vary from point to point, and the flux density vector field would describe how this chemical species is being transported through the solvent. Such transport processes also take place in solids, as when a dopant impurity is diffused into a semiconductor chip. In fact, even though *heat* is not a physical substance, its flow in a body is a diffusive process described by a flux vector field, the "heat flux."

In all these cases the transport processes are physically driven by *nonuniformities* in the species concentration; thus solutes flow from regions of high concentrations to regions of low concentrations, and heat flows from high temperature points to lower ones. Experimentally one finds that a great many of these processes obey *Fick's law,* which states that the flux vectors are proportional to the gradients of the species concentrations:

$$\mathbf{F}(x,y,z) = -K\nabla f(x,y,z)$$

where $f(x,y,z)$ is the concentration of the species or the temperature, K is a physical parameter known as the diffusivity or thermal conductivity, and the minus sign emphasizes that the flow is *away* from regions of higher concentration. (Fick's law can also be justified by statistical mechanics.)

In section 3.3 we saw that the divergence of **F** measures the net outflux per unit volume. Thus, if the K in Fick's law is constant, this outflux is proportional to the laplacian of the species concentration:

$$\nabla \cdot \mathbf{F} = -K\nabla^2 f$$

When the laplacian is negative, there is a positive outflux of the species at the point, and the concentration (or temperature) must go down; positive laplacians indicate influxes and increasing concentrations (assuming there are no sources or sinks maintaining the imbalances). *When an "unpumped" diffusive process is at equilibrium the laplacian must be zero everywhere* and the concentration satisfies *Laplace's equation,*

$$\nabla^2 f = 0 \tag{3.17}$$

while if sources are present the concentration satisfies *Poisson's equation,*

$$\nabla^2 f = g(x,y,z)$$

where g is a measure of the source at (x,y,z). Solutions of Laplace's equation are known as *harmonic functions.*

(Appendix D describes some occurrences of the laplacian in electromagnetics.)

We can conjecture some of the mathematical properties of the laplacian from its form:

$$\nabla^2 f = \frac{\partial^2 f}{\partial x^2} + \frac{\partial^2 f}{\partial y^2} + \frac{\partial^2 f}{\partial z^2}$$

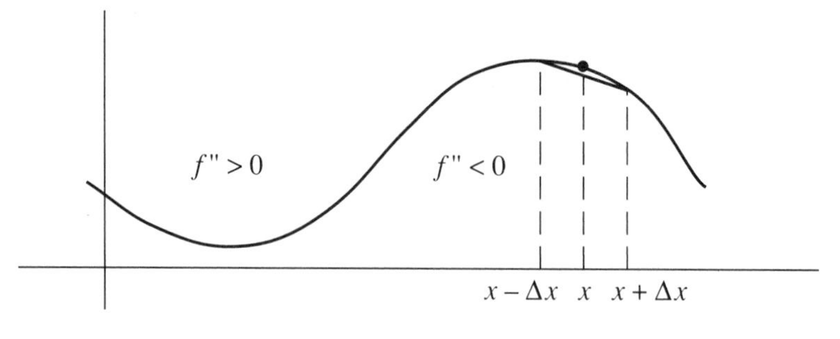

Figure 3.17

Certainly $\nabla^2 f$ is a generalization of the second derivative of a function of one variable. Recall that the second derivative determines the *convexity* of a function; if f'' is positive, the graph of f "holds water," and if it is negative it spills water (fig. 3.17).

In other words *in a region where $f'' < 0$ the value of f at x lies above the secant line connecting the values of f at $x + \Delta x$ and $x - \Delta x$,*

$$f(x) > \frac{f(x + \Delta x) + f(x - \Delta x)}{2} \tag{3.18}$$

and when $f'' > 0$ the opposite occurs. The extent to which f exceeds the average of its neighbors, as quantified by eq. (3.18), appears to be proportional to the negative of the second derivative in the figure. [In fact the expression

$$-\frac{2}{\Delta x^2}\left\{f(x) - \frac{f(x + \Delta x) + f(x - \Delta x)}{2}\right\}$$

is a well-known approximation for $f''(x)$.]

Thus we might reasonably expect that some variant of this property would hold for the "three-dimensional second derivative," $\nabla^2 f$; it should measure (in a negative sense) the extent to which $f(x,y,z)$ exceeds the averaged neighboring values $\bar{f}$:

$$f(x,y,z) - \bar{f} \approx -M\nabla^2 f \tag{3.19}$$

In the next section we shall show that an approximation like eq. (3.19) holds when f is averaged over a small cube centered at (x,y,z), with the constant M equal to $\frac{1}{24}$ the square of the length of one side. (The reader will be invited to demonstrate that if f is averaged over a sphere, eq. (3.19) holds with M equal to $\frac{1}{10}$ the square of the radius.)

In section 5.2 we will derive a very specific equality expressing this property of the laplacian in exact terms. For the present let us proceed with the study of the algebraic properties of the operator.

The formal differential operator

$$\Delta = \nabla^2 = \frac{\partial^2}{\partial x^2} + \frac{\partial^2}{\partial y^2} + \frac{\partial^2}{\partial z^2}$$

may also be applied to *vector* fields to obtain new vector fields, since if **F** is a *vector* field, $(\partial^2\mathbf{F}/\partial x^2) + (\partial^2\mathbf{F}/\partial y^2) + (\partial^2\mathbf{F}/\partial z^2)$ makes perfectly good sense. For example, if

$$\mathbf{F} = x^2y\mathbf{i} + y^2z^3\mathbf{j} + xyz^4\mathbf{k}$$

then we have

$$\frac{\partial^2\mathbf{F}}{\partial x^2} = 2y\mathbf{i}$$

$$\frac{\partial^2\mathbf{F}}{\partial y^2} = 2z^3\mathbf{j}$$

$$\frac{\partial^2\mathbf{F}}{\partial z^2} = 6y^2z\mathbf{j} + 12xyz^2\mathbf{k}$$

whence $\nabla^2\mathbf{F} = 2y\mathbf{i} + (2z^3 + 6y^2z)\mathbf{j} + 12xyz^2\mathbf{k}$. When used in this sense, to operate on vector fields to produce vector fields, ∇^2 is sometimes called the *vector* laplacian operator.

A summary of the vector operators appears in table 3.1.

Table 3.1 Vector Operators

$$\nabla = \mathbf{i}\frac{\partial}{\partial x} + \mathbf{j}\frac{\partial}{\partial y} + \mathbf{k}\frac{\partial}{\partial z}$$

Name; Symbol	Interpretation
grad $\phi = \nabla\phi$	Maximum rate of change of ϕ, in the maximal direction
div $\mathbf{F} = \nabla \cdot \mathbf{F}$	Net outflux of **F** per unit volume
curl $\mathbf{F} = \nabla \times \mathbf{F}$	Swirl of **F** per unit area
Laplacian $\phi = \nabla^2\phi = \Delta\phi$	A measure of difference between $\phi(\mathbf{R})$ and the average of ϕ around **R**

EXERCISES

1. Find $\nabla^2 f$, given that $f(x,y,z) = x^5yz^3$.
2. Find $\nabla^2 f$, given that $f(x,y,z) = 1/(x^2 + y^2 + z^2)^{1/2}$.
3. Find $\nabla^2\mathbf{F}$, given that $\mathbf{F}(x,y,z) = 3\mathbf{i} + \mathbf{j} - x^2y^3z^4\mathbf{k}$.
4. Which of the following functions satisfies Laplace's equation?
   (a) $f(x,y,z) = e^z \sin y$
   (b) $f(x,y,z) = \sin x \sinh y + \cos x \cosh z$
   (c) $f(x,y,z) = \sin px \sinh qy$ (p and q are constants)
5. Tell whether each of the following is a vector field or a scalar field, given that f is a scalar field and $\mathbf{F}$ is a vector field. Two of the expressions are meaningless; determine which two.
   (a) ∇f
   (b) $\nabla \cdot \mathbf{F}$
   (c) $\nabla \times \mathbf{F}$
   (d) $\nabla \cdot (\nabla f)$
   (e) $\nabla \times (\nabla f)$
   (f) $\nabla \times f$
   (g) $\nabla^2\mathbf{F}$
   (h) $\nabla \times (\nabla^2\mathbf{F})$
   (i) $\nabla \times (\nabla^2 f)$
   (j) $\nabla(\nabla^2 f)$
6. (a) Show that, if f and g satisfy Laplace's equation, $f + g$ does also.
   (b) Find a function satisfying Laplace's equation and also the following identities: $f(0,y,z) = 0$, $f(x,0,z) = 0$, $f(\pi,y,z) = 0$, $f(x,5,z) = \sin x + \sin 2x$. [*Hint:* Use exercises 6(a) and 4(c) to guess an answer.]
7. Given $f(x,y,z) = 2x^2 + y$ and $\mathbf{R} = x\mathbf{i} + y\mathbf{j} + z\mathbf{k}$, find
   (a) ∇f
   (b) $\nabla \cdot \mathbf{R}$
   (c) $\nabla^2 f$
   (d) $\nabla \times (f\mathbf{R})$
8. If $\mathbf{F} = x^2\mathbf{i} + xy\mathbf{j} + z\mathbf{k}$, evaluate each of the following at the point $(-1,2,3)$:
   (a) $\nabla^2\mathbf{F}$
   (b) $\nabla \times \mathbf{F}$
   (c) $\nabla \cdot \mathbf{F}$

3.7 Optional Reading: Dyadics: Taylor Polynomials

Notice that when ∇^2 is used as an operator on vector fields, as in $\nabla^2\mathbf{F}$, its interpretation as div **grad** is rather strained. After all, there is no meaning to **grad** $\mathbf{F}$ in our scheme of things.

However, in some areas of physics and engineering it proves convenient to use such strange symbols. To see how this might come about, suppose we want to express the vector component of a vector $\mathbf{F}$ in the direction of a unit vector $\mathbf{n}$. The answer is given by $\mathbf{n}(\mathbf{n} \cdot \mathbf{F})$ (recall section 1.9). This formula tempts one to define an operator, the *projection operator in the direction* $\mathbf{n}$, and to denote it as $\mathbf{nn}$; then the projection of $\mathbf{F}$ in the direction $\mathbf{n}$ is

$$(\mathbf{nn}) \cdot \mathbf{F} \equiv \mathbf{nn} \cdot \mathbf{F}$$

Generalizing, given any two vectors $\mathbf{A}$ and $\mathbf{B}$, one formally defines the *dyadic* $\mathbf{AB}$ as an operator acting as follows: for any vector $\mathbf{F}$,

$$(\mathbf{AB}) \cdot \mathbf{F} \equiv \mathbf{AB} \cdot \mathbf{F}$$

and

$$\mathbf{F} \cdot (\mathbf{AB}) \equiv \mathbf{F} \cdot \mathbf{AB}$$

Thus the dyadic $\mathbf{ii}$ projects a vector onto the x axis. As another example, observe that the dyadic $\mathbf{ii} + \mathbf{jj} + \mathbf{kk}$ is an identity operator because

$$(\mathbf{ii} + \mathbf{jj} + \mathbf{kk}) \cdot \mathbf{F} = \mathbf{F} = \mathbf{F} \cdot (\mathbf{ii} + \mathbf{jj} + \mathbf{kk})$$

[Recall eq. (1.15).]

In this context, we can consider **grad F**, or $\nabla \mathbf{F}$, as a dyadic:

$$\begin{aligned}
\nabla \mathbf{F} &= \left(\mathbf{i}\frac{\partial}{\partial x} + \mathbf{j}\frac{\partial}{\partial y} + \mathbf{k}\frac{\partial}{\partial z}\right)(F_1\mathbf{i} + F_2\mathbf{j} + F_3\mathbf{k}) \\
&= \frac{\partial F_1}{\partial x}\mathbf{ii} + \frac{\partial F_1}{\partial y}\mathbf{ji} + \frac{\partial F_1}{\partial z}\mathbf{ki} \\
&\quad + \frac{\partial F_2}{\partial x}\mathbf{ij} + \frac{\partial F_2}{\partial y}\mathbf{jj} + \frac{\partial F_2}{\partial z}\mathbf{kj} \\
&\quad + \frac{\partial F_3}{\partial x}\mathbf{ik} + \frac{\partial F_3}{\partial y}\mathbf{jk} + \frac{\partial F_3}{\partial z}\mathbf{kk}
\end{aligned}$$

(Notice that we do not equate $\mathbf{ij}$ with $\mathbf{ji}$.) Then the dyadic interpretation of div **grad F** becomes

$$\begin{aligned}
\nabla \cdot (\nabla \mathbf{F}) &= \left(\mathbf{i}\frac{\partial}{\partial x} + \mathbf{j}\frac{\partial}{\partial y} + \mathbf{k}\frac{\partial}{\partial z}\right) \cdot \left(\frac{\partial F_1}{\partial x}\mathbf{ii} + \cdots + \frac{\partial F_3}{\partial z}\mathbf{kk}\right) \\
&= \frac{\partial^2 F_1}{\partial x^2}\mathbf{i} + \frac{\partial^2 F_2}{\partial x^2}\mathbf{j} + \frac{\partial^2 F_3}{\partial x^2}\mathbf{k} \\
&\quad + \frac{\partial^2 F_1}{\partial y^2}\mathbf{i} + \frac{\partial^2 F_2}{\partial y^2}\mathbf{j} + \frac{\partial^2 F_3}{\partial y^2}\mathbf{k} \\
&\quad + \frac{\partial^2 F_1}{\partial z^2}\mathbf{i} + \frac{\partial^2 F_2}{\partial z^2}\mathbf{j} + \frac{\partial^2 F_3}{\partial z^2}\mathbf{k}
\end{aligned}$$

which is the same as $\nabla^2 \mathbf{F}$.

In calculus it is shown that if a function $f(x)$ has at least n derivatives at the point x_0, then the *Taylor polynomial of order n,*

$$p_n(x) = f(x_0) + f'(x_0)(x - x_0)/1! + f''(x_0)(x - x_0)^2/2! + \cdots + f^{(n)}(x_0)(x - x_0)^n/n! \tag{3.20}$$

provides a "best fit" to f near x_0, in that the first n derivatives (and the zeroth) match at x_0. Various forms of the "error" term, $f(x) - p(x)$, are known (Lagrange, Cauchy, integral) and can be used to estimate the accuracy of the approximation. Roughly speaking, this error goes to zero faster than $(x - x_0)^n$ as x approaches x_0, if $f^{(n+1)}(x)$ is continuous.

The Taylor *series* for f at x_0 is the power series generated by letting n be infinite in eq. (3.20); it exists only if f has derivatives of all orders at x_0, and the question of whether it converges to $f(x)$ is the essence of analytic function theory. Some students get the impression that the convergence of the Taylor series is the whole story. This is quite wrong; even if the Taylor *series* fails to converge (or doesn't exist!), the Taylor *polynomials* provide good approximations to f in the neighborhood of x_0.

In this section we are going to see how to express the Taylor polynomials for a scalar field, that is, a function of three variables, $f(x,y,z)$. We shall take $\mathbf{R}_0$ to be the "base point" (corresponding to x_0 above) and we seek the formula for the Taylor polynomials evaluated at the point $\mathbf{R}$.

It is convenient to express the difference $\mathbf{R} - \mathbf{R}_0$ as $s\mathbf{u}$, where $\mathbf{u}$ is a unit vector. Thus

$$\mathbf{R} = \mathbf{R}_0 + s\mathbf{u} \tag{3.21}$$

where s is the distance between $\mathbf{R}$ and $\mathbf{R}_0$.

Now let $g(s)$ be the single-variable function defined by

$$g(s) = f(\mathbf{R}_0 + s\mathbf{u}) \tag{3.22}$$

where we regard $\mathbf{u}$, as well as $\mathbf{R}_0$, as fixed. Expanding f around $\mathbf{R}_0$ corresponds to expanding g around zero. The Taylor series for g looks like

$$g(0) + g'(0)s/1! + g''(0)s^2/2! + \cdots \tag{3.23}$$

What is $g'(0)$? It is the rate of change of $g(s) = f(\mathbf{R}_0 + s\mathbf{u}) = f(\mathbf{R})$ with respect to s, which measures distance in the direction $\mathbf{u}$, at the point $s = 0$. Well, that is precisely the directional derivative of f at $\mathbf{R}_0$! In other words,

$$g'(0) = \mathbf{u} \cdot \nabla f(\mathbf{R}_0)$$

By the same token, $g''(0)$, being the derivative of the first derivative, is the directional derivative of the above quantity:

$$g''(0) = \mathbf{u} \cdot \nabla[\mathbf{u} \cdot \nabla f(\mathbf{R}_0)]$$

Remembering that the dot product is commutative, and that ∇ commutes with constants, we rewrite this in dyadic form:

$$g''(0) = \mathbf{u} \cdot \nabla\nabla f(\mathbf{R}_0) \cdot \mathbf{u}$$

where the dyadic in the middle, which is known as the *Hessian* of f at $\mathbf{R}_0$, takes the form

$$\begin{aligned} \nabla\nabla f &= \left(\mathbf{i}\frac{\partial}{\partial x} + \mathbf{j}\frac{\partial}{\partial y} + \mathbf{k}\frac{\partial}{\partial z}\right)\left(\mathbf{i}\frac{\partial f}{\partial x} + \mathbf{j}\frac{\partial f}{\partial y} + \mathbf{k}\frac{\partial f}{\partial z}\right) \\ &= \mathbf{ii}\frac{\partial^2 f}{\partial x^2} + \mathbf{ij}\frac{\partial^2 f}{\partial x \partial y} + \mathbf{ik}\frac{\partial^2 f}{\partial x \partial z} + \mathbf{ji}\frac{\partial^2 f}{\partial y \partial x} + \mathbf{jj}\frac{\partial^2 f}{\partial y^2} + \mathbf{jk}\frac{\partial^2 f}{\partial y \partial z} \\ &+ \mathbf{ki}\frac{\partial^2 f}{\partial z \partial x} + \mathbf{kj}\frac{\partial^2 f}{\partial z \partial y} + \mathbf{kk}\frac{\partial^2 f}{\partial z^2} \end{aligned} \tag{3.24}$$

Expressing the variables in terms of $\mathbf{R}$ and $\mathbf{R}_0$, we obtain the second-order Taylor polynomial for f around $\mathbf{R}_0$:

$$\begin{aligned} P_2(\mathbf{R}) &= f(\mathbf{R}_0) + (\mathbf{R} - \mathbf{R}_0) \cdot \nabla f(\mathbf{R}_0) \\ &+ \tfrac{1}{2}(\mathbf{R} - \mathbf{R}_0) \cdot \nabla\nabla f(\mathbf{R}_0) \cdot (\mathbf{R} - \mathbf{R}_0) \end{aligned} \tag{3.25}$$

In matrix notation this takes the form

$$\begin{aligned} P_2(x,y,z) &= f(x_0,y_0,z_0) + \left|(x - x_0)\ (y - y_0)\ (z - z_0)\right| \begin{vmatrix} f_x \\ f_y \\ f_z \end{vmatrix} \\ &+ \tfrac{1}{2}\left|(x - x_0)\ (y - y_0)\ (z - z_0)\right| \begin{vmatrix} f_{xx} & f_{xy} & f_{xz} \\ f_{yx} & f_{yy} & f_{yz} \\ f_{zx} & f_{zy} & f_{zz} \end{vmatrix} \begin{vmatrix} x - x_0 \\ y - y_0 \\ z - z_0 \end{vmatrix} \end{aligned} \tag{3.26}$$

where the subscripts denote differentiation and the Hessian matrix, which is symmetric, is evaluated at $\mathbf{R}_0$.

The reader will be invited in exercise 1 to verify directly that all partial derivatives of P_2 and f, up to order 2, agree at $\mathbf{R}_0$.

Example 3.23 Compute the second-order Taylor polynomial for $\sin x \sin y + \sin z$ around (0,0,0).

Solution We have $f(0,0,0) = 0$. The gradient matrix is

$$\nabla f = \begin{vmatrix} 0 \\ 0 \\ 1 \end{vmatrix}$$

The Hessian is

$$\begin{vmatrix} 0 & 1 & 0 \\ 1 & 0 & 0 \\ 0 & 0 & 0 \end{vmatrix}$$

Thus $P_2(x,y,z) = z + xy$.

Example 3.24 Compute the second-order Taylor polynomials for the function $-x^4 - 2y^2 - 4z^4 + 2z^2$ at the points $(0,0,0)$ and $(0,0,\frac{1}{2})$.

Solution It is easy to see that the gradient is zero at the two points in question, and that $f(0,0,0) = 0$ while $f(0,0,\frac{1}{2}) = \frac{1}{4}$. The Hessians at the two points are

$$\begin{vmatrix} 0 & 0 & 0 \\ 0 & -4 & 0 \\ 0 & 0 & 4 \end{vmatrix} \quad \text{and} \quad \begin{vmatrix} 0 & 0 & 0 \\ 0 & -4 & 0 \\ 0 & 0 & -8 \end{vmatrix}$$

respectively. Thus we derive the Taylor polynomials $-2y^2 + 2z^2$ and $\frac{1}{4} - 2y^2 - 4(z - \frac{1}{2})^2$.

We can learn a lot about the *extremal* points of the function from the second-order Taylor polynomials. The polynomial $-2y^2 + 2z^2$, for example, will increase in the z direction. Thus $(0,0,0)$ could not have been a *maximum* point in the above example; $(0,0,\frac{1}{2})$ is a better bet, because the polynomial $\frac{1}{4} - 2y^2 - 4(z - \frac{1}{2})^2$ does not increase from this point.

In general, when the second-order Taylor polynomial is expanded about a maximum point, the linear terms are missing because the gradient vanishes. Thus the quadratic terms describe the behavior of f for small deviations from $\mathbf{R}_0$ (unless they, too, are zero). From the display in eq. (3.25), then, we would agree that *at a maximum point* $\mathbf{R}_0$ *of* f *the Hessian must have the property that* $\mathbf{u} \cdot \nabla\nabla f(\mathbf{R}_0) \cdot \mathbf{u}$ *is negative or zero for every direction* $\mathbf{u}$. This statement, called the "negative semidefinite" condition, generalizes the second-derivative criterion for a maximum of a function of a single variable.

Example 3.25 Using the second-order Taylor polynomial, estimate the difference between the value of a function $f(x,y,z)$ at (0,0,0) and the average of its values throughout the interior of a small cube of side a centered at (0,0,0).

Solution The Taylor polynomial around the origin takes the form

$$f(0,0,0) + xf_x + yf_y + zf_z + \frac{x^2}{2}f_{xx} + \frac{y^2}{2}f_{yy} + \frac{z^2}{2}f_{zz} + xyf_{xy} + yzf_{yz} + xzf_{xz}$$

where subscripts denote the derivatives evaluated at (0,0,0). Let's average these terms over the cube. The first term is constant, and thus equals its average. By symmetry, the average of x, y, and z is zero, as is the average of the cross-terms xy, yz, and xz.

The average of x^2 is given by

$$\frac{1}{a^3}\int_{-a/2}^{a/2}\int_{-a/2}^{a/2}\int_{-a/2}^{a/2} x^2\,dx\,dy\,dz = \frac{a^2}{12},$$

and similarly for y^2 and z^2. Thus the difference between $f(0,0,0)$ and its average value throughout the small cube is given approximately by

$$f(0,0,0) - f_{av} \approx -\frac{a^2}{24}(f_{xx} + f_{yy} + f_{zz}) = -\frac{a^2}{24}\nabla^2 f(0,0,0)$$

We alluded to this property of the laplacian in section 3.6.

EXERCISES

1. Verify that the zeroth, first, and second partial derivatives of f and its second-degree Taylor polynomial agree at the base point $\mathbf{R}_0$.
2. Work out the second-order Taylor polynomials for the function $e^{xy} + z^2$ expanded around (0,0,0) and (1,1,1).
3. Work out the third-order Taylor polynomial for a function of two variables, $f(x,y)$.
4. The Taylor (Maclauren) series for e^x is

$$1 + x + \frac{x^2}{2!} + \frac{x^3}{3!} + \cdots$$

   Use this to interpret the expression $e^{s\mathbf{u}\cdot\nabla}f$; what do you get?
5. Show that the difference between $f(0,0,0)$ and its average over a small sphere of radius R is given approximately by $-(R^2/10)\nabla^2 f(0,0,0)$. (*Hint:* The average value of x^2 over the sphere is $R^2/5$.)
6. Another way of deriving the Taylor polynomials for $f(x,y,z)$ is as follows. Regard y and z as fixed and expand f in powers of $(x - x_0)$. The coefficients in this expansion will be functions of y and z. Regard z as fixed and expand each coefficient in powers of $(y - y_0)$. Now the coefficients are functions of z. Expand them in powers of $(z - z_0)$.

   Carry out this program to second order and verify that the results agree with the display in eq. (3.26).

7. Show that $\nabla^2 f \leq 0$ at a local maximum of f, and $\nabla^2 f \geq 0$ at a minimum. (*Hint:* consider f as a function of x, y, and z separately.) Note that this statement agrees with the interpretation of $\nabla^2 f$ as a measure of the difference of the averaged value of f and its local value.

3.8 Vector Identities

Although we continue to use the del notation, formally manipulating

$$\nabla = \mathbf{i}\frac{\partial}{\partial x} + \mathbf{j}\frac{\partial}{\partial y} + \mathbf{k}\frac{\partial}{\partial z}$$

as though it were a vector, this practice has certain hazards. Keep in mind that the derivative operators appearing in the del operator act only on functions appearing to the right of the del operator.

For example, supposing that

$$\mathbf{F} = x^3y\mathbf{i} + y^2\mathbf{j} + x^2z\mathbf{k} \qquad \text{and} \qquad \mathbf{R} = x\mathbf{i} + y\mathbf{j} + z\mathbf{k},$$

let us compare the two expressions $(\nabla \cdot \mathbf{R})\mathbf{F}$ and $(\mathbf{R} \cdot \nabla)\mathbf{F}$. For the first of these we have

$$(\nabla \cdot \mathbf{R})\mathbf{F} = 3\mathbf{F} = 3x^3y\mathbf{i} + 3y^2\mathbf{j} + 3x^2z\mathbf{k}$$

On the other hand, in the second expression, $\mathbf{R}$ is to the left of ∇, and therefore the derivatives in the del operator do not act on $\mathbf{R}$. We have

$$\begin{aligned}(\mathbf{R} \cdot \nabla)\mathbf{F} &= \left(x\frac{\partial}{\partial x} + y\frac{\partial}{\partial y} + z\frac{\partial}{\partial z}\right)(x^3y\mathbf{i} + y^2\mathbf{j} + x^2z\mathbf{k}) \\ &= x(3x^2y\mathbf{i} + 2xz\mathbf{k}) + y(x^3\mathbf{i} + 2y\mathbf{j}) + z(x^2\mathbf{k}) \\ &= 4x^3y\mathbf{i} + 2y^2\mathbf{j} + 3x^2z\mathbf{k} \\ &\neq (\nabla \cdot \mathbf{R})\mathbf{F}\end{aligned}$$

Also, it is common practice to omit parentheses in a vector expression when there is only one interpretation of the expression that makes sense, in the context of ordinary vector analysis (i.e., excluding dyadics). For example, $\nabla \cdot \mathbf{RF}$ and $\mathbf{R} \cdot \nabla\mathbf{F}$ must mean $(\nabla \cdot \mathbf{R})\mathbf{F}$ and $(\mathbf{R} \cdot \nabla)\mathbf{F}$, respectively, since $\nabla \cdot (\mathbf{RF})$ and $\mathbf{R} \cdot (\nabla\mathbf{F})$ do not make sense in this context.

Similarly, $\nabla \cdot f\mathbf{F}$ means $\nabla \cdot (f\mathbf{F})$, simply the divergence of $f\mathbf{F}$, since $\nabla \cdot f$, and hence $(\nabla \cdot f)\mathbf{F}$, is meaningless.

In some cases where parentheses are omitted, two interpretations are possible, both of which make sense. For example, if $\mathbf{A} = A_1\mathbf{i} + A_2\mathbf{j} + A_3\mathbf{k}$ is a vector field and f is a scalar field, both $(\mathbf{A} \cdot \nabla)f$ and $\mathbf{A} \cdot (\nabla f)$ are meaningful and are sometimes written $\mathbf{A}\cdot\nabla f$. This is legitimate because both interpretations lead to exactly the same result. We have

$$(\mathbf{A}\cdot\nabla)f = \left(A_1\frac{\partial}{\partial x} + A_2\frac{\partial}{\partial y} + A_3\frac{\partial}{\partial z}\right)f = A_1\frac{\partial f}{\partial x} + A_2\frac{\partial f}{\partial y} + A_3\frac{\partial f}{\partial z}$$

and also

$$\mathbf{A}\cdot(\nabla f) = \mathbf{A}\cdot\left(\frac{\partial f}{\partial x}\mathbf{i} + \frac{\partial f}{\partial y}\mathbf{j} + \frac{\partial f}{\partial z}\mathbf{k}\right) = A_1\frac{\partial f}{\partial x} + A_2\frac{\partial f}{\partial y} + A_3\frac{\partial f}{\partial z}$$

Because of the convention adopted above, it is especially important to preserve order in working with ∇. For instance, $\nabla\cdot\mathbf{A}$ is a scalar field, simply the divergence of $\mathbf{A}$, but $\mathbf{A}\cdot\nabla$ is the differential operator

$$A_1\frac{\partial}{\partial x} + A_2\frac{\partial}{\partial y} + A_3\frac{\partial}{\partial z}$$

— a horse of quite a different color.

We list sixteen vector identities in table 3.2.

Table 3.2 Vector Operator Identities

$\mathbf{F}$ and $\mathbf{G}$ denote vector fields, ϕ denotes a scalar field, and $\mathbf{R} = x\mathbf{i} + y\mathbf{j} + z\mathbf{k}$. $\mathbf{A}$ is any constant vector, and f is any differentiable function of a single variable.

$$\nabla(\phi_1\phi_2) = \phi_1\nabla\phi_2 + \phi_2\nabla\phi_1 \tag{3.27}$$

$$\nabla\cdot\phi\mathbf{F} = \phi\nabla\cdot\mathbf{F} + \mathbf{F}\cdot\nabla\phi \tag{3.28}$$

$$\nabla\times\phi\mathbf{F} = \phi\nabla\times\mathbf{F} + \nabla\phi\times\mathbf{F} \tag{3.29}$$

$$\nabla f(\phi) = \frac{df}{d\phi}\nabla\phi \tag{3.30}$$

$$\nabla\cdot(\mathbf{R}-\mathbf{A}) = 3 \tag{3.31}$$

$$\nabla\times(\mathbf{R}-\mathbf{A}) = \mathbf{0} \tag{3.32}$$

$$\nabla(|\mathbf{R}-\mathbf{A}|^n) = n|\mathbf{R}-\mathbf{A}|^{n-2}(\mathbf{R}-\mathbf{A}) \tag{3.33}$$

$$\mathbf{F}\cdot\nabla(\mathbf{R}-\mathbf{A}) = \mathbf{F} \tag{3.34}$$

$$\nabla(\mathbf{A}\cdot\mathbf{R}) = \mathbf{A} \tag{3.35}$$

$$\nabla\cdot(\mathbf{F}\times\mathbf{G}) = \mathbf{G}\cdot(\nabla\times\mathbf{F}) - \mathbf{F}\cdot(\nabla\times\mathbf{G}) \tag{3.36}$$

$$\nabla\times(\mathbf{F}\times\mathbf{G}) = (\mathbf{G}\cdot\nabla)\mathbf{F} - (\mathbf{F}\cdot\nabla)\mathbf{G} + (\nabla\cdot\mathbf{G})\mathbf{F} - (\nabla\cdot\mathbf{F})\mathbf{G} \tag{3.37}$$

$$\nabla\times(\nabla\times\mathbf{F}) = \nabla(\nabla\cdot\mathbf{F}) - \nabla^2\mathbf{F} \tag{3.38}$$

$$\nabla(\mathbf{F}\cdot\mathbf{G}) = (\mathbf{F}\cdot\nabla)\mathbf{G} + (\mathbf{G}\cdot\nabla)\mathbf{F} + \mathbf{F}\times(\nabla\times\mathbf{G}) + \mathbf{G}\times(\nabla\times\mathbf{F}) \tag{3.39}$$

$$\nabla\times\nabla(\phi) = \mathbf{0} \tag{3.40}$$

$$\nabla\cdot(\nabla\times\mathbf{F}) = 0 \tag{3.41}$$

$$\nabla\cdot(\nabla\phi_1\times\nabla\phi_2) = 0 \tag{3.42}$$

Identities (3.27) through (3.29) are very simple. They are based on the formula expressing the derivative of a product as the sum of two terms, each containing the derivative of one factor. Any of these is easy to verify componentwise. For instance, the z component of $\nabla \times \phi\mathbf{F}$ is

$$\frac{\partial}{\partial x}(\phi F_2) - \frac{\partial}{\partial y}(\phi F_1)$$

Breaking this up, we see that this is ϕ times the z component of $\nabla \times \mathbf{F}$, plus the z component of $(\nabla\phi) \times \mathbf{F}$.

Identity (3.30) expresses the chain rule; its x component merely says

$$\frac{\partial}{\partial x} f(u) = \frac{df}{du}\frac{\partial u}{\partial x}$$

It can be generalized to functions of more than one variable. For example, if u_1 and u_2 are functions of x, y, and z, and if f is a function of u_1 and u_2, then we have

$$\nabla f(u_1,u_2) = \frac{\partial f}{\partial u_1}\nabla u_1 + \frac{\partial f}{\partial u_2}\nabla u_2$$

Identities (3.31) through (3.35), which involve the vector $\mathbf{R}$, are quite trivial but extremely useful. Note that $\mathbf{R} - \mathbf{A}$ locates (x,y,z) "relative to" $\mathbf{A}$ (section 1.6).

Identities (3.36) through (3.39) involve the interplay of the vector and differential properties of ∇, and they are quite complex. Any of them can be verified by laboriously working out the components, and we cheerfully invite the devoted student to do so. In the next (optional) section on tensor notation we will use some heavy notational machinery to derive these equations more efficiently. However, we would like to mention a heuristic device for guessing at the form of the identities.

Let's take identity (3.36), and go to work on

$$\nabla \cdot (\mathbf{F} \times \mathbf{G}) \tag{3.43}$$

We know that, as far as the *vector* nature of the triple scalar product is concerned, we can interchange the dot and the cross. Thus we suspect that the expression (3.43) is equal to

$$(\nabla \times \mathbf{F}) \cdot \mathbf{G} \tag{3.44}$$

However, we must interpret (3.44) in an unconventional manner, namely, the operator ∇ must continue to differentiate *both* $\mathbf{F}$ *and* $\mathbf{G}$ [and not merely $\mathbf{F}$, as eq. (3.44) dictates]. So to be correct we must split eq. (3.44) into two terms, analogous to the splitting in differentiating a product. The term where $\mathbf{F}$ alone is differentiated can be expressed unambiguously as

$$\mathbf{G} \cdot (\nabla \times \mathbf{F}) \tag{3.45}$$

To get the term where $\mathbf{G}$ is differentiated, we "rewrite" eq. (3.44) as

$$-(\nabla \times \mathbf{G}) \cdot \mathbf{F} \tag{3.46}$$

which is consistent with the *vector* nature of the triple scalar product. Clearly, from eq. (3.46) we can display the part of the formula in which **G** is differentiated as

$$-\mathbf{F} \cdot (\nabla \times \mathbf{G}) \tag{3.47}$$

Thus we are led to guess that eq. (3.43) equals eqs. (3.45) plus (3.47), in agreement with identity (3.36)!

Let us try this out again on formula (3.37). Using our old rule for $\mathbf{A} \times (\mathbf{B} \times \mathbf{C})$, we first write, *incorrectly,*

$$\nabla \times (\mathbf{F} \times \mathbf{G}) = (\nabla \cdot \mathbf{G})\mathbf{F} - (\nabla \cdot \mathbf{F})\mathbf{G} \tag{3.48}$$

This is incorrect because we must interpret ∇ as differentiating both **F** and **G** in each expression on the right. To break up this compound derivative and get a correct expression for "$(\nabla \cdot \mathbf{G})\mathbf{F}$," we observe that $(\nabla \cdot \mathbf{G})\mathbf{F}$, interpreted conventionally, gives the term in which **G** is differentiated, while $(\mathbf{G} \cdot \nabla)\mathbf{F}$ gives the term where **G** is treated as constant and we differentiate **F**. Handling the other term in eq. (3.48) similarly, we propose that

$$\nabla \times (\mathbf{F} \times \mathbf{G}) = (\nabla \cdot \mathbf{G})\mathbf{F} + (\mathbf{G} \cdot \nabla)\mathbf{F} - (\nabla \cdot \mathbf{F})\mathbf{G} - (\mathbf{F} \cdot \nabla)\mathbf{G}$$

This is identity (3.37).

Clearly the above reasoning is tricky, but it can be very helpful in suggesting "which way to turn" in the derivation of complicated vector equations (e.g., those of electromagnetic theory). Suffice it to say that in practice one always breathes easier after verifying any such "identity" in a reliable reference.

Identities (3.40) through (3.42) are based on the appearance, in each case, of differences of mixed second derivatives. For example, the z component of $\nabla \times (\nabla\phi)$ is

$$\frac{\partial}{\partial x}\frac{\partial \phi}{\partial y} - \frac{\partial}{\partial y}\frac{\partial \phi}{\partial x}$$

It is well known from advanced calculus that such mixed derivatives are the same when taken in either order; hence these terms cancel. [To be rigorous, when applying eqs. (3.40) through (3.42) we should stipulate that ϕ and **F** possess *continuous second derivatives.*] A proof of the equality of the mixed partial derivatives appears in Appendix B.

The labor of computing a gradient can be reduced by using (3.30) and a little common sense, and the following example should be studied carefully.

Example 3.26 Find the gradient of the scalar field f given by $f(R) = 1/R$, where R is the distance from the origin, $R = |\mathbf{R}| = \sqrt{x^2 + y^2 + z^2}$.

Solution Since R is the distance from the origin, ∇R can be computed by using properties 3.1 through 3.4 (sec. 3.1). Obviously, R increases most rapidly in the direction away from the origin, so the direction of ∇R is the same as the direction of the position vector **R**, which also points away from the origin. When we move in this direction, the rate of increase of R per unit distance is simply $dR/dR = 1$. So ∇R is a unit vector directed away from the origin, and hence equals $\mathbf{R}/|\mathbf{R}|$, which is the position vector divided by its own magnitude. That is,

$$\nabla R = \nabla|\mathbf{R}| = \frac{\mathbf{R}}{|\mathbf{R}|} = \frac{x\mathbf{i} + y\mathbf{j} + z\mathbf{k}}{(x^2 + y^2 + z^2)^{1/2}}$$

Applying eq. (3.30) with $\phi = |\mathbf{R}|$ and $f(R) = 1/R$, we have $\nabla f(R) = f'(R)\nabla R = (-1/R^2)\nabla R$, and, therefore,

$$\nabla\left(\frac{1}{R}\right) = \left(-\frac{1}{R^2}\right)\nabla R = -\frac{1}{R^2}\frac{\mathbf{R}}{|\mathbf{R}|} = -\frac{x\mathbf{i} + y\mathbf{j} + z\mathbf{k}}{(x^2 + y^2 + z^2)^{3/2}} = -\frac{\mathbf{R}}{R^3}$$

[which agrees with eq. (3.33) when $n = -1$]. Readers familiar with electric fields will recognize this expression. Except for some physical constants, it is the electric field intensity due to a point charge located at the origin.

As in section 1.15, the reader is advised to attach a permanent bookmark to this section for future referencing.

EXERCISES

1. Verify eqs. (3.27) and (3.28).
2. Verify (3.31) through (3.35).
3. Verify (3.40) through (3.42).
4. "Derive" (3.38) heuristically.
5. Why is the following "identity" obviously not valid? (*Hint:* Check the symmetry.)

$$\nabla \cdot (\mathbf{F} \times \mathbf{G}) = \mathbf{G} \cdot (\nabla \times \mathbf{F}) + \mathbf{F} \cdot (\nabla \times \mathbf{G})$$

6. If $\mathbf{A} = a_1\mathbf{i} + a_2\mathbf{j} + a_3\mathbf{k}$ is a constant vector, and if $\mathbf{R} = x\mathbf{i} + y\mathbf{j} + z\mathbf{k}$ and $R = |\mathbf{R}|$, show that

$$\nabla \cdot \frac{\mathbf{A} \times \mathbf{R}}{R} = 0$$

7. If $\mathbf{V}(\mathbf{R})$ can be expressed as $\mathbf{V}(\mathbf{R}) = \mathbf{A}f(\mathbf{R} \cdot \mathbf{B})$, where $\mathbf{A}$ and $\mathbf{B}$ are constant, prove that **curl V** is perpendicular to both **A** and **B.**
8. Evaluate $\nabla^2[(\mathbf{i} + \mathbf{j} + \mathbf{k}) \times \nabla(\mathbf{R} \cdot \mathbf{R})^2]$.
9. Evaluate $\mathbf{A} \cdot \nabla\mathbf{R} + \nabla(\mathbf{A} \cdot \mathbf{R}) + \mathbf{A} \times \nabla \times \mathbf{R}$, where $\mathbf{A}$ is a constant vector field.
10. If $R^2 = x^2 + y^2 + z^2$, $\mathbf{R} = x\mathbf{i} + y\mathbf{j} + z\mathbf{k}$, and $\mathbf{A}$ is a constant vector field, find
    (a) $\nabla \cdot (R^2\mathbf{A})$
    (b) $\nabla \times (R^2\mathbf{A})$
    (c) $\mathbf{R} \cdot \nabla(R^2\mathbf{A})$
    (d) $\nabla(\mathbf{A} \cdot \mathbf{R})^4$
    (e) $\nabla \cdot (R\mathbf{A})$
    (f) $\mathbf{R} \cdot \nabla(\mathbf{A} \cdot R\mathbf{A})$
    (g) $\nabla \cdot (\mathbf{A} \times \mathbf{R})$
    (h) $\nabla \times (\mathbf{A} \times \mathbf{R})$
    (i) $\nabla^2(\mathbf{R} \cdot \mathbf{R})$

11. Evaluate

$$\nabla\left(\mathbf{A}\cdot\nabla\frac{1}{R}\right)+\nabla\times\left(\mathbf{A}\times\nabla\frac{1}{R}\right)$$

12. For what value of the constant C is the vector field $V = (x + 4y)\mathbf{i} + (y - 3z)\mathbf{j} + Cz\mathbf{k}$ the curl of some vector field **F**? [*Hint:* use eq. (3.41).]

13. Derive the identity

$$\nabla|\mathbf{F}|^2 = 2\,\mathbf{F}\cdot\nabla\,\mathbf{F} + 2\,\mathbf{F}\times(\nabla\times\mathbf{F})$$

14. The time derivative, $\partial\mu/\partial t$, in the equation of continuity (eq. (3.9)) measures the time rate of change of the fluid density (μ) at a fixed location **R**:

$$\frac{\partial\mu}{\partial t} = \lim_{\Delta t\to 0}\frac{\mu(\mathbf{R}, t+\Delta t) - \mu(\mathbf{R}, t)}{\Delta t}.$$

This is sometimes called the Eulerian derivative. However, in formulating Newton's laws of motion analysts need to express the rate of change of density as measured by an observer moving with the fluid—the so-called convective or Lagrangian derivative.

(a) Argue that the convective derivative for any fluid property η (such as density, temperature, etc.) moving with the fluid at the velocity $\mathbf{v}$ is expressed by

$$\frac{d\eta}{dt} = \lim_{\Delta t\to 0}\frac{\eta(\mathrm{R} + \mathrm{v}\,\Delta t, t+\Delta t) - \eta(\mathrm{R}, t)}{\Delta t}.$$

(b) Evaluate the limit of this expression and show that it is given by

$$\frac{d\eta}{dt} = \frac{\partial\eta}{\partial t} + \mathbf{v}\cdot\nabla\eta.$$

(c) Use identity (2.38) to show that if η is the fluid density μ, the equation of continuity (eq. 3.9) becomes

$$\frac{d\mu}{dt} = -\mu\nabla\cdot\mathbf{v}$$

in terms of the convective derivative.

3.9 Optional Reading: Tensor Notation

The operator ∇, considered as a vector operator, has components $\partial/\partial x$, $\partial/\partial y$, and $\partial/\partial z$. In tensor notation we adopt two conventions which enable us to absorb ∇ into our system painlessly. First, we designate coordinates by the triple (x_1,x_2,x_3) instead of (x,y,z); this makes the ith component of ∇ equal $\partial/\partial x_i$. Second, we abbreviate $\partial/\partial x_i$ by ∂_i. Now let us write down the tensor expressions for the concepts introduced in this chapter.

The ith component of the gradient of ϕ is $\partial_i\phi$.

The divergence of **F** is the scalar $\partial_i F_i$ (remember summation).

The ith component of the curl, $\nabla \times \mathbf{F}$, is $\varepsilon_{ijk}\,\partial_j\,F_k$ (recall the determinant expression for curl).

The laplacian of ϕ is $\partial_i\partial_i\phi$. We may write this as $\partial_i^2\,\phi$ if we stipulate that the summation convention applies to *squared* terms, since they would have repeated subscripts if written out.

Now the proof of the identities of the last section can be carried out easily. To check identity (3.29), observe that the ith component of $\nabla \times \phi\mathbf{F}$ is

$$\varepsilon_{ijk}\,\partial_j\,(\phi F_k) = \varepsilon_{ijk}(\partial_j\phi)F_k + \varepsilon_{ijk}\phi\,\partial_j\,F_k$$

These terms we identify as the ith components of $\nabla\phi \times \mathbf{F}$ and $\phi\nabla \times \mathbf{F}$.

To check (3.34), observe that the ith component of $\mathbf{F} \cdot \nabla\mathbf{R}$ is $F_j\partial_j x_i$ (summing over j). But $\partial_j x_i = \delta_{ij}$, the Kronecker delta; so this expression is $F_j\delta_{ij} = F_i$, the ith component of **F**.

The proof of formula (3.37) proceeds as follows:

$$\begin{aligned}
\varepsilon_{ijk}\,\partial_j\,(\mathbf{F} \times \mathbf{G})_k &= \varepsilon_{ijk}\,\partial_j\,(\varepsilon_{klm}F_lG_m) \\
&= \varepsilon_{ijk}\varepsilon_{klm}\partial_j\,(F_lG_m) \\
&= (\delta_{il}\delta_{jm} - \delta_{im}\delta_{jl})\,\partial_j\,(F_lG_m) \\
&= \partial_j\,(F_iG_j) - \partial_j\,(F_jG_i) \\
&= G_j\partial_jF_i + (\partial_jG_j)F_i - (\partial_jF_j)G_i - F_j\partial_jG_i \\
&= (\mathbf{G} \cdot \nabla)\mathbf{F}_i + (\nabla \cdot \mathbf{G})\mathbf{F}_i - (\nabla \cdot \mathbf{F})\mathbf{G}_i - (\mathbf{F} \cdot \nabla)\mathbf{G}_i
\end{aligned}$$

The proof of (3.39) is rather complicated. We begin by developing the obvious expression for the ith component of $\nabla(\mathbf{F} \cdot \mathbf{G})$:

$$\partial_i\,(F_jG_j) = F_j\partial_iG_j + G_j\partial_iF_j \tag{3.49}$$

Now we are stumped; the terms on the right seem to have no vector analogs. How can we identify the right-hand side of identity (3.39) here? The clue lies in the tensor expression for $\mathbf{F} \times (\nabla \times \mathbf{G})$; its ith component is

$$\begin{aligned}
\varepsilon_{ijk}F_j(\varepsilon_{klm}\partial_lG_m) &= \varepsilon_{ijk}\varepsilon_{klm}F_j\partial_lG_m \\
&= (\delta_{il}\delta_{jm} - \delta_{im}\delta_{jl})F_j\partial_lG_m \\
&= F_j\partial_iG_j - F_j\partial_jG_i
\end{aligned}$$

We observe the appearance of one of these "mystery" terms, $F_j\partial_i G_j$, plus the ith component of $-(\mathbf{F}\cdot\nabla)\mathbf{G}$. Transposing, we see that $F_j\partial_i G_j$ is the ith component of $\mathbf{F}\times(\nabla\times\mathbf{G})+\mathbf{F}\cdot\nabla\mathbf{G}$. Putting this into eq. (3.49) above, and using a similar expression for $G_j\partial_i F_j$, we get

$$\nabla(\mathbf{F}\cdot\mathbf{G}) = \mathbf{F}\times(\nabla\times\mathbf{G}) + \mathbf{F}\cdot\nabla\mathbf{G} + \mathbf{G}\times(\nabla\times\mathbf{F}) + \mathbf{G}\cdot\nabla\mathbf{F}$$

We have derived the identity.

The equality of mixed second derivatives of any (twice continuously differentiable) function can be expressed in tensor notation by the equation

$$\partial_j\partial_i\phi = \partial_i\partial_j\phi$$

or simply

$$\partial_j\partial_i = \partial_i\partial_j$$

That is, the components of ∇ *commute* with each other. (Of course, they do not commute with functions: $\partial_i\phi$ is very different from $\phi\partial_i$.) This makes the verification of identities (3.40) through (3.42) simple. For eq. (3.42), we use ψ and χ for ϕ_1 and ϕ_2 respectively, in order not to confuse subscripts. We then have

$$\partial_i[\varepsilon_{ijk}(\partial_j\psi)(\partial_k\chi)] = \varepsilon_{ijk}(\partial_i\partial_j\psi)(\partial_k\chi) + \varepsilon_{ijk}(\partial_j\psi)(\partial_i\partial_k\chi)$$

Because of the antisymmetric nature of ε_{ijk}, as we sum over i and j the terms $\partial_i\partial_j\psi$ and $\partial_j\partial_i\psi$ come in with opposite signs for $i \neq j$, and with coefficient zero if $i = j$. Thus all the addends in the first term cancel, as do those in the second, and we get zero, in accordance with the identity (3.42).

EXERCISES

Using the tensor notation, prove the following vector identities:

1. $\nabla\cdot\phi\mathbf{F} = \phi\nabla\cdot\mathbf{F} + \mathbf{F}\cdot\nabla\phi$
2. $\nabla(\mathbf{A}\cdot\mathbf{R}) = \mathbf{A}$ if $\mathbf{A}$ is constant
3. $\nabla\times\mathbf{R} = \mathbf{0}$
4. $\nabla\cdot(\mathbf{F}\times\mathbf{G}) = \mathbf{G}\cdot(\nabla\times\mathbf{F}) - \mathbf{F}\cdot(\nabla\times\mathbf{G})$
5. $\nabla\times(\nabla\times\mathbf{F}) = \nabla(\nabla\cdot\mathbf{F}) - \nabla^2\mathbf{F}$
6. $\nabla\times(\nabla\phi) = \mathbf{0}$
7. $\nabla\cdot(\nabla\times\mathbf{F}) = 0$

3.10 Cylindrical and Spherical Coordinates

Recall (section 2.5) that many two-dimensional problems can be expressed more conveniently in polar coordinates than in cartesian coordinates. This is the case, for instance, if some type of circular symmetry is present. Analogous situations arise in three dimensions, of course, and therefore one is led to construct generalizations of the polar coordinate system. The two generalizations that have proved most useful are *cylindrical coordinates* and *spherical coordinates*. In this section we shall study how the various vector relationships are expressed in these systems.

Cylindrical coordinates are the most direct generalization of polar coordinates. To see this, we observe first that the *cartesian* system can be described in the following manner: the third coordinate, z, gives the (signed) height of the point above the xy plane; and the first two coordinates, x and y, are the two-dimensional cartesian coordinates of the projection of the point on that plane.

For the cylindrical coordinate system, the third coordinate, z, is again the height above the xy plane, but the first two coordinates are the *polar* coordinates ρ and θ of the projection of the point on that plane (see fig. 3.18). Notice that ρ in cylindrical coordinates plays the role of r in polar coordinates; the reason for the change in terminology will be seen later. However, be aware that *there is no standard terminology* among authors for these coordinate systems!

The equations relating cartesian and polar coordinates are as follows:

$$
\begin{aligned}
x &= \rho\cos\theta & \rho &= (x^2+y^2)^{1/2} \\
y &= \rho\sin\theta & \theta &= \sin^{-1}\frac{y}{\sqrt{x^2+y^2}} = \cos^{-1}\frac{x}{\sqrt{x^2+y^2}} \\
z &= z & z &= z
\end{aligned}
\tag{3.50}
$$

The extra equation for θ serves to remind us that we use the value θ appropriate to the quadrant of (x,y), not necessarily the principal value.

The angle θ in cylindrical coordinates is not defined on the z axis, when $\rho = 0$, but otherwise the equations of (3.50) specify a one-to-one correspondence between the two systems.

The nomenclature "cylindrical" comes from the fact that the surfaces $\rho = constant$ are cylinders. The surfaces $\theta = constant$ are "half-planes" extending out from the z axis and, of course, $z = constant$ defines a family of horizontal planes. Normals to these surfaces are given by **grad** ρ, **grad** θ, and **grad** z, respectively. From figure 3.19, we can see that **grad** ρ points away from the z axis, **grad** θ points counterclockwise in the horizontal plane, and **grad** z points upward. In the order **grad** ρ, **grad** θ, and **grad** z, these vectors form a right-handed orthogonal system.

Any two surfaces $\rho = constant$ and $\theta = constant$ intersect in a vertical line, which is a curve along which only z varies. It is called a *coordinate curve* for z. Coordinate curves for ρ are horizontal rays extending from the z axis. Coordinate curves for θ are horizontal circles. Notice that **grad** z, **grad** ρ, and **grad** θ are everywhere tangent to their respective coordinate curves.

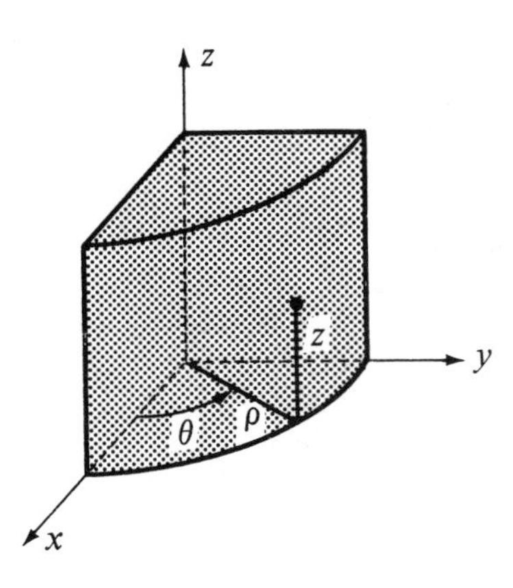

Figure 3.18

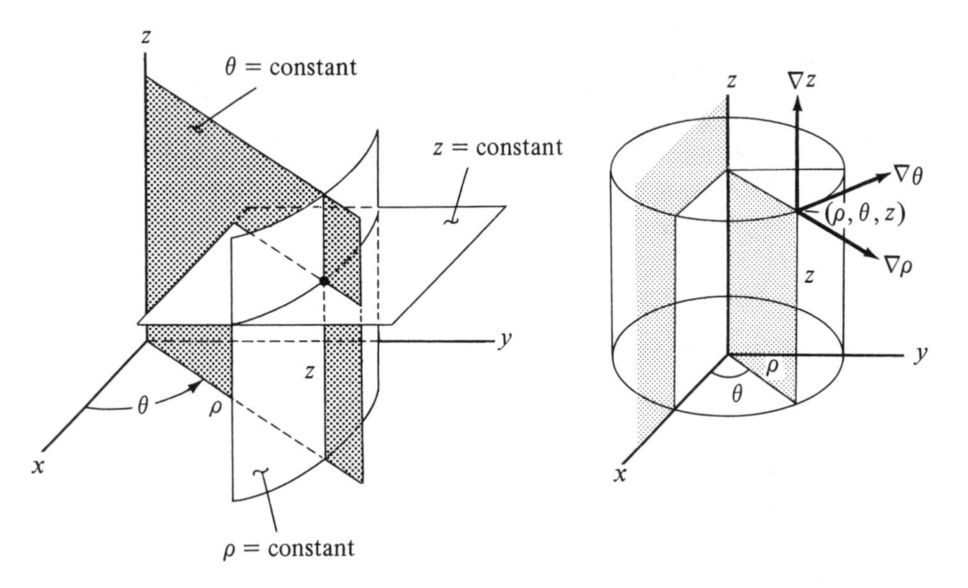

Figure 3.19

These features make it convenient to introduce unit vectors in the directions of **grad** z, **grad** ρ, and **grad** θ. Accordingly, we define

$$\mathbf{e}_z = \frac{\mathbf{grad}\ z}{|\mathbf{grad}\ z|}$$

$$\mathbf{e}_\rho = \frac{\mathbf{grad}\ \rho}{|\mathbf{grad}\ \rho|}$$

$$\mathbf{e}_\theta = \frac{\mathbf{grad}\ \theta}{|\mathbf{grad}\ \theta|} \tag{3.51}$$

Readers should convince themselves that $\mathbf{e}_z$ is the same as $\boldsymbol{k}$, and $\mathbf{e}_\rho$ and $\mathbf{e}_\theta$ are the three-dimensional analogs of $\mathbf{u}_r$ and $\mathbf{u}_\theta$ in section 2.5. In fact, recalling that $|\mathbf{grad}\, f| = df/ds$ when s measures distance in the direction of $\mathbf{grad}\, f$, one can simplify these equations. Along the coordinate curves of z, $ds = |dz|$. Hence, $|\mathbf{grad}\, z| = dz/dz = 1$. Along the coordinate curves of ρ, $ds = |d\rho|$. Hence also, $|\mathbf{grad}\, \rho| = d\rho/d\rho = 1$. Along coordinate curves of θ, however, $ds = \rho|d\theta|$. Therefore, $|\mathbf{grad}\, \theta| = d\theta/\rho\, d\theta = 1/\rho$. This results in

$$\mathbf{e}_z = \mathbf{grad}\, z$$
$$\mathbf{e}_\rho = \mathbf{grad}\, \rho$$
$$\mathbf{e}_\theta = \rho\, \mathbf{grad}\, \theta$$

The reader should observe that the position vector of a point can be expressed in cylindrical coordinates as

$$\mathbf{R} = x\mathbf{i} + y\mathbf{j} + z\mathbf{k} = \rho\mathbf{e}_\rho + z\mathbf{e}_z$$

To compute arc length in cylindrical coordinates, observe that in figure 3.20 the displacement $d\mathbf{R}$ can be expressed as the sum of three orthogonal displacements:

$$d\mathbf{R} = \mathbf{e}_\rho\, d\rho + \mathbf{e}_\theta\, \rho\, d\theta + \mathbf{e}_z\, dz \tag{3.52}$$

Hence *the element of arc length in cylindrical coordinates is given by*

$$ds = |d\mathbf{R}| = (d\rho^2 + \rho^2\, d\theta^2 + dz^2)^{1/2} \tag{3.53}$$

Example 3.27 Find the arc length along the helix

$$x = \sin t \qquad y = \cos t \qquad z = t$$

for $0 \leq t \leq 4\pi$.

Solution Transforming to cylindrical coordinates, we find from eqs. (3.50)

$$\rho = 1 \qquad \theta = \frac{\pi}{2} - t \qquad z = t$$

Hence

$$s = \int_0^{4\pi} \left[\left(\frac{d\rho}{dt}\right)^2 + \rho^2\left(\frac{d\theta}{dt}\right)^2 + \left(\frac{dz}{dt}\right)^2\right]^{1/2} dt$$
$$= \int_0^{4\pi} (1 + 1)^{1/2}\, dt = 4\sqrt{2}\pi$$

From figure 3.20 it is easy to see that *the element of volume in cylindrical coordinates is given by*

$$dV = d\rho\, \rho\, d\theta\, dz = \rho\, d\rho\, d\theta\, dz \tag{3.54}$$

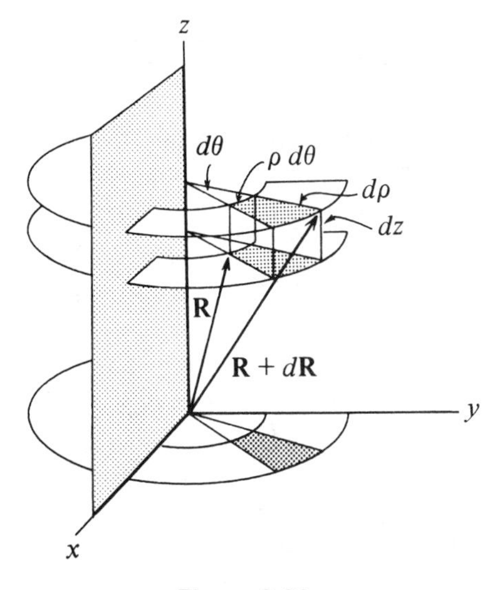

Figure 3.20

Now consider a scalar field f expressed in cylindrical coordinates, $f = f(\rho, \theta, z)$. To express **grad** f in these coordinates, we shall exploit the characteristics of the gradient as expounded in section 3.1. First observe that since $\mathbf{e}_\rho$, $\mathbf{e}_\theta$, and $\mathbf{e}_z$ are mutually orthogonal unit vectors, the coefficients in the expression $\nabla f = \alpha\, \mathbf{e}_\rho + \beta\, \mathbf{e}_\theta + \gamma\, \mathbf{e}_z$ are simply the dot products:

$$\nabla f = (\mathbf{e}_\rho \cdot \nabla f)\, \mathbf{e}_\rho + (\mathbf{e}_\theta \cdot \nabla f)\, \mathbf{e}_\theta + (\mathbf{e}_z \cdot \nabla f)\, \mathbf{e}_z.$$

Next recall that each of these coefficients gives the rate of change of f with respect to distance, df/ds, in the corresponding direction. Applying the expression (3.53) for ds, we find

$$\mathbf{e}_\rho \cdot \nabla f = \left.\frac{\partial f}{\partial s}\right|_{\theta,z \text{ constant}} = \frac{\partial f}{\partial \rho}$$

$$\mathbf{e}_\theta \cdot \nabla f = \left.\frac{\partial f}{\partial s}\right|_{\rho,z} = \frac{1}{\rho}\frac{\partial f}{\partial \theta}$$

$$\mathbf{e}_z \cdot \nabla f = \left.\frac{\partial f}{\partial s}\right|_{\rho,\theta} = \frac{\partial f}{\partial z}$$

Therefore *the expression for* **grad** f *in cylindrical coordinates is*

$$\mathbf{grad}\, f = \frac{\partial f}{\partial \rho}\mathbf{e}_\rho + \frac{1}{\rho}\frac{\partial f}{\partial \theta}\mathbf{e}_\theta + \frac{\partial f}{\partial z}\mathbf{e}_z \tag{3.55}$$

Example 3.28 Compute **grad** f in cylindrical coordinates if f is given in cartesian coordinates by (a) $f = z/(x^2 + y^2)$; (b) $f = x$.

Solution We first express f in cylindrical coordinates. From eqs. (3.50), we have $z/(x^2 + y^2) = z/\rho^2$ and $x = \rho \cos \theta$. Hence

$$\mathbf{grad}\,\frac{z}{\rho^2} = -\frac{2z}{\rho^3}\,\mathbf{e}_\rho + \frac{1}{\rho^2}\,\mathbf{e}_z,$$

$$\mathbf{grad}\,[\rho \cos \theta] = \cos \theta\,\mathbf{e}_\rho - \sin \theta\,\mathbf{e}_\theta.$$

Of course the last expression equals **i.**

The expressions for the divergence and curl of a vector field can be derived by heuristic reasoning with infinitesimals (as in sections 3.3 and 3.4), but one must be extra careful when the coordinate system is not cartesian.

Let us compute div **F** as flux per unit volume out of the box in figure 3.21. We start with the vector field **F** given in cylindrical coordinates:

$$\mathbf{F} = F_\rho(\rho,\theta,z)\mathbf{e}_\rho + F_\theta(\rho,\theta,z)\mathbf{e}_\theta + F_z(\rho,\theta,z)\mathbf{e}_z$$

The flux of **F** out of face I equals the outward-normal component of **F** times the surface area $(-F_\rho)(\rho\, d\theta\, dz)$. A similar expression holds for face IV, but with a different value of ρ. Hence the contribution from faces I and IV will be given, in the limit, by

$$(F_\rho \rho\, d\theta\, dz)_{\mathrm{IV}} - (F_\rho \rho\, d\theta\, dz)_{\mathrm{I}} = \frac{\partial(\rho F_\rho)}{\partial \rho}\, d\rho\, d\theta\, dz \tag{3.56}$$

Notice that we regard the dimensions $d\theta$ and dz as the same for faces I and IV, so they are constants in eq. (3.56).

The flux out of face II is $(-F_\theta)\, d\rho\, dz$ and, combining this with face V, we obtain a contribution

$$(F_\theta\, d\rho\, dz)_{\mathrm{V}} - (F_\theta\, d\rho\, dz)_{\mathrm{II}} = \frac{\partial F_\theta}{\partial \theta}\, d\theta\, d\rho\, dz$$

The flux out of faces III and VI contributes

$$(F_z \rho\, d\theta\, d\rho)_{\mathrm{VI}} - (F_z \rho\, d\theta\, d\rho)_{\mathrm{III}} = \frac{\partial F_z}{\partial z}\, dz\, \rho\, d\theta\, d\rho \tag{3.57}$$

The reader may feel a little queasy about this last expression since in figure 3.21 face III is not a genuine rectangle, having side li of length $\rho\, d\theta$ and the opposite side kj of length $(\rho + d\rho)\, d\theta$. To play it safe, we may replace ρ in eq. (3.57) by $\tilde{\rho}$, some intermediate value between ρ and $\rho + d\rho$. Adding all the contributions to the flux, we obtain

$$\frac{\partial(\rho F_\rho)}{\partial \rho}\, d\rho\, d\theta\, dz + \frac{\partial F_\theta}{\partial \theta}\, d\theta\, d\rho\, dz + \frac{\partial F_z}{\partial z}\, dz\, \tilde{\rho}\, d\theta\, d\rho$$

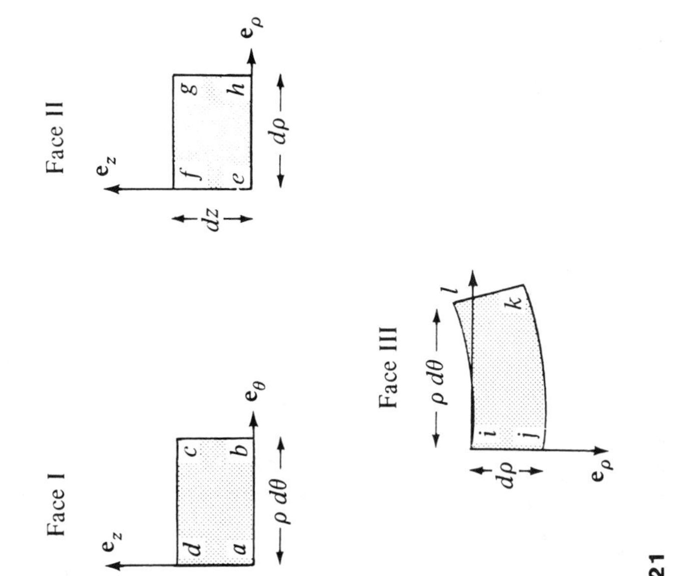
Face I
e_z
d
c
a
b
dz
ρ dθ
e_θ
Face II
e_z
f
g
e
h
dz
dρ
e_ρ
Face III
ρ dθ
i
j
k
l
dρ
e_ρ

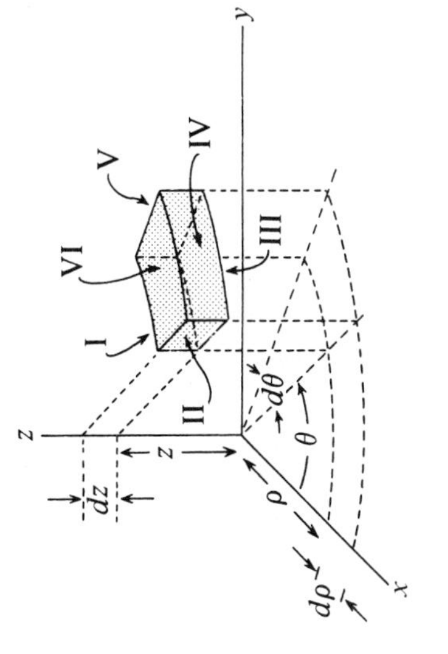
z
dz
z
I
VI
V
IV
III
II
dθ
θ
ρ
dρ
x
y

Figure 3.21

Next we divide by the volume element eq. (3.54) and, noticing that in the limit $\tilde{\rho}$ approaches ρ (so our precaution was unnecessary), we find that *the divergence of a vector field is given in cylindrical coordinates by*

$$\text{div}\,\mathbf{F} = \frac{1}{\rho}\frac{\partial(\rho F_\rho)}{\partial\rho} + \frac{1}{\rho}\frac{\partial F_\theta}{\partial\theta} + \frac{\partial F_z}{\partial z} \tag{3.58}$$

Example 3.29 Compute the divergence of

$$\mathbf{F}(\rho,\theta,z) = \rho\mathbf{e}_\rho + z\sin\theta\,\mathbf{e}_\theta + \rho z\mathbf{e}_z$$

Solution Applying eq. (3.58), we find

$$\nabla\cdot\mathbf{F} = \frac{1}{\rho}\frac{\partial(\rho^2)}{\partial\rho} + \frac{1}{\rho}\frac{\partial(z\sin\theta)}{\partial\theta} + \frac{\partial(\rho z)}{\partial z} = 2 + \frac{z\cos\theta}{\rho} + \rho$$

We compute **curl F** in cylindrical coordinates by employing the physical characterization of curl as "swirl per unit area" (section 3.4). To compute the $\mathbf{e}_\rho$ component, consider the sum of the lengths of the edges of the curvilinear rectangle *abcda* in figure 3.21, weighted by the counterclockwise components of **F**.

The contribution to the swirl from *ab* is $F_\theta\rho\,d\theta$; from *cd* it is $(-F_\theta)\rho\,d\theta$, but with a larger value of z. Since ρ and $d\theta$ are the same along these edges, we obtain a net contribution of

$$(F_\theta\rho\,d\theta)_{ab} - (F_\theta\rho\,d\theta)_{cd} = -\frac{\partial F_\theta}{\partial z}dz\,\rho\,d\theta$$

Similarly, edges *bc* and *da* contribute

$$(F_z\,dz)_{bc} - (F_z\,dz)_{da} = \frac{\partial F_z}{\partial\theta}d\theta\,dz$$

Hence the $\mathbf{e}_\rho$ component of **curl F** is the sum of these divided by the area $\rho\,d\theta\,dz$, or

$$\frac{1}{\rho}\frac{\partial F_z}{\partial\theta} - \frac{\partial F_\theta}{\partial z}$$

To find the $\mathbf{e}_\theta$ component, we add around the edge *efghe* of face II (because $\mathbf{e}_\theta$ points into the page). The total is

$$(F_z\,dz)_{ef} - (F_z\,dz)_{gh} + (F_\rho\,d\rho)_{fg} - (F_\rho\,d\rho)_{he}$$
$$= -\frac{\partial F_z}{\partial\rho}d\rho\,dz + \frac{\partial F_\rho}{\partial z}dz\,d\rho$$

and, dividing by the area $d\rho\, dz$, we find the $\mathbf{e}_\theta$ component to be

$$\frac{\partial F_\rho}{\partial z} - \frac{\partial F_z}{\partial \rho}$$

The $\mathbf{e}_z$ component is obtained by summing around the edge *ijkli* of face III:

$$\begin{aligned}(F_\rho\, d\rho)_{ij} &- (F_\rho\, d\rho)_{kl} + (F_\theta \rho\, d\theta)_{jk} - (F_\theta \rho\, d\theta)_{li} \\ &= -\frac{\partial F_\rho}{\partial \theta} d\theta\, d\rho + \frac{\partial(\rho F_\theta)}{\partial \rho} d\rho\, d\theta\end{aligned}$$

(keeping in mind that ρ on edge *li* is different from ρ on edge *jk*). Dividing by the area $\tilde{\rho}\, d\theta\, d\rho$, with $\tilde{\rho}$ between ρ and $\rho + d\rho$ as before, and then taking the limit, we find the $\mathbf{e}_z$ component of **curl F** to be

$$\frac{1}{\rho}\left[-\frac{\partial F_\rho}{\partial \theta} + \frac{\partial(\rho F_\theta)}{\partial \rho}\right]$$

Combining these components, we see that *the curl of a vector field is given in cylindrical coordinates by*

$$\begin{aligned}\mathbf{curl\ F} &= \left(\frac{1}{\rho}\frac{\partial F_z}{\partial \theta} - \frac{\partial F_\theta}{\partial z}\right)\mathbf{e}_\rho + \left(\frac{\partial F_\rho}{\partial z} - \frac{\partial F_z}{\partial \rho}\right)\mathbf{e}_\theta \\ &+ \frac{1}{\rho}\left(\frac{\partial(\rho F_\theta)}{\partial \rho} - \frac{\partial F_\rho}{\partial \theta}\right)\mathbf{e}_z \qquad (3.59)\end{aligned}$$

or, equivalently (exercise 3),

$$\mathbf{curl\ F} = \frac{1}{\rho}\begin{vmatrix} \mathbf{e}_\rho & \rho\mathbf{e}_\theta & \mathbf{e}_z \\ \dfrac{\partial}{\partial \rho} & \dfrac{\partial}{\partial \theta} & \dfrac{\partial}{\partial z} \\ F_\rho & \rho F_\theta & F_z \end{vmatrix} \qquad (3.60)$$

Example 3.30 Compute the curl of **F** given in example 3.29.

Solution Applying eq. (3.60), we find

$$\mathbf{curl\ F} = \frac{1}{\rho}\begin{vmatrix} \mathbf{e}_\rho & \rho\mathbf{e}_\theta & \mathbf{e}_z \\ \dfrac{\partial}{\partial \rho} & \dfrac{\partial}{\partial \theta} & \dfrac{\partial}{\partial z} \\ \rho & \rho z \sin\theta & \rho z \end{vmatrix}$$

$$= (0 - \rho \sin\theta)\frac{\mathbf{e}_\rho}{\rho} + (0 - z)\mathbf{e}_\theta + (z\sin\theta - 0)\frac{\mathbf{e}_z}{\rho}$$

$$= -\sin\theta\, \mathbf{e}_\rho - z\mathbf{e}_\theta + \frac{z \sin\theta}{\rho}\mathbf{e}_z$$

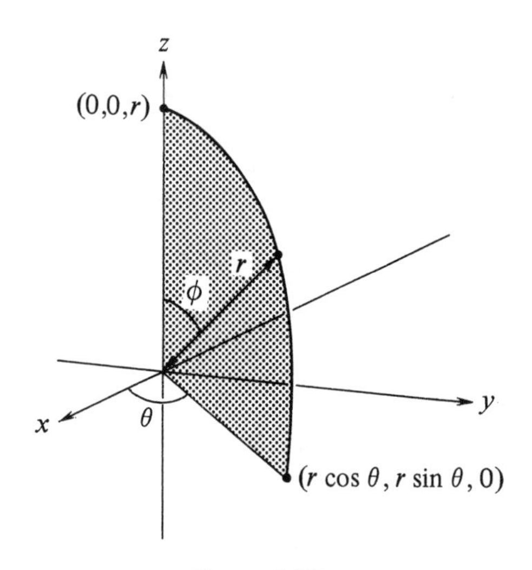

Figure 3.22

Spherical coordinates are also generalizations of polar coordinates in the plane. The first coordinate, r, is the distance of the point from the origin; hence it is a three-dimensional generalization of the two-dimensional r. The second coordinate, ϕ, is the angle between the positive z axis and the position vector $\mathbf{R}$ (see fig. 3.22). Therefore ϕ lies between 0 and π radians. The third coordinate, θ, is the same angle as in the cylindrical coordinate system, and lies between 0 and 2π radians.

Surfaces of constant r are, of course, spheres centered at the origin. Surfaces of constant ϕ are cones—right circular cones, in fact (see fig. 3.23). Surfaces of constant θ are half-planes, as in the case of cylindrical coordinates.

The equations of transformation between spherical and cartesian coordinates are easy to see once we recognize that the cylindrical coordinate ρ equals $r \sin \phi$, and z equals $r \cos \phi$. Then with the help of eqs. (3.50), we find

$$\begin{aligned} x &= r \sin \phi \cos \theta \qquad & r &= (x^2 + y^2 + z^2)^{1/2} \\ y &= r \sin \phi \sin \theta \qquad & \phi &= \cos^{-1} \frac{z}{(x^2 + y^2 + z^2)^{1/2}} \quad \text{(principal value)} \\ z &= r \cos \phi \qquad & \theta &= \sin^{-1} \frac{y}{\sqrt{x^2 + y^2}} = \cos^{-1} \frac{x}{\sqrt{x^2 + y^2}} \end{aligned} \tag{3.61}$$

The coordinate curves (curves where one coordinate varies and the other two are constant) are rays emanating from the origin (for r), vertical semicircles (for ϕ), and horizontal circles (for θ). If we consider the surface of the earth as a sphere, r = constant, the coordinate curves for ϕ are semicircles of constant longitude, and those for θ are circles of constant latitude: $\phi = \pi/2$ defines the equator (see fig. 3.24).

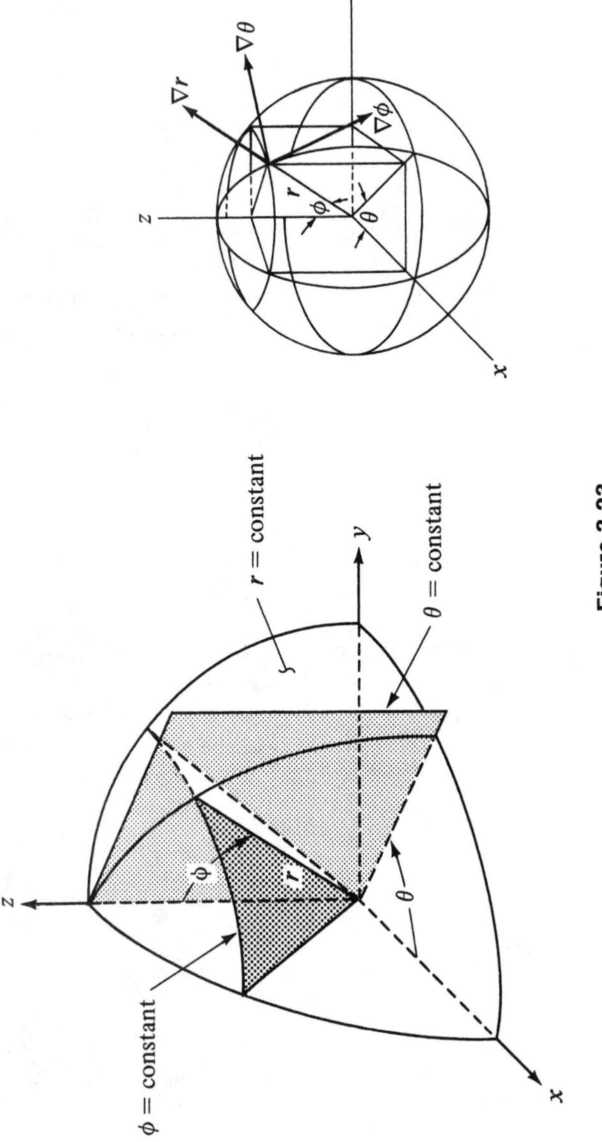

z
y
x
ϕ = constant
r = constant
θ = constant
ϕ
r
θ
∇r
$\nabla \theta$
$\nabla \phi$

Figure 3.23

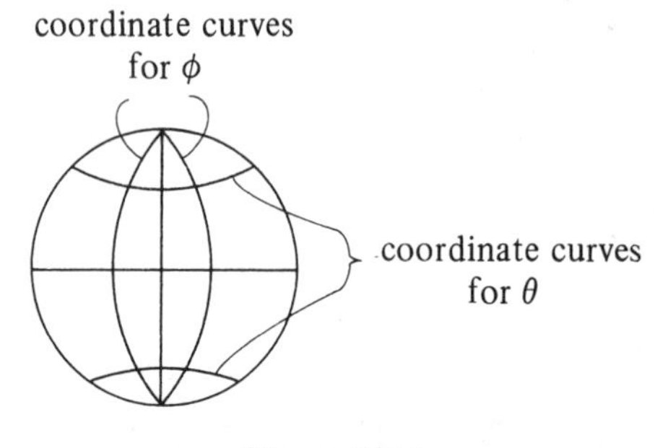

Figure 3.24

Staying with this earth analogy for a moment, we can see from the constant surfaces in figure 3.23 that **grad** r points along the *local vertical,* **grad** ϕ points due south, and **grad** θ points due east. These vectors are also tangent to their respective coordinate curves, and they are mutually orthogonal. Hence we define a set of unit vectors

$$\mathbf{e}_r = \frac{\mathbf{grad}\, r}{|\mathbf{grad}\, r|}$$

$$\mathbf{e}_\phi = \frac{\mathbf{grad}\, \phi}{|\mathbf{grad}\, \phi|}$$

$$\mathbf{e}_\theta = \frac{\mathbf{grad}\, \theta}{|\mathbf{grad}\, \theta|} \tag{3.62}$$

and observe that in the order $\mathbf{e}_r$, $\mathbf{e}_\phi$, and $\mathbf{e}_\theta$ they form a right-handed system. Again recalling that $|\mathbf{grad}\, f| = df/ds$, we can be more specific in the eqs. (3.62). Along coordinate curves of r, $ds = |dr|$. The coordinate curves of ϕ are semicircles of radius r, so $ds = |r\, d\phi|$. The coordinate curves of θ are circles of radius (be careful!) $r \sin \phi$, so $ds = |r \sin \phi\, d\theta|$ (see fig. 3.25). Therefore

$$\mathbf{e}_r = \mathbf{grad}\, r$$

$$\mathbf{e}_\phi = r\, \mathbf{grad}\, \phi$$

$$\mathbf{e}_\theta = r \sin \phi\, \mathbf{grad}\, \theta$$

The position vector in spherical coordinates is simply

$$\mathbf{R} = r\mathbf{e}_r$$

Now we can model the computations, made previously for cylindrical coordinates, to obtain the analogous expressions in spherical coordinates. From figure 3.25 we see that the displacement $d\mathbf{R}$ can be expressed

$$d\mathbf{R} = \mathbf{e}_r\, dr + \mathbf{e}_\phi r\, d\phi + \mathbf{e}_\theta r \sin \phi\, d\theta \tag{3.63}$$

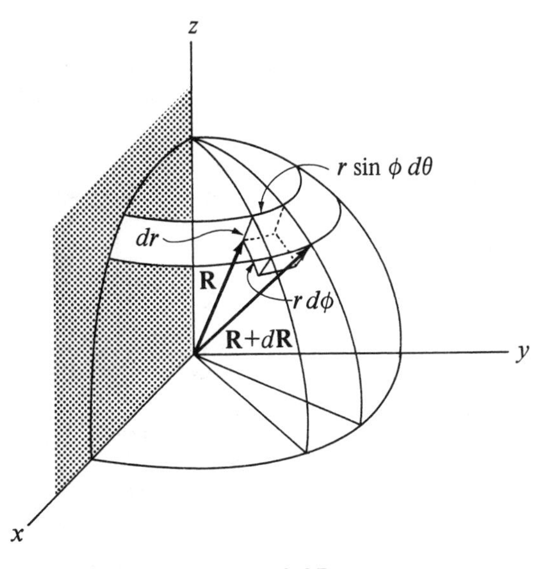

Figure 3.25

Thus *the element of arc length in spherical coordinates is*

$$ds = |d\mathbf{R}| = (dr^2 + r^2\,d\phi^2 + r^2 \sin^2\phi\,d\theta^2)^{1/2} \tag{3.64}$$

From figure 3.25 we see that *the volume element in spherical coordinates is given by*

$$dV = (dr)(r\,d\phi)(r \sin\phi\,d\theta) = r^2 \sin\phi\,dr\,d\phi\,d\theta \tag{3.65}$$

The component of the gradient of $f(r,\phi,\theta)$ in the direction $\mathbf{e}_r$ is the rate of change of f with respect to distance along the r-coordinate curve; and similarly for the $\mathbf{e}_\phi$ and $\mathbf{e}_\theta$ components. Using eq. (3.64) for distances, we find, analogous to eq. (3.55), that *the expression for the gradient in spherical coordinates is*

$$\mathbf{grad}\, f(r,\phi,\theta) = \frac{\partial f}{\partial r}\mathbf{e}_r + \frac{1}{r}\frac{\partial f}{\partial \phi}\mathbf{e}_\phi + \frac{1}{r\sin\phi}\frac{\partial f}{\partial \theta}\mathbf{e}_\theta \tag{3.66}$$

If $\mathbf{F}$ is a vector field given in spherical coordinates by

$$\mathbf{F}(r,\phi,\theta) = F_r\mathbf{e}_r + F_\phi\mathbf{e}_\phi + F_\theta\mathbf{e}_\theta$$

we can compute its divergence as before by reasoning on the infinitesimal parallelepiped in figure 3.26. The total flux out of all the faces can be expressed

$$\begin{aligned}(F_r r \sin\phi\,d\theta\,r\,d\phi)_{\mathrm{IV}} &- (F_r r \sin\phi\,d\theta\,r\,d\phi)_{\mathrm{I}} + (F_\theta r\,d\phi\,dr)_{\mathrm{V}} - (F_\theta r\,d\phi\,dr)_{\mathrm{II}} \\ &+ (F_\phi r \sin\phi\,d\theta\,dr)_{\mathrm{VI}} - (F_\phi r \sin\phi\,d\theta\,dr)_{\mathrm{III}} \\ &= \frac{\partial(r^2F_r)}{\partial r}dr \sin\phi\,d\theta\,d\phi + \frac{\partial F_\theta}{\partial\theta}d\theta\,r\,d\phi\,dr + \frac{\partial(F_\phi \sin\phi)}{\partial\phi}d\phi\,r\,d\theta\,dr\end{aligned}$$

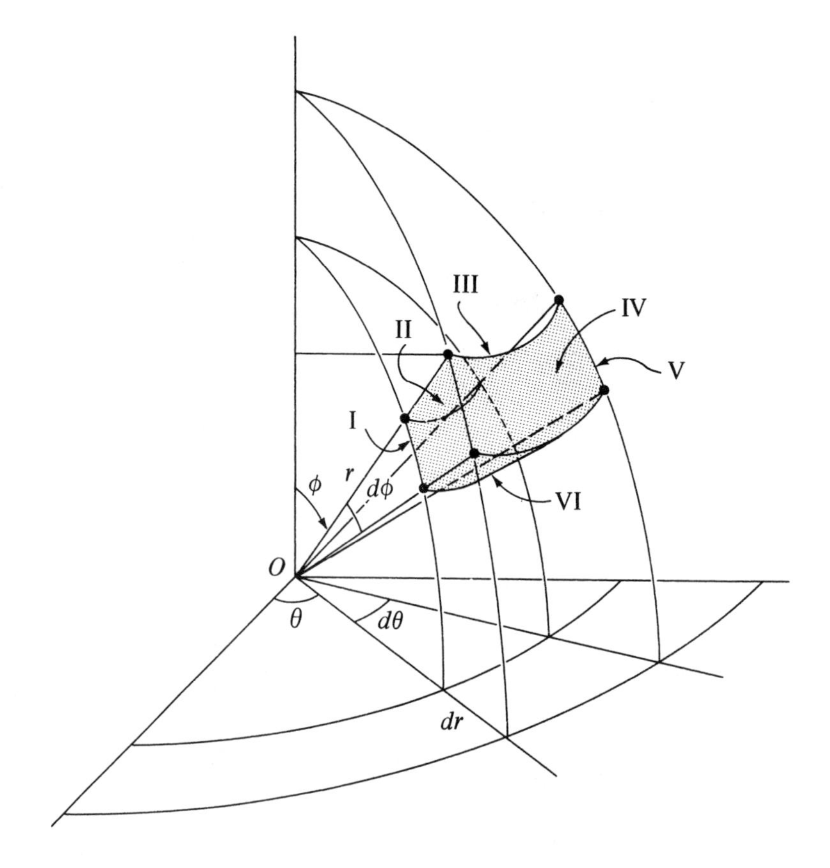
III
II
IV
V
I
VI
ϕ
r
$d\phi$
O
θ
$d\theta$
dr

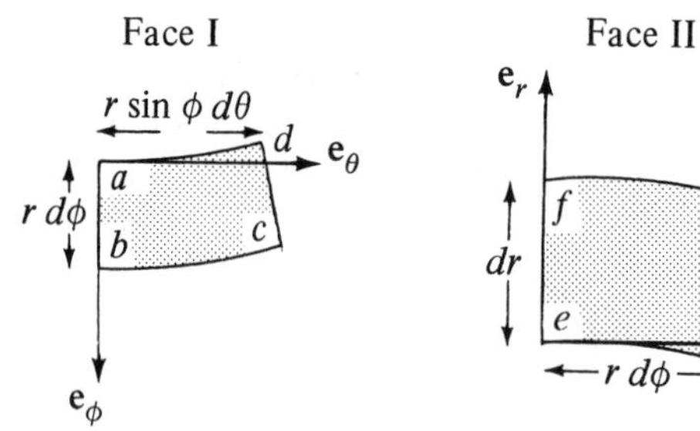
Face I
$r \sin \phi \, d\theta$
d
$\mathbf{e}_\theta$
a
$r\, d\phi$
b
c
$\mathbf{e}_\phi$

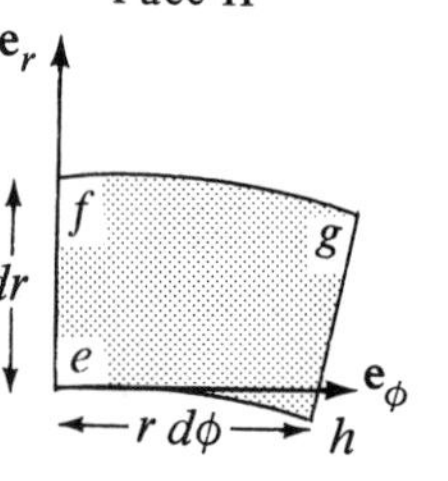
Face II
$\mathbf{e}_r$
f
g
dr
e
$\mathbf{e}_\phi$
$r\, d\phi$
h

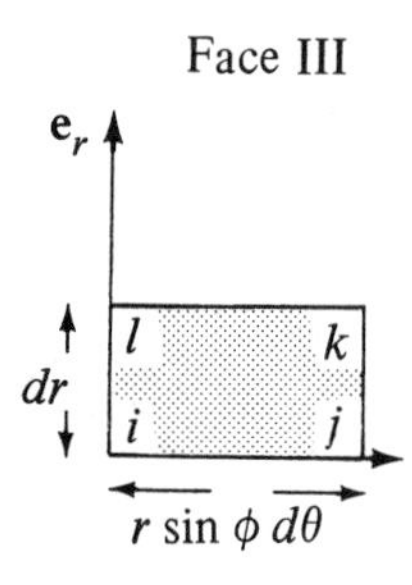
Face III
$\mathbf{e}_r$
l
k
dr
i
j
$r \sin \phi \, d\theta$

Figure 3.26

(keeping careful track of which variable changes from face to face). Dividing by the volume element eq. (3.65), we find that *the expression for the divergence of a vector field in spherical coordinates is given by*

$$\text{div}\,\mathbf{F} = \frac{1}{r^2}\frac{\partial(r^2F_r)}{\partial r} + \frac{1}{r\sin\phi}\frac{\partial F_\theta}{\partial \theta} + \frac{1}{r\sin\phi}\frac{\partial(F_\phi \sin\phi)}{\partial\phi} \tag{3.67}$$

Analogous reasoning on figure 3.26 yields the expression for the curl. The swirl around the properly oriented edge of face I is

$$\begin{aligned}(F_\phi r\,d\phi)_{ab} &- (F_\phi r\,d\phi)_{cd} + (F_\theta r\sin\phi\,d\theta)_{bc} - (F_\theta r\sin\phi\,d\theta)_{da}\\ &= -\frac{\partial F_\phi}{\partial\theta}d\theta\,r\,d\phi + \frac{\partial(F_\theta\sin\phi)}{\partial\phi}d\phi\,r\,d\theta\end{aligned}$$

and dividing by the area $r^2 \sin\phi\, d\theta\, d\phi$, we obtain the $\mathbf{e}_r$ component of **curl F**:

$$\frac{1}{r\sin\phi}\left[\frac{\partial(F_\theta\sin\phi)}{\partial\phi} - \frac{\partial F_\phi}{\partial\theta}\right]$$

The swirl around face II produces

$$\begin{aligned}(F_r\,dr)_{ef} &- (F_r\,dr)_{gh} + (F_\phi r\,d\phi)_{fg} - (F_\phi r\,d\phi)_{he}\\ &= -\frac{\partial F_r}{\partial\phi}d\phi\,dr + \frac{\partial(rF_\phi)}{\partial r}dr\,d\phi\end{aligned}$$

and dividing by the area $r\,d\phi\,dr$, we obtain the $\mathbf{e}_\theta$ component:

$$\frac{1}{r}\left[\frac{\partial(rF_\phi)}{\partial r} - \frac{\partial F_r}{\partial\phi}\right]$$

The swirl around face III produces

$$\begin{aligned}(F_\theta r\sin\phi\,d\theta)_{ij} &- (F_\theta r\sin\phi\,d\theta)_{kl} + (F_r\,dr)_{jk} - (F_r\,dr)_{li}\\ &= -\frac{\partial(rF_\theta)}{\partial r}dr\sin\phi\,d\theta + \frac{\partial F_r}{\partial\theta}d\theta\,dr\end{aligned}$$

and dividing by the area $r\sin\phi\,d\theta\,dr$ yields the $\mathbf{e}_\phi$ component:

$$\frac{1}{r\sin\phi}\left[\frac{\partial F_r}{\partial\theta} - \frac{\partial(rF_\theta)}{\partial r}\sin\phi\right]$$

The reader should verify that these results can be summarized thus: *the expression for the curl of a vector field in spherical coordinates is given by*

$$\textbf{curl F} = \frac{1}{r^2\sin\phi}\begin{vmatrix}\mathbf{e}_r & r\mathbf{e}_\phi & r\sin\phi\,\mathbf{e}_\theta\\ \dfrac{\partial}{\partial r} & \dfrac{\partial}{\partial\phi} & \dfrac{\partial}{\partial\theta}\\ F_r & rF_\phi & r\sin\phi\,F_\theta\end{vmatrix} \tag{3.68}$$

Throughout this chapter we have adopted the point of view that the gradient, divergence, and curl are more than mathematical combinations of derivatives. They represent physical characteristics of the fields—characteristics which can be described without reference to the coordinates. Thus observe that the following "definitions" make no reference to coordinates:

(*i*) the gradient of f is a vector pointing in the direction of maximal rate of increase of f with respect to distance, and has magnitude equal to this maximal rate;
(*ii*) the divergence of **F** is the net outflux of **F** per unit volume, in the limit of diminishing volumes;
(*iii*) the curl of **F** is twice the local angular velocity in the associated fluid velocity field.

This is the reason that the scale factors like ρ, $r \sin \phi$, etc. come into play in the curvilinear-coordinate formulas for the vector operators. If we simply took the cylindrical-coordinate gradient, for example, to be given by the array of partials,

$$\frac{\partial f}{\partial \rho}\mathbf{e}_\rho + \frac{\partial f}{\partial \theta}\mathbf{e}_\theta + \frac{\partial f}{\partial z}\mathbf{e}_z,$$

then it wouldn't possess the directional derivative property. Physicists, in fact, would no longer call it a *vector!*

In this light, then, it is reassuring to show directly that if we first compute the gradient in cartesian coordinates—where we fully comprehend its properties—and then express it in terms of $\mathbf{e}_\rho$, $\mathbf{e}_\theta$, and $\mathbf{e}_z$, the result is consistent with formula (3.55).

Example 3.31 Transform the cartesian coordinate expression for ∇f into cylindrical coordinates.

Solution This is a long computation. We start with $\nabla f = \partial f/\partial x\, \mathbf{i} + \partial f/\partial y\, \mathbf{j} + \partial f/\partial z\, \mathbf{k}$. By the chain rule for calculus we have

$$\frac{\partial f}{\partial x} = \frac{\partial f}{\partial \rho}\frac{\partial \rho}{\partial x} + \frac{\partial f}{\partial \theta}\frac{\partial \theta}{\partial x} + \frac{\partial f}{\partial z}\frac{\partial z}{\partial x}$$

with similar formulas for $\partial f/\partial y$ and $\partial f/\partial z$.

From the transformation equations (3.50) we compute (exercise 15)

$$\frac{\partial \rho}{\partial x} = \frac{x}{\rho} = \cos\theta, \quad \frac{\partial \theta}{\partial x} = -\frac{y}{\rho^2} = -\frac{\sin\theta}{\rho}, \quad \frac{\partial z}{\partial x} = 0;$$

$$\frac{\partial \rho}{\partial y} = \frac{y}{\rho} = \sin\theta, \quad \frac{\partial \theta}{\partial y} = \frac{x}{\rho^2} = \frac{\cos\theta}{\rho}, \quad \frac{\partial z}{\partial y} = 0;$$

$$\frac{\partial \rho}{\partial z} = 0, \quad \frac{\partial \theta}{\partial z} = 0, \quad \frac{\partial z}{\partial z} = 1;$$

and with a little trigonometry we can express

$$\mathbf{i} = \cos\theta\,\mathbf{e}_\rho - \sin\theta\,\mathbf{e}_\theta,\ \mathbf{j} = \sin\theta\,\mathbf{e}_\rho + \cos\theta\,\mathbf{e}_\theta,\ \mathbf{k} = \mathbf{e}_z$$

(recall example 3.28). Assembling all this we find

$$\frac{\partial f}{\partial x}\mathbf{i} + \frac{\partial f}{\partial y}\mathbf{j} + \frac{\partial f}{\partial z}\mathbf{k} =$$

$$\left\{\frac{\partial f}{\partial \rho}\cos\theta + \frac{\partial f}{\partial \theta}\left(-\frac{\sin\theta}{\rho}\right)\right\}\left\{(\cos\theta\,\mathbf{e}_\rho - \sin\theta\,\mathbf{e}_\theta\right\}$$
$$+\left\{\frac{\partial f}{\partial \rho}\sin\theta + \frac{\partial f}{\partial \theta}\frac{\cos\theta}{\rho}\right\}\left\{\sin\theta\,\mathbf{e}_\rho + \cos\theta\,\mathbf{e}_\theta\right\} + \frac{\partial f}{\partial z}\mathbf{e}_z,$$

which simplifies to

$$\frac{\partial f}{\partial x}\mathbf{i} + \frac{\partial f}{\partial y}\mathbf{j} + \frac{\partial f}{\partial z}\mathbf{k} = \frac{\partial f}{\partial \rho}\mathbf{e}_\rho + \frac{1}{\rho}\frac{\partial f}{\partial \theta}\mathbf{e}_\theta + \frac{\partial f}{\partial z}\mathbf{e}_z,$$

in keeping with formula (3.55).

This same type of verification can be carried out for each of the vector operators in cylindrical, spherical, or other curvilinear coordinate systems (described in the next section). As a matter of fact, in abstract texts, vectors (and tensors) are *defined* in terms of their transformation properties! This requirement of transformation "co-variance," when expressed in a four-dimensional space-time framework, lies at the heart of the formulation of the physics of relativity.

For reference purposes the formulas derived in this section will be listed, together with their generalizations, at the end of the next section.

EXERCISES

1. Derive the equations of transformation between cylindrical and spherical coordinates.
2. Use eqs. (3.50) and (3.51) to derive

$$\begin{aligned}\mathbf{e}_z &= \mathbf{k}\\ \mathbf{e}_\rho &= \frac{x\mathbf{i} + y\mathbf{j}}{(x^2+y^2)^{1/2}}\\ \mathbf{e}_\theta &= \frac{-y\mathbf{i} + x\mathbf{j}}{(x^2+y^2)^{1/2}}\end{aligned} \tag{3.69}$$

3. Verify eq. (3.60).

4. Use eqs. (3.61) and (3.62) to derive

$$\mathbf{e}_r = \frac{x\mathbf{i} + y\mathbf{j} + z\mathbf{k}}{(x^2 + y^2 + z^2)^{1/2}}$$

$$\mathbf{e}_\phi = \frac{z(x\mathbf{i} + y\mathbf{j}) - (x^2 + y^2)\mathbf{k}}{(x^2 + y^2)^{1/2}(x^2 + y^2 + z^2)^{1/2}}$$

$$\mathbf{e}_\theta = \frac{-y\mathbf{i} + x\mathbf{j}}{(x^2 + y^2)^{1/2}}$$

5. Verify eq. (3.68).
6. Compute the laplacian $\nabla^2 f$ in cylindrical and spherical coordinates. (*Hint:* Use $\nabla^2 = \text{div}\ \textbf{grad}$.)
7. Show that if f is a function of r only, then

$$\nabla^2 f(r) = f''(r) + \frac{2}{r}f'(r)$$

8. Change to cylindrical coordinates and find the divergence and curl of
   (a) $\mathbf{F} = \dfrac{x\mathbf{i} + y\mathbf{j}}{x^2 + y^2}$
   (b) $\mathbf{F} = \dfrac{-y\mathbf{i} + x\mathbf{j}}{x^2 + y^2}$
   [*Hint:* Observe eq. (3.69).]
9. What is the arc length of the curve $r = \sin\phi$, $\theta = \pi/2$, for $0 \le \phi \le \pi$?
10. Evaluate $\nabla(r^n)$.
11. Compute the gradient, in spherical coordinates, of $f(r, \phi, \theta) = \cos\phi/r^2$.
12. Compute the divergence and curl, in spherical coordinates, of $\mathbf{F}(r, \phi, \theta) = \mathbf{e}_r + r\mathbf{e}_\phi + r\cos\phi\ \mathbf{e}_\theta$.
13. For what values of n does $\nabla \cdot (r^n\mathbf{e}_r) = 0$?
14. For what values of n does $\nabla \times (r^n\mathbf{e}_r) = \mathbf{0}$?
15. Verify the calculations in example 3.31. [eqs. (2.47) may be helpful.]
16. Verify the expression (3.58) for the divergence by transforming the cartesian expression as in example 3.31.

3.11 Optional Reading: Orthogonal Curvilinear Coordinates

Our experience with the cylindrical and spherical coordinate systems places us in a good position to analyze general coordinate systems, or *curvilinear coordinates.*

The general situation is this: each point in a certain region of space is specified by three numbers (u_1,u_2,u_3), called the curvilinear coordinates of the point. Possibly the numbers can be interpreted as lengths or angles, but no such geometric visualization is required. All that is needed are the transformation equations between the curvilinear coordinates and cartesian coordinates, which we represent by

$$\begin{aligned} x &= x(u_1,u_2,u_3) \qquad & u_1 &= u_1(x,y,z) \\ y &= y(u_1,u_2,u_3) \qquad & u_2 &= u_2(x,y,z) \\ z &= z(u_1,u_2,u_3) \qquad & u_3 &= u_3(x,y,z) \end{aligned} \tag{3.70}$$

Equations (3.70) include eqs. (3.50) and (3.61) as special instances.

Observe that it is not feasible to choose the functions u_1, u_2, and u_3 arbitrarily. For example, the system $u_1 = x^2$, $u_2 = y - z$, $u_3 = 2y - 2z$ is unsatisfactory because one cannot invert the equations; in fact, the points $(x,y,z) = (1,2,3)$ and $(x,y,z) = (1,3,4)$ would have identical curvilinear coordinates $(1, -1, -2)$. Therefore, we stipulate that the functions defining u_1, u_2, and u_3 assign different ordered triples to different points in the region of interest. We also assume that they possess continuous partial derivatives of all orders and that at every point P the gradients of these functions are nonzero.

(Sometimes we do not require that the coordinates satisfy these requirements at every point in space. For example, if we pass through the z axis along a line parallel to the x axis, the spherical coordinate θ undergoes a discontinuous jump from 0 to π. We shall generally ignore this difficulty or work only in a subdomain where the conditions are satisfied.)

These conditions ensure that through any point P in the domain, having curvilinear coordinates equal to (c_1,c_2,c_3), will pass three isotimic surfaces $u_1(x,y,z) = c_1$, $u_2(x,y,z) = c_2$, $u_3(x,y,z) = c_3$. As illustrated in figure 3.27, these surfaces intersect in pairs to give three curves passing through P, along each of which only one coordinate varies; these are the coordinate curves. The normal to the surface $u_i = c_i$ is the gradient

$$\nabla u_i = \frac{\partial u_i}{\partial x}\mathbf{i} + \frac{\partial u_i}{\partial y}\mathbf{j} + \frac{\partial u_i}{\partial z}\mathbf{k} \tag{3.71}$$

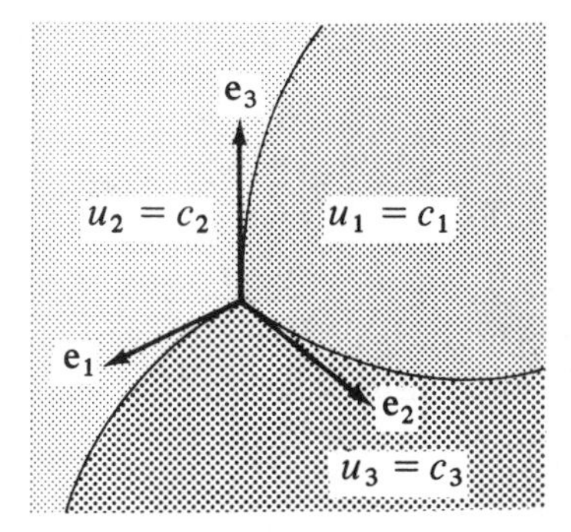

Figure 3.27

and a tangent to the coordinate curve for u_i is the vector

$$\frac{\partial \mathbf{R}}{\partial u_i} = \frac{\partial x}{\partial u_i}\mathbf{i} + \frac{\partial y}{\partial u_i}\mathbf{j} + \frac{\partial z}{\partial u_i}\mathbf{k} \qquad (3.72)$$

Recall that in the cases of cylindrical and spherical coordinates, the three normals to the isotimic surfaces were mutually perpendicular. In general, whenever the vectors ∇u_1, ∇u_2, and ∇u_3 are mutually orthogonal at every point, we say that (u_1,u_2,u_3) comprise *orthogonal curvilinear coordinates.* In this section we shall restrict our analysis to such systems. Moreover, we assume that the u_i's are numbered so that ∇u_1, ∇u_2, and ∇u_3 (in that order) form a right-handed system.

Another feature of the special coordinate systems we studied in the last section is that each gradient vector ∇u_i was seen to be parallel to the tangent vector $\partial \mathbf{R}/\partial u_i$ for the corresponding coordinate curve. This is always true for orthogonal curvilinear coordinates: *any coordinate curve for u_i intersects the isotimic surface $u_i = c_i$ at right angles when* (u_1,u_2,u_3) *form orthogonal curvilinear coordinates.* To see this, consider, say, a coordinate curve for u_1.

(*i*) This curve is the intersection of two surfaces $u_2 = c_2$ and $u_3 = c_3$. Hence its tangent $\partial \mathbf{R}/\partial u_1$ is perpendicular to both surface normals ∇u_2 and ∇u_3.
(*ii*) The vector ∇u_1 is also perpendicular to ∇u_2 and ∇u_3, by definition of orthogonal curvilinear coordinates.
(*iii*) This implies that $\partial \mathbf{R}/\partial u_1$ is parallel or, perhaps, antiparallel to ∇u_1.

Since both point in the direction of increasing u_1, they are parallel.

It follows, of course, that the vectors $\partial \mathbf{R}/\partial u_1$, $\partial \mathbf{R}/\partial u_2$, and $\partial \mathbf{R}/\partial u_3$ also form a right-handed system of mutually orthogonal vectors. In fact, by the chain rule

$$\begin{aligned}(\nabla u_i)\cdot\left(\frac{\partial \mathbf{R}}{\partial u_j}\right) &= \frac{\partial u_i}{\partial x}\frac{\partial x}{\partial u_j} + \frac{\partial u_i}{\partial y}\frac{\partial y}{\partial u_j} + \frac{\partial u_i}{\partial z}\frac{\partial z}{\partial u_j} \\ &= \frac{\partial u_i}{\partial u_j} = \begin{cases} 1 & \text{if } i = j \\ 0 & \text{if } i \neq j \end{cases}\end{aligned}$$

Thus it is natural to define the right-handed system of mutually orthogonal unit vectors $(\mathbf{e}_1,\mathbf{e}_2,\mathbf{e}_3)$ by

$$\mathbf{e}_i = \frac{\nabla u_i}{|\nabla u_i|} = \frac{\partial \mathbf{R}/\partial u_i}{|\partial \mathbf{R}/\partial u_i|} \qquad (i = 1, 2, 3) \qquad (3.73)$$

The vectors $(\mathbf{e}_\rho,\mathbf{e}_\theta,\mathbf{e}_z)$ and $(\mathbf{e}_r,\mathbf{e}_\phi,\mathbf{e}_\theta)$ are special instances of eq. (3.73).

To express the vector operations in general orthogonal curvilinear coordinates, we need to evaluate three functions h_i known as the *scale factors. The scale factor h_i is defined to be the rate at which arc length increases on the ith coordinate curve,* *with respect to* u_i. In other words, if s_i denotes arc length on the ith coordinate curve, measured in the direction of increasing u_i, then

$$h_1 = \frac{ds_1}{du_1} \qquad h_2 = \frac{ds_2}{du_2} \qquad h_3 = \frac{ds_3}{du_3} \tag{3.74}$$

Since arc length in general can be expressed

$$ds = |d\mathbf{R}| = \left| \frac{\partial \mathbf{R}}{\partial u_1} du_1 + \frac{\partial \mathbf{R}}{\partial u_2} du_2 + \frac{\partial \mathbf{R}}{\partial u_3} du_3 \right|$$

we see that

$$h_i = \left| \frac{\partial \mathbf{R}}{\partial u_i} \right| \qquad (i = 1, 2, 3) \tag{3.75}$$

Combining the last two equations shows that the displacement vector can be expressed in terms of the scale factors by

$$d\mathbf{R} = h_1\, du_1\, \mathbf{e}_1 + h_2\, du_2\, \mathbf{e}_2 + h_3\, du_3\, \mathbf{e}_3 \tag{3.76}$$

We can get another formula for the scale factor h_i by the following observations:

(*i*) $|\nabla u_i|$ is the rate of change of u_i with respect to distance in the direction of ∇u_i.
(*ii*) The direction of ∇u_i is the direction of the coordinate curve for u_i.
(*iii*) s_i measures distance along the coordinate curve for u_i.

It follows that

$$|\nabla u_i| = \frac{du_i}{ds_i} = \frac{1}{h_i}$$

Therefore

$$h_1 = \frac{1}{|\nabla u_1|} \qquad h_2 = \frac{1}{|\nabla u_2|} \qquad h_3 = \frac{1}{|\nabla u_3|} \tag{3.77}$$

Example 3.32 Consider the curvilinear coordinate system defined for $z \geq 0$ by

$$\begin{aligned} x &= u_1 - u_2 \\ y &= u_1 + u_2 \\ z &= u_3^2 \end{aligned} \tag{3.78}$$

Verify that the system is orthogonal and right-handed, and compute the unit vectors $\mathbf{e}_i$ and the scale factors h_i.

Solution We do not need the inverse equations for this example, but they are easy to derive:

$$u_1 = \frac{x + y}{2}$$

$$u_2 = \frac{y - x}{2}$$

$$u_3 = z^{1/2} \qquad \textit{(Take the positive square root for definiteness.)}$$

We use the right-hand expression in eq. (3.73) to compute the $\mathbf{e}_i$:

$$\mathbf{e}_1 = \frac{\mathbf{i} + \mathbf{j}}{\sqrt{2}}$$

$$\mathbf{e}_2 = \frac{-\mathbf{i} + \mathbf{j}}{\sqrt{2}}$$

$$\mathbf{e}_3 = \frac{2u_3\mathbf{k}}{|2u_3|} = \mathbf{k} \tag{3.79}$$

Clearly this set is right-handed and orthogonal, so eqs. (3.78) do, in fact, define orthogonal curvilinear coordinates. The h_i have already been computed in the denominators of eqs. (3.79):

$$h_1 = \sqrt{2} \qquad h_2 = \sqrt{2} \qquad h_3 = 2u_3 \tag{3.80}$$

Example 3.33 Compute the scale factors for cylindrical and spherical coordinates.

Solution Comparing eqs. (3.76), (3.53), and (3.64), respectively, we have

$$h_\rho = 1 \qquad h_\theta = \rho \qquad h_z = 1$$

$$h_r = 1 \qquad h_\phi = r \qquad h_\theta = r \sin \phi$$

Recall that these were obtained simply by examining the diagrams; the computations in eq. (3.75) for the h_i are often unnecessary when the curvilinear coordinates can be visualized.

The scale factors will allow us to write general formulas for arc length, volume, gradient, divergence, and curl in terms of curvilinear coordinates. We will present heuristic derivations for these expressions.

From the discussion above, we can say that ds_i is the arc length along the ith coordinate curve, corresponding to a change in the ith coordinate from u_i to $u_i + du_i$. Since an arbitrary displacement $d\mathbf{R}$ is generated by changes du_1, du_2, and du_3, each in mutually perpendicular directions, we can express the element of arc length $|d\mathbf{R}|$ by the pythagorean theorem as

$$|d\mathbf{R}|^2 = ds_1^2 + ds_2^2 + ds_3^2$$

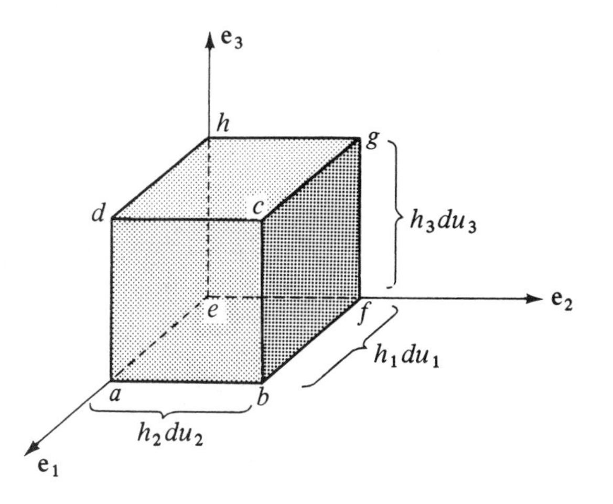

Figure 3.28

Using the scale factors from eq. (3.74), we find that *the arc length along a curve C is given by the line integral*

$$\int_C |d\mathbf{R}| = \int ds = \int \sqrt{(h_1\, du_1)^2 + (h_2\, du_2)^2 + (h_3\, du_3)^2} \qquad (3.81)$$

generalizing the formula in section 2.2.

From figure 3.28, we observe that the solid generated by displacements du_1, du_2, and du_3 is approximately a rectangular parallelepiped with edges $h_1\, du_1$, $h_2\, du_2$, and $h_3\, du_3$. Its volume is therefore

$$dV = h_1 h_2 h_3\, du_1\, du_2\, du_3 \qquad (3.82)$$

In spherical coordinates, $dV = r^2 \sin\phi\, dr\, d\theta\, d\phi$; in cylindrical coordinates, $dV = \rho\, d\rho\, d\theta\, dz$.

In section 3.1 it was shown that the component of **grad** f in the $\mathbf{e}_1$ direction is given by df/ds_1, the rate of change of f with respect to distance in the $\mathbf{e}_1$ direction. Since $\mathbf{e}_1$, $\mathbf{e}_2$, and $\mathbf{e}_3$ are mutually orthogonal unit vectors, we can immediately express **grad** f in terms of these:

$$\mathbf{grad}\, f = \frac{df}{ds_1}\mathbf{e}_1 + \frac{df}{ds_2}\mathbf{e}_2 + \frac{df}{ds_3}\mathbf{e}_3$$

or, introducing the scale factors,

$$\mathbf{grad}\, f = \frac{1}{h_1}\frac{\partial f}{\partial u_1}\mathbf{e}_1 + \frac{1}{h_2}\frac{\partial f}{\partial u_2}\mathbf{e}_2 + \frac{1}{h_3}\frac{\partial f}{\partial u_3}\mathbf{e}_3 \qquad (3.83)$$

Example 3.34 Compute **grad** f in the coordinate system in eq. (3.78), for $f(u_1,u_2,u_3) = u_1u_2 + u_3^2$.

Solution From eqs. (3.83) and (3.80),

$$\begin{aligned}\nabla f &= \frac{1}{\sqrt{2}}u_2\mathbf{e}_1 + \frac{1}{\sqrt{2}}u_1\mathbf{e}_2 + \frac{1}{2u_3}2u_3\mathbf{e}_3 \\ &= (u_2\mathbf{e}_1 + u_1\mathbf{e}_2)\frac{\sqrt{2}}{2} + \mathbf{e}_3\end{aligned}$$

The expression for divergence is more complicated. Let

$$\mathbf{F} = F_1\mathbf{e}_1 + F_2\mathbf{e}_2 + F_3\mathbf{e}_3$$

be the vector field, given in terms of the unit vectors $\mathbf{e}_1$, $\mathbf{e}_2$, and $\mathbf{e}_3$. We will calculate div **F** as the flux of **F** out of the sides of the box in figure 3.28, divided by the volume of the box, in accordance with the interpretation of divergence given in section 3.3.

The flux density normal to the face *abcd* is $\mathbf{F} \cdot \mathbf{e}_1 = F_1$, and the area of this face is $h_2h_3\,du_2\,du_3$. Therefore the flux *out*ward is $F_1h_2h_3\,du_2\,du_3$. The unit *out*ward normal to face *efgh* is $-\mathbf{e}_1$, so that the flux outward from that face is $-F_1h_2h_3\,du_2\,du_3$. Since F_1, h_2, and h_3 are functions of u_1 as we move along the u_1-coordinate curve, the sum of these two is approximately

$$\left[\frac{\partial}{\partial u_1}(F_1h_2h_3)\,du_1\right]du_2\,du_3$$

From this and similar expressions for the other two pairs of faces, we see that the net flux outward from the parallelepiped is approximately

$$\left[\frac{\partial}{\partial u_1}(F_1h_2h_3) + \frac{\partial}{\partial u_2}(F_2h_1h_3) + \frac{\partial}{\partial u_3}(F_3h_1h_2)\right]du_1\,du_2\,du_3$$

and so the flux output per unit volume is this expression divided by the volume $h_1h_2h_3\,du_1\,du_2\,du_3$. Hence

$$\text{div}\,\mathbf{F} = \frac{1}{h_1h_2h_3}\left[\frac{\partial}{\partial u_1}(F_1h_2h_3) + \frac{\partial}{\partial u_2}(F_2h_1h_3) + \frac{\partial}{\partial u_3}(F_3h_1h_2)\right] \tag{3.84}$$

Combining eqs. (3.83) and (3.84), we have the expression for the laplacian:

$$\begin{aligned}\nabla^2 f &= \text{div}\,\mathbf{grad}\,f \\ &= \frac{1}{h_1h_2h_3}\left[\frac{\partial}{\partial u_1}\left(\frac{h_2h_3}{h_1}\frac{\partial f}{\partial u_1}\right) + \frac{\partial}{\partial u_2}\left(\frac{h_1h_3}{h_2}\frac{\partial f}{\partial u_2}\right) + \frac{\partial}{\partial u_3}\left(\frac{h_1h_2}{h_3}\frac{\partial f}{\partial u_3}\right)\right]\end{aligned} \tag{3.85}$$

Now let us find the expression for **curl F.** We shall use the "swirl" characterization described in section 3.4. The component of **curl F** in the direction $\mathbf{e}_1$ will be the swirl of $\mathbf{F}$ around the curve *efghe* in figure 3.28, divided by the area enclosed by this curve. The contribution of the edge *ef* is approximately

$$F_2(u_1,u_2,u_3)h_2(u_1,u_2,u_3)\,du_2$$

Along *gh* we are proceeding in the opposite direction; furthermore, the third coordinate is now $u_3 + du_3$, so the contribution is

$$-F_2(u_1, u_2, u_3 + du_3)h_2(u_1, u_2, u_3 + du_3)\,du_2$$

Thus the net contribution from *ef* and *gh* is given by

$$-\frac{\partial}{\partial u_3}(F_2h_2)\,du_3\,du_2$$

Similarly, the contribution from *fg* and *he* is

$$\frac{\partial}{\partial u_2}(F_3h_3)\,du_2\,du_3$$

Dividing by the area $h_2\,du_2\,h_3\,du_3$, we have

$$(\mathbf{curl\ F})\cdot\mathbf{e}_1 = \frac{1}{h_2h_3}\left(\frac{\partial}{\partial u_2}(F_3h_3) - \frac{\partial}{\partial u_3}(F_2h_2)\right)$$

Reasoning similarly for the other components, we find that the curl is given by

$$\begin{aligned}\mathbf{curl\ F} &= \frac{1}{h_2h_3}\left(\frac{\partial}{\partial u_2}(F_3h_3) - \frac{\partial}{\partial u_3}(F_2h_2)\right)\mathbf{e}_1 \\ &\quad + \frac{1}{h_1h_3}\left(\frac{\partial}{\partial u_3}(F_1h_1) - \frac{\partial}{\partial u_1}(F_3h_3)\right)\mathbf{e}_2 \\ &\quad + \frac{1}{h_1h_2}\left(\frac{\partial}{\partial u_1}(F_2h_2) - \frac{\partial}{\partial u_2}(F_1h_1)\right)\mathbf{e}_3 \\ &= \frac{1}{h_1h_2h_3}\begin{vmatrix} h_1\mathbf{u}_1 & h_2\mathbf{u}_2 & h_3\mathbf{u}_3 \\ \dfrac{\partial}{\partial u_1} & \dfrac{\partial}{\partial u_2} & \dfrac{\partial}{\partial u_3} \\ F_1h_1 & F_2h_2 & F_3h_3 \end{vmatrix} \end{aligned} \tag{3.86}$$

Example 3.35 Compute the divergence and curl of the vector field

$$\mathbf{F}(u_1,u_2,u_3) = u_3u_1\mathbf{e}_1 + u_3u_2\mathbf{e}_2 + u_1u_2\mathbf{e}_3$$

in the coordinate system of eq. (3.78).

Solution From eqs. (3.84) and (3.80),

$$\nabla \cdot \mathbf{F} = \frac{1}{4u_3}\left[\frac{\partial}{\partial u_1}(u_3 u_1 2\sqrt{2}u_3) + \frac{\partial}{\partial u_2}(u_3 u_2 2\sqrt{2}u_3) + \frac{\partial}{\partial u_3}(2u_1 u_2)\right]$$
$$= \sqrt{2}u_3$$

From eq. (3.86),

$$\nabla \times \mathbf{F} = \frac{1}{4u_3}\begin{vmatrix} \sqrt{2}\mathbf{e}_1 & \sqrt{2}\mathbf{e}_2 & 2u_3\mathbf{e}_3 \\ \dfrac{\partial}{\partial u_1} & \dfrac{\partial}{\partial u_2} & \dfrac{\partial}{\partial u_3} \\ \sqrt{2}u_3 u_1 & \sqrt{2}u_3 u_2 & 2u_1 u_2 u_3 \end{vmatrix}$$
$$= \frac{(2u_1u_3 - \sqrt{2}u_2)\sqrt{2}}{4u_3}\mathbf{e}_1 + \frac{(\sqrt{2}u_1 - 2u_2u_3)\sqrt{2}}{4u_3}\mathbf{e}_2$$

For reference purposes we list the various vector operations in general orthogonal curvilinear coordinates, cylindrical coordinates, and spherical coordinates together here. The reader is advised to attach a third, and final, permanent bookmark to this page.

General Orthogonal Curvilinear Coordinates

Scale factors:

$$h_i = \left|\frac{\partial \mathbf{R}}{\partial u_i}\right| = \frac{1}{|\nabla u_i|} \qquad (i = 1, 2, 3)$$

Displacement vector:

$$d\mathbf{R} = h_1\, du_1\, \mathbf{e}_1 + h_2\, du_2\, \mathbf{e}_2 + h_3\, du_3\, \mathbf{e}_3$$

Arc length:

$$ds = (h_1^2\, du_1^2 + h_2^2\, du_2^2 + h_3^2\, du_3^2)^{1/2}$$

Volume element:

$$dV = h_1 h_2 h_3\, du_1\, du_2\, du_3$$

Gradient:

$$\nabla f = \frac{1}{h_1}\frac{\partial f}{\partial u_1}\mathbf{e}_1 + \frac{1}{h_2}\frac{\partial f}{\partial u_2}\mathbf{e}_2 + \frac{1}{h_3}\frac{\partial f}{\partial u_3}\mathbf{e}_3$$

Divergence:

$$\nabla \cdot \mathbf{F} = \frac{1}{h_1h_2h_3}\left[\frac{\partial}{\partial u_1}(F_1h_2h_3) + \frac{\partial}{\partial u_2}(F_2h_1h_3) + \frac{\partial}{\partial u_3}(F_3h_1h_2)\right]$$

Curl:

$$\nabla \times \mathbf{F} = \frac{1}{h_1h_2h_3}\begin{vmatrix} h_1\mathbf{e}_1 & h_2\mathbf{e}_2 & h_3\mathbf{e}_3 \\ \dfrac{\partial}{\partial u_1} & \dfrac{\partial}{\partial u_2} & \dfrac{\partial}{\partial u_3} \\ F_1h_1 & F_2h_2 & F_3h_3 \end{vmatrix}$$

Laplacian:

$$\nabla^2 f = \frac{1}{h_1h_2h_3}\left[\frac{\partial}{\partial u_1}\left(\frac{h_2h_3}{h_1}\frac{\partial f}{\partial u_1}\right) + \frac{\partial}{\partial u_2}\left(\frac{h_1h_3}{h_2}\frac{\partial f}{\partial u_2}\right) + \frac{\partial}{\partial u_3}\left(\frac{h_1h_2}{h_3}\frac{\partial f}{\partial u_3}\right)\right]$$

Cylindrical Coordinates

Displacement vector:

$$d\mathbf{R} = d\rho\, \mathbf{e}_\rho + \rho\, d\theta\, \mathbf{e}_\theta + dz\, \mathbf{e}_z$$

Arc length:

$$ds = (d\rho^2 + \rho^2\, d\theta^2 + dz^2)^{1/2}$$

Volume element:

$$dV = \rho\, d\rho\, d\theta\, dz$$

Gradient:

$$\nabla f = \frac{\partial f}{\partial \rho}\mathbf{e}_\rho + \frac{1}{\rho}\frac{\partial f}{\partial \theta}\mathbf{e}_\theta + \frac{\partial f}{\partial z}\mathbf{e}_z$$

Divergence:

$$\nabla \cdot \mathbf{F} = \frac{1}{\rho}\frac{\partial(\rho F_\rho)}{\partial \rho} + \frac{1}{\rho}\frac{\partial F_\theta}{\partial \theta} + \frac{\partial F_z}{\partial z}$$

Curl:

$$\nabla \times \mathbf{F} = \frac{1}{\rho}\begin{vmatrix} \mathbf{e}_\rho & \rho\mathbf{e}_\theta & \mathbf{e}_z \\ \dfrac{\partial}{\partial \rho} & \dfrac{\partial}{\partial \theta} & \dfrac{\partial}{\partial z} \\ F_\rho & \rho F_\theta & F_z \end{vmatrix}$$

Laplacian:

$$\nabla^2 f = \frac{1}{\rho}\frac{\partial}{\partial \rho}\left(\rho\frac{\partial f}{\partial \rho}\right) + \frac{1}{\rho^2}\frac{\partial^2 f}{\partial \theta^2} + \frac{\partial^2 f}{\partial z^2}$$

Spherical Coordinates

Displacement vector:

$$d\mathbf{R} = dr\,\mathbf{e}_r + r\,d\phi\,\mathbf{e}_\phi + r\sin\phi\,d\theta\,\mathbf{e}_\theta$$

Arc length:

$$ds = (dr^2 + r^2\,d\phi^2 + r^2\sin^2\phi\,d\theta^2)^{1/2}$$

Volume element:

$$dV = r^2\sin\phi\,dr\,d\phi\,d\theta$$

Gradient:

$$\nabla f = \frac{\partial f}{\partial r}\mathbf{e}_r + \frac{1}{r}\frac{\partial f}{\partial \phi}\mathbf{e}_\phi + \frac{1}{r\sin\phi}\frac{\partial f}{\partial \theta}\mathbf{e}_\theta$$

Divergence:

$$\nabla\cdot\mathbf{F} = \frac{1}{r^2}\frac{\partial}{\partial r}(r^2F_r) + \frac{1}{r\sin\phi}\frac{\partial}{\partial\phi}(F_\phi\sin\phi) + \frac{1}{r\sin\phi}\frac{\partial F_\theta}{\partial\theta}$$

Curl:

$$\nabla\times\mathbf{F} = \frac{1}{r^2\sin\phi}\begin{vmatrix} \mathbf{e}_r & r\mathbf{e}_\phi & (r\sin\phi)\mathbf{e}_\theta \\ \dfrac{\partial}{\partial r} & \dfrac{\partial}{\partial\phi} & \dfrac{\partial}{\partial\theta} \\ F_r & rF_\phi & (r\sin\phi)F_\theta \end{vmatrix}$$

Laplacian:

$$\nabla^2 f = \frac{1}{r^2}\frac{\partial}{\partial r}\left(r^2\frac{\partial f}{\partial r}\right) + \frac{1}{r^2\sin\phi}\frac{\partial}{\partial\phi}\left(\sin\phi\frac{\partial f}{\partial\phi}\right) + \frac{1}{r^2\sin^2\phi}\frac{\partial^2 f}{\partial\theta^2}$$

EXERCISES

1. Verify that the formulas for the vector operations in cylindrical and spherical coordinates, as computed in section 3.10, are instances of the general formulas derived in this section when the scale factors from Example 3.33 are inserted.

2. Verify Example 3.34 by expressing f in cartesian coordinates, applying ∇, and transforming.
3. Explain why curvilinear coordinates defined by functions of the form

$$u_1 = u_1(z) \qquad u_2 = u_2(x) \qquad u_3 = u_3(y)$$

are automatically orthogonal. What other combinations have this property? How about the following?

$$u_1 = u_1(\rho) \qquad u_2 = u_2(\theta) \qquad u_3 = u_3(z)$$

4. What is the element of volume relative to the coordinate system $u_1 = e^x$, $u_2 = y$, $u_3 = z$?
5. Compute $\nabla^2 g$ if $g = u_1^3 + u_2^3 + u_3^3$ in the coordinate system in eq. (3.78).
6. Let $u_1 = x + y$, $u_2 = x - y$, and $u_3 = 2z$.
   (a) Is this an orthogonal coordinate system?
   (b) Solve for x, y, and z in terms of u_1, u_2, and u_3.
   (c) Find ds^2 and hence determine h_1, h_2, and h_3 for this coordinate system.
   (d) What is the laplacian relative to this coordinate system?
   (e) Let $f(u_1,u_2,u_3) = u_1 + u_2 + 2u_3$. Find **grad** f.
7. Let $u_1 = x + y$, $u_2 = x - 2y$, and $u_3 = 2z$.
   (a) Solve for x, y, and z in terms of u_1, u_2, and u_3.
   (b) Attempt to determine the scale factors h_1, h_2, and h_3.
   (c) What is "wrong"?
8. Consider the transformation

$$x = u_1^2 - u_2^2$$
$$y = 2u_1u_2$$
$$z = u_3$$

   (a) Show that (u_1, u_2, u_3) form right-handed orthogonal curvilinear coordinates.
   (b) Compute the scale factors.
   (c) Express $\nabla^2 g(u_1, u_2, u_3)$.
   (d) Find the divergence and curl of the vector field $\mathbf{F} = u_3\mathbf{e}_1 + u_1\mathbf{e}_2 + u_2\mathbf{e}_3$.
9. Consider the transformation

$$x = u_3$$
$$y = \exp(u_2) \cos u_1$$
$$z = \exp(u_2) \sin u_1$$

   (a) Show that (u_1, u_2, u_3) constitute orthogonal curvilinear coordinates.
   (b) Compute the scale factors.
   (c) Find $\nabla^2 g$ if $g = u_1^2 + u_2^2 + u_3^2$.
   (d) Find the divergence and curl of the vector field $\mathbf{F} = -\exp(u_2)\, \mathbf{e}_3 + u_3\, \mathbf{e}_1$.

10. Consider the coordinate system $u_1 = y$, $u_2 = x$, $u_3 = z$. The scale factors are all equal to unity, so that eq. (3.86) takes an especially simple form.

(a) Let $\mathbf{F} = -u_2\mathbf{e}_1 + u_1\mathbf{e}_2$. Show that eq. (3.86) gives **curl** $\mathbf{F} = 2\mathbf{e}_3$.

(b) Obviously $\mathbf{e}_1 = \mathbf{j}$, $\mathbf{e}_2 = \mathbf{i}$, and $\mathbf{e}_3 = \mathbf{k}$, so that $\mathbf{F} = y\mathbf{i} - x\mathbf{j}$ and, by part (a), **curl** $\mathbf{F} = 2\mathbf{k}$. But direct calculation of **curl F** in cartesian coordinates shows that **curl** $\mathbf{F} = -2\mathbf{k}$, not $2\mathbf{k}$. What is "wrong"?

11. Suppose that u, v, w are orthogonal curvilinear coordinates for which $ds^2 = v^2\,du^2 + u^2\,dv^2 + dw^2$.

(a) Calculate the divergence of $\mathbf{u}$, where $\mathbf{u}$ is the unit vector tangent to a u curve.

(b) Determine the laplacian of the function $f = uvw$.

12. Parabolic cylindrical coordinates (u,v,z) are defined by $x = \frac{1}{2}(u^2 - v^2)$, $y = uv$, $z = z$, where $-\infty < u < \infty$, $v \geq 0$, $-\infty < z < \infty$. To make use of the formulas in section 3.11, it is necessary to know the scale factors h_u, h_v, and h_z. Determine these scale factors.

13. What is the element of volume in parabolic cylindrical coordinates?

14. (a) Write div $\mathbf{A}$ in parabolic cylindrical coordinates.

(b) Write Laplace's equation $\nabla^2\phi = 0$ in parabolic cylindrical coordinates.

4 Line, Surface, and Volume Integrals

4.1 Line Integrals

In this chapter we are going to study the integration of vector fields over one, two, and three-dimensional objects. Actually, the latter are the easiest to treat; for our purposes, three-dimensional objects are simply regions of three-dimensional space—"blobs," as it were. The two-dimensional objects are curved surfaces in three-dimensional space, and they are very difficult to characterize rigorously; even a heuristic treatment like ours turns up some surprises such as the Möbius strip (section 4.6). Integrals over one-dimensional objects—curves—have a level of difficulty intermediate between the other two.

Nonetheless we shall study these integrals in the natural order: one, two, and three dimensions. In this section we are going to address the operation of integrating a vector field along a curve in space—a construction that has found considerable utility in mathematics and physics.

Let us give some thought to the meaning of integration along a curve. By analogy with the theory of (Riemann) integration in elementary calculus, one would suspect that the curve is partitioned into short arcs, then some sort of sum is formed over the partition; and, finally, the "integral" emerges as the limit of these sums as the partitions are made finer and finer.

In fact, we have already gained some experience with this type of process in section 2.2, where we computed arc length for a smooth arc. Figure 4.1 (a replica of fig. 2.15, repeated for convenience) illustrates how the points $Q_0, Q_1, \ldots, Q_n$ (with position vectors $\mathbf{R}_0, \mathbf{R}_1, \ldots, \mathbf{R}_n$, respectively) partition the curve and generate the inscribed polygonal path, whose length we calculate by summing the lengths

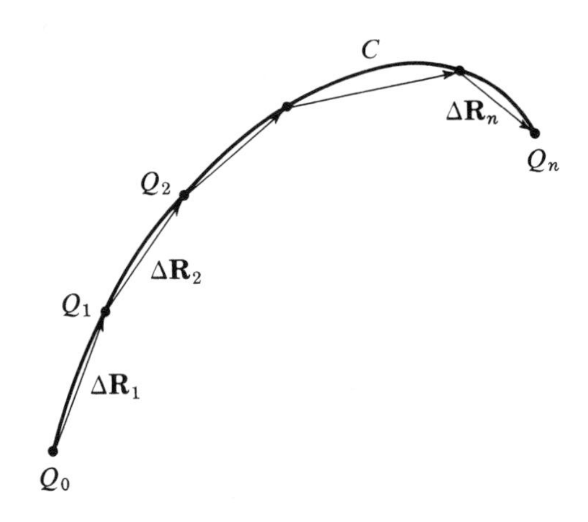

Figure 4.1

of the sides $|\Delta\mathbf{R}_k| = |\mathbf{R}_k - \mathbf{R}_{k-1}|$. The length of the *curve* is then taken to be the limit of these sums, as the partitions are refined in such a manner that the largest length $|\Delta\mathbf{R}_k|$ goes to zero.

In that same section we saw how this limit can be evaluated if the curve is parametrized by $\mathbf{R} = \mathbf{R}(t)$, $a \leq t \leq b$. Recall the essentials of the technique: the interval $[a,b]$ is partitioned $a = t_0 < t_1 < t_2 < \cdots < t_n = b$ to correspond with the points $\mathbf{R}_k = \mathbf{R}(t_k)$, and the approximation

$$\Delta\mathbf{R}_k \approx \frac{d\mathbf{R}}{dt}\,\Delta t_k$$

is used to argue that

$$\int_C |d\mathbf{R}| \equiv \lim \sum_{k=1}^{n} |\Delta\mathbf{R}_k| = \lim \sum_{k=1}^{n} \left|\frac{d\mathbf{R}}{dt}\,\Delta t_k\right| = \int_a^b \left|\frac{d\mathbf{R}}{dt}\right| dt$$

as the Δt_k go to zero. This final expression is an ordinary integral.

Using this as a model, we now turn to the definition of the line integral. (As will be seen, the terminology "curve integral" would be a more accurate term.) Imagine that the curve C is imbedded in a continuous vector field $\mathbf{F}$, as in figure 4.2. The line integral of $\mathbf{F}$ along C is a measure of the degree to which C "lines up" with $\mathbf{F}$. It is highest for curves that are everywhere parallel to $\mathbf{F}$, as in figure 4.2*a* (recall that we called such curves "flow lines" in section 3.3); it is zero if C is orthogonal to $\mathbf{F}$ as in figure 4.2*b;* and it is lowest if C runs counter to the field, as in figure 4.2*c.*

To express this notion quantitatively we subdivide C into n smaller arcs and approximate it by a polygonal path, as in figure 4.3. If $\mathbf{F}_{k+1}$ denotes the value of $\mathbf{F}$ at the point Q_k in the figure, then clearly $\mathbf{F}_k \cdot \Delta\mathbf{R}_k$ measures the alignment of the field with the curve, and we formally define the line integral $\int_C \mathbf{F} \cdot d\mathbf{R}$ to be the

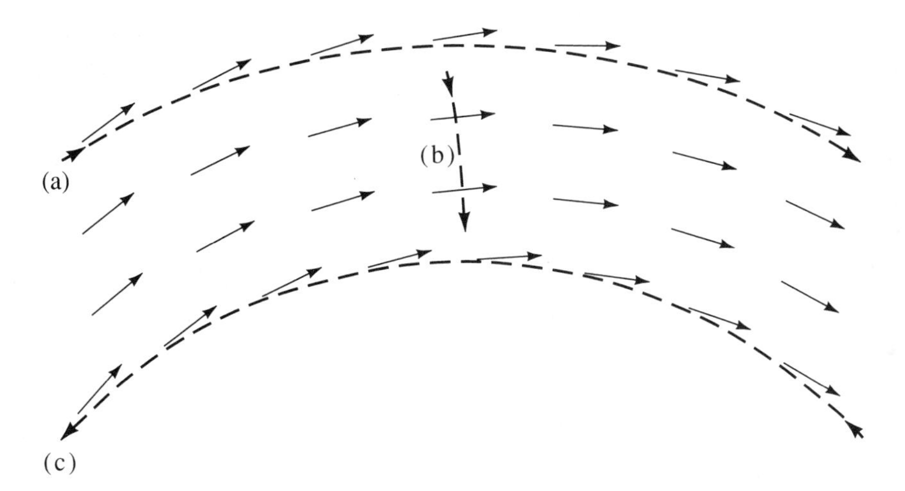

Figure 4.2

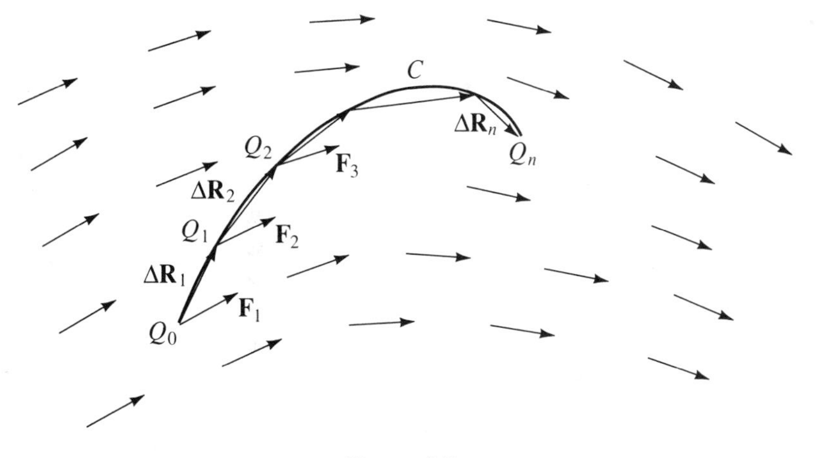

Figure 4.3

limit of sums of this form, when the approximating polygonal paths are obtained by taking increasingly small subdivisions:

$$\int_C \mathrm{F} \cdot d\mathbf{R} = \lim_{(max\ |\Delta R_k| \to 0)} \sum_k \mathbf{F}_k \cdot \Delta\mathbf{R}_k \tag{4.1}$$

If **F** is continuous and C is smooth, it can be shown that this limit exists and is independent of the particular subdivision scheme chosen.

In example 1.15 of section 1.9, it was pointed out that the *work* done by a force **F** acting through a displacement $\Delta\mathbf{R}$ is given by $\mathbf{F} \cdot \Delta\mathbf{R}$. Thus *the line integral*

$\int_C \mathbf{F} \cdot d\mathbf{R}$ *equals the work done by the force field* **F** *on a particle as it moves along the curve C.* It measures the extent to which the vector field abets the particle's motion.

Note that the necessary ingredients for a line integral are a vector field and an oriented curve, and the result is a scalar. The direction of the vectors $\Delta\mathbf{R}_k$ is taken to be consistent with the orientation of C, which in figure 4.3 is *from* Q_0 *to* Q_n. If C were oriented the opposite way, from Q_n to Q_0, each of the vectors $\Delta\mathbf{R}_k$ would point in the opposite direction and the line integral would change sign.

Other notations can be used to denote line integrals. If **T** is a unit vector tangent to the path, in the direction determined by the orientation, then $\mathbf{T} = d\mathbf{R}/ds$, and eq. (4.1) can be written

$$\int_C \mathbf{F} \cdot \mathbf{T}\, ds \tag{4.2}$$

where s, the arc length measured along C, is taken to be *increasing* in the direction determined by the orientation of C.

If $F_t = \mathbf{F} \cdot \mathbf{T}$ is the scalar component of **F** in the direction of the unit tangent, the line integral can also be written

$$\int_C F_t\, ds \tag{4.3}$$

In vector language, we sometimes speak of "the line integral of the tangential component of **F** over the oriented curve C." If we wish to be sloppier, we just say "the integral of **F** along C."

In books on advanced calculus that do not use vector notation, yet another form is used:

$$\int_C (F_1dx + F_2dy + F_3dz) \tag{4.4}$$

We obtain this from eq. (4.1) by taking $\mathbf{F} = F_1\mathbf{i} + F_2\mathbf{j} + F_3\mathbf{k}$. Then, since $d\mathbf{R} = dx\,\mathbf{i} + dy\,\mathbf{j} + dz\,\mathbf{k}$, we have $\mathbf{F} \cdot d\mathbf{R} = F_1dx + F_2dy + F_3dz$.

An expression such as $F_1dx + F_2dy + F_3dz$, where F_1, F_2, and F_3 are functions of x, y, and z, respectively, is called a *differential form*. We call eq. (4.4) the line integral of the differential form over the oriented curve C.

In a moment we are going to see how the line integral can be easily evaluated when the curve is parametrized. Just for the experience, however, we first present an example that computes a line integral directly from the definition. Keep in mind that example 4.1 is gimmicked to work out nicely, and is atypical in this respect.

Example 4.1 Let C be the curve $y = \sqrt{x}$ in the xy plane extending from (0,0,0) to (1,1,0), and let $\mathbf{F} = xy^2\mathbf{i} + y^2\mathbf{k}$. Find $\int_C \mathbf{F} \cdot d\mathbf{R}$ directly from the definition of the integral as the limit of a sum.

Solution For convenience, let all Δx's equal $1/n$, so that

$$Q_k = (x_k, y_k, z_k) = \left(\frac{k}{n}, \sqrt{\frac{k}{n}}, 0\right)$$

$$\Delta\mathbf{R}_k = \frac{1}{n}\mathbf{i} + \left(\sqrt{\frac{k}{n}} - \sqrt{\frac{k-1}{n}}\right)\mathbf{j}$$

$$\mathbf{F}_k = x_k y_k^2 \mathbf{i} + y_k^2 \mathbf{k} = \frac{k^2}{n^2}\mathbf{i} + \frac{k}{n}\mathbf{k}$$

$$\sum_{k=1}^{n} \mathbf{F}_k \cdot \Delta\mathbf{R}_k = \sum_{k=1}^{n} \frac{k^2}{n^3} = \frac{1}{n^3}\sum_{k=1}^{n} k^2$$

$$= \frac{1}{n^3}\left(\frac{1}{6}n(n+1)(2n+1)\right) = \frac{1}{6}\left(1 + \frac{1}{n}\right)\left(2 + \frac{1}{n}\right)$$

which tends to $\frac{1}{3}$ as n approaches infinity. Therefore

$$\int_C \mathbf{F} \cdot d\mathbf{R} = \frac{1}{3}$$

In the usual situation one has a parameterization $\mathbf{R} = \mathbf{R}(t)$, $a \leq t \leq b$, for the smooth curve C, and by analogy with the arc length technique we propose

$$\int_C \mathbf{F} \cdot d\mathbf{R} = \lim \sum_{k=1}^{n} \mathbf{F}_k \cdot \Delta\mathbf{R}_k$$

$$= \lim \sum_{k=1}^{n} \mathbf{F}_k \cdot \frac{d\mathbf{R}}{dt}\Delta t_k = \int_a^b \mathbf{F} \cdot \frac{d\mathbf{R}}{dt}\,dt \tag{4.5}$$

The final form is an ordinary definite integral. To see this, observe that we have the continuous vector function $\mathbf{F} = \mathbf{F}(x,y,z) = F_1\mathbf{i} + F_2\mathbf{j} + F_3\mathbf{k}$, and the continuously differentiable parametric functions $x = x(t)$, $y = y(t)$, and $z = z(t)$, so that plugging the latter into the former produces, in eq. (4.5), an ordinary integral of a function of t:

$$\int_C \mathbf{F} \cdot d\mathbf{R} = \int_a^b \left[F_1(x(t),y(t),z(t))\frac{dx}{dt} + F_2(x(t),y(t),z(t))\frac{dy}{dt} + F_3(x(t),y(t),z(t))\frac{dz}{dt}\right]dt$$

For instance, the curve in example 4.1 can be parametrized $x = t$, $y = \sqrt{t}$, $z = 0$, $0 \leq t \leq 1$, and eq. (4.5) becomes

$$\int_0^1 \left(xy^2 \frac{dx}{dt} + y^2 \frac{dz}{dt} \right) dt = \int_0^1 t^2\, dt = \frac{1}{3}$$

Observe that the line integral has been defined without reference to the parameterization of the curve, so its value will depend only on the field **F** and the oriented curve C, not on the choice of the parameter t. Sometimes *arc length* is a convenient parameter; sometimes it is better to use an *angle* or the *time*, or one of the variables x, y, z. Examples are given below; study them carefully!

The integrals $\int_a^b f(x)\, dx$ that occur in elementary calculus can be regarded as very special kinds of line integrals. Indeed, let us suppose that **F** is always directed parallel to the x axis, so that $\mathbf{F} = f(x)\mathbf{i}$, and suppose that C is a segment of the x axis, $a \leq x \leq b$, oriented in the direction of increasing x. Then $d\mathbf{R} = dx\,\mathbf{i}$, and $\int_C \mathbf{F} \cdot d\mathbf{R} = \int_a^b f(x)\, dx$. So you already have had some experience in evaluating line integrals! *Caution:* In general, line integrals do not represent areas under curves, nor arc length.

Example 4.2 Compute the line integral $\int \mathbf{F} \cdot d\mathbf{R}$ from (0,0,0) to (1,2,4) if

$$\mathbf{F} = x^2\mathbf{i} + y\mathbf{j} + (xz - y)\mathbf{k}$$

(a) along the line segment joining these two points.
(b) along the curve given parametrically by $x = t^2$, $y = 2t$, $z = 4t^3$.

Solutions

(a) Parametric equations for the line segment joining (0,0,0) to (1,2,4) are $x = t$, $y = 2t$, $z = 4t$ (section 1.8). We have

$$\begin{aligned} \int_C \mathbf{F} \cdot d\mathbf{R} &= \int_C x^2\, dx + y\, dy + (xz - y)\, dz \\ &= \int_0^1 t^2 dt + (2t)(2\, dt) + (4t^2 - 2t)(4\, dt) \\ &= \int_0^1 (17t^2 - 4t)\, dt = \frac{11}{3} \end{aligned}$$

(b) In this case we have

$$\begin{aligned} \int_C \mathbf{F} \cdot d\mathbf{R} &= \int_0^1 (t^4)(2t\, dt) + (2t)(2\, dt) + (4t^5 - 2t)(12t^2 dt) \\ &= \int_0^1 (2t^5 + 4t + 48t^7 - 24t^3)\, dt = \frac{7}{3} \end{aligned}$$

Example 4.3 Find the line integral of the tangential component of $\mathbf{F} = x\mathbf{i} + x^2\mathbf{j}$ from $(-1,0)$ to $(1,0)$ in the xy plane,

(a) along the x axis.
(b) along the semicircle $y = \sqrt{1 - x^2}$.
(c) along the dotted polygonal path shown in figure 4.4.

Solutions

(a) Along the x axis, $y = 0$; hence $dy = 0$, and

$$\int \mathbf{F} \cdot d\mathbf{R} = \int (x\,dx + x^2\,dy)$$

$$= \int_{-1}^{1} x\,dx = \frac{1}{2} x^2 \bigg|_{-1}^{1} = 0$$

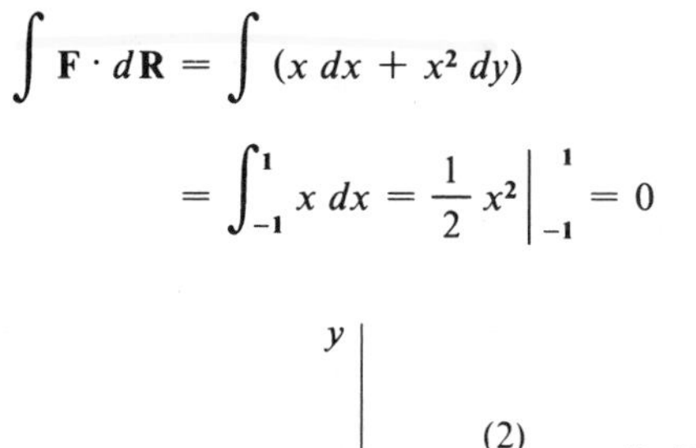

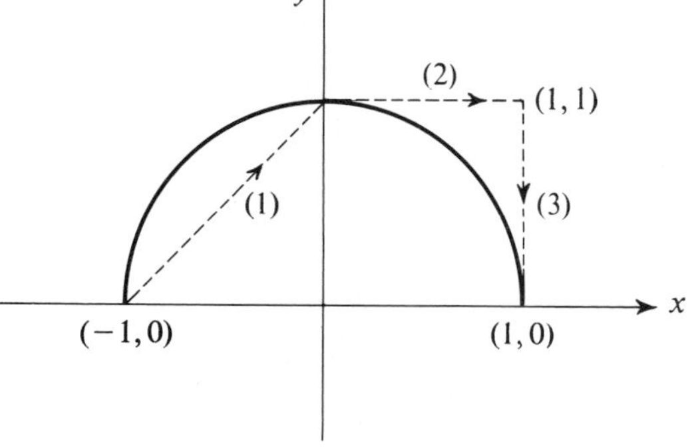

Figure 4.4

(b) Along the semicircle, the convenient parameter is the polar coordinate θ. Since the radius of the circle is unity, we have, for points (x,y) on this path, $x = \cos\theta$, $y = \sin\theta$; hence $dx = -\sin\theta\,d\theta$, $dy = \cos\theta\,d\theta$, and θ runs from π down to zero:

$$\int \mathbf{F} \cdot d\mathbf{R} = \int (x\,dx + x^2\,dy)$$

$$= \int_{\pi}^{0} [(\cos\theta)(-\sin\theta\,d\theta) + (\cos^2\theta)(\cos\theta\,d\theta)]$$

$$= \int_{\pi}^{0} (-\sin\theta\cos\theta + \cos^3\theta)\,d\theta$$

$$= \left[-\frac{\sin^2\theta}{2} + \sin\theta - \frac{\sin^3\theta}{3}\right]_{\pi}^{0} = 0$$

(c) Along the path labeled (1) in figure 4.4, $y = x + 1$, so that $dy = dx$, and

$$\int (x\,dx + x^2\,dy) = \int_{-1}^{0} (x\,dx + x^2\,dx) = -\frac{1}{6}$$

Along path (2), $y = 1$, so that $dy = 0$, and

$$\int (x\,dx + x^2\,dy) = \int_{0}^{1} x\,dx = \frac{1}{2}$$

Along path (3), $x = 1$, so that $dx = 0$, and

$$\int (x\,dx + x^2\,dy) = \int_{1}^{0} dy = -1$$

[Note that we use y instead of x as the parameter along path (3).] The value of the integral is $-\frac{1}{6} + \frac{1}{2} - 1 = -\frac{2}{3}$.

Note: If a curve is *closed,* that is, its initial and final points coincide, the notation $\oint \mathbf{F} \cdot d\mathbf{R}$ is frequently used. The line integral of $\mathbf{F}$ around a closed curve C is called the *circulation* of $\mathbf{F}$ about C.

EXERCISES

1. Find $\int \mathbf{F} \cdot d\mathbf{R}$, where $\mathbf{F} = x^2\mathbf{i} + \mathbf{j} + yz\mathbf{k}$, along C: $x = t$, $y = 2t^2$, $z = 3t$, $0 \leq t \leq 1$.
2. Let C be the curve given by

$$\mathbf{R}(t) = \cos t\,\mathbf{i} + \sin t\,\mathbf{j} + e^t\,\mathbf{k} \qquad (0 \leq t \leq \pi/2)$$

and let

$$\mathbf{F} = -\frac{2x}{x^2 + y^2}\mathbf{i} + \frac{2y}{x^2 + y^2}\mathbf{j} + 2z\mathbf{k}$$

Express $\int_C \mathbf{F} \cdot d\mathbf{R}$ in terms of t and evaluate the resulting integral.
3. Find $\int \mathbf{F} \cdot d\mathbf{R}$ from (1,0,0) to (1,0,4), if $\mathbf{F} = x\mathbf{i} - y\mathbf{j} + z\mathbf{k}$,
   (a) along the line segment joining (1,0,0) and (1,0,4).
   (b) along the helix $x = \cos 2\pi t$, $y = \sin 2\pi t$, $z = 4t$.
4. Find the value of $\oint [(3x + 4y)\,dx + (2x + 3y^2)\,dy]$ around the circle $x^2 + y^2 = 4$.
5. Find the line integral $\int \mathbf{F} \cdot d\mathbf{R}$ along the line segment from (1,0,2) to (3,4,1), where $\mathbf{F} = 2xy\mathbf{i} + (x^2 + z)\mathbf{j} + y\mathbf{k}$.
6. Find the integral $\oint \mathbf{F} \cdot d\mathbf{R}$ around the circumference of the circle $x^2 - 2x + y^2 = 2$, $z = 1$, where $\mathbf{F} = y\mathbf{i} + x\mathbf{j} + xyz^2\mathbf{k}$.

7. Let

$$\mathbf{F} = \frac{y}{x^2 + y^2}\mathbf{i} - \frac{x}{x^2 + y^2}\mathbf{j}$$

Find the line integral of the tangential component of **F**, from $(-1,0)$ to $(1,0)$,
(a) along the semicircle $y = \sqrt{1 - x^2}$.
(b) along the dotted polygonal path shown in figure 4.4.

8. By changing to polar coordinates, find the answers to exercise 7 by inspection.
9. Evaluate the following line integrals over the straight-line segment C joining the point $(2,1,4)$ to the point $(3,3,4)$:
(a) $\int_C [3xy\,dx + 3\,dy + yz\,dz]$
(b) $\int_C e^{xyz}(yz\,dx + xz\,dy + xy\,dz)$
10. Evaluate

$$\oint [(y + yz\cos xyz)\,dx + (x^2 + xz\cos xyz)\,dy + (z + xy\cos xyz)\,dz]$$

along the ellipse $x = 2\cos\theta$, $y = 3\sin\theta$, $z = 1$, $0 \le \theta \le 2\pi$.

11. Evaluate $\oint_C [(\sin x + y^2)\,dx + (x - e^{-y})\,dy]$, where C is the boundary of the semicircular region $x^2 + y^2 \le 4$, $y \ge 0$.
12. Compute the line integral $\int_C \mathbf{F} \cdot d\mathbf{R}$, where C is the intersection of the plane $x + y + z = 1$ with the cylinder $x^2 + y^2 = 1$ and $\mathbf{F} = (x + y)\mathbf{i} + (y + z)\mathbf{j} + (z + x)\mathbf{k}$. Orient C clockwise as viewed from above.
13. Find $\int \mathbf{R} \cdot d\mathbf{R}$ from $(1,2,2)$ to $(3,6,6)$, along the line segment joining these points,
(a) in the manner described in the text.
(b) by observing that $\mathbf{R} \cdot d\mathbf{R} = s\,ds$, where $s = (x^2 + y^2 + z^2)^{1/2}$ is the distance from the origin, and computing $\int_3^9 s\,ds$.
14. Let $\mathbf{F} = \boldsymbol{\omega} \times \mathbf{R}$, where $\boldsymbol{\omega}$ is a constant. Compute $\int \mathbf{F} \cdot d\mathbf{R}$ along the straight line from $(0,0,0)$ to $(2,2,2)$. (*Hint:* Use a little thought, and you can avoid any work.)
15. For example 4.3 (refer to fig. 4.4), determine **T**,
(a) along path (1), in the direction shown, in terms of **i** and **j**.
(b) along path (2), in the direction shown.
(c) along path (3), in the direction shown.
16. For example 4.3, determine ds, in terms of dx or dy,
(a) along path (1).
(b) along path (2).
(c) along path (3).
17. Show that $d\mathbf{R} = dx\,\mathbf{i} + dy\,\mathbf{j}$ is the same as $\mathbf{T}\,ds$ in each of the three special cases referred to in exercises 15 and 16. (This illustrates the general rule that, in practice, it is easier to find $d\mathbf{R}$ directly than to find **T** and ds separately and multiply.)
18. Let $\mathbf{F} = (x^2/y)\mathbf{i} + y\mathbf{j} + \mathbf{k}$.
(a) Find the equation for the flow line for **F** that passes through the point $(1,1,0)$.
(b) Show that this flow line passes also through the point $(e,e,1)$.
(c) Evaluate $\int_C \mathbf{F} \cdot d\mathbf{R}$, where C is the path, along the given flow line, from $(1,1,0)$ to $(e,e,1)$.

19. Let $\mathbf{F}(x,y) = (x^2 + y^2)(\mathbf{i} + \mathbf{j})$, and let C be a directed straight-line segment of unit length, with one endpoint at the origin, (0,0). Find the direction of C such that the line integral $I = \int_C \mathbf{F} \cdot d\mathbf{R}$ is
(a) a maximum (give the direction of C and the value of I).
(b) a minimum (give the direction of C and the value of I).
(c) zero (give the direction of C).

20. If the vector field $\mathbf{F}(x,y,z)$ is everywhere parallel to $\mathbf{R}$ and C is a curve drawn on a sphere with center at the origin, then $\int_C \mathbf{F} \cdot d\mathbf{R} = 0$; why?

4.2 Domains: Simply Connected Domains

We recall from elementary calculus that many of the functions that arise are not defined for all values of x, but only for certain intervals. For example, the function $f(x) = 1/x$ is not defined at $x = 0$, and the function $f(x) = \csc x$ is not defined when x is an integral multiple of π.

Similarly, the vector fields that arise in practice are frequently not defined at all points (x,y,z) in space, but only in certain regions of space.

For instance, we learn in elementary physics that the magnitude of the magnetic field intensity due to a current flowing along a straight line varies inversely with the distance from that line. As we get nearer to the line, the magnetic intensity increases in magnitude. The magnetic field is not defined along the line itself. The region of definition consists of all points in space except those along the line.

Similarly, the electric intensity due to a system of n point charges is defined everywhere in space except at the n points in question.

To be sure, the fields that arise in elementary physics are extremely idealized (is a charge really concentrated at a *point?*), but they are useful in theoretical discussions and their study is essential to more advanced work.

The reader with limited knowledge of electric or magnetic field theory may imagine instead that the fields we consider are the velocity fields of fluids that are in some container. Obviously, it is nonsense to speak of the velocity vector at any point outside the container. The region of definition in this case consists of all points within the container.

The vector fields that usually arise, both in theory and in practice, have two important properties. First, such a field is defined in the interior of a given region but not on the boundary of the region. Second, if the field is defined at two points P and Q, it is possible to find a smooth arc C joining P to Q along which the field is everywhere defined.

For instance, the velocity of a fluid in a container is not defined for points on the surface of the container, but only for points in the interior of the container. Moreover, it is unusual to consider a container with separate compartments; we usually assume that if there is fluid at two points P and Q, it is possible to move from P to Q without passing through any separating walls. Motivated by these ideas, we now give several precise definitions.

Figure 4.5

If P is any given point and ϵ is any positive number (zero is excluded), we say that an ϵ *neighborhood of P* is the set of all points that are *less* than ϵ in distance away from P. Thus, if we are speaking of points in the plane, an ϵ neighborhood of a point P consists of all points in the interior (but not on the circumference) of a circle of radius ϵ and center at P. If we are speaking of points in space, an ϵ neighborhood of P consists of all points in the interior (but not on the surface) of a sphere of radius ϵ and center at P.

Given a region R, we say that P is an *interior point* of R if it is possible to find an ϵ neighborhood of P that lies completely within R. We say that P is a *boundary point* of R if, no matter how small we take the positive number ϵ, the ϵ neighborhood of P contains at least one point in R and one point not in R. So, by definition, an interior point cannot be a boundary point, nor can a boundary point be an interior point. (The other points are called *exterior.*)

A region is said to be *open* if every point in the region is an interior point of the region. Thus, if the region of definition of a vector field is an open region, we can say that, if the field is defined at a point P, it will also be defined in some ϵ neighborhood of P. Of course, if P is very near the boundary of the region, ϵ may have to be very small.

By definition, an open region does not include its boundary. (For example, the set of all points *within* a cube is an open region in space, but the set consisting of all those points either within or on the surface of a cube is not an open region.) If we say that an arc C lies in an open region, then by definition C cannot intersect or even touch the boundary of the region.

Henceforth, we shall consider only open regions.

An open region R is said to be *connected* if, given any two points P and Q in R, there can be found a smooth arc in R that joins P to Q.

In figure 4.5 we show a region in the plane that is *not* connected. Obviously we cannot join P to Q by a smooth arc that lies completely within the region. We will have no occasion to consider such regions; henceforth we consider only connected regions.

A region that is both open and connected is called a *domain.*

The region of definition of the magnetic field due to a steady line of current flowing along the z axis consists of all points except those on the z axis. The region of definition of the electric field due to a system of n fixed point charges consists of all points other than the given n points. It is easy to see that in either case the region is both open and connected, so that the word "domain" applies.

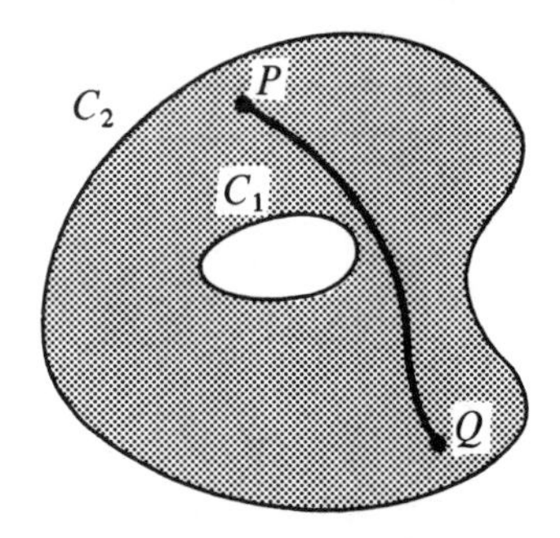

Figure 4.6

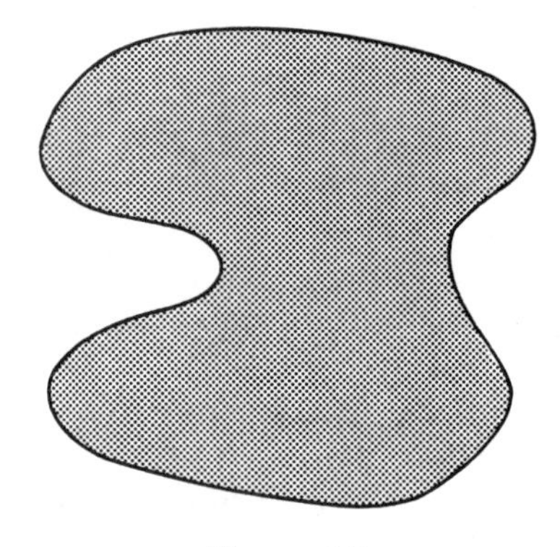

Figure 4.7

In figure 4.6 we give an example of a region in the plane. If we let D denote the set of points within the shaded region, not including any points on either of the curves C_1 and C_2, then D is a domain. The points on the curves C_1 and C_2 constitute the boundary of the domain. In the figure we give an example of a smooth arc joining two points P and Q.

Of special importance are those domains that are simply connected. In figure 4.7 we show a region in the plane that is simply connected, but the regions indicated in figures 4.6 and 4.8 are not simply connected. Roughly speaking, a domain is said to be *simply connected* if every closed curve lying in the domain can be continuously shrunk to a point in the domain without any part of the curve passing through regions outside the domain. The plane regions indicated in figures 4.6 and 4.8 are not simply connected because no closed curve surrounding one of the "holes" could be shrunk to a point while still always remaining in the domain. Thus, in the special case of a domain of points in the plane, this simply means that, given any closed curve in the domain, all points within the closed curve are also in the domain. In other words, there are no "holes" in the domain.

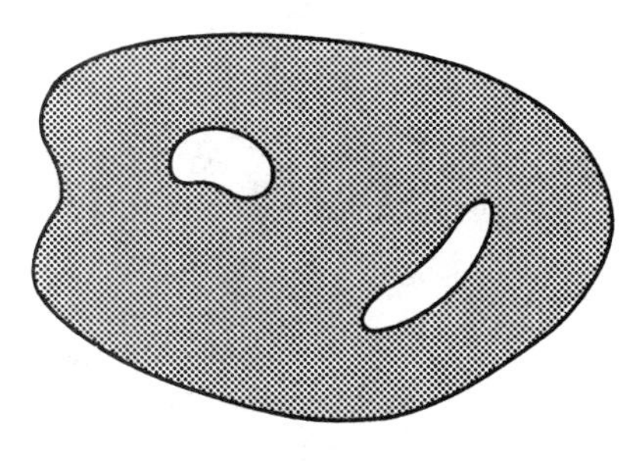

Figure 4.8

Simply connected domains in space are, very roughly speaking, those domains through which no holes have been bored. Thus, the set of points in the interior of a torus (doughnut) is not simply connected, since a closed curve within the torus surrounding the hole cannot be shrunk to a point while remaining always within the torus.

A closed curve C, in the process of being shrunk to a point, will generate a surface having the original curve C as its boundary. Thus, another way of wording the definition is as follows: a domain is simply connected if, given any closed curve lying in the domain, there can be found a surface within the domain that has that curve as its boundary.

The domain consisting of all points in the interior of a sphere is simply connected. As another example, suppose we are given two concentric spheres; then the set of points outside the inner sphere but inside the outer sphere constitutes a simply connected domain.

As a further example, consider the cylinder $x^2 + y^2 = 1$. This is a cylinder of radius 1, concentric with the z axis. Every point outside the cylinder has coordinates (x,y,z) satisfying the inequality $x^2 + y^2 > 1$ (z arbitrary), and the set of all such points is a domain that *is not* simply connected. The set of points in the interior of the cylinder $x^2 + y^2 < 1$ is simply connected.

Vector fields defined in simply connected regions have much simpler properties, in general, than those having domains of definition that are not simply connected. Domains that are not simply connected may be very complicated; the reader may wish to contemplate the region of space within an old-fashioned steam radiator, which is very far indeed from being simply connected.

In this chapter we shall have occasion to refer to a *star-shaped domain.* A domain is called star-shaped if there is a point P in the domain such that, for any other point Q in the domain, the entire line segment PQ lies in the domain. Sometimes we say the domain is star-shaped *with respect to P.* Figure 4.9 illustrates some star-shaped domains.

A star-shaped domain is simply connected; indeed, any curve can be shrunk to the point P.

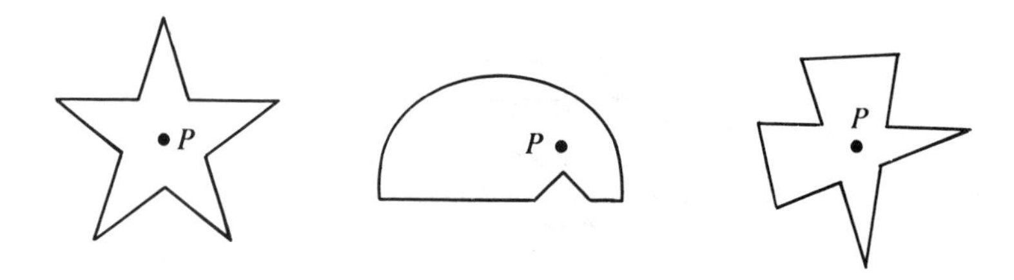

Figure 4.9

EXERCISES

In each of the following cases, a region D is defined. Tell whether the region is a domain. If it is a domain, determine whether or not it is simply connected. If it is not a domain, explain why not.

1. The region of definition of a magnetic field due to a steady current flowing along the z axis [i.e., the region consisting of all points (x,y,z) such that $x^2 + y^2 > 0$].
2. The region of definition of an electric field due to n point charges.
3. The region consisting of all points above the xy plane [i.e., all points (x,y,z) such that $z > 0$].
4. The region D consisting of all points (x,y,z) for which $z \geq 0$.
5. The region D consisting of all points (x,y,z) such that $x^2 + y^2 + z^2 > 4$.
6. The region D consisting of all points (x,y,z) for which $1 < x^2 + y^2 < 4$ (i.e., all points outside a cylinder of radius 1 and within a cylinder of radius 2, both cylinders concentric with the z axis).
7. The region D consisting of all points (x,y,z) for which $1 < x < 2$ (i.e., all points between the planes $x = 1$ and $x = 2$).
8. The region D consisting of all points (x,y,z) for which $z \neq 0$.
9. The region in the plane between two concentric circles.
10. The region in space between two concentric spheres.

4.3 Conservative Fields: The Potential Function

In this section we let **F** denote a vector field that is defined and continuous throughout a domain D. Then

$$\mathbf{F} = F_1\mathbf{i} + F_2\mathbf{j} + F_3\mathbf{k} \tag{4.6}$$

where F_1, F_2, and F_3 are scalar-valued functions, each of which is continuous throughout D. If these three functions have partial derivatives (there will be nine such derivatives, $\partial F_1/\partial x$, $\partial F_1/\partial y$, . . . , $\partial F_3/\partial z$) all of which are continuous throughout D, then $\mathbf{F}$ is said to be *continuously differentiable* in D. It follows from these definitions that, if $\mathbf{F}$ is continuously differentiable in D, then **curl** $\mathbf{F}$ is a vector field that is continuous in D, and div $\mathbf{F}$ is a scalar field that is continuous in D.

A vector field $\mathbf{F}$ is said to be *conservative* in a domain D if there can be found some scalar field ϕ defined in D such that $\mathbf{F} = \mathbf{grad}\ \phi$. If this is possible, then ϕ is called a *potential function*, or simply a *potential*, for $\mathbf{F}$.

Notice that the potential function for a conservative field is not unique, since one can always add an arbitrary constant to ϕ to obtain a new potential whose gradient is also $\mathbf{F}$. (Physicists conventionally choose potentials to satisfy certain natural boundary conditions; e.g., they may choose the constant so that the potential function for a gravitational field is zero along the laboratory floor, or so that the potential function for an electric field tends to zero at infinity.)

The following theorem may indicate why conservative fields are so important.

THEOREM 4.1 *A vector field* $\mathbf{F}$ *continuous in a domain D is conservative if and only if the line integral of* $\mathbf{F}$ *along every regular curve in D depends only on the endpoints of the curve. In that case, the line integral is simply the difference in potential of the endpoints. That is, we have*

$$\int_C \mathbf{F} \cdot d\mathbf{R} = \phi(Q) - \phi(P)$$

where P and Q are initial and terminal points of C respectively.

Before we continue, let us be sure we understand this theorem. We are given a vector field $\mathbf{F}$ defined and continuous in a domain D. The theorem says this field is conservative if and only if the following condition holds: if we are given any two points P and Q in D, and any regular curve C within the domain extending from P to Q, then

$$\int_P^Q \mathbf{F} \cdot d\mathbf{R}$$

depends only on the location of the endpoints P and Q and not in any way on the choice of the curve C that joins them. (We summarize this condition by stating that the line integral is independent of path.) Moreover, if this condition holds, then we can *evaluate* this line integral by first finding a function ϕ such that $\mathbf{F} = \mathbf{grad}\ \phi$, and then subtracting the value of ϕ at P from its value at Q.

This is the major theorem of vector analysis. We strongly urge the student to study the following outline of the proof.

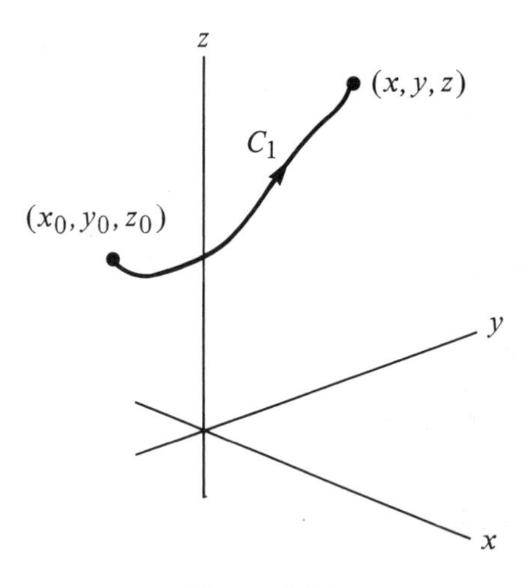

Figure 4.10

Proof The phrase "if and only if" requires that we prove the implication in both directions. We break up the proof into four steps. First, we assume that the line integral of **F** depends only on the endpoints, and (*i*) define a function ϕ in a certain manner, (*ii*) show that ϕ is a potential for **F**, and (iii) demonstrate that

$$\int_P^Q \mathbf{F} \cdot d\mathbf{R} = \phi(Q) - \phi(P)$$

Finally, we complete the argument by proving the converse; (*iv*) assuming that **F** is conservative, we show that the line integral is given by $\phi(Q) - \phi(P)$ and hence is independent of path. Here we go.

(*i*) *Definition of the potential function*

We choose, once and for all, an arbitrary point (x_0,y_0,z_0) in D, which we call the "point of zero potential." Given any other point (x,y,z) in D, we choose some smooth arc C_1 in D extending from (x_0,y_0,z_0) to (x,y,z); this is possible since we assume that D is a domain (fig. 4.10). We define $\phi(x,y,z)$ to be

$$\phi(x,y,z) = \int_{(x_0,y_0,z_0)}^{(x,y,z)} \mathbf{F} \cdot d\mathbf{R}$$

where we integrate along C_1. By hypothesis, this integral is independent of path, and so this definition of $\phi(x,y,z)$ does not depend on the particular arc C_1 that we choose. In other words, $\phi(x,y,z)$ depends on the point (x,y,z) in an unambiguous manner.

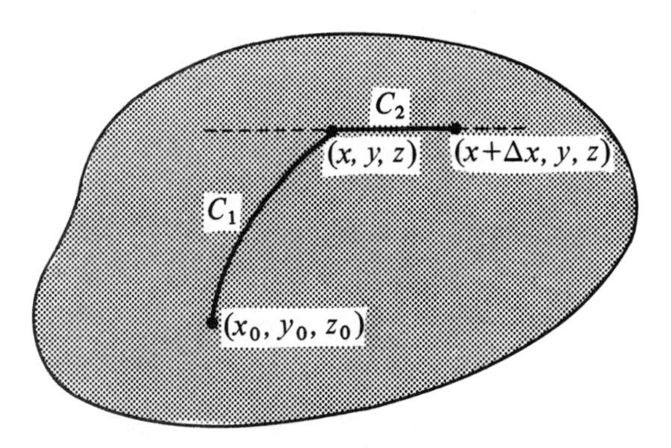

Figure 4.11

(*ii*) *Proof that* $\mathbf{F} = \nabla\phi$

We begin by computing $\partial\phi/\partial x$ at (x,y,z). By definition, this is

$$\lim_{\Delta x \to 0} \frac{\phi(x + \Delta x, y, z) - \phi(x,y,z)}{\Delta x} \tag{4.7}$$

Since D is open (every domain is) there is some ϵ neighborhood of (x,y,z) that is within D. Let us consider a line segment, parallel to the x axis and passing through (x,y,z), that is within this ϵ neighborhood. For a point $(x + \Delta x, y, z)$ along this line segment, let C_2 denote that part of the segment extending from (x,y,z) to $(x + \Delta x, y, z)$. Then C_2, being a line segment, is *a fortiori* a smooth arc, and the path from (x_0,y_0,z_0) to $(x + \Delta x, y, z)$ obtained by joining C_2 to C_1 consists of two smooth arcs and is therefore a regular curve (fig. 4.11). We integrate along this curve to find $\phi(x + \Delta x, y, z)$ by first integrating along C_1 and then along C_2. Since the first integral gives $\phi(x,y,z)$, we have

$$\phi(x + \Delta x, y, z) = \phi(x,y,z) + \int_{(x,y,z)}^{(x+\Delta x,y,z)} \mathbf{F} \cdot d\mathbf{R}$$

from which it follows that the numerator of eq. (4.7) is simply the integral

$$\int_{(x,y,z)}^{(x+\Delta x,y,z)} \mathbf{F} \cdot d\mathbf{R}$$

taken along C_2. Since y and z are constant along this line segment, we have $d\mathbf{R} = dx\,\mathbf{i}$, and hence $\mathbf{F} \cdot d\mathbf{R} = F_1\,dx$. Thus eq. (4.7) becomes

$$\lim_{\Delta x \to 0} \frac{\int_{(x,y,z)}^{(x+\Delta x,y,z)} F_1\,dx}{\Delta x} \tag{4.8}$$

Only one variable is involved in eq. (4.8) since y and z are constant along C_2; in other words, one can treat the numerator just like any integral one meets

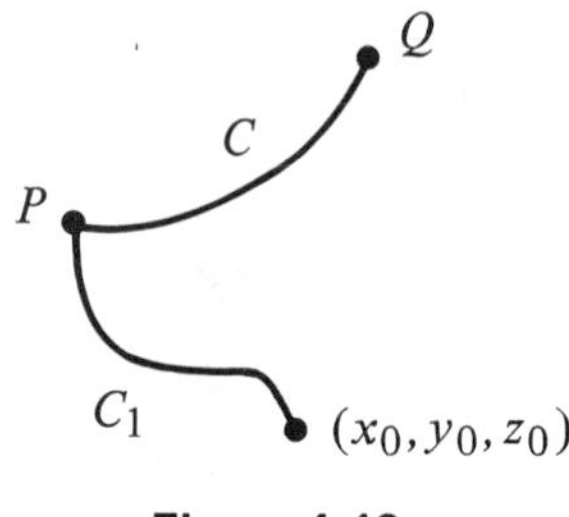

Figure 4.12

in elementary calculus. The reader will recognize this integral, divided by Δx, as simply the average value of F_1 along the line segment C_2. Since F_1 is continuous, this average value tends to $F_1(x,y,z)$ as Δx tends to zero. It follows that, at any point (x,y,z), we have $\partial\phi/\partial x = F_1$.

Similarly, we can show (taking line segments parallel to the y and z axes, respectively) that $\partial\phi/\partial y = F_2$ and $\partial\phi/\partial z = F_3$. Therefore

$$\mathbf{grad}\,\phi = \frac{\partial\phi}{\partial x}\mathbf{i} + \frac{\partial\phi}{\partial y}\mathbf{j} + \frac{\partial\phi}{\partial z}\mathbf{k} = F_1\mathbf{i} + F_2\mathbf{j} + F_3\mathbf{k} = \mathbf{F}$$

which proves that ϕ is a potential function for **F**.

(*iii*) *Proof that* $\int_P^Q \mathbf{F}\cdot d\mathbf{R} = \phi(Q) - \phi(P)$

Let P and Q be two distinct points in D, and let C denote any regular curve extending from P to Q. Let C_1 be a smooth arc extending from (x_0,y_0,z_0) to P. Since the integral is independent of path, $\phi(Q)$ must equal the integral taken along the regular curve obtained by attaching C_1 and C together (fig. 4.12). Thus

$$\begin{aligned}\phi(Q) &= \int_{C_1}\mathbf{F}\cdot d\mathbf{R} + \int_C \mathbf{F}\cdot d\mathbf{R}\\ &= \phi(P) + \int_C \mathbf{F}\cdot d\mathbf{R}\end{aligned}$$

from which it follows that

$$\int_C \mathbf{F}\cdot d\mathbf{R} = \phi(Q) - \phi(P)$$

(*iv*) *Proof of the converse*

To prove the converse, we *assume* **F** to be conservative, that is, that there exists ϕ such that $\mathbf{F} = \mathbf{grad}\,\phi$. Then along any smooth arc we have **F** and $d\mathbf{R}$ expressed in terms of some parameter t and its differential dt. Therefore we compute

$$\int_P^Q \mathbf{F} \cdot d\mathbf{R} = \int_P^Q \left(\frac{\partial \phi}{\partial x} dx + \frac{\partial \phi}{\partial y} dy + \frac{\partial \phi}{\partial z} dz\right)$$
$$= \int_P^Q \left(\frac{\partial \phi}{\partial x}\frac{dx}{dt} + \frac{\partial \phi}{\partial y}\frac{dy}{dt} + \frac{\partial \phi}{\partial z}\frac{dz}{dt}\right) dt$$

Now we make use of the chain rule; if ϕ is a function having continuous partial derivatives with respect to x, y, and z, where x, y, and z are differentiable functions of a single parameter t, then

$$\frac{d\phi}{dt} = \frac{\partial \phi}{\partial x}\frac{dx}{dt} + \frac{\partial \phi}{\partial y}\frac{dy}{dt} + \frac{\partial \phi}{\partial z}\frac{dz}{dt}$$

It follows that

$$\int_P^Q \mathbf{F} \cdot d\mathbf{R} = \int_P^Q \frac{d\phi}{dt} dt = \phi(Q) - \phi(P)$$

This completes the proof. Note that if the path C is *closed,* that is, if P and Q coincide, then $\oint_C \mathbf{F} \cdot d\mathbf{R} = 0$, since $\phi(P) - \phi(P) = 0$. Conversely, if $\oint_C \mathbf{F} \cdot d\mathbf{R} = 0$ around every regular closed curve in the domain, then $\mathbf{F}$ must be conservative (see exercise 1 for the proof).

THEOREM 4.2 *A vector field* **F** *continuous in a domain D is conservative if and only if around every regular closed curve in D the line integral of* **F** (*i.e., its circulation*) *is zero.*

Example 4.4 Show that $\mathbf{F} = xy^2\mathbf{i} + x^3y\mathbf{j}$ is *not* conservative.

Solution A quick way of solving such problems will be given in example 4.5. However, we can prove that a field is not conservative by showing that its line integral does depend on the path. In this case, for instance, let us compute the integral along two paths joining (0,0) to (1,1) in the xy plane (fig. 4.13). Along the line $y = x$, we have

$$\int_{(0,0)}^{(1,1)} (xy^2\, dx + x^3y\, dy) = \int_{x=0}^{x=1} (x^3 + x^4)\, dx = \frac{9}{20}$$

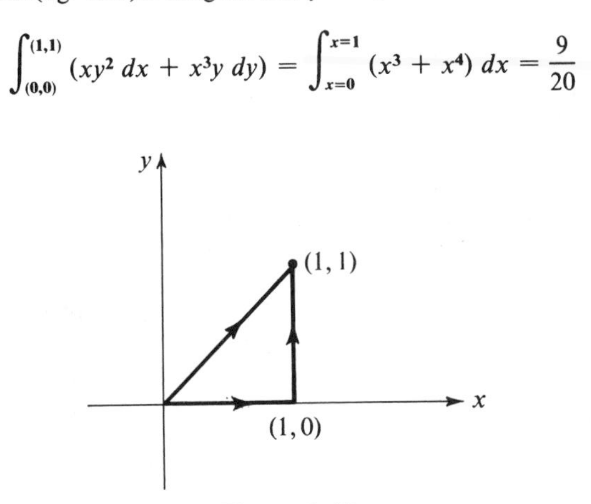

Figure 4.13

Now let us move along the regular path consisting of two line segments, the first joining (0,0) to (1,0) and the second joining (1,0) to (1,1). Along the first line segment, $y = 0$, so that the line integral is zero. Along the second line segment, $x = 1$, so that $dx = 0$, and the integral becomes

$$\int_{y=0}^{y=1} y\,dy = \frac{1}{2}$$

The total of the two integrals is thus $\frac{1}{2}$, differing from $\frac{9}{20}$. Hence the field is not conservative.

It is important to notice that if these two line integrals had turned out to be equal, we would not have been able to draw any conclusions from that alone. Such a result could have happened by coincidence even though the field $\mathbf{F}$ was not conservative. Since it is obviously impossible to compute $\int \mathbf{F} \cdot d\mathbf{R}$ along every conceivable regular curve, the theorem does not provide a practical way of showing that a given field is conservative.

Example 4.5 Show that $\mathbf{F} = xy^2\mathbf{i} + x^3y\mathbf{j}$ is not conservative, without computing any integrals.

Solution This can be done by contradiction. Suppose $\mathbf{F}$ were conservative. Then $\mathbf{F} = \mathbf{grad}\ \phi$ for some function ϕ. Since

$$\mathbf{grad}\ \phi = \frac{\partial \phi}{\partial x}\mathbf{i} + \frac{\partial \phi}{\partial y}\mathbf{j} + \frac{\partial \phi}{\partial z}\mathbf{k}$$

we must have $\partial\phi/\partial x = xy^2$ and $\partial\phi/\partial y = x^3y$. But this is impossible, since the mixed derivatives $\partial^2\phi/\partial y\partial x$ and $\partial^2\phi/\partial x\partial y$ would be $2xy$ and $3x^2y$ respectively, whereas the theory of partial differentiation requires these derivatives to be equal. This contradiction shows that such a function ϕ cannot exist, and so $\mathbf{F}$ is not conservative.

Example 4.6 Show that $\mathbf{F} = 3x^2y\mathbf{i} + (x^3 + 1)\mathbf{j} + 9z^2\mathbf{k}$ is conservative.

Solution Again, a routine way of solving such problems will be given later. At this point, we have no alternative but to try to find a function ϕ such that $\mathbf{F} = \mathbf{grad}\ \phi$. As we have remarked already, the theorems of this section are not useful in proving that a field *is* conservative since we would have to compute an infinite number of integrals. (If we were to take two points and compute line integrals along a dozen or so paths joining these points, the equality of these numbers might lead us to suspect the field to be conservative, but the experiment would not provide a rigorous proof.)

If $\mathbf{F} = \mathbf{grad}\ \phi$, then $\partial\phi/\partial x = 3x^2y$, $\partial\phi/\partial y = x^3 + 1$, and $\partial\phi/\partial z = 9z^2$. In computing $\partial\phi/\partial x$, one differentiates while holding y and z constant, and so evidently

$$\phi = x^3y + (\text{either a constant term or a term involving only } y \text{ and } z)$$

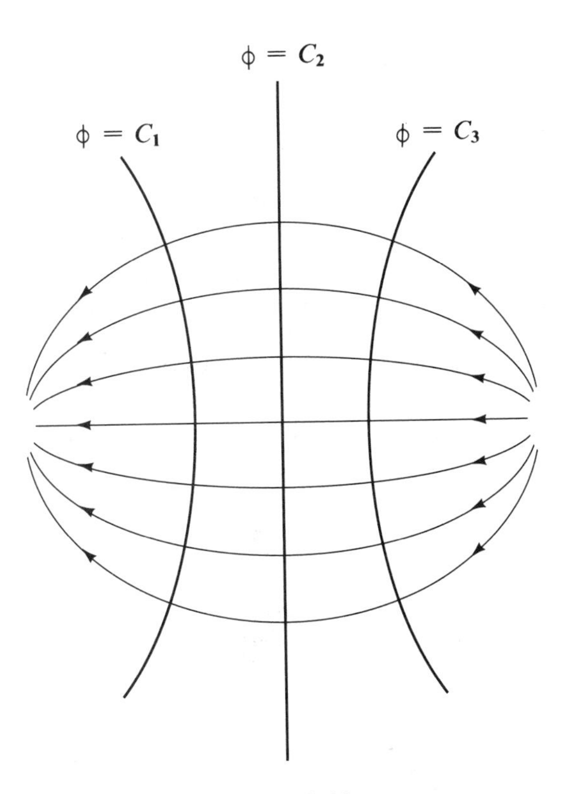

Figure 4.14

Let us write this as $\phi = x^3y + g(y,z)$, where g is a function not yet determined. Differentiating, we have $\partial\phi/\partial y = x^3 + (\partial g/\partial y)$. Comparing this with $\partial\phi/\partial y$ above, we see that $\partial g/\partial y = 1$. Since g is a function of y and z, evidently

$$g(y,z) = y + (\text{either a constant term or a term involving } z \text{ alone})$$

Therefore we have $\phi = x^3y + y + h(z)$, where h depends only on z (or may possibly be a constant). Differentiating, this time with respect to z, we have $\partial\phi/\partial z = h'(z)$, and comparison with the above gives $h'(z) = 9z^2$. It follows that $h(z) = 3z^3 + C$, where C is a constant that may be chosen arbitrarily. Now we have $\phi = x^3y + y + 3z^3 + C$, and it is easy to check this to see that **grad** ϕ = **F**. Hence **F** is conservative.

Remark: A common error is to integrate separately and add the results. Since $\partial\phi/\partial x = 3x^2y$, $\phi = x^3y$; since $\partial\phi/\partial y = x^3 + 1$, $\phi = x^3y + y$; since $\partial\phi/\partial z = 9z^2$, $\phi = 3z^3$. Adding these, we obtain $\phi = 2x^3y + y + 3z^3$, which is incorrect. (It doesn't check.)

It is easy to visualize the relation between a conservative field and its potential function; recall that **grad** ϕ is normal to the surface ϕ = constant. Thus the flow lines of **F** = **grad** ϕ pierce its equipotential surfaces at right angles (see fig. 4.14).

We close this section with some comments on the physical applications of the potential function. When **F** is a *force* field, it is common to choose the opposite sign for the potential ϕ:

$$\phi(x,y,z) = -\int_{(x_0,y_0,z_0)}^{(x,y,z)} \mathbf{F} \cdot d\mathbf{R} \qquad \mathbf{F} = -\,\mathbf{grad}\,\phi$$

As is demonstrated in physics textbooks, this convention enables one to associate ϕ with *potential energy.* The reader should have little trouble making this adjustment.

Also, the divergence of a conservative field is sometimes regarded as the "source" of the field, and it is important to have formulas for computing the potential ϕ from $\nabla \cdot \mathbf{F}$ rather than from **F** itself. This will be discussed in a subsequent section.

EXERCISES

1. Show that, if $\oint_C \mathbf{F} \cdot d\mathbf{R} = 0$ for every regular closed curve C, then for any two points P and Q, $\int_P^Q \mathbf{F} \cdot d\mathbf{R}$ is independent of path. (*Hint:* Let C_1 and C_2 be two paths extending from P to Q, and construct a closed curve out of these.)
2. Using the method of example 4.4, or some similar method, show that the following fields are not conservative:
   (a) $\mathbf{F} = -y\mathbf{i} + x\mathbf{j}$
   (b) $\mathbf{F} = y\mathbf{i} + y(x - 1)\mathbf{j}$
   (c) $\mathbf{F} = y\mathbf{i} + x\mathbf{j} + x^2\mathbf{k}$ [*Suggestion:* Consider two different paths extending from (0,0,0) to (1,1,1).]
   (d) $\mathbf{F} = z\mathbf{i} + z\mathbf{j} + (y - 1)\mathbf{k}$
   (e) $\mathbf{F} = \dfrac{x\mathbf{i} + x\mathbf{j}}{x^2 + y^2}$ (not defined at the origin)
3. Using methods similar to that of example 4.5, show that the fields of exercise 2 are not conservative.
4. Compute $\oint \mathbf{F} \cdot d\mathbf{R}$ around the closed path consisting of a circle of radius r, centered at the origin, in the xy plane, taking

   $$\mathbf{F} = (-y\mathbf{i} + x\mathbf{j})/(x^2 + y^2)$$

   (*Hint:* Change to polar coordinates.)
5. If you worked correctly, you obtained a nonzero answer to exercise 4. Yet it appears that $\mathbf{F} = \mathbf{grad}\,\phi$, where $\phi = \tan^{-1}(y/x)$, and this would contradict theorem 4.2. Investigate this mystery.
6. Find a potential for the force field $\mathbf{F} = (y + z\cos xz)\mathbf{i} + x\mathbf{j} + (x\cos xz)\mathbf{k}$.
7. Show that the field $\mathbf{F} = 2xy\mathbf{i} + (x^2 + z)\mathbf{j} + y\mathbf{k}$ is conservative.
8. Under what conditions can two scalar fields have the same gradient?

4.4 Conservative Fields: Irrotational Fields

In section 4.3 we saw that a continuously differentiable vector field **F** defined in a domain D is conservative if and only if it possesses any one (and hence all) of the following properties:

(i) It is the gradient of a scalar function.
(ii) Its integral around any regular closed curve is zero.
(iii) Its integral along any regular curve extending from a point P to a point Q is independent of the path.

Note that we are using slightly sloppy language here. When we say "its integral" we mean "the line integral of the tangential component," and when we say "any regular closed curve" or "any regular curve" we really do not mean *any* such curve, since we require the curve to lie completely within the domain D.

If the domain D in which **F** is defined is *simply connected,* we can add a fourth property, equivalent to any one of the other three:

$$\mathbf{curl\ F} = \mathbf{0} \tag{4.9}$$

This is of practical utility since, if we are given a vector field **F** defined in a simply connected domain D, we can quickly test to determine whether it is conservative by computing its curl. In terms of components, the test to determine whether

$$\mathbf{F} = F_1\mathbf{i} + F_2\mathbf{j} + F_3\mathbf{k}$$

is conservative consists of checking to see whether *all* the following equations are valid:

$$\frac{\partial F_1}{\partial y} = \frac{\partial F_2}{\partial x} \qquad \frac{\partial F_2}{\partial z} = \frac{\partial F_3}{\partial y} \qquad \frac{\partial F_1}{\partial z} = \frac{\partial F_3}{\partial x} \tag{4.10}$$

Equations (4.10) will be valid if and only if **curl F** = **0**, as one sees easily from the definition of **curl F.**

Some of the problems of section 4.3 may be solved quite easily by using this test. For instance, consider the vector field $\mathbf{F} = y\mathbf{i} + x\mathbf{j} + x^2\mathbf{k}$. Equations (4.10) written would demand

$$\frac{\partial}{\partial y}(y) = \frac{\partial}{\partial x}(x) \qquad \frac{\partial}{\partial z}(x) = \frac{\partial}{\partial y}(x^2) \qquad \frac{\partial}{\partial z}(y) = \frac{\partial}{\partial x}(x^2)$$

The first two of these equations are valid but the third is not, and so the vector field is not conservative.

A vector field whose curl vanishes everywhere is said to be *irrotational.* Now let us turn to the theorem that justifies our claim.

THEOREM 4.3 *A vector field* **F** *defined and continuously differentiable in a simply connected domain D is conservative if and only if* **curl F** = **0** *throughout D.*

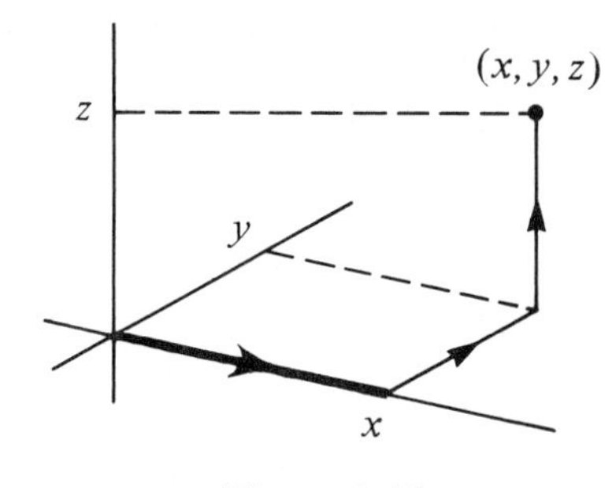

Figure 4.15

The "only if" statement is trivial; $\nabla \times \nabla\phi = \mathbf{0}$ always. So we need to show that irrotational fields are conservative in simply connected domains. We shall give a proof here for the special case when D is all of space. In exercise 12 the reader is invited to extend the argument to spherical and rectangular domains. At the end of the section, a proof for star-shaped domains is given, as optional reading. These restricted forms suffice for most practical applications. The extension of the result to arbitrary, simply connected domains requires some topological gymnastics, and we shall not go into this.

Proof of "if" part when D is all space We are given the information that $\mathbf{curl\,F} = \mathbf{0}$ everywhere, and our goal is to show that a scalar field ϕ exists such that $\mathbf{F} = \mathbf{grad}\,\phi$. Thinking ahead, we know that the line integrals of $\mathbf{F}$ can be computed in terms of ϕ; thus let us *define* ϕ by a line integral and then try to prove the theorem.

Specifically, we define $\phi(x,y,z)$ to be the line integral of $\mathbf{F}$ from (0,0,0) to (x,y,z) along the following curve (fig. 4.15):

(*i*) from (0,0,0) to $(x,0,0)$ along the x axis,
(*ii*) from $(x,0,0)$ to $(x,y,0)$ parallel to the y axis,
(*iii*) from $(x,y,0)$ to (x,y,z) parallel to the z axis.

The parameterization is trivial, and we have

$$\phi(x,y,z) = \int_0^x F_1(t,0,0)\,dt + \int_0^y F_2(x,t,0)\,dt + \int_0^z F_3(x,y,t)\,dt \qquad (4.11)$$

At this point we cannot assume that line integrals are independent of path. The function $\phi(x,y,z)$, however, is computed in terms of a *specific* path; it therefore is well defined.

Now we proceed to show **grad** ϕ = **F**, by components. The z component is easy:

$$\frac{\partial\phi}{\partial z} = \lim_{\Delta z\to 0}\frac{\phi(x, y, z + \Delta z) - \phi(x,y,z)}{\Delta z}$$

$$= \lim_{\Delta z\to 0}\frac{\int_z^{z+\Delta z} F_3(x,y,t)\,dt}{\Delta z} = F_3(x,y,z)$$

[by reasoning as in eq. (4.8)].

For the y component we have

$$\frac{\partial\phi}{\partial y} = \lim_{\Delta y\to 0}\frac{\phi(x, y + \Delta y, z) - \phi(x,y,z)}{\Delta y}$$

$$= \lim_{\Delta y\to 0}\frac{1}{\Delta y}\left(\int_0^{y+\Delta y} F_2(x,t,0)\,dt + \int_0^z F_3(x, y + \Delta y, t)\,dt - \int_0^y F_2(x,t,0)\,dt - \int_0^z F_3(x,y,t)\,dt\right)$$

$$= \lim_{\Delta y\to 0}\frac{\int_y^{y+\Delta y} F_2(x,t,0)\,dt}{\Delta y} + \lim_{\Delta y\to 0}\int_0^z \frac{F_3(x, y + \Delta y, t) - F_3(x,y,t)}{\Delta y}\,dt$$

$$= F_2(x,y,0) + \int_0^z \frac{\partial F_3(x,y,t)}{\partial y}\,dt$$

(Here we have interchanged a limiting process and an integral; the justification relies on continuity arguments and is omitted.)

Now we use the second of eqs. (4.10), keeping in mind that in the above formula the third coordinate is named t, not z:

$$\frac{\partial\phi}{\partial y} = F_2(x,y,0) + \int_0^z \frac{\partial F_2(x,y,t)}{\partial t}\,dt$$

$$= F_2(x,y,0) + F_2(x,y,z) - F_2(x,y,0)$$

$$= F_2(x,y,z)$$

We sketch the computation for the x component, allowing the reader to supply the details:

$$\frac{\partial\phi}{\partial x} = F_1(x,0,0) + \int_0^y \frac{\partial F_2(x,t,0)}{\partial x}\,dt + \int_0^z \frac{\partial F_3(x,y,t)}{\partial x}\,dt$$

$$= F_1(x,0,0) + \int_0^y \frac{\partial F_1(x,t,0)}{\partial t}\,dt + \int_0^z \frac{\partial F_1(x,y,t)}{\partial t}\,dt$$

$$= F_1(x,0,0) + F_1(x,y,0) - F_1(x,0,0) + F_1(x,y,z) - F_1(x,y,0)$$

$$= F_1(x,y,z)$$

This completes the proof that **grad** ϕ = **F**.

Example 4.7 Show that $\mathbf{F} = 2xy\mathbf{i} + (x^2 + 1)\mathbf{j} + 6z^2\mathbf{k}$ is conservative, and find a scalar potential ϕ.

Solution We use the test of eqs. (4.10), which is acceptable since this field **F** is defined and continuously differentiable throughout space:

$$\frac{\partial F_1}{\partial y} = \frac{\partial F_2}{\partial x} = 2x$$

$$\frac{\partial F_2}{\partial z} = \frac{\partial F_3}{\partial y} = 0$$

$$\frac{\partial F_1}{\partial z} = \frac{\partial F_3}{\partial x} = 0$$

The curl is zero; hence the field is conservative.

The potential may be found by the method of example 4.6, or we may use the line integral of eq. (4.11). Let us try the latter technique.

Inserting the given expression for **F**, we find

$$\phi(x,y,z) = \int_0^x 0\,dt + \int_0^y (x^2 + 1)\,dt + \int_0^z 6t^2\,dt$$

$$= 0 + (x^2 t + t)\Big|_0^y + 2t^3\Big|_0^z = x^2 y + y + 2z^3$$

As a check, let us compute ϕ by integrating **F** along the straight-line segment from (0,0,0) to (x,y,z) parametrized by

$$\mathbf{R}(t) = tx\mathbf{i} + ty\mathbf{j} + tz\mathbf{k} \qquad (0 \le t \le 1)$$

As before, we find

$$\phi(x,y,z) = \int_0^1 [2(tx)(ty)\mathbf{i} + (t^2x^2 + 1)\mathbf{j} + 6(t^2z^2)\mathbf{k}] \cdot [x\mathbf{i} + y\mathbf{j} + z\mathbf{k}]\,dt$$

$$= 2x^2y \int_0^1 t^2\,dt + x^2y \int_0^1 t^2\,dt + y \int_0^1 dt + 6z^3 \int_0^1 t^2\,dt$$

$$= x^2y + y + 2z^3$$

Example 4.8 Use eq. (4.11) to find a potential for

$$\mathbf{F} = (3x^2yz + y + 5)\mathbf{i} + (x^3z + x - z)\mathbf{j} + (x^3y - y + 7)\mathbf{k}$$

which has the value 10 at the origin.

Solution By eq. (4.11) we have

$$\begin{aligned}\phi(x,y,z) &= 10 + \int_0^x 5\,dt + \int_0^y x\,dt + \int_0^z (x^3y - y + 7)\,dt \\ &= 10 + 5t\Big|_0^x + xt\Big|_0^y + (x^3yt - yt + 7t)\Big|_0^z \\ &= 10 + 5x + xy + x^3yz - yz + 7z\end{aligned}$$

As a rule, the reader should use the *test* given in this section to determine whether a given field **F** is conservative, but use the *method* of example 4.6 to construct the potential ϕ. The reason we do not advise using (4.11) of this section is that it may be a little tricky for most students to use correctly.

It is very enlightening to investigate the vector field

$$\mathbf{F} = \frac{-y\mathbf{i} + x\mathbf{j}}{x^2 + y^2}$$

The reader should verify that **curl** $\mathbf{F} = \mathbf{0}$, but **F** is not conservative since its line integrals around closed paths are not always zero, as shown in exercise 4 of section 4.3. This does not contradict the theorem, because **F** fails to be defined on the z axis, where $x^2 + y^2 = 0$, and thus the domain D is not simply connected.

In exercise 5 of section 4.3, we tried to shake the reader up by suggesting that this **F** was a gradient, namely, the gradient of $\phi = \tan^{-1}(y/x)$. This definition is subject, of course, to quadrant ambiguities, so we might try $\mathbf{F} = \mathbf{grad}\,\theta$, where θ is the polar angle defined in section 2.5. However, the polar angle jumps from π to $-\pi$ as we cross the negative x axis, so it is discontinuous there and its gradient is not defined (or is infinite!). Thus the theorem remains unchallenged.

Optional Reading: Proof of Theorem 4.3 for Star-shaped Domains

As before, we have to prove only the "if" part. Assuming D is star-shaped with respect to the point P, we can *define* ϕ at the point Q to be the line integral of **F** taken along *the straight-line segment from P to Q*. Again notice that the function $\phi(Q)$ is computed *in terms of a specific path;* therefore it is not ambiguous.

Now we proceed to show $\mathbf{grad}\,\phi = \mathbf{F}$. First we parametrize the path of integration. Let (x_0,y_0,z_0) be the coordinates of P, and (x,y,z) be coordinates for Q. Since we customarily use **R** to denote the vector (x,y,z), which now is an *endpoint* in our integral, we shall write

$$\phi(x,y,z) = \int_P^Q \mathbf{F} \cdot d\mathbf{r} \tag{4.12}$$

using $\mathbf{r}(t)$ to designate the path of integration. Then a parametrization for the segment is

$$\begin{aligned}\mathbf{r}(t) &= [x_0 + t(x - x_0)]\mathbf{i} + [y_0 + t(y - y_0)]\mathbf{j} + [z_0 + t(z - z_0)]\mathbf{k} \\ &= \mathbf{R}_0 + t(\mathbf{R} - \mathbf{R}_0) \qquad (0 \leq t \leq 1)\end{aligned} \tag{4.13}$$

The explicit dependence of $\mathbf{F}$ on the parameter t in the integral (4.12) is given by

$$\begin{aligned}\mathbf{F} &= \mathbf{F}[x_0 + t(x - x_0), y_0 + t(y - y_0), z_0 + t(z - z_0)] \\ &= \mathbf{F}(X,Y,Z)\end{aligned}$$

where we have abbreviated the first argument, $x_0 + t(x - x_0)$, of $\mathbf{F}$ by X, the second by Y, and the third by Z.

Now we compute the gradient of ϕ. Since ∇ operates only on the variables x, y, and z,

$$\nabla = \mathbf{i}\frac{\partial}{\partial x} + \mathbf{j}\frac{\partial}{\partial y} + \mathbf{k}\frac{\partial}{\partial z}$$

(and *not* on t), we can bring the differential operator inside the integral. Using identity (3.41), we find

$$\begin{aligned}\nabla\phi &= \int_0^1 \nabla\left(\mathbf{F}\cdot\frac{d\mathbf{r}}{dt}\right) dt \\ &= \int_0^1 \left[(\mathbf{F}\cdot\nabla)\frac{d\mathbf{r}}{dt} + \left(\frac{d\mathbf{r}}{dt}\cdot\nabla\right)\mathbf{F} + \mathbf{F}\times\left(\nabla\times\frac{d\mathbf{r}}{dt}\right) + \frac{d\mathbf{r}}{dt}\times(\nabla\times\mathbf{F})\right] dt\end{aligned} \tag{4.14}$$

Here we must be very careful in our interpretation. As we said, ∇ operates on x, y, and z; but the arguments of $\mathbf{F}$ are X, Y, and Z. Thus we cannot identify, for example, $\nabla \times \mathbf{F}(X,Y,Z)$ as the curl of $\mathbf{F}$, evaluated at (X,Y,Z). This latter would be

$$\nabla^* \times \mathbf{F}(X,Y,Z)$$

where ∇^* denotes the operator

$$\nabla^* = \mathbf{i}\frac{\partial}{\partial X} + \mathbf{j}\frac{\partial}{\partial Y} + \mathbf{k}\frac{\partial}{\partial Z}$$

However, we have the following relation, because of the definition of X in terms of x:

$$\frac{\partial \mathbf{F}}{\partial x} = \frac{\partial X}{\partial x}\frac{\partial \mathbf{F}}{\partial X} = t\frac{\partial \mathbf{F}}{\partial X} \tag{4.15}$$

Similarly,

$$\begin{aligned}\frac{\partial \mathbf{F}}{\partial y} &= t\frac{\partial \mathbf{F}}{\partial Y} \\ \frac{\partial \mathbf{F}}{\partial z} &= t\frac{\partial \mathbf{F}}{\partial Z}\end{aligned} \tag{4.15$'$}$$

It follows from these identities that

$$\nabla \times \mathbf{F}(X,Y,Z) = t\,\nabla^* \times \mathbf{F}(X,Y,Z) = t\,\mathbf{curl}\,\mathbf{F}(X,Y,Z)$$

By hypothesis, the curl of $\mathbf{F}$ is zero; hence, from the above,

$$\nabla \times \mathbf{F} = \mathbf{0}$$

in expression (4.14).

Furthermore, since

$$\frac{d\mathbf{r}}{dt} = \mathbf{R} - \mathbf{R}_0$$

we have

$$\nabla \times \frac{d\mathbf{r}}{dt} = \mathbf{0}$$

and

$$(\mathbf{F} \cdot \nabla)\frac{d\mathbf{r}}{dt} = \mathbf{F}$$

[Recall identities (3.34) and (3.36).]

Combining these data in (4.14), we have shown

$$\nabla\phi(x,y,z) = \int_0^1 \left[\mathbf{F} + \left(\frac{d\mathbf{r}}{dt} \cdot \nabla\right)\mathbf{F}\right] dt \tag{4.16}$$

One more simplification is possible. If we differentiate $t\mathbf{F}$ with respect to t along the curve, using the chain rule we find

$$\frac{d}{dt}\{t\mathbf{F}[x_0 + t(x - x_0), y_0 + t(y - y_0), z_0 + t(z - z_0)]\}$$

$$= t\frac{\partial \mathbf{F}}{\partial X}(x - x_0) + t\frac{\partial \mathbf{F}}{\partial Y}(y - y_0) + t\frac{\partial \mathbf{F}}{\partial Z}(z - z_0) + \mathbf{F}$$

By eq. (4.15) this can be written

$$\begin{aligned}\frac{d(t\mathbf{F})}{dt} &= (x - x_0)\frac{\partial \mathbf{F}}{\partial x} + (y - y_0)\frac{\partial \mathbf{F}}{\partial y} + (z - z_0)\frac{\partial \mathbf{F}}{\partial z} + \mathbf{F} \\ &= \left(\frac{d\mathbf{r}}{dt} \cdot \nabla\right)\mathbf{F} + \mathbf{F}\end{aligned} \tag{4.17}$$

Using this in eq. (4.16), we get

$$\nabla\phi(x,y,z) = \int_0^1 \frac{d}{dt}(t\mathbf{F})\,dt = t\mathbf{F}\Big|_0^1 = \mathbf{F}(x,y,z)$$

We have succeeded in proving that the gradient of ϕ is **F**, that is, that **F** is conservative!

EXERCISES

1. Use the zero curl test to determine whether the following fields are conservative:
   (a) $\mathbf{F} = (12xy + yz)\mathbf{i} + (6x^2 + xz)\mathbf{j} + xy\mathbf{k}$
   (b) $\mathbf{F} = ze^{xz}\mathbf{i} + xe^{xz}\mathbf{k}$
   (c) $\mathbf{F} = \sin x\,\mathbf{i} + y^2\mathbf{j} + e^z\mathbf{k}$
   (d) $\mathbf{F} = 3x^2yz^2\mathbf{i} + x^3z^2\mathbf{j} + x^3yz\mathbf{k}$
   (e) $\mathbf{F} = \dfrac{2x}{x^2 + y^2}\mathbf{i} + \dfrac{2y}{x^2 + y^2}\mathbf{j} + 2z\mathbf{k}$

   For which one of the fields is the test not applicable? How, then, can you test this field to determine whether it is conservative in its domain of definition?
2. Let **F** and **G** be conservative vector fields with potentials ϕ and ψ respectively. Is the vector field **F** + **G** conservative? If so, determine a potential for it.
3. Show that the scalar field $\phi = -1/|\mathbf{R}|$, which is defined everywhere except at the origin, is a potential function for the vector field $\mathbf{R}/|\mathbf{R}|^3$, where $\mathbf{R} = x\mathbf{i} + y\mathbf{j} + z\mathbf{k}$,
   (a) by writing ϕ in terms of x, y, and z and computing its gradient.
   (b) by inspection, using the second and third fundamental properties of the gradient listed in section 3.1.
4. A force field is defined by

   $$\mathbf{F} = \frac{x\mathbf{i} + y\mathbf{j} + z\mathbf{k}}{(x^2 + y^2 + z^2)^{3/2}}$$

   at all points in space except the origin. A particle is moved along the straight-line segment from the point (1,2,3) to the point (2,3,5). What is the work done by the force on the particle? [*Hint:* Avoid a lot of work (!) by making use of the statement of exercise 3.]
5. Would your answer to exercise 4 be any different if the path extending from (1,2,3) to (2,3,5) were not straight?
6. Let $\mathbf{F} = (6x - 2e^{2x}y^2)\mathbf{i} - 2ye^{2x}\mathbf{j} + \cos z\,\mathbf{k}$.
   (a) Determine whether **F** is conservative or not. Explain.
   (b) Evaluate the line integral $\int_C \mathbf{F} \cdot d\mathbf{R}$ along the path parametrized by

   $$\mathbf{R}(t) = t\mathbf{i} + (t - 1)(t - 2)\mathbf{j} + \frac{\pi}{2}t^3\mathbf{k} \qquad (0 \le t \le 1)$$

   (c) Evaluate $\int_C \mathbf{F} \cdot d\mathbf{R}$ along

   $$\mathbf{R}(t) = \frac{1}{2}(t - 1)\mathbf{i} + t(3 - t)\mathbf{j} + \frac{\pi}{4}(t - 1)\mathbf{k} \qquad (1 \le t \le 3)$$

7. Let

$$\mathbf{F} = [(1 + x)e^{x+y}]\mathbf{i} + [xe^{x+y} + 2y]\mathbf{j} - 2z\mathbf{k}$$
$$\mathbf{G} = [(1 + x)e^{x+y}]\mathbf{i} + [xe^{x+y} + 2z]\mathbf{j} - 2y\mathbf{k}$$

(a) Show that **F** is conservative by finding a potential ϕ for **F**.
(b) Evaluate $\int_C \mathbf{G} \cdot d\mathbf{R}$, where C is the path given by

$$x = (1 - t)e^t \qquad y = t \qquad z = 2t \qquad (0 \le t \le 1)$$

(*Hint:* Take advantage of the similarity between **F** and **G**.)

8. (a) Show that the field $\mathbf{F} = xe^y\mathbf{i} + ye^z\mathbf{j} + ze^x\mathbf{k}$ is not conservative.
(b) Find a potential ϕ for the field **G**, when

$$\begin{aligned}\mathbf{G} = {} & [xe^{-x^2} + (xyz + y)e^{zx} + y^2ze^{xy} + ze^{yz}]\mathbf{i} \\ & + [ye^{-y^2} + (yzx + z)e^{xy} + z^2xe^{yz} + xe^{zx}]\mathbf{j} \\ & + [ze^{-z^2} + (zxy + x)e^{yz} + x^2ye^{zx} + ye^{xy}]\mathbf{k}\end{aligned}$$

9. (a) Find a potential ϕ for the field

$$\mathbf{F} = (2xyz + z^2 - 2y^2 + 1)\mathbf{i} + (x^2z - 4xy)\mathbf{j} + (x^2y + 2xz - 2)\mathbf{k}$$

(b) The field

$$\mathbf{G} = \frac{x}{(x^2 + z^2)^2}\mathbf{i} + \frac{z}{(x^2 + z^2)^2}\mathbf{k}$$

satisfies the condition that $\nabla \times \mathbf{G} = \mathbf{0}$ at all points except on the y axis. Is **G** conservative?

10. Evaluate $\int_{(0,0)}^{(1,2)} (15x^4 - 3x^2y^2)\,dx - 2x^3y\,dy$ along the path $2x^4 - 6xy^3 + 23y = 0$.

11. If **F** and **G** are conservative fields, is $\mathbf{F} \times \mathbf{G}$ necessarily conservative?

12. Show that the proof of theorem 4.3 in the text can be adapted for some other domains D, specifically
(a) the interior of a sphere.
(b) the interior of a parallelepiped with edges parallel to the axes.
(*Hint:* You must verify that all the line integrals are well defined.)

13. A function $f(x,y,z)$ is said to be *homogeneous of degree k* if $f(tx,ty,tz) = t^kf(x,y,z)$. Suppose that the components F_1, F_2, F_3 of the vector field $\mathbf{F}(x,y,z)$ are each homogeneous of degree k, and **curl F** = **0**. Prove

$$\mathbf{F} = \nabla\left(\frac{xF_1 + yF_2 + zF_3}{k + 1}\right)$$

4.5 Optional Reading: Vector Potentials and Solenoidal Fields

In section 4.4 we discussed a *partial converse* to identity (3.42), which states that the curl of a gradient is zero. It is a *converse* because it states that, *if* the curl of a field is zero, the field is a gradient; but it is only a *partial* converse because it is valid only in simply connected domains.

The astute reader will wonder if there is also a converse, or at least a partial converse, to identity (3.43), which asserts that the divergence of a curl is zero. If the divergence of a vector field is zero, is that field necessarily the curl of another vector field? The answer is yes, provided the domain of definition is star-shaped (or simply connected).

A vector field whose divergence is everywhere zero is called *solenoidal.* If $\mathbf{F} = \nabla \times \mathbf{G}$, $\mathbf{G}$ is called a *vector potential* for $\mathbf{F}$. Notice that $\mathbf{G}$ is not unique; in fact, according to eq. (3.42), we can add the gradient of any scalar to $\mathbf{G}$.

Now let us prove the statement about the existence of a vector potential.

THEOREM 4.4 *A vector field* $\mathbf{F}$ *continuously differentiable in a star-shaped domain* D *is solenoidal if and only if there is a vector field* $\mathbf{G}$ *such that* $\mathbf{F} = \mathbf{curl\ G}$ *throughout* D.

Proof The "if" part follows from identity (3.43). For the "only if" statement, we assume that $\mathbf{F}$ is solenoidal in a domain D that is star-shaped with respect to the point P. We wish to find $\mathbf{G}(x,y,z)$ such that $\mathbf{F} = \nabla \times \mathbf{G}$.

The proof is quite similar to that in section 4.4. (In fact, both theorems are special cases of a result known as *Poincaré's lemma.*) We again parameterize the straight-line segment from $P(x_0,y_0,z_0)$ to $Q(x,y,z)$ by eq. (4.13).

$$\begin{aligned}\mathbf{r}(t) &= [x_0 + t(x - x_0)]\mathbf{i} + [y_0 + t(y - y_0)]\mathbf{j} + [z_0 + t(z - z_0)]\mathbf{k} \\ &= X\mathbf{i} + Y\mathbf{j} + Z\mathbf{k} \\ &= \mathbf{R}_0 + t(\mathbf{R} - \mathbf{R}_0)\end{aligned}$$

(using the same notation as in section 4.5).

Now we define $\mathbf{G}(x,y,z)$:

$$\mathbf{G}(x,y,z) = \int_0^1 t\mathbf{F} \times \frac{d\mathbf{r}}{dt}\, dt \tag{4.18}$$

where the dependence of $\mathbf{F}$ and $\mathbf{r}$ on t is exactly as in the previous section. Equation (4.18) is, as it stands, simply the integral of a vector function of t; it is not a line integral. However, it lends an obvious interpretation to an expression like $\int_P^Q t\mathbf{F} \times d\mathbf{r}$.

We compute the curl of $\mathbf{G}$. Again, ∇ may be taken inside the integral, and by identity (3.39) we have

$$\begin{aligned}\nabla \times \mathbf{G} &= \int_0^1 t\, \nabla \times \left(\mathbf{F} \times \frac{d\mathbf{r}}{dt}\right) dt \\ &= \int_0^1 \left[\left(\frac{d\mathbf{r}}{dt} \cdot \nabla\right) \mathbf{F} - (\mathbf{F} \cdot \nabla) \frac{d\mathbf{r}}{dt} \right. \\ &\qquad \left. + \left(\nabla \cdot \frac{d\mathbf{r}}{dt}\right) \mathbf{F} - (\nabla \cdot \mathbf{F}) \frac{d\mathbf{r}}{dt}\right] t\, dt\end{aligned} \tag{4.19}$$

As in section 4.4,

$$\frac{d\mathbf{r}}{dt} = \mathbf{R} - \mathbf{R}_0$$

so that

$$\nabla \cdot \frac{d\mathbf{r}}{dt} = 3$$

and

$$(\mathbf{F} \cdot \nabla) \frac{d\mathbf{r}}{dt} = \mathbf{F}$$

For the reasons stated in section 4.4, $\nabla \cdot \mathbf{F}(X,Y,Z)$ is not the divergence of $\mathbf{F}$, but because

$$\nabla \cdot \mathbf{F}(X,Y,Z) = t\, \nabla^* \cdot \mathbf{F}(X,Y,Z)$$

with ∇^* defined as before, we see that $\nabla^* \cdot \mathbf{F} = 0$ implies $\nabla \cdot \mathbf{F} = 0$.

Incorporating all of this into eq. (4.19), we get

$$\nabla \times \mathbf{G} = \int_0^1 \left[\left(\frac{d\mathbf{r}}{dt} \cdot \nabla \right) \mathbf{F} - \mathbf{F} + 3\mathbf{F} \right] t\, dt \tag{4.20}$$

Now using eq. (4.17) we find that

$$\frac{d(t^2\mathbf{F})}{dt} = t\,\frac{d(t\mathbf{F})}{dt} + (t\mathbf{F})\frac{dt}{dt} = 2t\mathbf{F} + t\left(\frac{d\mathbf{r}}{dt} \cdot \nabla\right)\mathbf{F}$$

This is precisely what we have in eq. (4.20); therefore

$$\nabla \times \mathbf{G} = \int_0^1 \frac{d}{dt}(t^2\mathbf{F})\, dt = t^2\mathbf{F}\Big|_0^1 = \mathbf{F}(x,y,z)$$

That is, **F** is the curl of **G**.

Example 4.9 Find a vector potential for the uniform (constant) vector field $\mathbf{F}(x,y,z) = \mathbf{A}$.

Solution Obviously **F** is solenoidal (as well as irrotational). Taking P to be the origin in eq. (4.18), we have the parameterization of the segment PQ:

$$\mathbf{r}(t) = t\mathbf{R}$$

Therefore

$$\mathbf{G}(x,y,z) = \int_0^1 t\mathbf{F} \times \frac{d\mathbf{r}}{dt}\, dt = \mathbf{A} \times \mathbf{R} \int_0^1 t\, dt = \frac{\mathbf{A} \times \mathbf{R}}{2}$$

and $\mathbf{A} = \nabla \times \left\{\dfrac{\mathbf{A} \times \mathbf{R}}{2}\right\}$. (For future reference we note also that **A** has a scalar potential given by $\phi = \mathbf{A} \cdot \mathbf{R}$.)

Example 4.10 Find a vector potential for $\mathbf{F} = \mathbf{A} \times \mathbf{R}$, where **A** is a constant vector (recall example 3.25, where **F** was identified as a fluid velocity field with uniform angular velocity).

Solution The verification that $\nabla \cdot \mathbf{F} = 0$ is immediate. Taking P to be the origin in the above equations, we have the parameterization of the segment PQ:

$$\mathbf{r}(t) = t\mathbf{R}$$

Therefore, since $\mathbf{F} = \mathbf{A} \times \mathbf{r}(t)$ in eq. (4.18),

$$\begin{aligned} \mathbf{G}(x,y,z) &= \int_0^1 t(\mathbf{A} \times t\mathbf{R}) \times \mathbf{R}\, dt \\ &= (\mathbf{A} \times \mathbf{R}) \times \mathbf{R} \int_0^1 t^2\, dt \end{aligned}$$

So we obtain

$$\mathbf{G}(x,y,z) = \frac{1}{3}(\mathbf{A} \times \mathbf{R}) \times \mathbf{R}$$

In exercise 1, the reader is invited to verify that $\nabla \times \mathbf{G} = \mathbf{F}$.

As we saw in figure 4.14, the scalar potential for an irrotational field is easy to visualize; its equipotential surfaces intersect the field's flow lines orthogonally. The vector potential, however, seems rather more forbidding. We can't draw "equipotential surfaces" for a *vector* field. We have to visualize it through its own flow lines. Now if we were given a diagram of the *potential* field $\mathbf{G}(x,y,z)$, it would be straightforward to sketch its curl, $\mathbf{F} = \nabla \times \mathbf{G}$; the flow lines of **F** would simply be the *vortex lines* for **G**. So, reversing the argument, given **F** we visualize the flow lines of **G** as being *wrapped around* those of **F**.

Figure 4.16 depicts the flow lines for the vector potential for the constant field derived in example 4.9 (as well as the equipotential surfaces for the scalar potential). The flow lines for the angular velocity field discussed in example 4.10 and its vector potential are depicted in figure 4.17.

This visualization, however, is slightly flawed. Just as the scalar potential is only determined up to a constant, any *gradient* can be added to the vector potential **G** without affecting its relationship to **F**:

$$\mathbf{F} = \nabla \times \mathbf{G} = \nabla \times \{\mathbf{G} + \nabla\psi\}$$

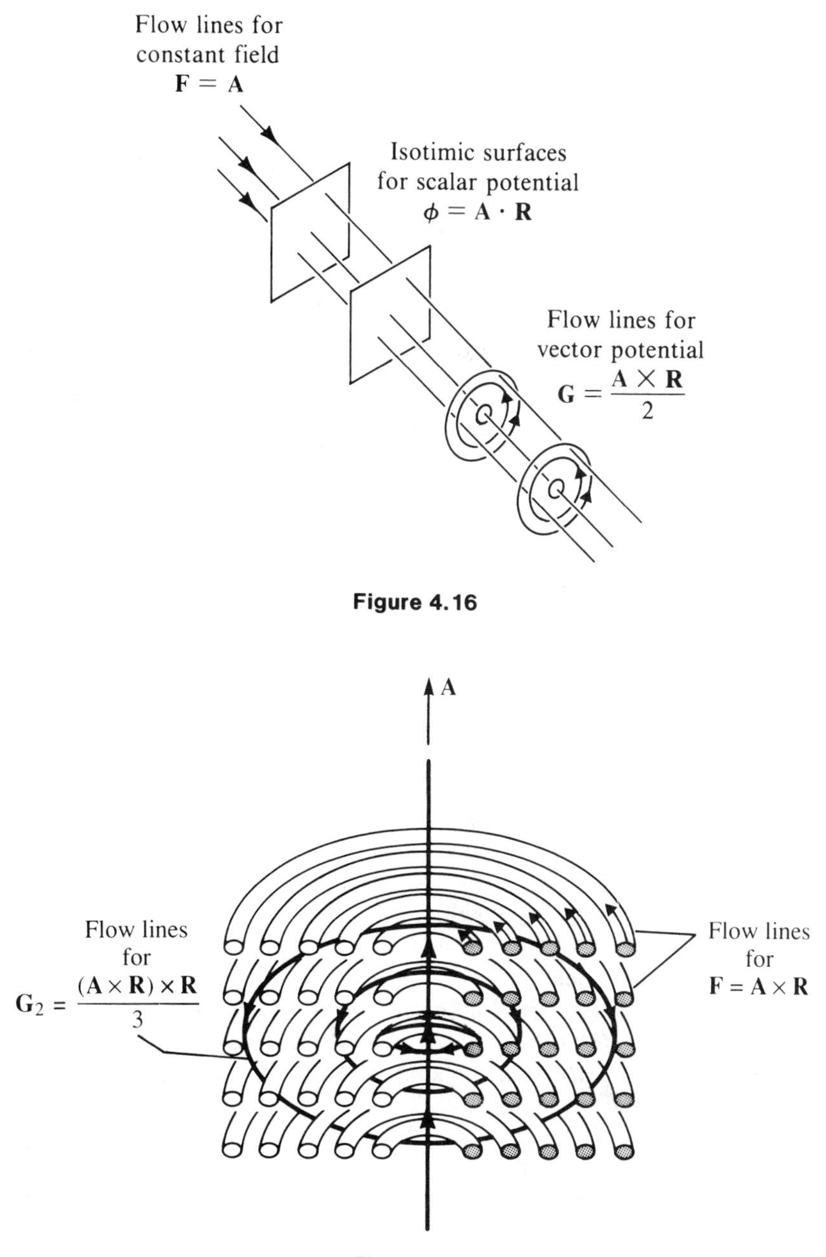

Figure 4.16

Figure 4.17

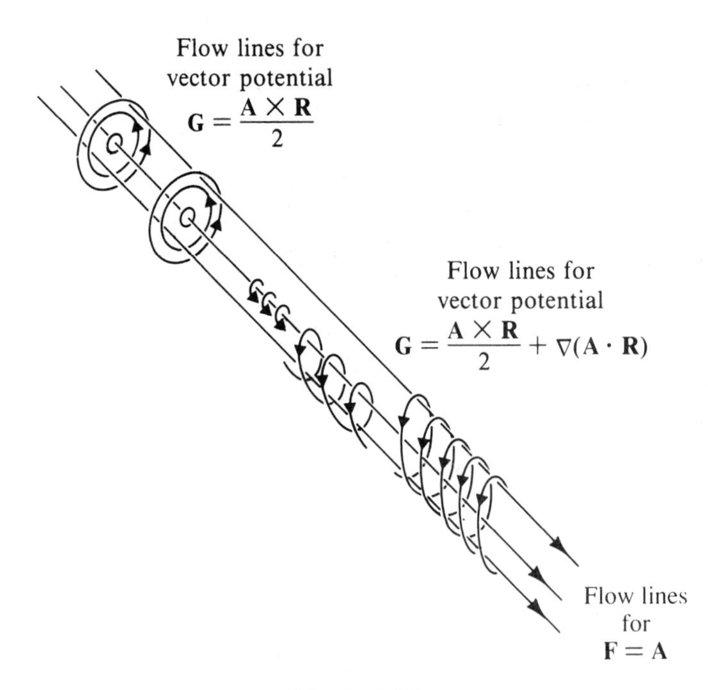

Figure 4.18

We will elaborate on this shortly, but for now observe that if we take the vector potential

$$\mathbf{G}_1 = \frac{\mathbf{A} \times \mathbf{R}}{2} + \nabla(\mathbf{A} \cdot \mathbf{R}) = \frac{\mathbf{A} \times \mathbf{R}}{2} + \mathbf{A}$$

for the constant field **A**, then its flow lines acquire a "downstream" component as depicted in figure 4.18.

If one adds the field $\nabla(|\mathbf{R}|^2\mathbf{A} \cdot \mathbf{R})/3$ to the vector potential $\mathbf{G}_1 = (\mathbf{A} \times \mathbf{R}) \times \mathbf{R}/3$ for the rotating fluid field $\mathbf{F} = \mathbf{A} \times \mathbf{R}$, one obtains an equivalent vector potential $\mathbf{G}_2 = (\mathbf{A} \cdot \mathbf{R})\mathbf{R}$ with flow lines depicted in figure 4.19. Still another valid potential, $\mathbf{G}_3 = -|\mathbf{R}|^2\mathbf{A}/2$, is obtained by subtracting $\nabla(|\mathbf{R}|^2\mathbf{R} \cdot \mathbf{A})/6$ from $\mathbf{G}_1$, with flow lines depicted in figure 4.20. It takes a bit of imagination to see how the flow lines of $\mathbf{G}_2$ and $\mathbf{G}_3$ "wrap around" those of **F** (but recall a similar situation in section 3.4, fig. 3.15!).

In this and the previous section we have seen that the vector identities **curl grad** = **0**, div **curl** = 0, have converses in simply connected domains. Table 4.1 may aid in remembering these results.

In many physical applications one must *compute* a vector field $\mathbf{F}(x,y,z)$ from its sources and boundary values. The scalar and vector potentials can then play important roles in simplifying the formulations of the problems.

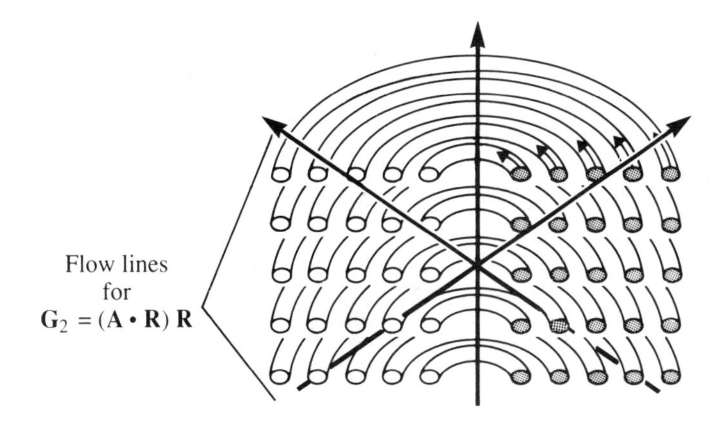

Figure 4.19

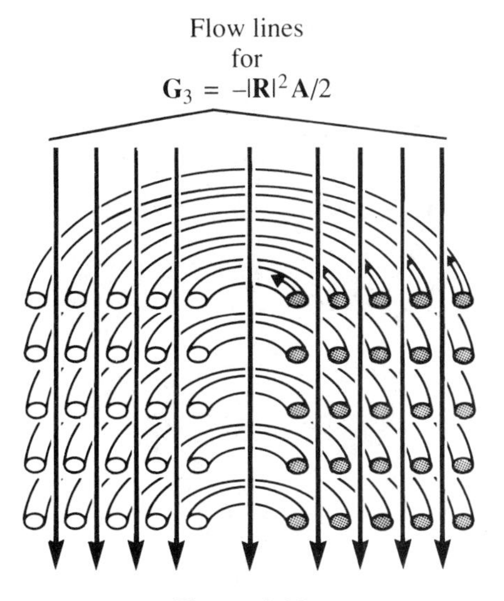

Figure 4.20

TABLE 4.1 The Potential Theorems

Vector Identity	Associated Potential Theorem (Simply Connected Domain)
$\nabla \times \nabla\phi = 0$	$\nabla \times \mathbf{F} = 0$ implies $\mathbf{F} = \nabla\phi$
$\nabla \cdot \nabla \times \mathbf{G} = 0$	$\nabla \cdot \mathbf{F} = 0$ implies $\mathbf{F} = \nabla \times \mathbf{G}$

The use of the scalar potential in solving for an irrotational field is an obvious economy, since it reduces the number of unknowns from three (components of **F**) to one (ϕ). It is enlightening to see how this comes about informally by accounting for the number of constraints and number of "degrees of freedom."

At first glance it might seem that the equation $\nabla \times \mathbf{F} = 0$ imposes three independent conditions on the three components of **F**, and thus there should be no degrees of freedom left in the choice for **F**. And yet we know there is *one*, as expressed by $\mathbf{F} = \nabla\phi$, $\phi(x,y,z)$ unconstrained. This apparent paradox is resolved by the observation that the three conditions specifying $\nabla \times \mathbf{F}$ are *not* independent. Since mathematically any curl must have zero divergence,

$$\nabla \cdot \nabla \times \mathbf{F} = 0 = \frac{\partial}{\partial x}(\nabla \times \mathbf{F})_x + \frac{\partial}{\partial y}(\nabla \times \mathbf{F})_y + \frac{\partial}{\partial z}(\nabla \times \mathbf{F})_z$$

we can see that if (say) only the first two components of $\nabla \times \mathbf{F}$ are constrained to be zero, the third component is also constrained (to be independent of z). Thus the condition that it, too, is zero is partially redundant. **F** has *three* degrees of freedom, the condition $\nabla \times \mathbf{F} = 0$ removes *two* of them, and there is *one* left ($\nabla\phi$).

For the solenoidal case **F** has three degrees of freedom and $\nabla \cdot \mathbf{F} = 0$ blatantly removes one of them, so there should be two left. Since the vector potential **G** has three components, one might expect that the latitude in **G** implied by

$$\mathbf{F} = \nabla \times \mathbf{G} = \nabla \times \{\mathbf{G} + \nabla\psi\}$$

could be exploited to render one of the components superfluous. Actually this is quite easy to accomplish. We simply select $\psi(x,y,z)$ so that $\partial\psi/\partial z$ cancels the z component of **G**:

$$\psi(x,y,z) = -\int^z G_3(x,y,\zeta)d\zeta$$

Example 4.11 Replace the vector potential $\mathbf{G} = \mathbf{A} \times \mathbf{R}/2$ for the constant field $\mathbf{F} = \mathbf{A}$ by one with zero z component.

Solution We have

$$\psi(x,y,z) = -\int^z [A_1 y - A_2 x]/2 \, d\zeta = -A_1 yz/2 + A_2 xz/2$$

Thus

$$\mathbf{G}^*(x,y,z) = \mathbf{A} \times \mathbf{R}/2 + \nabla\psi = (A_2 z - A_3 y/2)\mathbf{i} + (-A_1 z + A_3 x/2)\mathbf{j}$$

The verification that $\nabla \times \mathbf{G}^* = \mathbf{A}$ is left to the reader.

In physics the latitude in the choice of the gradient term $\nabla\psi$ for the vector potential is known as *gauge invariance.* The "gauge condition" described above is rarely used in practice. In electrodynamics, for instance, one has to solve for *two* coupled vector fields—one solenoidal and one irrotational. The gauge is chosen so that the scalar potential becomes superfluous.

The vector potential acquires new significance for *two-dimensional vector fields.* For such fields $\mathbf{F}(x,y,z)$ has no z component, and its x and y components are independent of z:

$$\mathbf{F} = F_1(x,y)\mathbf{i} + \mathbf{F}_2(x,y)\mathbf{j}$$

The condition that $\mathbf{F}$ be solenoidal

$$\nabla \cdot \mathbf{F} = \partial F_1/\partial x + \partial F_2/\partial y = 0$$

is exactly the same as the condition that $\mathbf{k} \times \mathbf{F}$ be irrotational

$$\nabla \times (\mathbf{k} \times \mathbf{F}) = \nabla \times \{-F_2\mathbf{i} + F_1\mathbf{j}\} = \{\partial F_1/\partial x + \partial F_2/\partial y\}\mathbf{k} = 0$$

Thus $\mathbf{k} \times \mathbf{F}$ is derivable from a scalar potential

$$\mathbf{k} \times \mathbf{F} = \nabla\chi : -F_2 = \partial\chi/\partial x \qquad F_1 = \partial\chi/\partial y$$

which implies that $\mathbf{G} = \chi\mathbf{k}$ is a vector potential for $\mathbf{F}$:

$$\nabla \times (\chi\mathbf{k}) = \partial\chi/\partial y\, \mathbf{i} - \partial\chi/\partial x\, \mathbf{j} = F_1\mathbf{i} + F_2\mathbf{j} = \mathbf{F}$$

Since $\mathbf{F}$ only has two components, the single constraint $\nabla \cdot \mathbf{F} = 0$ reduces the number of degrees of freedom to one, and $\mathbf{G}$ has a z component only.

The level curves of the vector potential $|\mathbf{G}| = |\chi| =$ constant are easy to interpret in the x,y plane; they coincide with the flow lines of $\mathbf{F}$! After all,

(*i*) $\mathbf{k} \times \mathbf{F}$ is perpendicular to $\mathbf{F}$;
(*ii*) $\mathbf{k} \times \mathbf{F} = \nabla\chi$; and
(*iii*) $\nabla\chi$ is perpendicular to the curve $|\chi| =$ constant.

Thus the curve $|\chi| =$ constant is everywhere parallel to $\mathbf{F}$. χ is sometimes known as the *stream function.*

Example 4.12 Compute scalar and vector potentials for the vector field $\mathbf{F}(x,y) = x\mathbf{i} - y\mathbf{j}$.

Solution Clearly $\nabla \cdot \mathbf{F} = 0$ and $\nabla \times \mathbf{F} = 0$. By inspection $\mathbf{F} = \nabla\{(x^2 - y^2)/2\} = \nabla\phi$. Similarly $\mathbf{k} \times \mathbf{F} = y\mathbf{i} + x\mathbf{j} = \nabla(xy)$; thus $\mathbf{F} = \nabla \times \{xy\mathbf{k}\} = \nabla \times \{\chi\mathbf{k}\}$. The level curves are shown in figure 4.21.

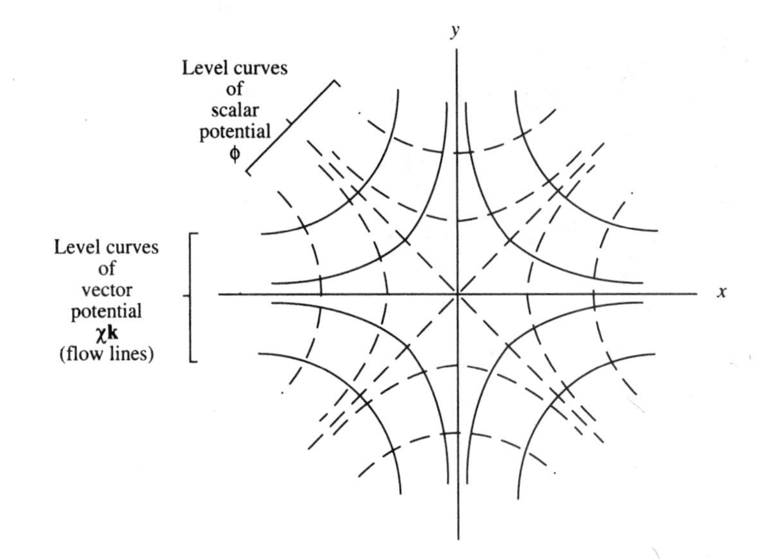

Figure 4.21

EXERCISES

1. Verify that $\mathbf{F} = \nabla \times \mathbf{G}$ in examples 4.9 and 4.10.
2. Find the vector potential for $\mathbf{F} = x\mathbf{j}$.
3. Verify directly that $\mathbf{G} = (\mathbf{A} \times \mathbf{R}) \times \mathbf{R}/|\mathbf{R}|^2$ is a vector potential for $\mathbf{F} = \mathbf{A} \times \mathbf{R}/|\mathbf{R}|^2$. Convince yourself that the corresponding flow line pattern is identical with that in figure 4.17.
4. Verify directly that the field $\mathbf{G} = -\{\log(x^2 + y^2)\}\mathbf{k}$ is a vector potential for $\mathbf{F} = \{2/(x^2 + y^2)\}\mathbf{k} \times \mathbf{R}$. Convince yourself that the flow lines of $\mathbf{F}$ are the same as for the rotating fluid, but the flow lines for $\mathbf{G}$ are vertical as in figure 4.20.
5. Verify the claims made for the various vector potentials $\mathbf{G}_1$, $\mathbf{G}_2$, and $\mathbf{G}_3$ for the rotating fluid field. Compute the divergence of each one.
6. Derive the following expression for the divergence of the vector potential defined by eq. (4.18):

$$\nabla \cdot \mathbf{G}(\mathbf{R}) = (\mathbf{R} - \mathbf{R}_0) \cdot \int_0^1 t^2\,[\mathbf{curl\ F}](t\mathbf{R})\,dt$$

7. Prove: if $\mathbf{F}$ and $\mathbf{G}$ are irrotational, then $\mathbf{F} \times \mathbf{G}$ is solenoidal. Can you find the vector potential for $\mathbf{F} \times \mathbf{G}$? (*Hint:* The problem is considerably easier if you have mastered tensor notation.)
8. Construct a vector potential $\mathbf{G}$ for the rotating fluid field $\mathbf{F} = \mathbf{A} \times \mathbf{R}$ with $G_3 = 0$.

9. Compute vector potentials for the following two-dimensional fields:

(a) $\mathbf{F} = \mathbf{A} = \text{constant}$ (b) $\mathbf{F} = xy\mathbf{i} - \frac{y^2}{2}\mathbf{j}$

(c) $[-y/(x^2 + y^2)]\,\mathbf{i} + [x/(x^2 + y^2)]\,\mathbf{j}$ (*Hint:* Consult exercise 4.)

Sketch the level curves for $|\mathbf{G}|$.

10. Compute scalar and vector potentials for the vector field $\mathbf{F}(x,y) = (x^2 - y^2)\mathbf{i} - 2xy\mathbf{j}$. Sketch the level curves.

4.6 Oriented Surfaces

In chapter 2 we considered, in some detail, the geometry of space curves. We now turn our attention to a study of surfaces. Just as the basic properties of curves hinge on the *tangent vectors,* the behavior of a surface is characterized in terms of the *normal vectors* at each point. [Recall that we have one method for computing normals; namely, $\mathbf{grad}\, f(x,y,z)$ is normal to the surface $f(x,y,z) = C$. Another method will be derived below.]

Keeping in mind that a smooth arc has a continuously turning tangent, we say that a surface S is *smooth* if it is possible to choose a unit normal vector $\mathbf{n}$ at every point of S in such a way that $\mathbf{n}$ varies continuously on S. It is said to be *piecewise smooth* if it consists of a finite number of smooth parts "joined together" (we refrain from specifying this precisely). For example, the surface of a sphere is smooth, whereas the surface of a cube is piecewise smooth (consisting of six smooth surfaces joined together).

At every point of a smooth surface there will, of course, be two choices for the unit normal $\mathbf{n}$. There will therefore be two ways in which we can define a field of unit normal vectors continuous on S. [If, for instance, the surface is given by an equation of the form $f(x,y,z) = C$, then the two fields are

$$\frac{\mathbf{grad}\, f}{|\mathbf{grad}\, f|} \quad \text{and} \quad \frac{-\mathbf{grad}\, f}{|\mathbf{grad}\, f|}$$

(section 3.1).] In choosing one of these two possibilities we *orient* the surface. Thus, there are always two possible orientations of a smooth surface. We have already discussed orientation for the special case of a plane (section 1.11). The situation is somewhat the same for more general surfaces. When a smooth surface has been oriented by choosing a particular unit normal field $\mathbf{n}$, then a positive direction for angles is determined at each point of the surface (fig. 4.22). If the surface is bounded by a regular closed curve C, the orientation also determines what we mean by the *positive direction* along C, by the following rule: an observer on the positive side of the surface (i.e., the side on which $\mathbf{n}$ emerges), walking along the boundary C with the surface at his left, is moving in the positive direction along C.

To produce an orientation on a piecewise smooth surface, we have to orient its smooth parts *consistently*. This means that along every edge shared by two smooth

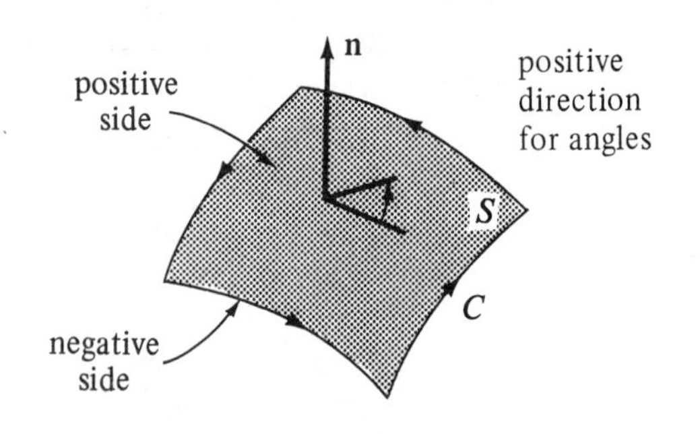

Figure 4.22

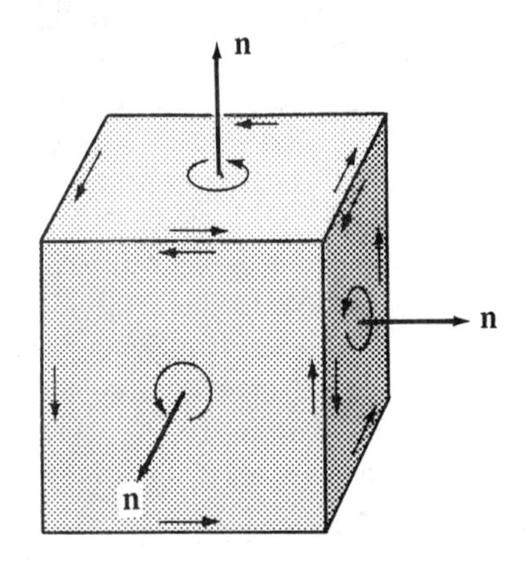

Figure 4.23

parts, the positive direction (on the edge) relative to one of the smooth surfaces is *opposite* to the positive direction relative to the other. Study figures 4.23 and 4.24 to see why this definition is chosen.

A *closed surface* is a surface that has no boundary. Thus the surfaces of figures 4.23 and 4.24 are closed, whereas the surface in figure 4.22 has a boundary and is not closed. It is conventional to take the orientation of a closed surface, which encloses a region of space, to be such that the unit normal **n** always points *away* from the enclosed region, as illustrated in figures 4.23 and 4.24.

A surface can be oriented only if it has two sides; the process of orientation consists essentially in choosing which side we will call "positive" and which "negative." (If the surface is closed, it is more natural to speak of the "outside" and the "inside.")

An example of a nonorientable surface is the *Möbius strip,* obtained by twisting and pasting together the ends of a strip of paper (fig. 4.25). This surface is nonorientable because it has only one side. If **n** is a unit vector normal to the surface at a point P, then as it moves around the strip its direction is reversed by the time

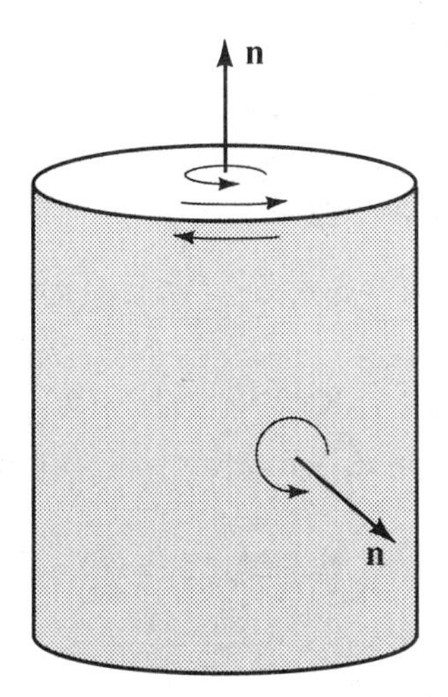

Figure 4.24

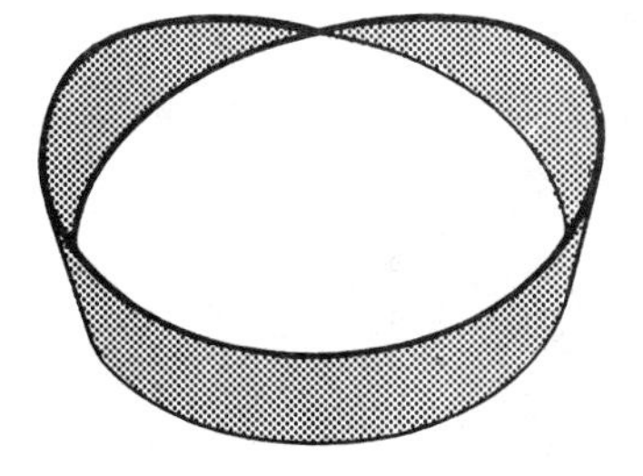

Figure 4.25

it reaches P again. This contradicts the requirement that $\mathbf{n}$ be unambiguous at every point and still vary continuously.

Readers may amuse themselves by taking two strips of paper and preparing two bands, one with a twist and one without. Cut along a central line of the cylindrical band and the Möbius strip. The cylindrical band will separate into two cylindrical bands, but the Möbius strip will not separate into distinct portions. Can you predict the result? What happens if you use a double twist?

Nonorientable surfaces have other mathematical properties that are rather amazing; so amazing, in fact, that we exclude them from further consideration! Henceforth, whenever we say "surface" we mean an orientable surface.

Just as it is possible to write the equation of a space curve in parametric form, giving x, y, and z as functions of a *single* parameter t (because the curve is a *one*-dimensional beast), it is also possible to represent these (*two*-dimensional) surfaces parametrically by giving x, y, and z as functions of two parameters u and v:

$$x = x(u,v) \qquad y = y(u,v) \qquad z = z(u,v) \tag{4.21}$$

In vector notation, we write eq. (4.21) as

$$\mathbf{R} = \mathbf{R}(u,v) \tag{4.21'}$$

Example 4.13 Construct a parametric representation of a plane.

Solution In section 1.10 we saw that points **R** on the plane through $\mathbf{R}_0$ and parallel to **A** and **B** could be expressed as

$$\mathbf{R} = \mathbf{R}_0 + s\mathbf{A} + t\mathbf{B}$$

This is a parametric representation, with u and v taken as s and t. If the plane were described by a nonparametric equation

$$ax + by + cz = d$$

then a parameterization would be

$$x = u \qquad y = v \qquad z = \frac{d - au - bv}{c}$$

(as long as $c \neq 0$).

Example 4.14 Construct a parametric representation of a sphere.

Solution As we saw in section 3.10, spheres are surfaces of constant r in spherical coordinates. For a sphere of radius a, then, the parameterization in terms of the coordinates ϕ and θ is given by the transformation eqs. (3.61):

$$x = a \sin\phi \cos\theta \qquad y = a \sin\phi \sin\theta \qquad z = a \cos\phi$$

See figure 4.26.

Example 4.15 Construct a parametric representation of a right circular cylinder.

Solution The cylinders are surfaces of constant ρ in cylindrical coordinates. The cylinder of radius a in figure 4.27 is thus parameterized in terms of θ and z by [eqs. (3.50)]

$$x = a \cos\theta \qquad y = a \sin\theta \qquad z = z$$

Example 4.16 Construct a parametric representation of a cone.

Solution Cones were characterized as surfaces of constant ϕ in spherical coordinates (fig. 3.23). Thus the transformation equations (3.61) provide the parametrization in terms of r and θ:

$$x = r \sin\phi_0 \cos\theta \qquad y = r \sin\phi_0 \sin\theta \qquad z = r \cos\phi_0$$

See figure 4.28.

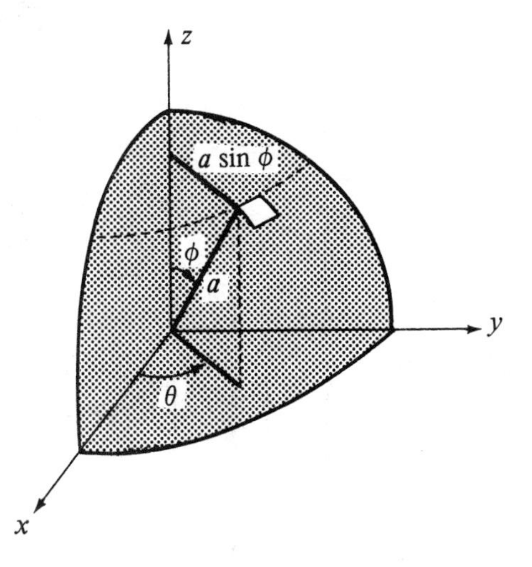

Figure 4.26

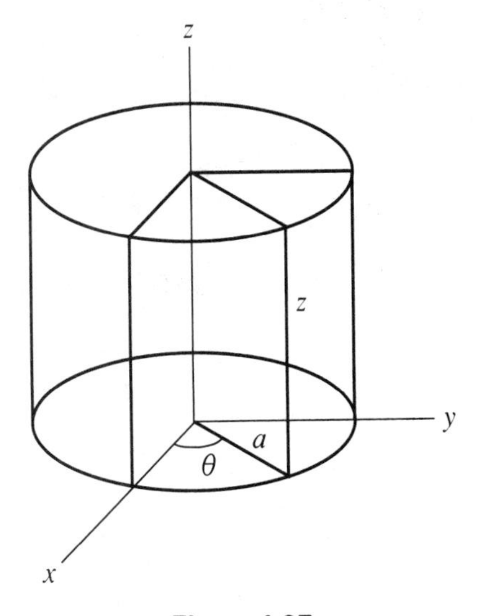

Figure 4.27

In fact it should be clear that the parametric equations of any surface of revolution—or "axially symmetric" surface—can be expressed as

$$x = v \cos \theta \qquad y = v \sin \theta \qquad z = f(v)$$

when the surface is generated by rotating the graph of $z = f(x)$ about the z axis (fig. 4.29).

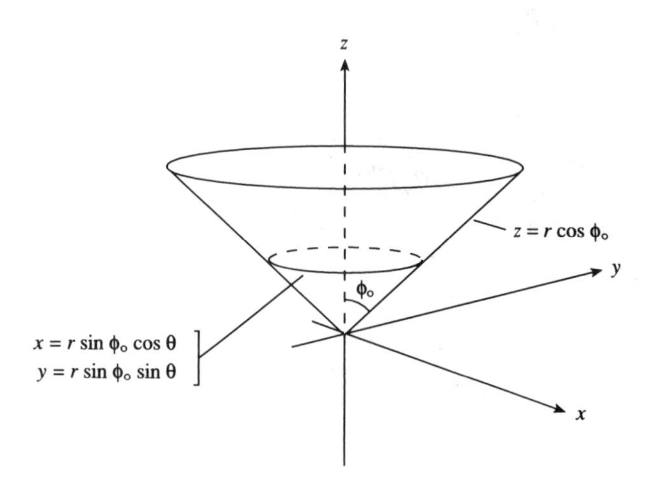

Figure 4.28

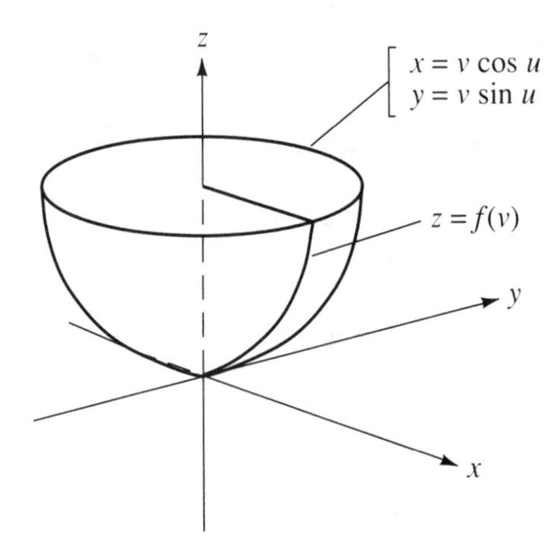

Figure 4.29

As the parameters u and v vary, the tip of the position vector $\mathbf{R}(u,v)$ generates the surface. In particular, if we fix the value of, say, v, and let u vary, then $\mathbf{R}(u,v)$ traces out a *one*-dimensional subset of points in the surface, that is, a curve lying in the surface. For a different fixed value of v, $\mathbf{R}(u,v)$ traces out a different curve in the surface. In fact, we can think of the surface itself, defined by eq. (4.21) with u and v varying independently, as being composed of "ribs" that are the curves given by (4.21) for variable u, fixed v.

Of course, it is equally valid to picture the surface as composed of ribs $\mathbf{R}(u,v)$ with v varying, u constant. These two families criss-cross and cover the surface like a fish net (see fig. 4.30).

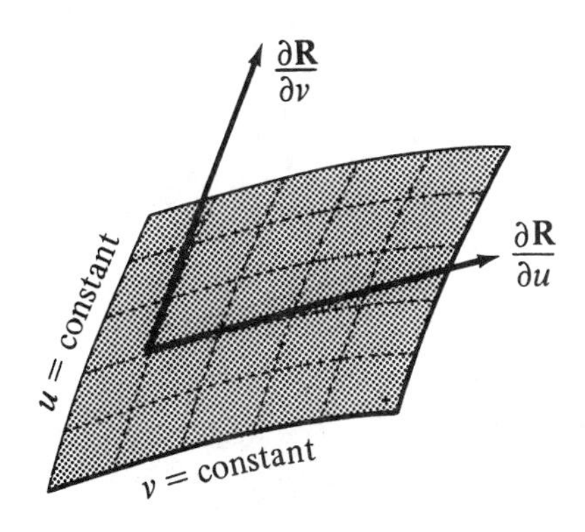

Figure 4.30

This is a very useful point of view because it allows us to apply the methods of chapter 2 to these curves, to learn about the underlying surface. For example, we know that a vector tangent to a curve with v = constant is given by

$$\frac{\partial \mathbf{R}}{\partial u} = \frac{\partial x}{\partial u}\mathbf{i} + \frac{\partial y}{\partial u}\mathbf{j} + \frac{\partial z}{\partial u}\mathbf{k} \tag{4.22}$$

and, similarly, $\partial\mathbf{R}/\partial v$ is tangent to a curve with u = constant (fig. 4.30). We assume here that the relevant derivatives exist, that $\partial\mathbf{R}/\partial u$ and $\partial\mathbf{R}/\partial v$ are nonzero and nonparallel at every point, and that these derivatives are continuous on the surface. Since both vectors are tangent to curves in the surface, they are tangent to the surface itself. Therefore, the vector

$$\frac{\partial \mathbf{R}}{\partial u} \times \frac{\partial \mathbf{R}}{\partial v} \tag{4.23}$$

is *normal* to the surface.

To recap, we have derived two ways of computing normals to a surface. If the surface is specified nonparametrically by $f(x,y,z) = C$, then $\mathbf{grad}\, f$ is a normal vector; if the surface is given parametrically through eqs. (4.21), then (4.23) is a normal vector. In both cases a *unit* normal $\mathbf{n}$ is obtained by dividing the normal by its length, and the induced orientation is reversed by changing the sign of $\mathbf{n}$ [or, in the parametric case, using $(\partial\mathbf{R}/\partial v) \times (\partial\mathbf{R}/\partial u)$ instead of eq. (4.23)].

Example 4.17 Write the equation for the plane tangent to the surface given by

$$x = u^2 \qquad y = uv \qquad z = v$$

at the point corresponding to $u = 1$, $v = 2$.

Solution The point corresponding to $u = 1$, $v = 2$ is $\mathbf{R}_0 = \mathbf{i} + 2\mathbf{j} + 2\mathbf{k}$. The tangents to a constant v, constant u curves are

$$\frac{\partial \mathbf{R}}{\partial u} = 2u\mathbf{i} + v\mathbf{j} = 2\mathbf{i} + 2\mathbf{j}$$

$$\frac{\partial \mathbf{R}}{\partial v} = u\mathbf{j} + \mathbf{k} = \mathbf{j} + \mathbf{k}$$

Hence a normal to the plane is given by the cross product:

$$\mathbf{n} = (2\mathbf{i} + 2\mathbf{j}) \times (\mathbf{j} + \mathbf{k}) = 2\mathbf{i} - 2\mathbf{j} + 2\mathbf{k}$$

The nonparametric equation of the plane is thus

$$(\mathbf{R} - \mathbf{R}_0) \cdot \mathbf{n} = 2(x - 1) - 2(y - 2) + 2(z - 2) = 0$$

A parametric form can be written using the tangent vectors:

$$\mathbf{R}(u,v) = u(2\mathbf{i} + 2\mathbf{j}) + v(\mathbf{j} + \mathbf{k}) + \mathbf{i} + 2\mathbf{j} + 2\mathbf{k}$$

Any portion of the surface that can be represented by equations of the form of eq. (4.21) in a manner such that to distinct ordered pairs (u,v) there correspond distinct points (x,y,z) on the surface, and satisfying the above differentiability and continuity requirements, is called a ***regular surface element***.

Recall that the arc length of a smooth arc was defined as the limit of the lengths of inscribed polygonal paths (section 2.2). The surface area of a regular surface element turns out to be a slightly trickier concept, so we will be content to present a heuristic argument that leads to the correct formula. Figure 4.31 shows a small patch of the surface bounded by curves of constant u and v. Notice that, for small Δu and Δv, the patch is well approximated by a parallelogram with sides PQ and PS represented by the displacement vectors $\mathbf{R}(u + \Delta u, v) - \mathbf{R}(u,v)$ and $\mathbf{R}(u, v + \Delta v) - \mathbf{R}(u,v)$, respectively. The area of this parallelogram is given by the cross product; so if we introduce the further approximations

$$\mathbf{R}(u + \Delta u, v) - \mathbf{R}(u,v) \approx \frac{\partial \mathbf{R}}{\partial u} \Delta u$$

$$\mathbf{R}(u, v + \Delta v) - \mathbf{R}(u,v) \approx \frac{\partial \mathbf{R}}{\partial v} \Delta v$$

we find that the area of the patch is given approximately by

$$\Delta S \approx \left| \frac{\partial \mathbf{R}}{\partial u} \times \frac{\partial \mathbf{R}}{\partial v} \right| \Delta u \, \Delta v$$

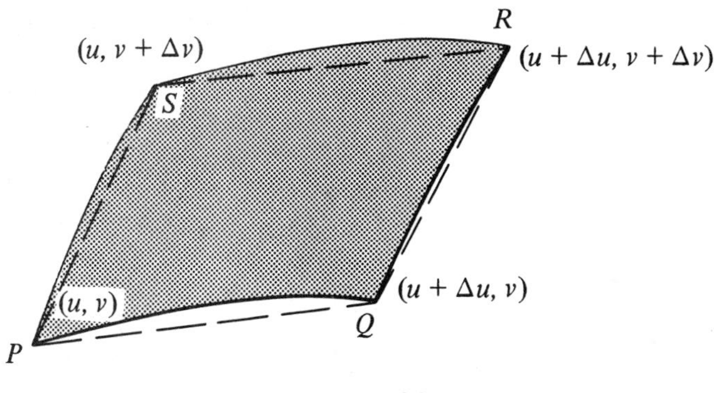

Figure 4.31

Summing these up over the surface and letting Δu and Δv go to zero, we argue that *the surface area of a regular surface element is given by*

$$S = \iint \left| \frac{\partial \mathbf{R}}{\partial u} \times \frac{\partial \mathbf{R}}{\partial v} \right| du\, dv \tag{4.24}$$

If we introduce the notation

$$d\mathbf{S} = \frac{\partial \mathbf{R}}{\partial u} du \times \frac{\partial \mathbf{R}}{\partial v} dv \tag{4.25}$$

then we see that $d\mathbf{S}$ is a vector normal to the surface at P, whose magnitude $dS = |d\mathbf{S}|$ is the element of area. The integral in eq. (4.24) may be written in the alternative forms

$$\iint |d\mathbf{S}|$$

or

$$\iint dS$$

or even

$$\iint \mathbf{n} \cdot d\mathbf{S}$$

where $\mathbf{n}$ is a unit normal in the same direction as $d\mathbf{S}$.

Example 4.18 Find the surface area of the cylinder defined by the equations

$$x = \cos\theta \qquad y = \sin\theta \qquad z = z$$

for $0 \le \theta \le 2\pi$, $0 \le z \le 1$.

Solution Identifying θ with u and z with v we have

$$\frac{\partial \mathbf{R}}{\partial u} = -\sin u\,\mathbf{i} + \cos u\,\mathbf{j}$$

$$\frac{\partial \mathbf{R}}{\partial v} = \mathbf{k}$$

$$d\mathbf{S} = \begin{vmatrix} \mathbf{i} & \mathbf{j} & \mathbf{k} \\ -\sin u & \cos u & 0 \\ 0 & 0 & 1 \end{vmatrix} du\,dv = (\cos u\,\mathbf{i} + \sin u\,\mathbf{j})\,du\,dv$$

$$\iint |d\mathbf{S}| = \int_0^1 \int_0^{2\pi} (\cos^2 u + \sin^2 u)^{1/2}\,du\,dv = 2\pi$$

The "ribs" of this cylinder for v constant, u varying are unit circles parallel to the xy plane. The u = constant ribs are vertical lines in the z direction. This right circular cylinder has unit radius and unit height. If it is cut along a "seam," it unfolds into a rectangle whose dimensions are 2π by 1.

We now consider a *special case* of eq. (4.24) that will illustrate further its geometrical significance. Let us suppose that the surface we consider is given in the form $z = f(x,y)$. In other words, we are told how far above the xy plane the surface is for each point (x,y) in the xy plane. Let us suppose that the projection of the surface element on the xy plane is bounded by the curves

$$y = y_1(x) \qquad y = y_2(x) \qquad x = a \qquad x = b$$

as shown in figure 4.32.

We can parametrize via the equations $x = u$, $y = v$, and $z = f(u,v)$ in order to make use of the preceding formulas. We have

$$\frac{\partial \mathbf{R}}{\partial u} = \frac{\partial \mathbf{R}}{\partial x} = \mathbf{i} + \left(\frac{\partial f}{\partial x}\right)\mathbf{k}$$

$$\frac{\partial \mathbf{R}}{\partial v} = \frac{\partial \mathbf{R}}{\partial y} = \mathbf{j} + \left(\frac{\partial f}{\partial y}\right)\mathbf{k}$$

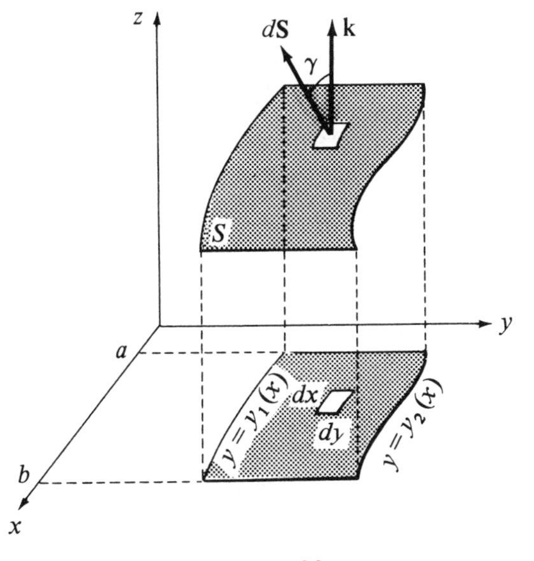

Figure 4.32

Taking the vector cross product, we obtain

$$d\mathbf{S} = \frac{\partial \mathbf{R}}{\partial x} \times \frac{\partial \mathbf{R}}{\partial y}\, dx\, dy = \begin{vmatrix} \mathbf{i} & \mathbf{j} & \mathbf{k} \\ 1 & 0 & \dfrac{\partial f}{\partial x} \\ 0 & 1 & \dfrac{\partial f}{\partial y} \end{vmatrix} dx\, dy = \left\{ \mathbf{k} - \frac{\partial f}{\partial x}\mathbf{i} - \frac{\partial f}{\partial y}\mathbf{j} \right\} dx\, dy$$

The magnitude of this vector is $\sqrt{1 + (\partial f/\partial x)^2 + (\partial f/\partial y)^2}$, so that the integral in eq. (4.24) is

$$S = \int_a^b \int_{y_1(x)}^{y_2(x)} \sqrt{1 + \left(\frac{\partial f}{\partial x}\right)^2 + \left(\frac{\partial f}{\partial y}\right)^2}\, dy\, dx \tag{4.26}$$

The geometrical significance of this is seen by considering the angle γ between $d\mathbf{S}$ and $\mathbf{k}$. By a simple calculation using scalar products we see that

$$|\cos \gamma| = \frac{|d\mathbf{S} \cdot \mathbf{k}|}{|d\mathbf{S}|} = \left[1 + \left(\frac{\partial f}{\partial x}\right)^2 + \left(\frac{\partial f}{\partial y}\right)^2 \right]^{-1/2}$$

so that eq. (4.26) is simply

$$S = \int_a^b \int_{y_1(x)}^{y_2(x)} \frac{dy\, dx}{|\cos \gamma|} \tag{4.27}$$

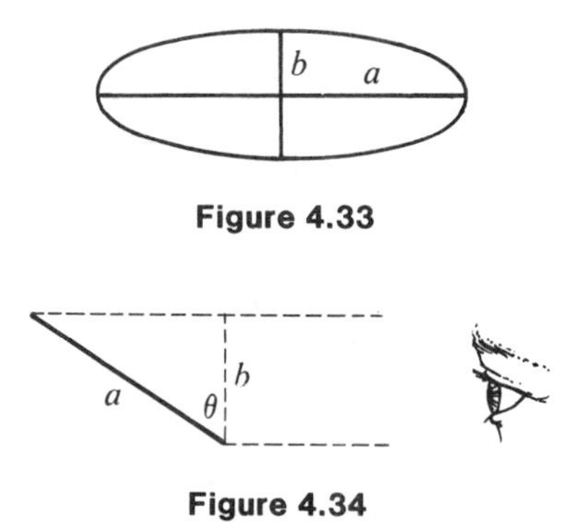

Figure 4.33

Figure 4.34

This integral could have been obtained heuristically by considering the *area cosine principle,* which says that if we look at a plane area A whose normal makes an acute angle θ with the line of sight, the area we appear to see is $A \cos \theta$. This is because distances in one direction will appear to be shorter by a factor of $\cos \theta$ and distances in a perpendicular direction will not change at all.

Let us digress for a moment to use this law to determine the area of the ellipse shown in figure 4.33. Pretend that this ellipse is really a circle of radius a that we are viewing at an angle. In other words, we imagine that this circle has been tipped in such a manner that vertical distances are shortened by a factor b/a (see fig. 4.34). The area we see will be $A \cos \theta = (\pi a^2)(b/a) = \pi ab$. Thus we find the area of this ellipse to be πab.

Returning to figure 4.32, we can consider an element of area in S whose projection on the xy plane has area $dx\, dy$. The angle between the normal to this area and the line of sight (imagine that you are below the xy plane looking up at the surface) is γ. By the area cosine principle, the area $dx\, dy$ that we see equals $dS|\cos \gamma|$. It follows that

$$dS = \frac{dx\, dy}{|\cos \gamma|}$$

The absolute value is unnecessary if γ is acute.

Frequently a judicious use of the area cosine principle makes use of eq. (4.24) unnecessary. The cosine of the relevant angle, in this case γ, is easily computed since we can find a normal to the surface by methods we have already learned and then use scalar products to find the desired cosine. For instance, the surface $z = f(x,y)$ can be represented by the equation $z - f(x,y) = 0$, and the gradient of the function $z - f(x,y)$ is $\mathbf{k} - (\partial f/\partial x)\mathbf{i} - (\partial f/\partial y)\mathbf{j}$. This is easier than computing the vector cross product given above. [*Caution:* Using gradients to give a normal vector $\mathbf{n}$ gives a vector that does not equal $d\mathbf{S}$ but is only a scalar multiple of it. However, this makes no difference since we are interested in computing only $\cos \gamma = \mathbf{n} \cdot \mathbf{k}/|\mathbf{n}|$ when using eq. (4.27).]

Example 4.19 Find the area of the hemisphere defined by

$$x^2 + y^2 + z^2 = 1 \qquad (x \geq 0)$$

Solution Let us use the area cosine principle that corresponds to a line of sight along the **i** direction, projecting the area onto the yz plane. A normal is given by the gradient

$$\nabla(x^2 + y^2 + z^2) = 2x\mathbf{i} + 2y\mathbf{j} + 2z\mathbf{k}$$

If α is the angle that the normal makes with **i**, then

$$\cos \alpha = \frac{(2x\mathbf{i} + 2y\mathbf{j} + 2z\mathbf{k}) \cdot \mathbf{i}}{(4x^2 + 4y^2 + 4z^2)^{1/2}} = \frac{2x}{2} = x$$

The projection of this hemisphere onto the yz plane is a unit circle; hence the area is

$$S = \iint \frac{dy\,dz}{\cos \alpha} = \int_{-1}^{1} \int_{-(1-z^2)^{1/2}}^{+(1-z^2)^{1/2}} x^{-1}\,dy\,dz$$

Since $x = (1 - y^2 - z^2)^{1/2}$,

$$S = 2\pi$$

(by standard methods). This checks with our expectations: a unit hemisphere would have area $4\pi/2 = 2\pi$.

Sometimes in practice, the formula (4.25) can be visualized. For example, suppose we parametrize a sphere of radius a as in example 4.14. Using the two parameters ϕ and θ we obtain

$$\frac{\partial \mathbf{R}}{\partial \phi} = a \cos \phi \cos \theta\, \mathbf{i} + a \cos \phi \sin \theta\, \mathbf{j} - a \sin \phi\, \mathbf{k} \tag{4.28}$$

$$\frac{\partial \mathbf{R}}{\partial \theta} = -a \sin \phi \sin \theta\, \mathbf{i} + a \sin \phi \cos \theta\, \mathbf{j} \tag{4.29}$$

whereupon we compute to show that (exercise 8)

$$\frac{\partial \mathbf{R}}{\partial \phi} \times \frac{\partial \mathbf{R}}{\partial \theta} = a^2 \sin^2 \phi \cos \theta\, \mathbf{i} + a^2 \sin^2 \phi \sin \theta\, \mathbf{j} + a^2 \sin \phi \cos \phi\, \mathbf{k} \tag{4.30}$$

The magnitude of the vector is $a^2 \sin \phi$; hence it follows that

$$dS = a^2 \sin \phi\, d\phi\, d\theta \tag{4.31}$$

This result can be visualized from figure 4.35. Holding θ fixed and varying ϕ by an amount $d\phi$, we trace out an arc of length $a\, d\phi$. Holding ϕ fixed and varying θ, we trace out an arc of a circle of radius $a \sin \phi$, the length of this arc being $a \sin \phi\, d\theta$. For small $d\phi$ and $d\theta$, this gives us very nearly a rectangle with area $a^2 \sin \phi\, d\phi\, d\theta$.

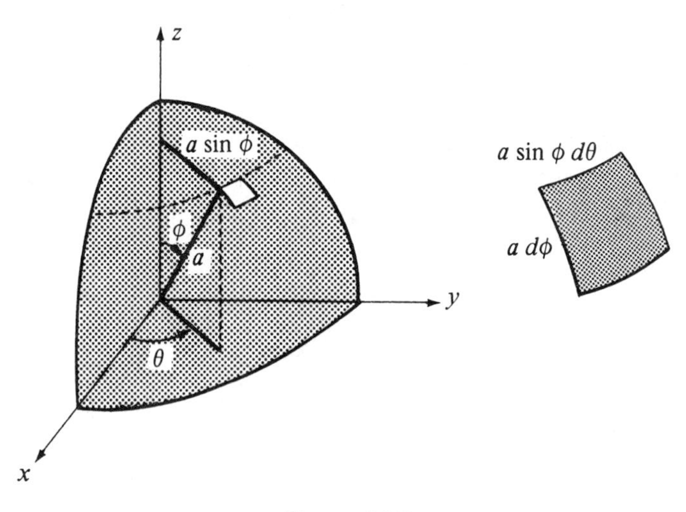

Figure 4.35

EXERCISES

1. Find the elements of surface area $d\mathbf{S}$ and dS, in terms of du and dv, for the surface S given parametrically by $x = u^2$, $y = \sqrt{2}uv$, and $z = v^2$.
2. Find $d\mathbf{S}$ and dS, in terms of $d\phi$ and $d\theta$, for the surface with parametric equations

$$x = (1 + \cos\theta)\cos\phi$$
$$y = (1 + \cos\theta)\sin\phi$$
$$z = \sin\theta$$

3. Determine the element of surface area dS for a right circular cylinder,

$$x = a\cos u \qquad y = a\sin u \qquad z = v$$

Interpret geometrically. (See fig. 4.27.)

4. Determine the element of surface area dS in the special case of the paraboloid of revolution $z = x^2 + y^2$.
5. Find the area of the section of the surface

$$x = u^2 \qquad y = uv \qquad z = \frac{1}{2}v^2$$

bounded by the curves $u = 0$, $u = 1$, $v = 0$, and $v = 3$.

6. Consider the triangle with vertices (1,0,0), (0,1,0), and (0,0,1).
   (a) Find a unit vector $\mathbf{n}$ normal to this triangle, pointing away from the origin.
   (b) Determine $\cos\gamma$ for this vector.

(c) Supply the appropriate limits for the integral

$$\iint \frac{dx\,dy}{|\cos \gamma|}$$

if it is to represent the area of this triangle.

(d) Evaluate the integral.

(e) Obtain the same answer by applying the area cosine principle to the projection of this triangle on the yz plane.

7. Draw a diagram similar to those of figures 4.23 and 4.24 for the surface of a tetrahedron.

8. (a) Derive eq. (4.30) from eqs. (4.28) and (4.29).

(b) Show that the magnitude of this vector is $a^2 \sin \phi$.

9. Show that the Möbius strip is piecewise smooth, and show why the smooth parts cannot be oriented consistently.

10. Derive the identity

$$dS = (EG - F^2)^{1/2}\, du\, dv$$

where

$$E = \left|\frac{\partial \mathbf{R}}{\partial u}\right|^2 \qquad F = \frac{\partial \mathbf{R}}{\partial u} \cdot \frac{\partial \mathbf{R}}{\partial v} \qquad G = \left|\frac{\partial \mathbf{R}}{\partial v}\right|^2$$

(The quantities E, F, and G are employed in differential geometry in developing the theory of surfaces. They constitute the "second fundamental form.")

4.7 Surface Integrals

Let S denote a smooth surface and let $f(x,y,z)$ be a function defined and continuous on S. The surface integral of f over S, denoted

$$\iint f\, dS$$

is defined by a construction that the reader can, no doubt, anticipate by now. We imagine the surface cut up into n pieces having area $\delta S_1, \delta S_2, \ldots, \delta S_n$. In each piece we choose a point (x_i,y_i,z_i), evaluate $f(x_i,y_i,z_i)$, and form the product $f(x_i,y_i,z_i)$ δS_i. We sum these numbers:

$$\sum_{i=1}^{n} f(x_i,y_i,z_i)\, \delta S_i \tag{4.32}$$

In this way we obtain a single number. Now let n tend to infinity, at the same time letting the pieces grow smaller so that the maximum dimension of the areas δS_1, $\delta S_2, \ldots, \delta S_n$ tends to zero. In other words, we are dividing the surface into smaller and smaller elements of area, each time forming a sum of type (4.32). If these sums tend to a limit, independent of the way we form the repeated subdivisions, that limit is called the *surface integral of f over S:*

$$\iint_S f(x,y,z)\,dS = \lim_{\substack{\max \delta S_i \to 0 \\ n \to \infty}} \sum_{i=1}^{n} f(x_i,y_i,z_i)\,\delta S_i \tag{4.33}$$

In most situations the function f arises from a scalar product involving a vector field **F**. In section 3.3 we defined the flux of **F** through the surface element δS to be $\mathbf{F}\cdot\mathbf{n}\,\delta S$, where **n** is the unit normal to the element. Thus *the total flux through a surface S equals*

$$\iint_S \mathbf{F}\cdot\mathbf{n}\,dS \tag{4.34}$$

Flux integrals are instances of eq. (4.33) with $f = \mathbf{F}\cdot\mathbf{n}$. Some physical examples of flux integrals will be discussed shortly, but first let us see how to compute them.

Using the notation of the previous section, we can write the flux as

$$\iint_S \mathbf{F}\cdot d\mathbf{S} \tag{4.34$'$}$$

Applying eq. (4.25), we convert this to a workable formula when the surface is parameterized by $\mathbf{R}(u,v)$:

$$\iint \mathbf{F}\cdot\frac{\partial\mathbf{R}}{\partial u}\times\frac{\partial\mathbf{R}}{\partial v}\,du\,dv \tag{4.34$''$}$$

If the surface is specified by giving z as a function of x and y, we would use

$$\iint \mathbf{F}\cdot\mathbf{n}\frac{dx\,dy}{|\cos\gamma|} \tag{4.34$'''$}$$

where the limits of integration are determined by the projection of S onto the xy plane [with an obvious modification for surfaces described by, say, $x = x(y,z)$].

If the surface S is only piecewise smooth, we integrate over each smooth part separately and add the numbers obtained.

Example 4.20 Compute the flux of the vector field $\mathbf{F} = \mathbf{i} + xy\mathbf{j}$ across the surface given by

$$x = u + v \qquad y = u - v \qquad z = u^2 \qquad (0 \le u \le 1, 0 \le v \le 1)$$

Solution Using eq. (4.34″),

$$\iint \mathbf{F} \cdot \frac{\partial \mathbf{R}}{\partial u} \times \frac{\partial \mathbf{R}}{\partial v}\, du\, dv = \int_0^1 \int_0^1 \begin{vmatrix} 1 & u^2 - v^2 & 0 \\ 1 & 1 & 2u \\ 1 & -1 & 0 \end{vmatrix} du\, dv$$

$$= \int_0^1 \int_0^1 (2u^3 - 2uv^2 + 2u)\, du\, dv$$

$$= \int_0^1 \left(\frac{1}{2} - v^2 + 1\right) dv = \frac{7}{6}$$

Example 4.21 Compute $\iint_S \mathbf{F} \cdot d\mathbf{S}$, where S is the surface of the sphere $x^2 + y^2 + z^2 = 4$ and $\mathbf{F} = x\mathbf{i} + y\mathbf{j} + z\mathbf{k}$.

Solution We recall that at a point (x,y,z) the vector $x\mathbf{i} + y\mathbf{j} + z\mathbf{k}$ points directly away from the origin. The outward normal $\mathbf{n}$ to this sphere also points away from the origin, since the center of the sphere is at the origin. Hence, for points on the surface,

$$\mathbf{F} \cdot \mathbf{n} = |\mathbf{F}||\mathbf{n}| \cos \theta = |\mathbf{F}| = (x^2 + y^2 + z^2)^{1/2} = 2$$

and

$$\iint_S \mathbf{F} \cdot d\mathbf{S} = \iint_S \mathbf{F} \cdot \mathbf{n}\, dS = \iint_S 2\, dS$$

$$= 2\text{ (total surface area)} = 2(4\pi r^2) = 32\pi$$

since $r = 2$ is the radius of the sphere.

In this example no integration was needed since $\mathbf{F} \cdot \mathbf{n}$ was constant over the entire surface.

Example 4.22 Compute $\iint_S \mathbf{F} \cdot d\mathbf{S}$, where S is the surface of the cube bounded by the planes $x = 0$, $x = 1$, $y = 0$, $y = 1$, $z = 0$, $z = 1$, and $\mathbf{F} = x\mathbf{i} + y\mathbf{j} + z\mathbf{k}$.

Solution We see from figure 4.36 that the unit normal to the front face of the cube is $\mathbf{n} = \mathbf{i}$, so

$$\mathbf{F} \cdot \mathbf{n} = \mathbf{i} \cdot (x\mathbf{i} + y\mathbf{j} + z\mathbf{k}) = x = 1$$

on this face. It follows that the integral over this face is

$$\iint \mathbf{F} \cdot d\mathbf{S} = \iint \mathbf{F} \cdot \mathbf{n}\, dS = \iint dS = 1$$

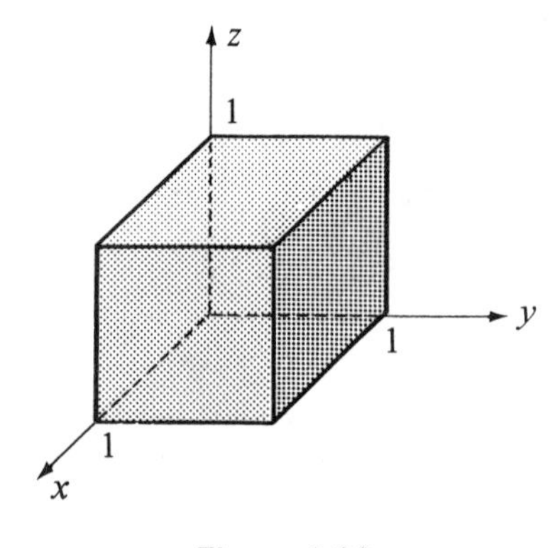

Figure 4.36

since the area of this face is unity. On the opposite face (in the yz plane) $\mathbf{n} = -\mathbf{i}$, so that $\mathbf{F} \cdot \mathbf{n} = -x$, but $x = 0$ for all points in this face, and hence

$$\iint \mathbf{F} \cdot \mathbf{n} \, dS = 0$$

On the top of the cube we have

$$\iint \mathbf{F} \cdot \mathbf{n} \, dS = \iint \mathbf{F} \cdot \mathbf{k} \, dS = \iint z \, dS = \iint dS = 1$$

and on the bottom we have

$$\iint \mathbf{F} \cdot \mathbf{n} \, dS = \iint (-z) \, dS = 0$$

since $z = 0$ in the xy plane. Along the right side we have $\mathbf{n} = \mathbf{j}$, so that the normal component of $\mathbf{F}$ is unity and the integral over this face is unity. Along the left side $\mathbf{n} = -\mathbf{j}$ and $\mathbf{F} \cdot \mathbf{n} = -y = 0$, so that the contribution to the integral is zero. Summing, we find that

$$\iint_S \mathbf{F} \cdot \mathbf{n} \, dS = 3$$

Example 4.23 Compute the surface integral of the normal component of $\mathbf{F} = x^2\mathbf{i} + yx\mathbf{j} + zx\mathbf{k}$ over the triangle with vertices (1,0,0), (0,2,0), (0,0,3). Consider the triangle oriented such that its positive side is that away from the origin (fig. 4.37).

Solution By the methods of section 1.12 we find easily that $\mathbf{n} = \frac{6}{7}\mathbf{i} + \frac{3}{7}\mathbf{j} + \frac{2}{7}\mathbf{k}$. Hence $\mathbf{F} \cdot \mathbf{n} = \frac{6}{7}x^2 + \frac{3}{7}yx + \frac{2}{7}zx$ and

$$\cos \gamma = \mathbf{n} \cdot \mathbf{k} = \frac{2}{7}$$

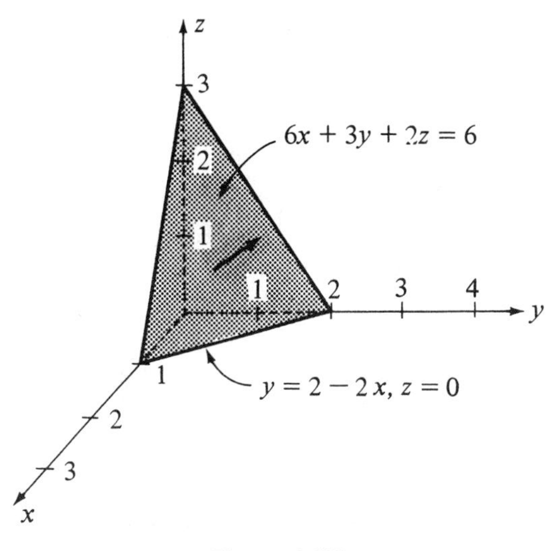

Figure 4.37

Using (4.34‴) we have

$$\iint_S \mathbf{F}\cdot\mathbf{n}\,dS = \int_0^1\int_0^{2-2x} \frac{7}{2}\left(\frac{6}{7}x^2 + \frac{3}{7}yx + \frac{2}{7}zx\right) dy\,dx$$
$$= \int_0^1\int_0^{2-2x}\left(3x^2 + \frac{3}{2}yx + zx\right) dy\,dx$$

On S we have $z = 3 - 3x - \frac{3}{2}y$, so that $zx = 3x - 3x^2 - \frac{3}{2}yx$, and the integral becomes

$$\int_0^1\int_0^{2-2x} 3x\,dy\,dx = \int_0^1 3x(2-2x)\,dx = 3x^2 - 2x^3\Big|_0^1 = 1$$

Example 4.24 Compute $\iint_S \mathbf{F}\cdot\mathbf{n}\,dS$ over the surface of the tetrahedron with vertices (1,0,0), (0,2,0), (0,0,3), (0,0,0), where $\mathbf{F} = x^2\mathbf{i} + yx\mathbf{j} + zx\mathbf{k}$ (fig. 4.37).

Solution We have already computed the integral over one surface in the previous example. Along the bottom face we have $\mathbf{n} = -\mathbf{k}$ and, hence, $\mathbf{F}\cdot\mathbf{n} = -zx$, but since $z = 0$ the integral over the bottom face is zero. On the face at the left we have $\mathbf{n} = -\mathbf{j}$ and $\mathbf{F}\cdot\mathbf{n} = -yx$, which is also zero since $y = 0$ there. On the rear face, in the yz plane, $\mathbf{n} = -\mathbf{i}$ and $\mathbf{F}\cdot\mathbf{n} = x^2 = 0$. It follows that

$$\iint_S \mathbf{F}\cdot d\mathbf{S} = 1$$

—the only nonzero contribution being the integral already computed.

Note that in examples 4.21, 4.22, and 4.24 we took **n** to be the *outward* normal, the usual convention for closed surfaces.

Now let's consider some physical examples of surface integrals. Suppose, for instance, that at any point (x,y,z) on a surface S, $f(x,y,z)$ gives the rate of flow of heat per unit area at that point, in units, say, of calories per second per square centimeter. Then $f(x_i,y_i,z_i)\ \delta S_i$ gives, approximately, the number of calories per second flowing across the element of area δS_i, and the sum in eq. (4.32) approximates the total number of calories per second flowing across the entire surface S. If $f(x,y,z)$ varies from point to point on the surface, this approximation can be improved by taking smaller elements of area (and hence more such elements). The limit

$$\iint_S f(x,y,z)\ dS$$

gives, exactly, the number of calories per second flowing across the surface.

Let us look a little deeper into the physics behind this phenomenon. If we assume a steady-state temperature distribution, where $T(x,y,z)$ denotes the temperature at each point in space, and if the region we consider is filled with a homogeneous material having coefficient of thermal conductivity k, then according to *Fourier's law* (a special instance of Fick's law, section 3.6),

$$\mathbf{Q} = -k\nabla T \tag{4.35}$$

gives, at each point in space, the direction in which the heat is flowing. The magnitude of **Q** gives the rate of heat flow per unit area across an area perpendicular to **Q**. More generally, we can say that the scalar component of **Q** in the direction of a unit vector **n** $(= \mathbf{Q} \cdot \mathbf{n})$ gives the number of calories per unit time and per unit area crossing an element of area perpendicular to **n.**

It follows that the function f is given by $(-k\nabla T) \cdot \mathbf{n}$, and the total number of calories per second flowing across a surface S equals

$$\iint_S (-k\nabla T) \cdot \mathbf{n}\ dS \tag{4.36}$$

The reason for the negative sign in eqs. (4.35) and (4.36) is that the temperature gradient ∇T points in the direction of maximum rate of increase of the temperature, whereas heat flows in the opposite direction, from hot to cold.

Let us consider another situation in physics in which surface integrals arise. If **v** denotes the velocity field of a fluid and μ its density, then as we saw in section 3.3, the amount of fluid crossing a patch of surface with area δS and unit normal **n,** per unit time, is approximately $\mu\mathbf{v} \cdot \mathbf{n}\ \delta S$. This formula becomes exact as δS goes to zero. Thus we can see that

$$\iint_S \mu\mathbf{v} \cdot \mathbf{n}\ dS \tag{4.37}$$

gives the rate of flow of liquid across the surface S, expressed as mass per unit time.

As yet another example, consider an electrostatic field **E** defined in a region of space. One can form

$$\iint_S \mathbf{n} \cdot \mathbf{E}\, dS$$

which is the surface integral of the normal component of **E** over the surface S. This integral arises in connection with Gauss' law of electrostatics (Appendix D), which states that if S is a closed surface,

$$\iint_S \mathbf{n} \cdot \mathbf{E}\, dS = \frac{q}{\epsilon_0} \tag{4.38}$$

where q is the total charge enclosed by the surface and ϵ_0 is a constant that depends on the system of units. The numerical value of the surface integral in eq. (4.38) is called the *flux across S* or the *number of flow lines of the vector field* **E** *crossing the surface*. This last phrase is not to be taken literally, since there will be a flow line crossing every point of S and therefore there are really an infinite number of flow lines crossing S. However, in drawing diagrams, it is impossible to draw an infinite number of flow lines, so it may be convenient to visualize eq. (4.38) as giving a measure of the number of flow lines we wish to picture crossing the surface. [This number is necessarily approximate since the value of eq. (4.38) may not be a whole number.]

Example 4.25 Use Gauss' law, eq. (4.38), to determine the magnitude of the electric field intensity at a point r units away from a point charge of magnitude q.

Solution Let S be a sphere of radius r with the charge q at its center. Symmetry considerations lead us to believe that $\mathbf{n} \cdot \mathbf{E}$ will be constant over the surface of this sphere, and that **E** will be normal to the surface. Hence we can bring $\mathbf{n} \cdot \mathbf{E}$ outside the integral, and we obtain from eq. (4.38)

$$\frac{q}{\epsilon_0} = \iint \mathbf{n} \cdot \mathbf{E}\, dS = \mathbf{n} \cdot \mathbf{E} \iint dS = 4\pi r^2(\mathbf{n} \cdot \mathbf{E})$$

It then follows that $\mathbf{n} \cdot \mathbf{E} = q/4\pi\epsilon_0 r^2$. Thus, if the charge is positive, $|\mathbf{E}| = q/4\pi\epsilon_0 r^2$ and **E** is directed away from q. If q is negative, **E** will be directed toward the charge q.

Example 4.26 Use Gauss' law, eq. (4.38), to determine the magnitude of the electric field intensity at a point r units away from an infinite sheet carrying a charge of density σ (charge per unit area).

Solution Let S be the surface of a right circular cylinder of length $2r$ and base area A, bisected by the charged sheet. We take the bases parallel to the sheet, so by symmetry we expect **E** to be perpendicular to the bases (fig. 4.38).

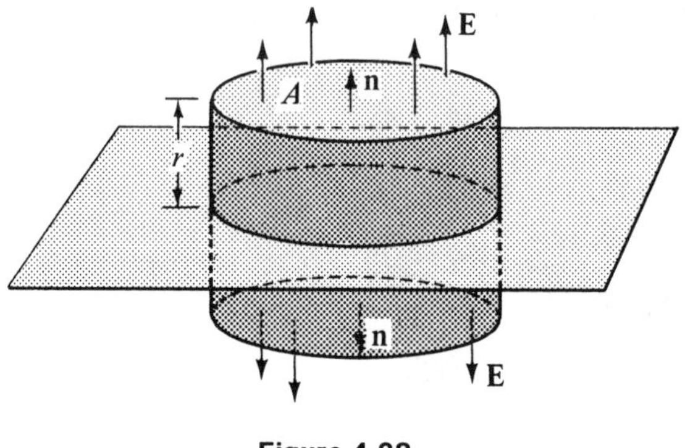

Figure 4.38

The charge within S is $q = \sigma A$. On each of the two bases we have $\mathbf{n} \cdot \mathbf{E} =$ constant by symmetry (since we assume the charged sheet is infinite in extent), and there will be no contribution to the integral around the curved surface of the cylinder because $\mathbf{E}$ is normal to the sheet (by symmetry), and thus parallel to this surface; therefore

$$\iint_S \mathbf{n} \cdot \mathbf{E}\, dS = \mathbf{n} \cdot \mathbf{E} \iint_S dS = (\mathbf{n} \cdot \mathbf{E})(2A)$$

By eq. (4.38) we have $(\mathbf{n} \cdot \mathbf{E})(2A) = \sigma A/\epsilon_0$, so $\mathbf{n} \cdot \mathbf{E} = \sigma/2\epsilon_0$. If σ is positive, this shows that $\mathbf{E}$ is in the same direction as $\mathbf{n}$ and $|\mathbf{E}| = \sigma/2\epsilon_0$, independent of r.

Example 4.27 Consider a cylindrical heat insulator surrounding a steam pipe. Let the inner and outer radii of the insulator be $r = a$ and $r = b$ respectively, and let T_a and T_b be the temperatures, respectively, of the inner and outer surfaces of the insulator. Find the equilibrium temperature T within the insulator as a function of r (fig. 4.39).

Solution A section of the insulator, of length L, is shown in the figure. By symmetry, we assume that T is a function of r alone, so that $\nabla T =$ **grad** T is directed radially toward the center of the pipe, with magnitude $-dT/dr$. (On the assumption that the pipe is hotter than the surroundings, dT/dr will be negative.) Let S be a cylindrical surface of radius r and length L within the insulator. By eq. (4.35), we have

$$\mathbf{Q} \cdot \mathbf{n} = (-k\nabla T) \cdot \mathbf{n} = -k\frac{dT}{dr}$$

(as usual, we take $\mathbf{n}$ to be outward, so $\nabla T \cdot \mathbf{n} = |\nabla T||\mathbf{n}| \cos 180° = -|\nabla T| = dT/dr$).

At equilibrium the number of calories of heat flowing across any such surface S will be the same as that across any other such surface, since otherwise the temperature would change with time. The quantity of heat flow per unit time across any such surface is

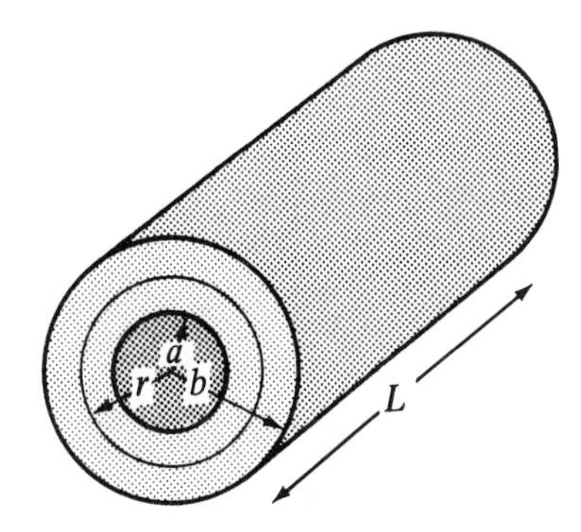

Figure 4.39

$$H = \iint_S \mathbf{Q} \cdot \mathbf{n}\, dS = \iint_S -k \frac{dT}{dr}\, dS = -k \frac{dT}{dr} \iint_S dS$$

$$= -k \frac{dT}{dr} (2\pi L r)$$

Here again we are using symmetry considerations in assuming that dT/dr is constant along any one surface S (but not necessarily the same as for surfaces with different r), and so dT/dr can be brought outside the integral sign.

Since H is independent of r, we treat it as a constant in solving the differential equation

$$H = -2\pi k L r \frac{dT}{dr}$$

Separating variables,

$$H \frac{dr}{r} = -2\pi k L\, dT$$

we integrate

$$H \int_a^b \frac{dr}{r} = -2\pi k L \int_{T_a}^{T_b} dT$$

which ultimately yields

$$H = \frac{2\pi L k (T_a - T_b)}{\ln(b/a)}$$

Substituting this value of H and integrating:

$$H \int_a^r \frac{dr}{r} = -2\pi k L \int_{T_a}^{T} dT$$

we finally obtain

$$T = T_a - (T_a - T_b) \frac{\ln(r/a)}{\ln(b/a)}$$

EXERCISES

1. If $\mathbf{F} = z\mathbf{k}$, find the surface integral of the normal component of $\mathbf{F}$ over the closed surface of the right circular cylinder with curved surface $x^2 + y^2 = 9$ and bases in the planes $z = 0$ and $z = 2$. (Mental arithmetic should suffice.)
2. Compute $\iint \mathbf{F} \cdot d\mathbf{S}$, where S is the surface of the cube bounded by the planes $x = \pm 1, y = \pm 1, z = \pm 1$, if
   (a) $\mathbf{F} = x\mathbf{i}$
   (b) $\mathbf{F} = x\mathbf{i} + y\mathbf{j}$
   (c) $\mathbf{F} = x\mathbf{i} + y\mathbf{j} + z\mathbf{k}$
   (d) $\mathbf{F} = x^2\mathbf{i} + y^2\mathbf{j} + z^2\mathbf{k}$
   (e) $\mathbf{F} = y\mathbf{i}$
   (f) $\mathbf{F} = z\mathbf{i}$
   (g) $\mathbf{F} = z^2\mathbf{i}$
3. Compute the surface integral of the normal component of $\mathbf{F} = x\mathbf{i}$ over the triangle with vertices (1,0,0), (0,2,0), (0,0,3), taking the normal on the side away from the origin.
4. Given $\mathbf{F} = x\mathbf{i} - y\mathbf{j}$, find the value of $\iint \mathbf{F} \cdot \mathbf{n}\, dS$ over the closed surface bounded by the planes $z = 0$, $z = 1$, and the cylinder $x^2 + y^2 = a^2$, where $\mathbf{n}$ is the unit outward normal,
   (a) by direct calculation. (*Hint:* The element of area is $dS = a\, d\theta\, dz$ in cylindrical coordinates on the curved surface.)
   (b) by symmetry considerations, without changing to cylindrical coordinates.
5. Given $\mathbf{F} = x\mathbf{i} + y\mathbf{j} + (z^2 - 1)\mathbf{k}$, find $\iint \mathbf{F} \cdot \mathbf{n}\, dS$ over the closed surface bounded by the planes $z = 0$, $z = 1$, and the cylinder $x^2 + y^2 = a^2$, where $\mathbf{n}$ is the unit outward normal.
6. Given that $\mathbf{F} = y\mathbf{i} + \mathbf{k}$, find the surface integral of the normal component of $\mathbf{F}$ over the box shown in figure 4.40, taking $\mathbf{n}$ to be the unit outward normal. Assume this box to have a bottom but no top, i.e., roughly like a shoe. (*Note:* Later on you will be asked to do the same problem by mental arithmetic, as a demonstration of the power of the divergence theorem. Take a furtive peek ahead at exercise 21, section 4.9.)

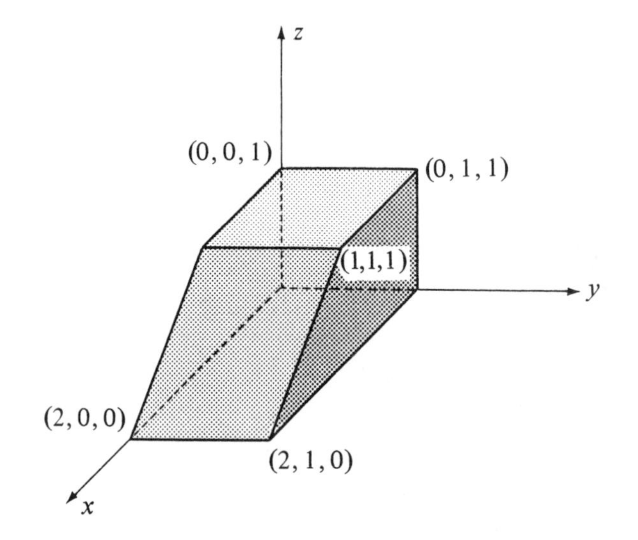

Figure 4.40

7. Let D be the region $x \geq 0$, $y \geq 0$, $z \geq 0$, $x + \frac{1}{2}y + \frac{1}{3}z \leq 1$.
   (a) Is this region a domain?
   (b) Is this region simply connected?
   (c) If $\mathbf{F} = 2x\mathbf{i} + y\mathbf{j} + z\mathbf{k}$, find the surface integral of the normal component of $\mathbf{F}$ over the boundary of this region, oriented by selecting the outward normal.
8. Calculate $\iint \mathbf{F} \cdot d\mathbf{S}$ over the section of surface described in exercise 5, section 4.6, for the vector field $\mathbf{F} = y\mathbf{i} - x\mathbf{j} + xy\mathbf{k}$.
9. Evaluate $\iint_S z^2\, dS$, where S is the part of the lateral surface of the right circular cylinder $x^2 + y^2 = 4$ between the planes $z = 0$ and $z = y + 3$.

In exercises 10–13 evaluate $\iint_S \mathbf{F} \cdot d\mathbf{S}$:

10. $\mathbf{F} = (x + 1)\mathbf{i} - (2y + 1)\mathbf{j} + z\mathbf{k}$; S is the triangle with vertices (1,0,0), (0,1,0), (0,0,1) and normal pointing away from the origin.
11. $\mathbf{F} = x^2\mathbf{i} + y^2\mathbf{j} + z^2\mathbf{k}$; S is the part of the cone $z^2 = x^2 + y^2$ for which $1 \leq z \leq 2$, with $\mathbf{n} \cdot \mathbf{k}$ positive.
12. $\mathbf{F} = \mathbf{j}$; S is the part of the cylinder lying above the curve in the xy plane expressed in polar coordinates by $r = 2 + \cos\theta$, and below the cone $z^2 = x^2 + y^2$, with normal pointing outward.
13. $\mathbf{F} = y^2\mathbf{i} + z\mathbf{j} - x\mathbf{k}$; S is the part of the cylinder generated by $y^2 = 1 - x$ between the planes $z = 0$ and $z = x$ for $x \geq 0$, with $\mathbf{n} \cdot \mathbf{i} > 0$.
14. Let $\mathbf{F} = y\mathbf{i} + (x + 2)\mathbf{j} + x^3 \sin(yz)\mathbf{k}$, and let S be the portion of the cylinder $x^2 + y^2 = 1$ that lies in the first octant and below $z = 1$.
15. Let $\mathbf{E} = -\mathbf{grad}\,(|\mathbf{R}|^{-1})$, where $\mathbf{R} = x\mathbf{i} + y\mathbf{j} + z\mathbf{k}$.
    (a) Show that $\mathbf{E} = \mathbf{R}/|\mathbf{R}|^3$.
    (b) Find $\int_C \mathbf{E} \cdot d\mathbf{R}$, when C is the line segment joining the points (0,1,0) and (0,0,1).
    (c) Compute $\iint_{S_1} \mathbf{E} \cdot d\mathbf{S}$, when S_1 is the sphere $x^2 + y^2 + z^2 = 9$.
16. A torus (doughnut) is shown in figure 4.41. Its major radius is A and its minor radius is a. Derive the parameterization $\mathbf{R}(u,v)$ in terms of the *toroidal angle u* and the *poloidal angle v*, where

$$x = A\cos u + a\cos u\cos v$$
$$y = A\sin u + a\sin u\cos v$$
$$z = a\sin v$$

    Show that the area of the torus is $4\pi^2 Aa$.
17. Consider a cone with vertex at the origin, as in figure 4.42. The *solid angle* Ω at the vertex is defined to be the surface area that this cone cuts out of the unit sphere centered at the origin.
    (a) If the cone is perfectly flat, that is, a plane, what is Ω?
    (b) What is Ω for the corner of a cube?
    (c) What is Ω for the $\phi = 45°$ cone?
    (d) What is the total solid angle around a point?

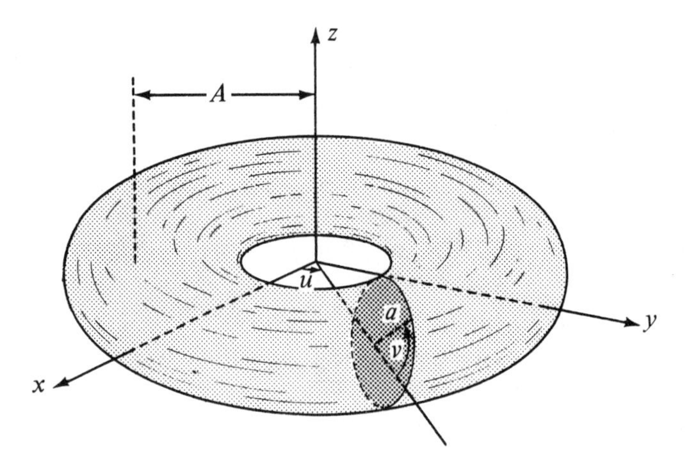

Figure 4.41

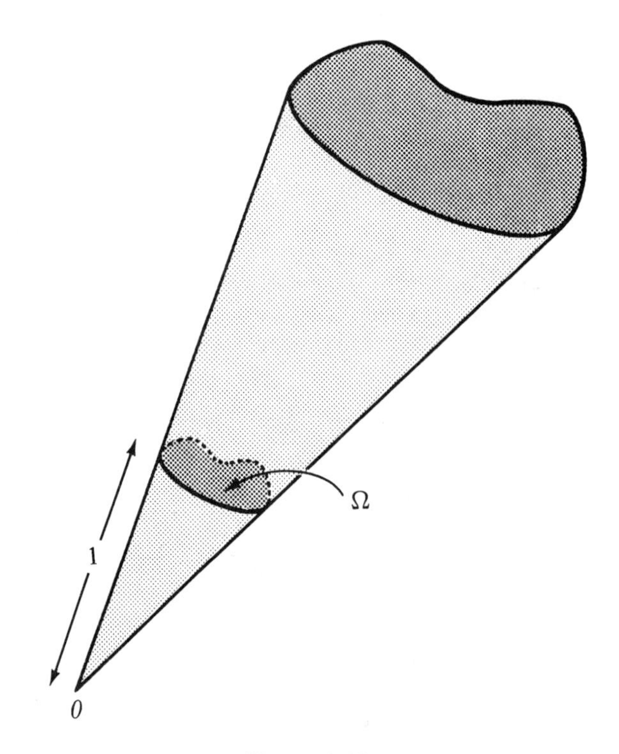

Figure 4.42

(e) Suppose the surface S, bounded by the simple closed curve C, has the property that every ray from the origin intersects S at most once. Then the *solid angle* Ω *subtended at the origin by* S is the solid angle at the vertex of the cone generated by the rays through C. Show that, if S is properly oriented,

$$\Omega = \iint_S \frac{\mathbf{R} \cdot d\mathbf{S}}{|\mathbf{R}|^3} \qquad (\mathbf{R} = x\mathbf{i} + y\mathbf{j} + z\mathbf{k})$$

Use the results to check Gauss' law, eq. (4.38), for a point charge at the origin. The expression for the electric field appears in example 4.25.

18. Use Gauss' law to determine the magnitude of the electric field intensity at a point r units away from an infinitely long thin wire carrying a charge of λ units per unit length. (Consider a cylinder of length L and radius r concentric with the wire.)

19. Consider a hollow sphere of homogeneous material, with inner radius a and outer radius b, and inner temperature T_a and outer temperature T_b.

(a) Find the steady-state temperature as a function of the distance r from the center, for values of r between a and b.

(b) For a value of r halfway between a and b, is T halfway between T_a and T_b?

20. Compute the surface area of the spiral ramp $\rho = u$, $\theta = \pi/2 - v$, $z = v$, for $0 \leq u \leq 1$, $0 \leq v \leq 2$. [*Hint:* Use eqs. (4.25) and (3.52).]

21. Compute the area of the cone $\phi = \text{constant} = \pi/6$, $0 \leq r \leq 2$. [*Hint:* Use eqs. (4.25) and (3.63).]

4.8 Volume Integrals

Volume integrals are defined through the familiar partition construction. We consider a function f (i.e., a scalar field) defined within and on the boundary of a domain V. We imagine that V is *bounded,* that is, that there exists a cube R sufficiently large that every point of V is within R. We imagine the cube R subdivided into rectangular parallelepipeds by planes parallel to the coordinate planes (fig. 4.43). Ignoring those parallelepipeds that contain no points of V, we let the volumes of the parallelepipeds that do overlap V be denoted $\delta V_1, \delta V_2, \ldots, \delta V_n$, and in each parallelepiped select a point (x_i,y_i,z_i) in V. We form the sum $\Sigma_{i=1}^{n} f(x_i,y_i,z_i)\,\delta V_i$ and define the volume integral of f over V, if it exists, to be

$$\iiint_V f(x,y,z)\,dV = \lim \sum_{i=1}^{n} f(x_i,y_i,z_i)\,\delta V_i \tag{4.39}$$

taking the limit as the dimensions of each volume δV_i tend to zero (which also makes n tend to infinity). For eq. (4.39) to make sense in an unambiguous way, we require that the limit exist independently of the particular manner of subdivision. It can be

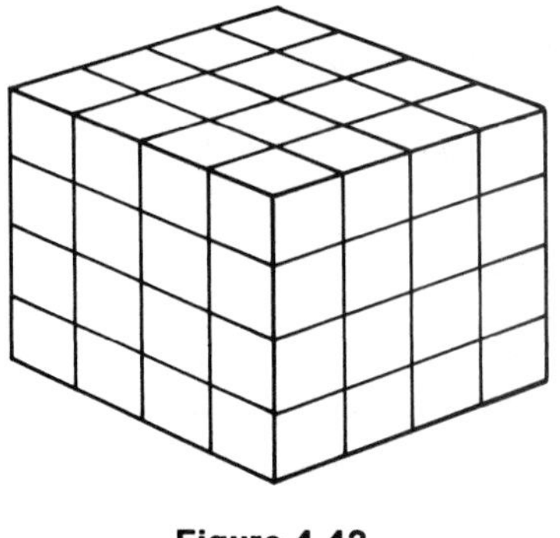

Figure 4.43

shown that this is the case if f is continuous within and on the boundary of V; we omit the proof.

Since the volume of a rectangular parallelepiped with edges dx, dy, and dz is $dV = dx\,dy\,dz$, we sometimes write

$$\iiint_V f(x,y,z)\,dx\,dy\,dz$$

instead of

$$\iiint_V f(x,y,z)\,dV$$

This suggests, and it can be proved, that a volume integral can be evaluated by triple integration, that is, by successively integrating with respect to x, then y, then z (the obvious extension of a double integral). In fact, a volume integral is almost always evaluated in this way—as an iterated integral. The only tricky part of volume integration is in supplying the limits of the "partial integrals"; we will give examples below.

One obvious application is that in which the function to be integrated is the mass density of a material. Let $\mu(x,y,z)$ denote the mass density of a material, say in grams per cubic centimeter, at a point (x,y,z). If μ is a constant, the mass of any material occupying a volume δV is precisely $\mu\;\delta V$. If μ varies from point to point, as may very well be the case for a compressible fluid, then if we take a point (x,y,z) in a small region of volume δV, we can say that $\mu(x,y,z)\;\delta V$ gives, approximately, the mass of the material within this region. We can then interpret the sum $\Sigma_{i=1}^{n}\,\mu(x_i,y_i,z_i)\;\delta V_i$ as giving an approximation to the mass within the entire domain V, and the integral in eq. (4.39) gives this mass precisely.

Similarly, if ρ represents the charge density (charge per unit volume), the volume integral of ρ over V gives the net total charge contained in the region V.

Of course, the volume of the domain V is defined by eq. (4.39), taking $f(x,y,z)$ to be identically equal to unity:

$$\text{volume of } V = \iiint_V dV = \iiint_V dx\, dy\, dz \tag{4.40}$$

Example 4.28 Find the volume integral of $f(x,y,z) = x + yz$ over the box bounded by the coordinate planes, $x = 1$, $y = 2$, and $z = 1 + x$.

Solution The region is illustrated in figure 4.44*a;* it can be described as a "four-walled house with a slant roof." The reader should take the time to identify the sides corresponding to $x = 0$, $y = 0$, $z = 0$, $y = 2$, $z = 1 + x$, and the wall $x = 1$, which has been "cut away."

Let us get the limits of integration. Consider a typical point (x,y,z) somewhere in the middle of the region. If we hold x and y fixed, z can slide down to 0 (the floor) and up to $1 + x$ (the slant roof), tracing out a *column.* If we now let x vary and hold y fixed, the columns trace out a *slice;* x goes back and forth from 0 to 1. The columns and slice are depicted in figure 4.44*b.* Now if we vary y from 0 to 2, these slices stack up and fill out the region. Integrating in this order we obtain

$$\begin{aligned}
\iiint f\, dz\, dx\, dy &= \int_0^2 \int_0^1 \int_0^{1+x} (x + yz)\, dz\, dx\, dy \\
&= \int_0^2 \int_0^1 \left(xz + \frac{1}{2} yz^2 \right) \Big|_0^{1+x} dx\, dy \\
&= \int_0^2 \int_0^1 \left(x + x^2 + \frac{1}{2} y + yx + \frac{1}{2} yx^2 \right) dx\, dy \\
&= \int_0^2 \left(\frac{5}{6} + \frac{7}{6} y \right) dy = 4
\end{aligned}$$

An alternative way of choosing limits arises if we vary z first, tracing out the vertical columns as before, but then vary y next. The vertical columns will trace out a slice from left to right, for $0 \leq y \leq 2$, and the slices will stack up as x goes from 0 to 1. This produces

$$\begin{aligned}
&\int_0^1 \int_0^2 \int_0^{1+x} (x + yz)\, dz\, dy\, dx \\
&= \int_0^1 \int_0^2 \left(x + x^2 + \frac{1}{2} y + yx + \frac{1}{2} yx^2 \right) dy\, dx \\
&= \int_0^1 \left(xy + x^2y + \frac{1}{4} y^2 + \frac{1}{2} y^2x + \frac{1}{4} y^2x^2 \right) \Big|_0^2 dx \\
&= \int_0^1 (1 + 4x + 3x^2)\, dx = 4
\end{aligned}$$

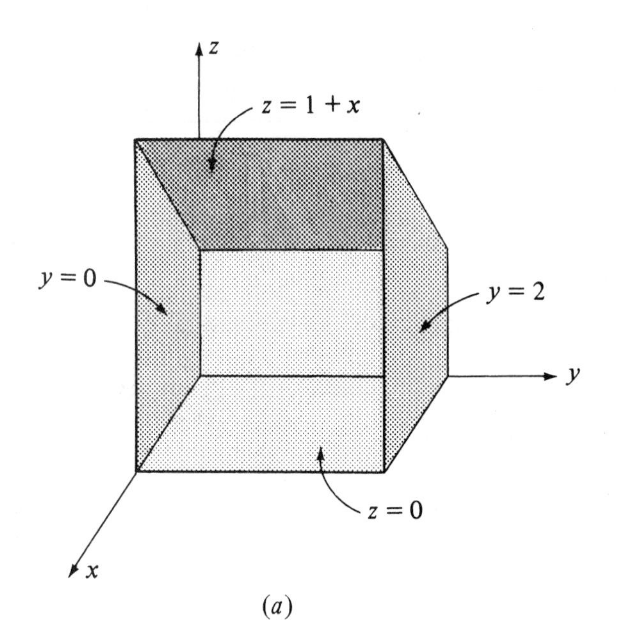

(a)

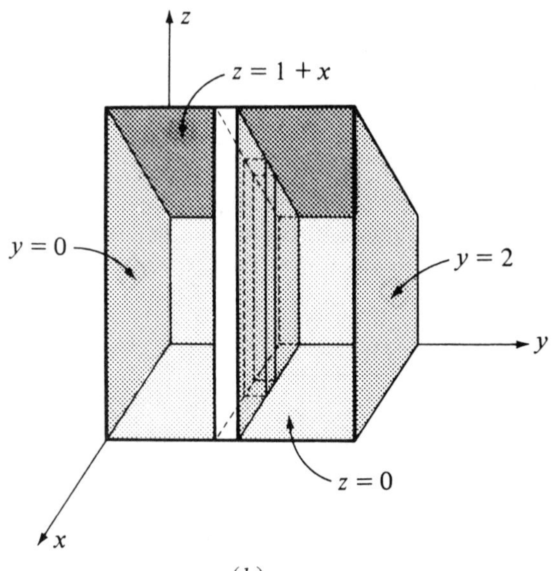

(b)

Figure 4.44

Now suppose, starting from our "typical point in the middle" (x,y,z) we let y vary first, holding x and z constant. Then we would trace out a horizontal column from left ($y = 0$) to right ($y = 2$). If we next vary z, holding x constant, these columns fill out a slice parallel to the yz plane, running from the floor ($z = 0$) to the ceiling ($z = 1 + x$). Stacking these slices for x between 0 and 1 fills out the region, and we have

$$\int_0^1 \int_0^{1+x} \int_0^2 (x + yz)\, dy\, dz\, dx = \int_0^1 \int_0^{1+x} \left(xy + \frac{1}{2} y^2 z\right)\Bigg|_0^2 dz\, dx$$

$$= \int_0^1 \int_0^{1+x} (2x + 2z)\, dz\, dx$$

$$= \int_0^1 (2xz + z^2)\Bigg|_0^{1+x} dx$$

$$= \int_0^1 (1 + 4x + 3x^2)\, dx = 4$$

What if, starting from our typical point in the middle, we let x vary first, holding y and z constant, which produces columns coming "out of the page" in figure 4.44? Then we have a complication. In the main part of the house, $z \leq 1$, the column runs from the back wall ($x = 0$) to the front ($x = 1$), but in the "attic," $z \geq 1$, the column only goes back to the slant roof, where $x = z - 1$. (Where does this equation come from?) Thus we have to break up the region, below and above the level $z = 1$, in order to establish consistent limits of integration. In both sections the columns can run left to right, $0 \leq y \leq 2$, generating horizontal slices that can be stacked, $0 \leq z \leq 1$ in the lower part and $1 \leq z \leq 2$ in the attic. This produces

$$\int_0^1 \int_0^2 \int_0^1 (x + yz)\, dx\, dy\, dz + \int_1^2 \int_0^2 \int_{z-1}^1 (x + yz)\, dx\, dy\, dz$$

$$= \int_0^1 \int_0^2 \left(\frac{1}{2} + yz\right) dy\, dz + \int_1^2 \int_0^2 \left(2yz - \frac{1}{2} z^2 + z - yz^2\right) dy\, dz$$

$$= \int_0^1 (1 + 2z)\, dz + \int_1^2 (6z - 3z^2)\, dz = 2 + 2 = 4$$

Example 4.28 demonstrates that volume integrals can be iterated in any order, but that some orders may be more complicated than others.

Example 4.29 Find the volume of the region of space above the xy plane and beneath the plane $z = 2 + x + y$, bounded by the planes $y = 0$, $x = 0$, and the surface $y = 1 - x^2$.

Solution Let us try to visualize the region. Its base, in the xy plane, is shown in figure 4.45*a*. Each of the shaded points is the base of a column reaching up to the "slanted roof," $z = 2 + x + y$ (fig. 4.45*b*).

To find the limits of integration, we start with the "typical point in the middle." If we let z vary first, with x and y fixed, we trace out a vertical column from $z = 0$ to

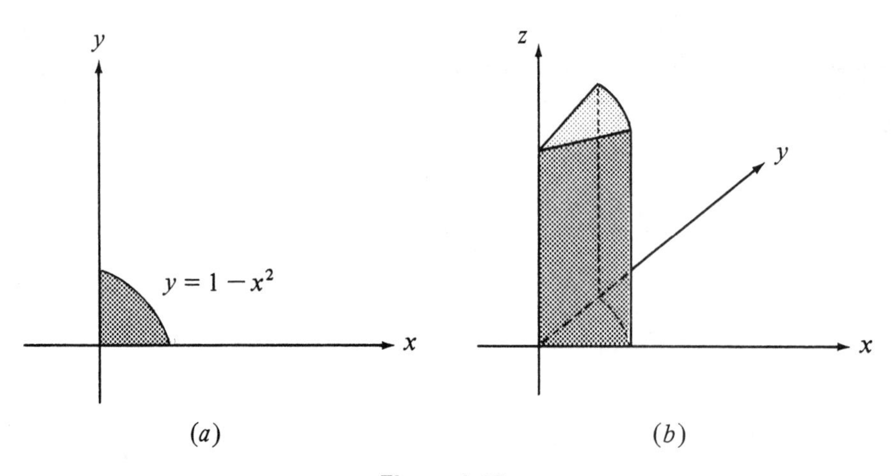

Figure 4.45

$z = 2 + x + y$. If, instead, we hold y and z fixed and vary x, the horizontal columns are troublesome; they run from 0 to $(1 - y)^{1/2}$ throughout most of the region, but in the upper right-hand corner (fig. 4.45) there is an "attic" where the slant roof cuts off the columns, on their left end. The same happens if we vary y first. Taking the easy way out, we choose the vertical columns.

If we hold x fixed and vary y, the columns generate slices from $y = 0$ to $y = 1 - x^2$. Stacking up these slices for $0 \leq x \leq 1$, we find

$$\begin{aligned}
\int_0^1 \int_0^{1-x^2} \int_0^{2+x+y} dz\, dy\, dx &= \int_0^1 \int_0^{1-x^2} (2 + x + y)\, dy\, dx \\
&= \int_0^1 \left(2y + xy + \frac{1}{2}y^2\right)\Bigg|_0^{1-x^2} dx \\
&= \int_0^1 \left(\frac{5}{2} + x - 3x^2 - x^3 + \frac{1}{2}x^4\right) dx \\
&= \frac{37}{20}
\end{aligned}$$

Example 4.30 Find the integral of $f(x,y,z) = y$ over the volume of the sphere contained inside $x^2 + y^2 + z^2 = 1$.

Solution Obviously, for a sphere we can integrate equally well in any order. From figure 4.46 we see that if we fix x and y, z traces out a column between the limits $\pm(1 - x^2 - y^2)^{1/2}$. These columns can be slid in the y direction, for fixed x, between the limits $y = \pm(1 - x^2)^{1/2}$, generating slices that can be stacked from $x = -1$ to $x = +1$.

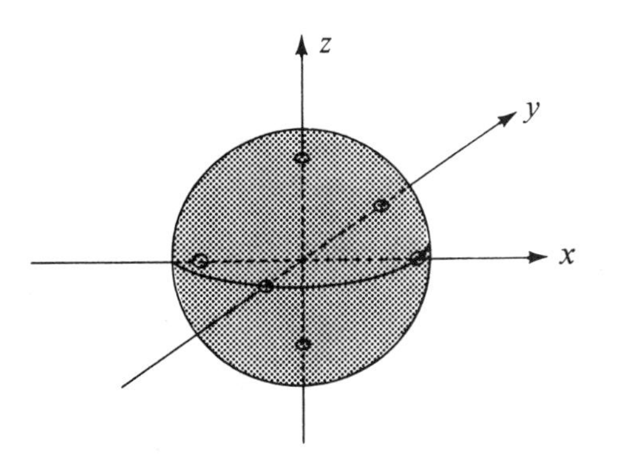

Figure 4.46

Hence

$$\iiint f\,dV = \int_{-1}^{1}\int_{-(1-x^2)^{1/2}}^{+(1-x^2)^{1/2}}\int_{-(1-x^2-y^2)^{1/2}}^{+(1-x^2-y^2)^{1/2}} y\,dz\,dy\,dx$$
$$= 2\int_{-1}^{1}\int_{-(1-x^2)^{1/2}}^{+(1-x^2)^{1/2}} y(1 - x^2 - y^2)^{1/2}\,dy\,dx$$
$$= 2\int_{-1}^{1} 0\,dx = 0$$

(We might have anticipated this answer by symmetry.)

Integrals over spheres are, as one would expect, easier to set up in spherical coordinates. For this example the interior of the sphere is generated as r runs from 0 to the outer radius (1), with ϕ and θ running through their full ranges. The volume element is given by eq. (3.65), and the integrand y is given by the transformation equations (3.61). Thus

$$\iiint f\,dV = \int_0^{2\pi}\int_0^{\pi}\int_0^{1} (r \sin\phi \sin\theta)\, r^2 \sin\phi\,dr\,d\phi\,d\theta$$
$$= 0. \text{ (Of course.)}$$

Example 4.31 Find the integral of the function $x^2 + y^2$ over the volume contained between the two cylinders $\rho = 1$ and $\rho = 2$, for $0 \leq z \leq 2$.

Solution In cylindrical coordinates the integral is given by [eqs. (3.50, 3.54)]

$$\iiint (x^2 + y^2)\,dV = \int_0^{2}\int_0^{2\pi}\int_1^{2} \rho^3\,d\rho\,d\theta\,dz = (2)(2\pi)\left(\frac{2^4}{4} - \frac{1^4}{4}\right) = 15\pi$$

EXERCISES

1. Compute the volume of the sphere of radius R by iterated integrals.
2. The volume of the region described in example 4.28 equals 3. Verify this four times by repeating each of the integrations given in example 4.28, taking $f(x,y,z) = 1$ instead of $f(x,y,z) = x + yz$.
3. Sketch the region whose volume is represented by the triple integral

$$\int_0^2 \int_0^3 \int_0^{\sqrt{9-y^2}} dx\, dy\, dz$$

4. In this exercise you will be asked to make a simple conjecture on the basis of carrying out the following computations.
   (a) Let $\mathbf{F}(x,y,z) = x^2\mathbf{i} + y\mathbf{j} + z\mathbf{k}$. Compute $\iint_S \mathbf{F} \cdot d\mathbf{S}$ over the surface of the cube bounded by the planes $x = 0$, $x = 1$, $y = 0$, $y = 1$, $z = 0$, $z = 1$ (fig. 4.47).

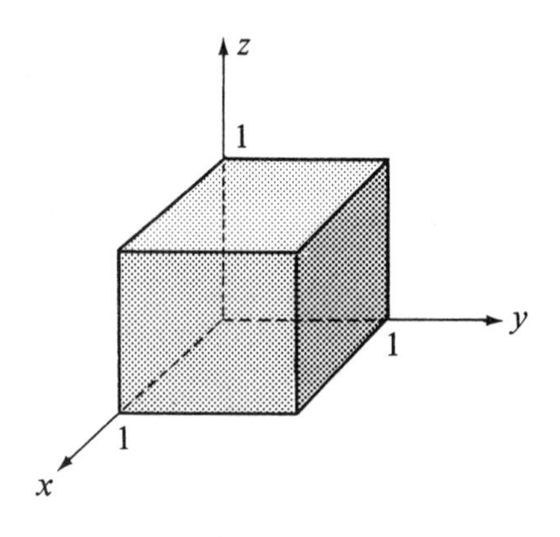

Figure 4.47

   (b) Let $f(x,y,z) = \nabla \cdot \mathbf{F}$, and compute

$$\iiint_V f(x,y,z)\, dV$$

   over the cube. Notice that here limits are no problem; we have simply

$$\int_0^1 \int_0^1 \int_0^1 f(x,y,z)\, dx\, dy\, dz$$

   (c) If your answers to (a) and (b) are not equal, check your work until you find the mistake.
   (d) Now invent another vector field $\mathbf{F}$ and repeat steps (a) and (b).
   (e) What do you conjecture from this?

5. Let V be a domain with volume v. Let $\mathbf{F} = x\mathbf{i} + y\mathbf{j} + z\mathbf{k}$.
   (a) What is $\iiint_V \nabla \cdot \mathbf{F}\, dV$?
   (b) On the basis of your answer to exercise 4, what do you conjecture is the value of $\iint_S \mathbf{F} \cdot d\mathbf{S}$, the surface integral of the normal component of $\mathbf{F}$ over the boundary of V?
6. Find the volume of the region bounded by the surface $z = e^{-(x^2+y^2)}$, the cylinder $x^2 + y^2 = 1$, and the plane $z = 0$.

4.9 Introduction to the Divergence Theorem and Stokes' Theorem

Having completed these preliminaries on higher-dimensional integration, we now turn to the most interesting part of our study. In this section we introduce two theorems of fundamental importance in vector analysis. Most of our work so far has been intended as preparation for these theorems. Their mathematically precise statements will be given in the next chapter; here we state them in rough form, omitting the rigorous conditions on continuity, differentiability, and so forth. And we give derivations for the theorems which are instructive and heuristic, deferring the more careful analysis until later.

First we present the divergence theorem.

THEOREM 4.5 *The volume integral of the divergence of a vector field, taken throughout a bounded domain D, equals the surface integral of the normal component of the vector field taken over the surface S bounding D.*

$$\iiint_D \nabla \cdot \mathbf{F}\, dV = \iint_S \mathbf{F} \cdot d\mathbf{S} \qquad (4.41)$$

In rough terms, the total divergence within D equals the net flux emerging through D's "skin." See figure 4.48. An example will clarify the statement.

Example 4.32 Compare both sides of eq. (4.41) when $\mathbf{F}$ is given in spherical coordinates by $\mathbf{F} = r^n\mathbf{e}_r$ and $\mathbf{D}$ is the upper hemisphere bounded above by $r = 3$, $0 \le \phi \le \pi/2$ and below by the equatorial plane ($z = 0$).

Solution In spherical coordinates the divergence of $\mathbf{F}$ is given by [eq. (3.67)]

$$\nabla \cdot \mathbf{F} = \frac{1}{r^2}\frac{\partial}{\partial r}(r^2 r^n) = (n + 2)r^{n-1}.$$

Integrating with respect to the spherical volume element [(eq. (3.65)] we calculate

$$\int_0^3 \int_0^{\pi/2} \int_0^{2\pi} (n + 2)r^{n-1}\, r^2 dr \sin\phi\, d\phi\, d\theta$$

$$= \{r^{n+2}\}_0^3 \{-\cos\phi\}_0^{\pi/2} \{\theta\}_0^{2\pi} = 3^{n+2} \times 2\pi.$$

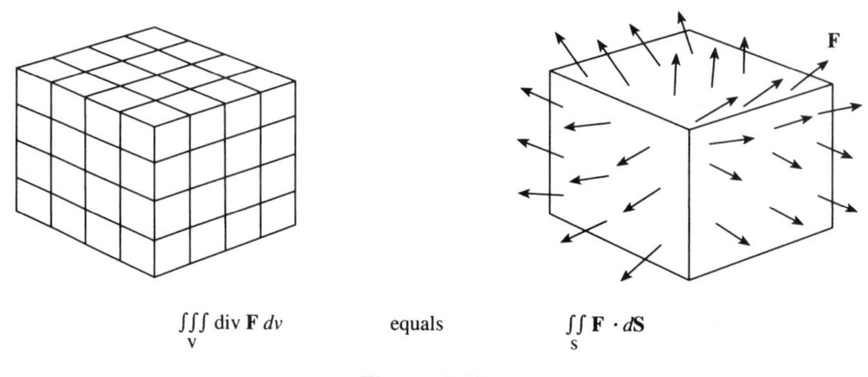

Figure 4.48

On the spherical part of the boundary, $\mathbf{F} \cdot d\mathbf{S} = r^n dS$. The element of area on a sphere was computed in eq. (4.31): inserting this into the flux integral produces

$$\int_0^{\pi/2} \int_0^{2\pi} r^n r^2 \sin \phi \, d\phi \, d\theta = 3^{n+2} \{-\cos \phi\}_0^{\pi/2} \{\theta\}_0^{\pi/2} = 3^{n+2} \times 2\pi.$$

On the equatorial plane $\mathbf{F}$ is perpendicular to the normal and there is no additional flux. Hence the divergence theorem is confirmed.

Example 4.33 Verify eq. (4.41) when $\mathbf{F} = \mathbf{R} \times \mathbf{k}$ and D is the cylinder described in cylindrical coordinates by $0 \leq \rho \leq 1$, $0 \leq z \leq 1$.

Solution In cartesian coordinates it is easy to verify that $\nabla \cdot \mathbf{F} = 0$ everywhere. In cylindrical coordinates $\mathbf{R} = \rho \mathbf{e}_\rho + z\mathbf{k}$, so $\mathbf{R} \times \mathbf{k} = -\rho \mathbf{e}_\theta$. The normals to the top and bottom faces of the cylinder are $\pm \mathbf{k}$, and the normal to the side wall is $\mathbf{e}_\rho$; hence $\mathbf{R} \times \mathbf{k}$ has no flux out of any part of the surface. Therefore eq. (4.41) merely states $0 = 0$ in this case.

Here is a heuristic proof of the divergence theorem.

Proof First let us consider a small rectangular parallelepiped bounded by planes of constant x, $x + dx$, y, $y + dy$, z, and $z + dz$. The surface integral of $\mathbf{F} \cdot \mathbf{n}$ over the six faces of this solid is the total flux of $\mathbf{F}$ out of the box. In section 3.3 we showed that this flux is given, in the limit, by

$$\nabla \cdot \mathbf{F} \, dx \, dy \, dz$$

Now let us divide the domain D into many small parallelepipeds, as if they were building blocks used in constructing D. See figure 4.48. What do we obtain if we sum up the flux out of all these blocks? If two such parallelepipeds are adjacent, the flux outward from one equals the flux inward to the other, over the face they have in common. Hence the only noncancelling contributions come from the blocks on the surface, and these terms add up to give the total flux of $\mathbf{F}$ out of the "brick structure." As we take smaller and smaller

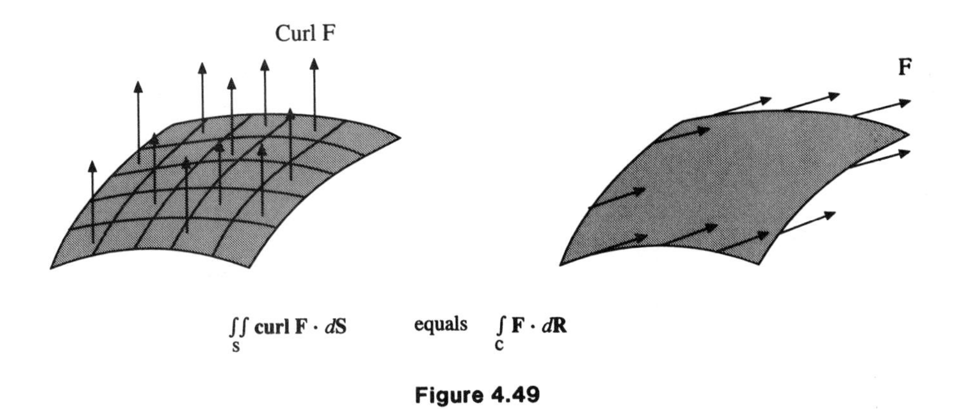

Figure 4.49

blocks, we expect that $\Sigma\, \nabla \cdot \mathbf{F}\, \delta V$ approaches the volume integral, and the flux out of the structure approaches the flux out of D. Hence

$$\iiint \nabla \cdot \mathbf{F}\, dV = \iint \mathbf{F} \cdot d\mathbf{S}$$

There are some obvious weaknesses in this proof. The passing to the limits must be considered much more carefully, particularly with regard to the surface integral. For example, it is not clear that one is justified in approximating, say, a spherical surface by a collection of little rectangles parallel to the coordinate planes (recall that in fig. 4.31 the rectangular patch is taken *tangent to the surface*). These heuristic arguments, however, are very valuable in helping one to remember theorems and conjecture new approaches.

Now let us turn to Stokes' theorem, the other crucial theorem in vector analysis.

THEOREM 4.6 *The surface integral of the normal component of the curl of a vector field, taken over a bounded surface S, equals the line integral of the tangential component of the field, taken over the closed curve C bounding the surface.*

$$\iint_S \nabla \times \mathbf{F} \cdot d\mathbf{S} = \int_C \mathbf{F} \cdot d\mathbf{R} \tag{4.42}$$

Here we assume that the surface S is oriented by a field of unit normals $\mathbf{n}$, and that the line integral around its boundary C is taken in the direction determined positive by this orientation (figure 4.49). Roughly speaking, then, Stokes' theorem states that the total flux of the curl equals the line integral around the edge.

Example 4.34 Verify Stokes' theorem for $\mathbf{F} = x\mathbf{j}$ and the upper hemispherical surface $r = 3, 0 \leq \phi \leq \pi/2$.

Solution In cartesian coordinates it is easy to see that $\nabla \times \mathbf{F} = \mathbf{k}$. The element of surface area on the hemisphere was computed in eq. (4.30) to be

$$d\mathbf{S} = \frac{\partial \mathbf{R}}{\partial \phi} \times \frac{\partial \mathbf{R}}{\partial \theta}\, d\phi\, d\theta$$
$$= \{r^2 \sin^2\phi \cos\theta\, \mathbf{i} + r^2 \sin^2\phi \sin\theta\, \mathbf{j} + r^2 \sin\phi \cos\phi\, \mathbf{k}\}\, d\phi\, d\theta$$

so the flux of **curl F** through the surface is given by

$$\int\int \nabla \times \mathbf{F} \cdot d\mathbf{S} = \int_0^{\pi/2} \int_0^{2\pi} 3^2 \sin\phi \cos\phi\, d\phi\, d\theta = 9\pi$$

Note that the direction of the normal $d\mathbf{S}$ is upwards, towards the positive z axis (as indicated by the sign of the coefficient of $\mathbf{k}$ for $0 \leq \phi \leq \pi/2$). Thus the boundary of the hemisphere—the circle of radius 3 in the xy plane—is oriented counterclockwise by the right-hand rule (section 4.6). On this circle $\mathbf{e}_r$ and $\mathbf{e}_\theta$ are identical with the polar vectors $\mathbf{u}_r$ and $\mathbf{u}_\theta$ of section 2.5, so for the element of arc length we combine eqs. (3.63) and (2.47) to write

$$d\mathbf{R} = \mathbf{e}_r\, dr + \mathbf{e}_\phi\, r\, d\phi + \mathbf{e}_\theta\, r \sin\phi\, d\theta = \mathbf{e}_\theta\, 3\, d\theta$$
$$= 3\,(-\sin\theta\, \mathbf{i} + \cos\theta\, \mathbf{j})\, d\theta$$

Thus the line integral around the boundary equals

$$\int \mathbf{F} \cdot d\mathbf{R} = \int_0^{2\pi} x\,(3\cos\theta)\, d\theta = \int_0^{2\pi} (3\cos\theta)\,(3\cos\theta)\, d\theta = 9\pi$$

and eq. (4.42) is confirmed.

To see the basis for Stokes' theorem, recall the "swirl per unit area" characterization of the curl that we derived in section 3.4 and exploited in sections 3.10 and 3.11. With reference to figure 3.16, we took the swirl of **F** around a small rectangle to be the sum of the lengths of the edges of the rectangle, each weighted by the counterclockwise component of **F** on that edge. As a result of our deliberations in section 4.1, we are now equipped to identify this calculation more precisely; it's the line integral of **F** around the rectangle! Thus in figure 4.50 as we calculate $\int\int$ (**curl F**) · **n** dS over the surface area, we are effectively computing the line integral of **F** around each infinitesimal rectangle and dividing by the area of the rectangle (to get (**curl F**) · **n**), then multiplying by this area and summing over the rectangles (to get the integral). This amounts to totaling up the line integrals around each rectangle. But these integrals cancel over the internal edges (since they go in opposite directions), and we are left with the line integral around the boundary curve C. That's the conclusion of Stokes' theorem.

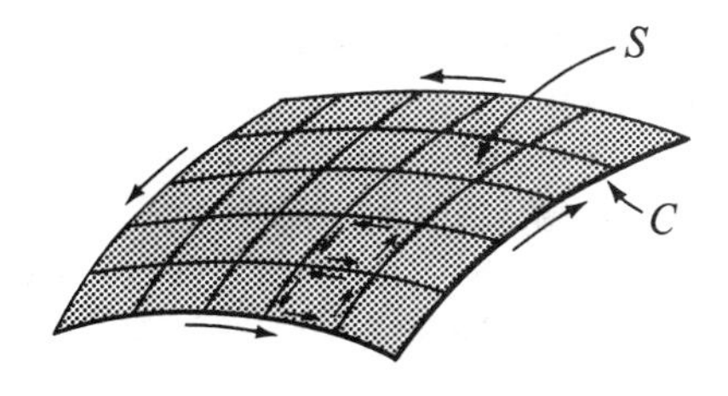

Figure 4.50

What subtleties are overlooked in this derivation? Obviously it is generally not possible to subdivide a surface into perfect rectangles. Thus we need to devise an argument that accounts more precisely for this approximation, or finesses it entirely. The more rigorous proof will be given in chapter 5.

In summary, we have stated the divergence theorem and Stokes' theorem, and we have given heuristic derivations of each. These derivations are motivated by the physical interpretations attributed to the concepts of "divergence" and "curl." However, there is another way of looking at these statements, based on a more *analytic* point of view. Note that the divergence theorem

$$\iiint_D \nabla \cdot \mathbf{F}\, dV = \iint_S \mathbf{F} \cdot \mathbf{n}\, dS$$

transforms a *three*-dimensional integral of a *differentiated* quantity into a *two*-dimensional integral, with the differential operator deleted. Stokes' theorem

$$\iint_S \nabla \times \mathbf{F} \cdot \mathbf{n}\, dS = \int_C \mathbf{F} \cdot d\mathbf{R}$$

changes a two-dimensional integral into a one-dimensional one, again dropping a differential operator. Now if we agree that isolated *points* are "zero-dimensional," then theorem 4.1 for conservative fields,

$$\int_P^Q \nabla\phi \cdot d\mathbf{R} = \phi(Q) - \phi(P)$$

fits this same pattern! In fact, Poincaré showed that all these theorems can be interpreted as higher-dimensional analogs of the fundamental theorem of calculus:

$$\int_{x_1}^{x_2} \frac{df}{dx}\, dx = f(x_2) - f(x_1)$$

In the next chapter we will present more rigorous analyses of the vector theorems, and the fundamental theorem will prove to be an important aid. Before proceeding to these sections, however, the reader is strongly urged to study the exercises below. The divergence and Stokes' theorems mark the apex of the subject of vector analysis, and the rest of our book will focus on a deeper study of the concepts we have introduced. Table 4.2 summarizes the important integral theorems; bear in mind that the divergence theorem applies to closed surfaces, and Stokes' theorem to open surfaces.

TABLE 4.2 Integral Theorems

Operator	Interpretation		Integral Theorem
grad ϕ	Maximum rate of change of ϕ, in the maximal direction		$\int_P^Q \nabla\phi \cdot d\mathbf{R} = \phi(Q) - \phi(P)$
div **F**	Net outflux of **F** per unit volume		$\iiint_D \nabla \cdot \mathbf{F}\, dV = \iint_S \mathbf{F} \cdot \mathbf{n}\, dS$
curl F	Swirl of **F** per unit area		$\iint_S \nabla \times \mathbf{F} \cdot \mathbf{n}\, dS = \int_C \mathbf{F} \cdot d\mathbf{R}$

EXERCISES

1. Use the divergence theorem to solve exercise 1, section 4.7.
2. Do all seven parts of exercise 2, section 4.7, by computing

$$\int_{-1}^{1}\int_{-1}^{1}\int_{-1}^{1} \nabla \cdot \mathbf{F}\, dx\, dy\, dz$$

   in each case.
3. Use the divergence theorem to solve
   (a) Exercise 4, section 4.7.
   (b) Exercise 5, section 4.7.
4. Find $\iiint_V \operatorname{div} \mathbf{F}\, dV$, when $\mathbf{F} = (x^2 + xy)\mathbf{i} + (y^2 + yz)\mathbf{j} + (z^2 + zx)\mathbf{k}$, and V is the cube centered at the origin and with faces on the planes $x = \pm 1$, $y = \pm 1$, $z = \pm 1$.
5. Use the divergence theorem to evaluate $\iint_S \mathbf{F} \cdot \mathbf{n}\, dS$, when $\mathbf{F} = y^2x\mathbf{i} + x^2y\mathbf{j} + z^2\mathbf{k}$, and when S is the complete surface of the region bounded by the cylinder $x^2 + y^2 = 4$ and by the planes $z = 0$ and $z = 2$.

6. (a) Find a vector field $\mathbf{F} = F_r(r)\mathbf{e}_r$ satisfying $\nabla \cdot \mathbf{F} = r^m$, m ≥ 0.
(b) Use the divergence theorem to prove

$$\iiint_D r^m \, dV = \frac{1}{m+3} \iint_S r^{m+1}\mathbf{e}_r \cdot d\mathbf{S}$$

7. Use Stokes' theorem to solve exercise 4, section 4.1.

8. Use Stokes' theorem to solve exercise 6, section 4.1.

9. Verify Stokes' theorem in the following special cases. Let C be the square in the xy plane with equation $|x| + |y| = 1$. Let $\mathbf{F}$ be as follows:
(a) $\mathbf{F} = x\mathbf{i}$
(b) $\mathbf{F} = y\mathbf{i}$
(c) $\mathbf{F} = -y\mathbf{i} + x\mathbf{j}$
(d) $\mathbf{F} = \mathbf{i} + \mathbf{j}$
(e) $\mathbf{F} = y^3\mathbf{i}$

10. By means of Stokes' theorem, find $\int \mathbf{F} \cdot d\mathbf{R}$ around the ellipse $x^2 + y^2 = 1$, $z = y$, where $\mathbf{F} = x\mathbf{i} + (x + y)\mathbf{j} + (x + y + z)\mathbf{k}$.

11. Evaluate $\iint_S (\nabla \times \mathbf{F}) \cdot d\mathbf{S}$, where $\mathbf{F} = y\mathbf{i} + (x - 2x^3z)\mathbf{j} + xy^3\mathbf{k}$ and S is the surface of a sphere $x^2 + y^2 + z^2 = a^2$ above the xy plane.

12. Let S be the portion of the paraboloid $z = 9 - x^2 - y^2$ that lies above the plane $z = 0$, and let $\mathbf{F} = (y - z)\mathbf{i} - (x + z)\mathbf{j} + (x + y)\mathbf{k}$. Find $\iint_S (\nabla \times \mathbf{F}) \cdot \mathbf{n}\, dS$.

13. Evaluate $\iint_S (\nabla \times \mathbf{F}) \cdot \mathbf{n}\, dS$, where $\mathbf{F} = 2y\mathbf{i} + (x - 2x^3z)\mathbf{j} + xy^3\mathbf{k}$, and where S is the curved surface of the hemisphere $x^2 + y^2 + z^2 = 1$, $z \geq 0$.

14. Use Stokes' theorem to evaluate

$$\int_C [x \sin y\, \mathbf{i} - y \sin x\, \mathbf{j} + (x + y)z^2\mathbf{k}] \cdot d\mathbf{R}$$

along the path consisting of straight-line segments successively joining the points $P_0 = (0,0,0)$ to $P_1 = (\pi/2,0,0)$ to $P_2 = (\pi/2,0,1)$ to $P_3 = (0,0,1)$ to $P_4 = (0,\pi/2,1)$ to $P_5 = (0,\pi/2,0)$, and back to $(0,0,0)$.

15. Let $\mathbf{F}$ be the field $\mathbf{F} = ye^x\mathbf{i} + (x + e^x)\mathbf{j} + z^2\mathbf{k}$, and let C be the curve given by

$$\mathbf{R}(t) = (1 + \cos t)\mathbf{i} + (1 + \sin t)\mathbf{j} + (1 - \sin t - \cos t)\mathbf{k}$$

for $0 \leq t \leq 2\pi$. Find $\int_C \mathbf{F} \cdot d\mathbf{R}$. (*Hint:* Use Stokes' theorem, observing that C is contained in a certain plane and that the projection of C on the xy plane is a circle.)

16. If $\mathbf{F} = xz\mathbf{i} - y\mathbf{j} + x^2y\mathbf{k}$, use Stokes' theorem to evaluate $\int_C \mathbf{F} \cdot d\mathbf{R}$, where C is the closed path consisting of the edges of the triangle with vertices at the points $P_1 = (1,0,0)$, $P_2 = (0,0,1)$, $P_3 = (0,0,0)$ transversed from P_1 to P_2 to P_3, and back to P_1.

17. Given $\phi(x,y,z) = xyz + 5$, find the surface integral of the normal component of **grad** ϕ over $x^2 + y^2 + z^2 = 9$.

18. (a) Show that, if ϕ is harmonic, $\nabla \cdot (\phi\nabla\phi) = |\nabla\phi|^2$.
(b) Given $\phi = 3x + 2y + 4z$, evaluate

$$\iint \phi \frac{\partial \phi}{\partial n} dS$$

over the surface $x^2 + y^2 + z^2 = 4$. Here, $\partial\phi/\partial n$ represents the normal derivative of ϕ (i.e., $\mathbf{n} \cdot \nabla\phi$).

19. Let $\mathbf{F} = \phi\nabla\phi$. Find the surface integral of the normal component of $\mathbf{F}$ over the surface of a sphere of radius 3 and center at the origin,
(a) if $\phi = x + y + z$.
(b) if $\phi = x^2 + y^2 + z^2$.

20. Let $\mathbf{F} = (x - yz)\mathbf{i} + (y + xz)\mathbf{j} + (z + 2xy)\mathbf{k}$, and let S_1 be the portion of the cylinder $x^2 + y^2 = 2$ that lies inside the sphere $x^2 + y^2 + z^2 = 4$. Let S_2 be the portion of the surface of the sphere $x^2 + y^2 + z^2 = 4$ that lies outside cylinder $x^2 + y^2 = 2$. Let V be the volume bounded by S_1 and S_2.
(a) Draw a diagram illustrating S_1, S_2, and V.
(b) Compute $\iint_{S_1} \mathbf{F} \cdot \mathbf{n}_1 \, dS_1$, with $\mathbf{n}_1$ pointing inward.
(c) Compute $\iiint_V (\nabla \cdot \mathbf{F}) \, dV$.
(d) Compute $\iint_{S_2} \mathbf{F} \cdot \mathbf{n}_2 \, dS_2$, with $\mathbf{n}_2$ pointing outward.

21. Despite the fact that the surface of exercise 6, section 4.7, is not closed, the divergence theorem can be used to reduce this to a problem in mental arithmetic. Show how to do this.

22. The moment of inertia about the z axis of a uniform solid is proportional to

$$\iiint_V (x^2 + y^2) \, dx \, dy \, dz$$

Express this as the flux of some vector field through the surface of the body.

23. One can compute the volume of a room by calculating the flux of the vector $\mathbf{R}$ through the walls. Show this.

24. If $\mathbf{R} = x\mathbf{i} + y\mathbf{j} + z\mathbf{k}$ and $R = |\mathbf{R}|$, find $\iint R \, \mathbf{R} \cdot \mathbf{n} \, dS$ over the surface of a sphere of radius b and centered at the origin,
(a) by interpreting the integrand geometrically.
(b) by using the divergence theorem.

25. Given

$$\mathbf{F} = \frac{x\mathbf{i} + y\mathbf{j} + z\mathbf{k}}{x^2 + y^2 + z^2}$$

find the surface integral of the normal component of $\mathbf{F}$ over the surface of the sphere $x^2 + y^2 + z^2 = 4$. Can you use the divergence theorem?

26. Let $\mathbf{F} = xyz\mathbf{i} + (y^2 + 1)\mathbf{j} + z^3\mathbf{k}$, and let S be the surface of the unit cube $0 \le x, y, z \le 1$. Evaluate the surface integral $\iint_S (\nabla \times \mathbf{F}) \cdot \mathbf{n} \, dS$ using
(a) the divergence theorem.
(b) Stokes' theorem.
(c) direct computation.

27. Stokes' theorem provides an interesting interpretation of theorem 4.3, which identifies irrotational fields with conservative fields in simply connected domains. Show that if **curl** $\mathbf{F} = \mathbf{0}$, then the line integral of $\mathbf{F}$ around any closed curve that bounds an oriented surface in the domain is zero. Where does simple connectedness come into play?

28. Explain: the flux of a solenoidal field through a surface depends only on the curve bounding the surface.

29. The abstract concept of a gooney sphere is derived from the shape of a gooney egg. A gooney bird is born with a pointed head and a prominent stubby tail; therefore the shape of the egg is roughly ellipsoidal but with pointed ends. Surface integrals over gooney spheres are difficult to compute; tables of gooney functions are needed, but these were tabulated during the war and are still classified top secret. All that is known is that a gooney sphere of minimal diameter $d = 1$ has volume approximately 0.7.

(a) Find the surface integral of the normal component of $\mathbf{F} = x\mathbf{i} + y\mathbf{j} + z\mathbf{k}$ over the surface of a gooney sphere with center at the origin and minimal diameter $d = 2$, making any assumptions you deem reasonable.

(b) Would your answer be the same if the gooney sphere had center at $(2,7,-3)$?

30. If electric field intensity is $\mathbf{E} = (x + 1)^2\mathbf{i} + y\mathbf{j} + z\mathbf{k}$, relevant to suitable choices of the units involved, what is the total charge within the cube bounded by the planes $x = 0$, $x = 1$, $y = 0$, $y = 1$, $z = 0$, and $z = 1$? Evaluate the left side of eq. (4.38),

(a) directly.

(b) by the divergence theorem.

31. Surface and volume integrals of vector-valued functions are defined as for numerically valued functions. Alternatively, they can be defined by simply integrating separately the x, y, and z components (which are numerical). Show formally that $\iiint_D \nabla\phi \, dV = \iint_S \phi\mathbf{n} \, dS$ by applying the divergence theorem to $\mathbf{F} = \phi\mathbf{C}$, where $\mathbf{C}$ is a constant vector field.

32. Derive the identity

$$\iiint_D \nabla \times \mathbf{F} \, dV = \iint_S \mathbf{n} \times \mathbf{F} \, dS$$

where $\mathbf{n}$ is the outer normal to S, the boundary of D.

33. Give a vector interpretation of each of the following.

(a) $\lim\limits_{V \to 0} \dfrac{\iint_S \mathbf{n}f \, dS}{V}$

(b) $\lim\limits_{V \to 0} \dfrac{\iint_S \mathbf{n} \times \mathbf{F} \, dS}{V}$

34. Indicate which of the following statements are true, and which are false. You may assume that all functions have continuous derivatives of all orders at all points.

(a) The divergence of $\nabla \times \mathbf{F}$ is zero, for every $\mathbf{F}$.

(b) In a simply connected region, $\int_C \mathbf{F} \cdot d\mathbf{R}$ depends only on the endpoints of C.

(c) If $\nabla f = \mathbf{0}$, then f is a constant function.

(d) If $\nabla \times \mathbf{F} = \mathbf{0}$, then $\mathbf{F}$ is a constant vector field.

(e) If div $\mathbf{F} = 0$, then $\iint_S \mathbf{F} \cdot d\mathbf{S} = 0$ for every closed surface S.

(f) If $\int_C \mathbf{F} \cdot d\mathbf{R} = 0$ for every closed contour C, then $\nabla \times \mathbf{F} = \mathbf{0}$.

35. Let $\mathbf{J}$ denote electric current density (a vector in the direction of the current, with magnitude in units of current per area) and $\mathbf{B}$ denote the magnetic field intensity. One of Maxwell's laws of electromagnetism states that, in the absence of a time-varying electric field,

$$\textbf{curl } \mathbf{B} = \mu_0 \mathbf{J}$$

where μ_0 is a constant. Use Stokes' theorem to derive

$$\int_C \mathbf{B} \cdot d\mathbf{R} = \mu_0 i$$

In words: the line integral of the tangential component of magnetic field intensity, around a closed loop, is proportional to the current i passing across any surface bounded by the loop.

4.10 Optional Reading: Introduction to the Transport Theorems

In some physics and engineering applications it is necessary to compute the time derivative of a surface or volume integral when the surface or volume of integration is in motion. For instance, in an electric generator a loop of wire is moved through a magnetic field in such a manner that the flux of the field through the surface bounded by the loop is changing. According to Faraday's law, an electromotive force is set up in the loop, proportional to the rate of change of this flux integral.

Similarly, one may wish to note the rate of change of some quantity, such as momentum or charge or stored energy, associated with a specific portion of a moving fluid. If this quantity is given as a volume integral of some density function, then the volume of interest is being transported downstream with the fluid as the time derivative is taken.

These two problems are related. First we treat the moving surface problem, then we use the answer in analyzing the transported volume problem.

The situation is as follows. We have a vector field **F** that changes with time, $\mathbf{F} = \mathbf{F}(\mathbf{R},t)$ and an oriented surface S that, together with its properly oriented boundary curve C, moves through space; we use the notation S_t and C_t to designate, respectively, the surface and curve at time t. Let $\Phi(t)$ be the flux of $\mathbf{F}(\mathbf{R},t)$ through S_t, at time t:

$$\Phi(t) = \iint_{S_t} \mathbf{F}(\mathbf{R},t) \cdot d\mathbf{S}$$

Notice that $\Phi(t)$ changes due to two effects, namely, the changing field **F** and the motion of the surface S_t. The problem is to compute $d\Phi/dt$.

We shall give a heuristic argument based on figure 4.51. A more rigorous argument using parametrizations appears in section 5.6.

Figure 4.51 shows the location of the surface at times t_1 and t_2, together with its boundary and the orientations. In figure 4.52, the displacements of corresponding points on S_{t_1} and S_{t_2} are shown. If we can associate a velocity field $\mathbf{v} = \mathbf{v}(\mathbf{R},t)$ on S that describes the pointwise motion of the surface, then for $t_2 - t_1 \equiv dt$ sufficiently small, the point **R** on S_{t_1} is carried to the point $\mathbf{R} + \mathbf{v}(\mathbf{R},t_1)\, dt$ on S_{t_2}. (The reader who is uneasy about the vagueness of these notions will feel more comfortable with the derivation in section 5.6.)

To compute $d\Phi/dt$, we must evaluate

$$\frac{d\Phi}{dt} = \lim_{t_2 \to t_1} \frac{1}{t_2 - t_1}\left(\iint_{S_{t_2}} \mathbf{F}(\mathbf{R},t_2) \cdot d\mathbf{S} - \iint_{S_{t_1}} \mathbf{F}(\mathbf{R},t_1) \cdot d\mathbf{S}\right) \tag{4.43}$$

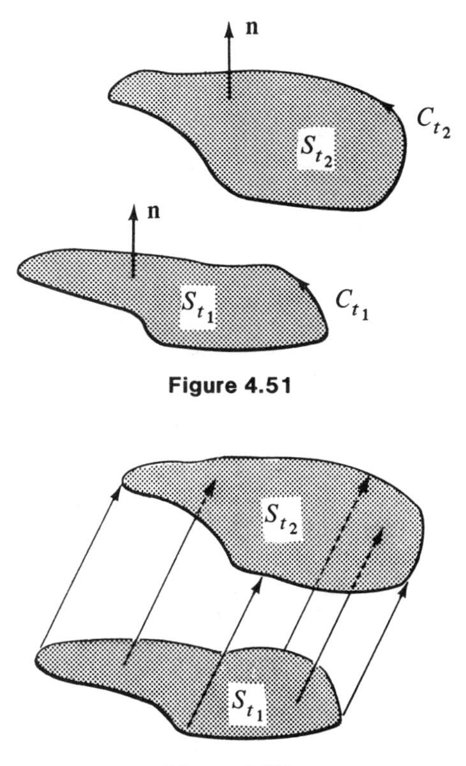

Figure 4.51

Figure 4.52

Visualizing these fluxes in figure 4.52 suggests that the divergence theorem, applied to the region D swept out by the surfaces in the intervening times, may be useful. Applying the theorem at time t_1, with due consideration for the distinction between the surface normal and the outward normal, we have

$$\iiint_D \nabla \cdot \mathbf{F}(\mathbf{R},t_1)\, dV = \iint_{S_{t_2}} \mathbf{F}(\mathbf{R},t_1) \cdot d\mathbf{S} - \iint_{S_{t_1}} \mathbf{F}(\mathbf{R},t_1) \cdot d\mathbf{S} + \iint_{\text{(sides)}} \mathbf{F}(\mathbf{R},t_1) \cdot d\mathbf{S} \tag{4.44}$$

Expressing

$$\mathbf{F}(\mathbf{R},t_2) \approx \mathbf{F}(\mathbf{R},t_1) + \frac{\partial \mathbf{F}}{\partial t} dt$$

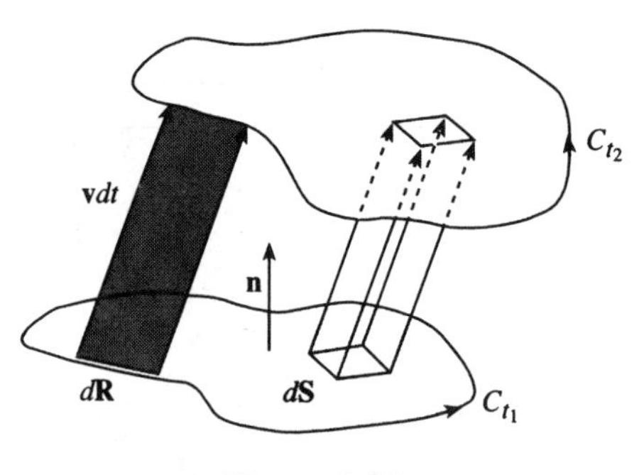

Figure 4.53

and using eq. (4.44) for the S_{t_1} integral, we find

$$\begin{aligned}\iint_{S_{t_2}} \mathbf{F}(\mathbf{R},t_2) \cdot d\mathbf{S} - \iint_{S_{t_1}} \mathbf{F}(\mathbf{R},t_1) \cdot d\mathbf{S} \\ = \iint_{S_{t_2}} \mathbf{F}(\mathbf{R},t_1) \cdot d\mathbf{S} + dt \iint_{S_{t_2}} \frac{\partial \mathbf{F}}{\partial t} \cdot d\mathbf{S} \\ - \iint_{S_{t_2}} \mathbf{F}(\mathbf{R},t_1) \cdot d\mathbf{S} - \iint_{\text{(sides)}} \mathbf{F}(\mathbf{R},t_1) \cdot d\mathbf{S} \\ + \iiint_D \nabla \cdot \mathbf{F}(\mathbf{R},t_1)\, dV \end{aligned} \tag{4.45}$$

Obviously, two of these integrals cancel. On the sides, figure 4.53 shows that the surface element $d\mathbf{S}$ equals $d\mathbf{R} \times \mathbf{v}\, dt$, where $d\mathbf{R}$ is taken along C_{t_1}. The element of volume in D has base $|d\mathbf{S}|$ and height $|\mathbf{v}\, dt \cdot \mathbf{n}|$, so that

$$dV = d\mathbf{S} \cdot \mathbf{v}\, dt$$

Consequently, eq. (4.45) becomes

$$\begin{aligned}\iint_{S_{t_2}} \mathbf{F}(\mathbf{R},t_2) \cdot d\mathbf{S} - \iint_{S_{t_1}} \mathbf{F}(\mathbf{R},t_1) \cdot d\mathbf{S} \\ = dt \iint_{S_{t_2}} \frac{\partial \mathbf{F}}{\partial t} \cdot d\mathbf{S} - dt \oint_{C_{t_1}} \mathbf{F}(\mathbf{R},t_1) \cdot d\mathbf{R} \times \mathbf{v} \\ + dt \iint_{S_{t_1}} (\nabla \cdot \mathbf{F})\mathbf{v} \cdot d\mathbf{S}\end{aligned}$$

Dividing by dt, we arrive at the *flux transport theorem:*

$$\frac{d\Phi}{dt} = \iint_{S_t} \left(\frac{\partial \mathbf{F}}{\partial t} + (\nabla \cdot \mathbf{F})\mathbf{v} \right) \cdot d\mathbf{S} + \oint_{C_t} \mathbf{F} \times \mathbf{v} \cdot d\mathbf{R} \tag{4.46}$$

We note that if the velocity field $\mathbf{v}$, defined on the surface S_t, can be extended as a continuously differentiable vector field throughout some region containing S_t, then Stokes' theorem can be employed to recast eq. (4.46) as

$$\frac{d\Phi}{dt} = \iint_{S_t} \left(\frac{\partial \mathbf{F}}{\partial t} + (\nabla \cdot \mathbf{F})\mathbf{v} + \nabla \times (\mathbf{F} \times \mathbf{v}) \right) \cdot d\mathbf{S} \tag{4.47}$$

This would be the case if, for instance, the surface were being transported inside a moving volume of fluid.

Now we turn to the transport theorem for volume integrals, known as *Reynold's transport theorem.* Let $\eta(\mathbf{R},t)$ be a continuously differentiable scalar field, and let V_t denote the volume of integration at time t. The points inside V_t move with velocity $\mathbf{v}(\mathbf{R},t)$, generating the motion of the volume. Our task is to compute

$$\frac{d}{dt} \iiint_{V_t} \eta(\mathbf{R},t)\, dV \tag{4.48}$$

The answer can be obtained heuristically as follows. If the volume is divided into small rectangular parallelepipeds, each of volume ΔV (as in section 4.8), then the integral in eq. (4.48) is approximated by

$$\sum \eta(\mathbf{R},t)\Delta V \tag{4.49}$$

Its derivative is approximated by

$$\sum \frac{d\eta(\mathbf{R},t)}{dt} \Delta V + \sum \eta(\mathbf{R},t) \frac{d\Delta V}{dt} \tag{4.50}$$

Now in exercise 12 of section 3.3 we argued that

$$\frac{1}{\Delta V} \cdot \frac{d\Delta V}{dt} = \nabla \cdot \mathrm{v} \tag{4.51}$$

Using (4.51) in equation (4.50) and letting $\Delta \mathrm{V} \rightarrow 0$, we derive the first form of Reynold's theorem:

$$\frac{d}{dt} \iiint_{V_t} \eta(\mathbf{R},\mathrm{t})\, dV = \iiint_{V_i} \left(\frac{d\eta}{dt} + \eta \nabla \cdot \mathbf{v} \right) dV \tag{4.52}$$

In exercise 14 of section 3.8 we derived the alternate expression for the *convective derivative*

$$\frac{d\eta(\mathbf{R},t)}{dt} = \frac{\partial \eta}{\partial t} + \mathbf{v} \cdot \nabla \eta \tag{4.53}$$

When this is substituted into (4.52) and simplified, the second form of Reynold's theorem results:

$$\begin{aligned} \frac{d}{dt}\iiint_{V_t} \eta \, dV &= \iiint_{V_t} \left(\frac{\partial \eta}{\partial t} + \mathbf{v} \cdot \nabla \eta + \eta \nabla \cdot \mathbf{v} \right) dV \\ &= \iiint_{V_t} \left(\frac{\partial \eta}{\partial t} + \nabla \cdot (\eta \mathbf{v}) \right) dV \end{aligned} \tag{4.54}$$

Applying the divergence theorem produces a third form:

$$\frac{d}{dt}\iiint_{V_t} \eta \, dV = \iiint_{V_t} \frac{\partial \eta}{\partial t} dV + \iint_{S_t} \eta \mathbf{v} \cdot d\mathbf{S} \tag{4.55}$$

Equation (4.55) can be interpreted as saying that the rate at which the volume integral of η increases when the volume is moving equals the rate of increase when the volume is stationary, plus the amount of η that is swept up through the moving sides.

Reynold's transport theorem plays a key role in the mathematical formulation of the laws of fluid motion. Usually in these applications η is a function of the mass density μ, which is constrained by the equation of continuity, expressed as

$$\frac{\partial \mu}{\partial t} = -\nabla \cdot \mu v \tag{4.56}$$

in terms of Eulerian derivatives [eq. (3.9)] or as

$$\frac{d\mu}{dt} = -\mu \, \nabla \cdot v \tag{4.57}$$

in terms of the moving, convective derivative (exercise 14, section 3.8).

For example if η is μ itself Reynold's equation (4.54), coupled with equation (4.56), implies

$$\frac{d}{dt} \iiint_V \mu \, dV = 0;$$

i.e., the net mass inside the volume remains constant. This, of course, is no surprise, since we are following a fixed group of fluid particles as they are transported by the flow.

A more interesting formula results if we let η represent the momentum density. Since the latter is a vector, we must interpret η as a particular component of the momentum. Thus if $\eta = \mu v_i$, then $\frac{d}{dt} \iiint_V \mu v_i \, dV$ is the rate of change of the ith component of the momentum of the fluid in the volume. According to Newton's second law, this balances the force on the fluid. By employing equation (4.57) in Reynold's equation (4.52) we can rewrite this momentum term as

$$\begin{aligned}
\frac{d}{dt} \iiint_V \mu v_i \, dV &= \iiint_V \left\{ \frac{d}{dt} (\mu v_i) + \mu v_i \nabla \cdot \mathbf{v} \right\} dV \\
&= \iiint_V \left\{ \frac{d\mu}{dt} v_i + \mu \frac{d}{dt} v_i + \mu v_i \nabla \cdot \mathbf{v} \right\} dV \\
&= \iiint_V \left\{ (-\mu \nabla \cdot \mathbf{v})\, v_i + \mu \frac{d}{dt} v_i + \mu v_i \nabla \cdot \mathbf{v} \right\} dV \\
&= \iiint_V \mu \frac{d}{dt} v_i \, dV
\end{aligned}$$

In full vector form, this becomes *Cauchy's expression for the rate of change of momentum:*

$$\frac{d}{dt} \iiint_V \mu \mathbf{v} \, dV = \iiint_V \mu \frac{d\mathbf{v}}{dt} \, dV. \tag{4.58}$$

The analysis of the forces in a fluid is too complicated to be included here, and we refer the reader to the textbooks on fluid dynamics for further study.

EXERCISES

1. Let S_t be a uniformly expanding hemisphere described by

$$x^2 + y^2 + z^2 = (vt)^2 \qquad (z \geq 0)$$

and let $\mathbf{F}$ be the vector field

$$\mathbf{F}(\mathbf{R},t) = \mathbf{R}t$$

Verify the flux transport theorem in this case.

2. Verify the flux transport theorem when S_t is the square with corners $(0,0,t)$, $(0,1,t)$, $(1,0,t)$, $(1,1,t)$, and $\mathbf{F}(\mathbf{R},t) = xz\mathbf{k}$.

3. Suppose the square $0 \leq x \leq 1$, $0 \leq y \leq 1$ is rotated about the x axis at a constant angular velocity. Verify the flux transport theorem with the uniform vector field $\mathbf{F}(\mathbf{R},t) = \mathbf{k}$.

4. Verify Reynold's transport theorem for the expanding sphere V_t:

$$x^2 + y^2 + z^2 \leq (vt)^2$$

and

$$\rho(\mathbf{R},t) = |\mathbf{R}|^2 t$$

5. Verify Reynold's theorem for a unit cube with edges parallel to the axes, sliding in the x direction at constant velocity and with $\rho(\mathbf{R},t) = xy$.

6. Show that if μ is the fluid density, then

$$\frac{d}{dt}\iiint_V \mu \mathbf{F}\, dV = \iiint_V \mu \frac{d\mathbf{F}}{dt}\, dV$$

is valid for any vector field **F**.

7. In a so-called *perfect fluid,* the *pressure* $p(\mathbf{R},t)$ exerts a force per unit area on a surface given by $-p\mathbf{n}$, where $\mathbf{n}$ is the unit surface normal. Gravity exerts a force per unit mass on the fluid given by the constant vector $\mathbf{g}$ (the gravitational acceleration).

   (a) Show that Newton's second law, applied to a volume of a perfect fluid, is expressed

   $$\frac{d}{dt}\iiint_V \mu\mathbf{v}\, dV = \iiint_V \mu\mathbf{g}\, dV - \iint_S p\mathbf{n} \cdot d\mathbf{S}$$

   where S is the surface boundary of V.

   (b) Modify the analysis of section 3.3 to rewrite the pressure force in terms of the gradient; then use Cauchy's expression for the momentum term and derive

   $$\iiint_V \mu \frac{d}{dt}\mathbf{v}\, dV = \iiint_V \{\mu\mathbf{g} - \nabla p\}\, dV. \tag{4.59}$$

   (c) Argue that (4.59) implies the equation of motion

   $$\mu \frac{d}{dt}\mathbf{v} = \mu\mathbf{g} - \nabla p.$$

   (d) For steady flow all the Eulerian time derivatives $(\partial/\partial t)$ are zero. Use exercise 14, section 3.8, to derive the equation for steady flow

   $$\mu\, \mathbf{v} \cdot \nabla\mathbf{v} = \mu\mathbf{g} - \nabla p.$$

5 Advanced Topics

This chapter contains a somewhat disjointed collection of topics which go beyond the customary realm of study for a first course in vector analysis. The divergence, Stokes', and transport theorems are reexamined with proofs which rely more on mathematical analysis and less on physical interpretation. An extensive exploration of the applications of the divergence theorem to partial differential equations is conducted. Finally we include some lesser-known applications of vector analysis to problems in matrix algebra.

5.1 The Divergence Theorem

As promised in section 4.9, we will now delve into a more careful, detailed analysis of the divergence theorem

$$\iiint_D \nabla \cdot \mathbf{F}\, dV = \iint_S \mathbf{F} \cdot \mathbf{n}\, dS \tag{5.1}$$

To fix ideas for the moment let us consider the vector field $\mathbf{F} = F_1\mathbf{i} + F_2\mathbf{j} + F_3\mathbf{k}$ to have continuous partial derivatives throughout a region. Let S denote the surface of a *sphere* located within the region, and let D be the set of points inside S. At each point on S we take $\mathbf{n}$ to be the *outward* normal, thus orienting S in the usual way.

Consider the single term arising in the left-hand side of eq. (5.1):

$$\iiint_D \frac{\partial F_3}{\partial z}\, dx\, dy\, dz$$

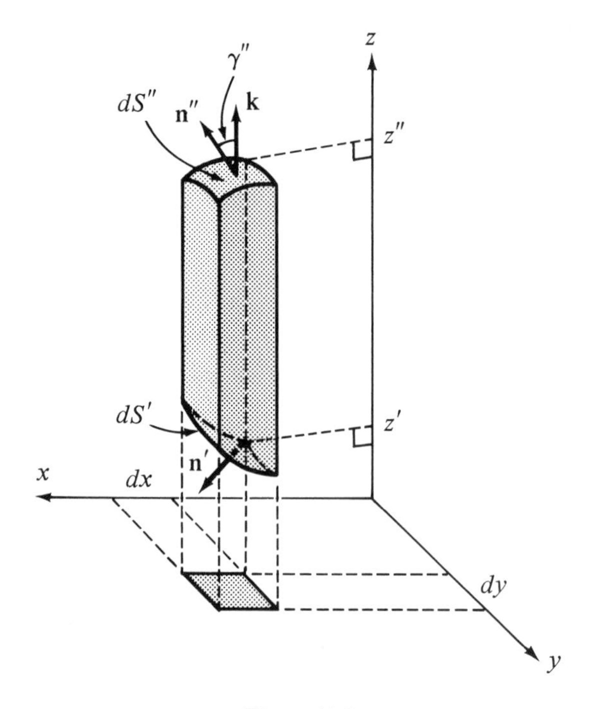

Figure 5.1

To evaluate this volume integral by the procedure of section 4.8, we can begin by integrating along a *column* parallel to the z axis (see fig. 5.1). We invoke the fundamental theorem of calculus to derive

$$\iiint \frac{\partial F_3}{dz}\, dz\, dx\, dy = \iint (F_3'' - F_3')\, dx\, dy \tag{5.2}$$

where $F_3''$and F_3' are, respectively, the values of F_3 at the top and bottom of the column. In the figure, $dx\,dy$ is the projection of the elemental area dS'' onto the xy plane:

$$dx\, dy = dS''\, |\cos \gamma''| = dS''\, \mathbf{k} \cdot \mathbf{n}'' \tag{5.3}$$

and, similarly,

$$dx\, dy = -dS'\, \mathbf{k} \cdot \mathbf{n}' \tag{5.4}$$

[since the *outward* normal $\mathbf{n}'$ points downward. Recall a similar analysis in section 4.6.] Insertion of eqs. (5.3) and (5.4) into eq. (5.2) yields

$$\iiint_D \frac{\partial F_3}{\partial z} dV = \iint_S (F_3\mathbf{k}) \cdot \mathbf{n}\, dS \tag{5.5}$$

Similarly, we can show that the volume integrals of $\partial F_1/\partial x$ and $\partial F_2/\partial y$ equal the fluxes of $F_1\mathbf{i}$ and $F_2\mathbf{j}$ through S, and adding these equations yields the divergence theorem, eq. (5.1).

In this way we prove the divergence theorem for a sphere. However, notice that the same proof applies when S is the surface of an ellipsoid, a cube, a right circular cylinder, or even a potato-shaped region of a fairly arbitrary nature. Here is a more precise statement of what we have shown.

THEOREM 5.1 *(The Divergence Theorem) Let D be any domain with the property that each straight line through any interior point of the domain cuts the boundary in exactly two points, and such that the boundary S is a piecewise smooth, closed, oriented surface with unit normal directed outward from the domain. Let $\mathbf{F}$ be a vector field, $\mathbf{F} = F_1\mathbf{i} + F_2\mathbf{j} + F_3\mathbf{k}$, continuous throughout a region containing D and its boundary, and such that the partial derivatives of F_1, F_2, and F_3 are also continuous in this region. Then*

$$\iiint_D \operatorname{div} \mathbf{F}\, dV = \iint_S \mathbf{F} \cdot \mathbf{n}\, dS$$

In proving this theorem, we made strong use of the idea of cutting up the sphere by filaments parallel to a coordinate axis. We assumed in figure 5.1 that any such filament cuts out two portions from the surface. Thus the proof does not apply without modification to a domain such as the dumbbell-shaped one in figure 5.2. Here, such a filament can cut out four portions from the surface. However, it is easy to see that the theorem still applies to such a domain, since the dumbbell can be cut in the middle and the theorem applied separately to the two parts. The volume integral over the whole domain equals the sum of the two separate volume integrals, and the corresponding surface integrals add up to give the surface integral over the dumbbell (there will be two contributions from the common boundary B, but they will cancel each other, since $\mathbf{n}$ will have opposite directions in the two integrals).

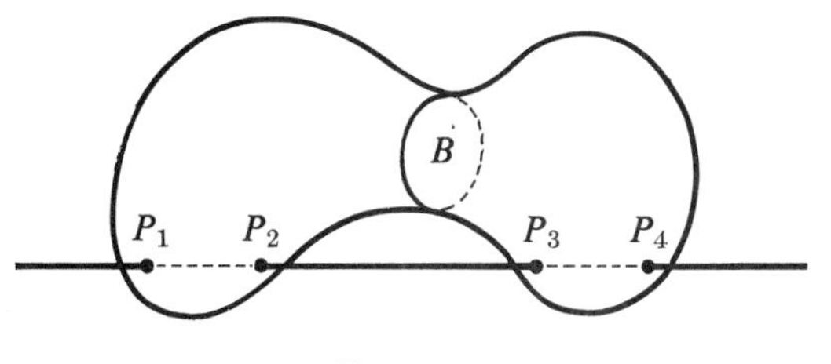

Figure 5.2

Let us now investigate one interesting consequence of the divergence theorem. Let us suppose that the domain D is a very small one surrounding a point P. If the domain is sufficiently small, div $\mathbf{F}$ will be approximately constant, and the volume integral of div **F** over the volume V will be approximately equal to the product (div **F**) V. More precisely, we have

$$\lim_{V \to 0} \frac{\iiint \text{div } \mathbf{F} \, dV}{V} = \text{div } \mathbf{F}$$

By the divergence theorem, we can replace the volume integral of div **F** by the surface integral of **F** over the boundary enclosing the volume, from which it follows that

$$\text{div } \mathbf{F} = \lim_{V \to 0} \frac{\iint_S \mathbf{F} \cdot \mathbf{n} \, dS}{V} \tag{5.6}$$

Of course this characterization of divergence as outflux per unit volume was our original motivation for defining the divergence, in section 3.3. There we restricted it to rectangular parallelepipeds with sides parallel to the axes and came up with the analytic formula

$$\text{div } \mathbf{F} = \frac{\partial F_1}{\partial x} + \frac{\partial F_2}{\partial y} + \frac{\partial F_3}{\partial z}, \tag{5.7}$$

which we took as the formal definition. But logically speaking, we never proved that the formula (5.7) would be consistent with the result of taking outflux per unit volume *for other shapes*. We *presumed* this in sections 3.10 and 3.11, trusting that the curvilinear coordinate expressions for the divergence, derived using the "definition" (5.6), gave the same vector as formula (5.7). Now thanks to the *analytic* proof of the divergence theorem—which is based only on the definition (5.7) and not on the heuristic picture—our calculations are justified. The divergence is, indeed, a coordinate-free concept. (As the physicists would say, it is "covariant.")

EXERCISES

1. At what point in the proof of the divergence theorem did we make use of the requirement that the partial derivative $\partial F_3/\partial z$ be a continuous function of z?
2. In the proof, we required that the three partial derivatives be continuous, that is, that each of them be continuous in all three variables. Why, for example, should we care whether or not the partial derivative $\partial F_2/\partial y$ is a continuous function of x?
3. Show, by a diagram similar to that of figure 5.1, that the volume integral of a function, taken over D, can be obtained by first integrating with respect to z and then integrating over the projection of S on the xy plane.
4. Outline a proof of the divergence theorem, taking exercise 3 as the starting point. Start with $\iiint_D \text{div } \mathbf{F} \, dV$, integrating first with respect to y. Your proof will differ only slightly

from that given in this section; that is, you will integrate first with respect to y rather than z. By using the definition of surface integral you can avoid use of such words as "approximately"; for simplicity, assume that S is a smooth surface.

5. Where, in your "proof" (exercise 4), did you make unconscious use of the fact that the points on S with normals parallel to the xz plane have a projection on the xz plane of zero area? [*Hint:* Look again at the definition of the area of a surface (section 4.6). What is $\cos \gamma$ for such points?]
6. What is the flux output per unit volume at $(3,1,-2)$ if $\mathbf{F} = x^3\mathbf{i} + yx\mathbf{j} - x^3\mathbf{k}$?
7. What is the flux output from an ellipsoid of volume v if $\mathbf{F} = 3x\mathbf{i} + y\mathbf{j} + z\mathbf{k}$?
8. If $\mathbf{F} = 3x^2\mathbf{i} + y\mathbf{j} + z\mathbf{k}$, would the flux output from an ellipsoid depend on the location of the ellipsoid as well as on its volume?
9. (a) Describe the oriented surface enclosing the region $1 \leq x^2 + y^2 + z^2 \leq 4$, assuming the usual convention concerning the orientation of a closed surface. (In section 4.6 it was mentioned that if a surface encloses a region of space, the unit normal points away from the enclosed region; in this problem, the surface has two disconnected parts.)

   (b) How would you compute the surface integral of the normal component of a vector field $\mathbf{F}$ over this surface?

   (c) If $\text{div } \mathbf{F} = 0$ except perhaps at the origin, what can you say about the flux $\iint \mathbf{F} \cdot \mathbf{n}\, dS$ over the two parts forming this surface, taking $\mathbf{n}$ to be the unit normal outward from the origin in each case?

   (d) Determine whether your answer to (c) would be any different if the region were that between the sphere $x^2 + y^2 + z^2 = 1$ and the ellipsoid

$$\frac{x^2}{4} + \frac{y^2}{9} + \frac{z^2}{16} = 1$$

   (e) Compute the surface integral of the normal component of

$$\mathbf{F} = \frac{x\mathbf{i} + y\mathbf{j} + z\mathbf{k}}{(x^2 + y^2 + z^2)^{3/2}}$$

   over the ellipsoid

$$\frac{x^2}{4} + \frac{y^2}{9} + \frac{z^2}{16} = 1$$

10. Letting $\mathbf{R} = x\mathbf{i} + y\mathbf{j} + z\mathbf{k}$, and $R = |\mathbf{R}|$, write the vector field $\mathbf{F} = \mathbf{R}/R^3$ in terms of R and $\mathbf{e}_r$.

    (a) Show that $\text{div } \mathbf{F}$ is identically zero throughout the domain of definition of $\mathbf{F}$.

    (b) Show that the surface integral of the normal component of $\mathbf{F}$ over the surface of the unit sphere $r = 1$ is 4π.

    (c) Explain why parts (a) and (b) do not contradict the divergence theorem.

    (d) What is the surface integral of the normal component of $\mathbf{F}$ over the surface of a unit sphere with center 4 units away from the origin?

5.2 Green's Formulas: Laplace's and Poisson's Equations

In this section we will use the divergence theorem

$$\iiint_D \nabla \cdot \mathbf{F}\, dV = \iint_S \mathbf{F} \cdot \mathbf{n}\, dS \tag{5.8}$$

to study the physical and mathematical significance of the laplacian, and to explore applications to some of the partial differential equations of mathematical physics.

As a start, we apply eq. (5.8) to a vector field concocted from a pair of scalar fields: $\mathbf{F} = \phi\nabla\psi$. From identity (3.30), we have

$$\nabla \cdot (\phi\nabla\psi) = \nabla\phi \cdot \nabla\psi + \phi\nabla^2\psi$$

Thus eq. (5.8) gives us

$$\begin{aligned}\iiint_D \nabla\phi \cdot \nabla\psi\, dV + \iiint_D \phi\nabla^2\psi\, dV &= \iiint_D \nabla \cdot (\phi\nabla\psi)\, dV \\ &= \iint_S \phi\nabla\psi \cdot \mathbf{n}\, dS\end{aligned}$$

which we rewrite as the *first Green formula:*

$$\iiint_D \phi\nabla^2\psi\, dV = \iint_S \phi\nabla\psi \cdot \mathbf{n}\, dS - \iiint_D \nabla\phi \cdot \nabla\psi\, dV \tag{5.9}$$

If we exchange the position of ϕ and ψ and subtract, the dot product cancels and we get the *second Green formula:*

$$\iiint_D (\phi\nabla^2\psi - \psi\nabla^2\phi)\, dV = \iint_S (\phi\nabla\psi - \psi\nabla\phi) \cdot \mathbf{n}\, dS \tag{5.10}$$

The third Green formula comes about by setting $\psi = 1/R$ in eq. (5.10), where R is the magnitude of the position vector $\mathbf{R} = x\mathbf{i} + y\mathbf{j} + z\mathbf{k}$. However, ψ approaches infinity at the origin, and this invalidates the divergence theorem there; so we must take extra pains to do the computation. To this end, we introduce a few technical lemmas.

LEMMA 5.1 *Let S be a sphere of radius b centered at the origin, and let $f(\mathbf{R})$ be a continuous scalar field. Then*

$$\lim_{b \to 0} \iint \frac{f}{R}\, dS = 0 \tag{5.11}$$

Proof On the sphere, $R = b$. The integral $\iint f\, dS$ equals the average value of f on the sphere times the area, $4\pi b^2$. By continuity, then, the limit is $\lim 4\pi b^2 f(0,0,0)/b = 0$.

LEMMA 5.2 *Let S and f be as in Lemma* 5.1. *Then*

$$\lim_{b \to 0} \iint_S f \nabla \left(\frac{1}{R}\right) \cdot \mathbf{n}\, dS = -4\pi f(0,0,0) \tag{5.12}$$

Proof The outward normal $\mathbf{n}$ on the sphere equals $\mathbf{R}/R$. From identity (3.33), $\nabla(1/R) = -\mathbf{R}/R^3$. Therefore

$$\nabla\left(\frac{1}{R}\right) \cdot \mathbf{n} = -\frac{R^2}{R^4} = -\frac{1}{b^2} \tag{5.13}$$

on the sphere. As before, the integral $\iint f\, dS$ is the average value of f on the sphere multiplied by $4\pi b^2$. Thus the left member of (5.12) is -4π times the limit of this average value, which by continuity approaches $f(0,0,0)$.

LEMMA 5.3 *The laplacian of* $1/R$ *is zero, except at the origin:*

$$\nabla^2\left(\frac{1}{R}\right) = 0 \qquad (R \neq 0) \tag{5.14}$$

Proof From identities (3.31) and (3.33),

$$\nabla^2\left(\frac{1}{R}\right) = \nabla \cdot \nabla\left(\frac{1}{R}\right) = \nabla \cdot \left(\frac{-\mathbf{R}}{R^3}\right) = -\frac{\nabla \cdot \mathbf{R}}{R^3} - \mathbf{R} \cdot \nabla\left(\frac{1}{R^3}\right)$$

$$= -\frac{3}{R^3} - \mathbf{R} \cdot \left(\frac{-3\mathbf{R}}{R^5}\right) = 0$$

Now we apply the second Green formula with $\psi = 1/R$ to a volume D enclosing the origin. We cautiously exclude a small sphere D' centered around the origin (fig. 5.3) and write

$$\iiint_{D-D'} \left[\phi \nabla^2\left(\frac{1}{R}\right) - \frac{1}{R}\nabla^2\phi\right] dV = \iint_S \left[\phi\nabla\left(\frac{1}{R}\right) - \frac{1}{R}\nabla\phi\right] \cdot \mathbf{n}\, dS$$
$$+ \iint_{S'}\left[\phi\nabla\left(\frac{1}{R}\right) - \frac{1}{R}\nabla\phi\right] \cdot \mathbf{n}\, dS$$

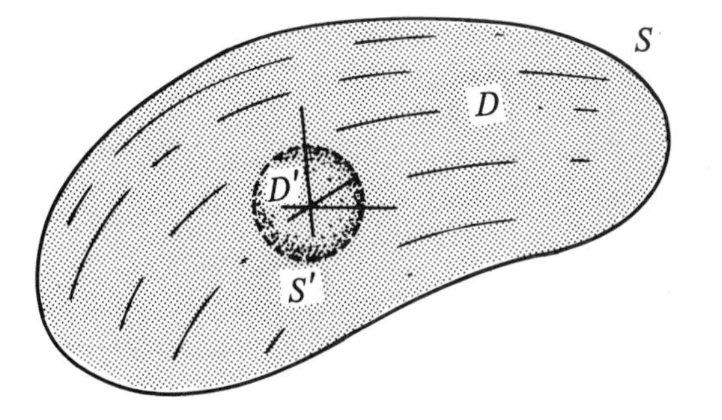

Figure 5.3

where S is the boundary of D and S' is the boundary of D'. Taking the limit as D' shrinks to a point, and noting that in the present context the *outward* normal on S' is oppositely directed to that in lemma 5.2, we use the lemmas to derive the *third Green formula:*

$$\phi(0,0,0) = \frac{-1}{4\pi}\iiint_D \frac{\nabla^2\phi}{R}\,dV + \frac{1}{4\pi}\iint_S \left[\frac{\nabla\phi}{R} - \phi\nabla\left(\frac{1}{R}\right)\right]\cdot \mathbf{n}\,dS \tag{5.15}$$

The derivation of this identity is equally valid if some other point $\mathbf{R}_0$ is used instead of the origin; we simply "relativize" and interpret R as $|\mathbf{R} - \mathbf{R}_0|$ in lemmas 5.1 through 5.3 (exercise 3). In fact, it will be helpful for future reference if we rewrite the result with $\mathbf{R}$ replacing the "origin," and $\mathbf{R}'$ the variable of integration:

$$\phi(\mathbf{R}) = \frac{-1}{4\pi}\iiint_D \frac{\nabla'^2\phi(\mathbf{R}')}{|\mathbf{R} - \mathbf{R}'|}\,dV' + \frac{1}{4\pi}\iint_S \left[\frac{\nabla'\phi(\mathbf{R}')}{|\mathbf{R} - \mathbf{R}'|} - \phi(\mathbf{R}')\nabla'\left(\frac{1}{|\mathbf{R} - \mathbf{R}'|}\right)\right]\cdot \mathbf{n}'\,dS' \tag{5.16}$$

Here we have used ∇' to denote the vector operator which differentiates with respect to the primed variables:

$$\nabla' = \mathbf{i}\frac{\partial}{\partial x'} + \mathbf{j}\frac{\partial}{\partial y'} + \mathbf{k}\frac{\partial}{\partial z'} \tag{5.17}$$

This is a very useful result, because it displays the values of any scalar field $\phi(\mathbf{R})$ in terms of the values of its laplacian throughout the domain, and the values of ϕ and $\nabla\phi$ on the boundary of the domain. In fact, we can regard the "field point" $\mathbf{R}$ as variable, and compute derivatives of ϕ directly from eq. (5.16)!

What will result if we take the laplacian ∇^2 (with respect to $\mathbf{R}$, of course) of every term in the third Green formula, eq. (5.16)? Since the variables of integration for the surface integrals are primed, the unprimed differential operators in ∇^2 can be moved inside the integrals. (Keep in mind that $\mathbf{R} \neq \mathbf{R}'$ since $\mathbf{R}'$ is constrained to the boundary surface S while $\mathbf{R}$ is inside D.) We know

$$\nabla^2\left(\frac{1}{|\mathbf{R} - \mathbf{R}'|}\right) = 0 \tag{5.18}$$

from lemma 5.3 (relativized version). Moreover, by taking the gradient of eq. (5.18) and noting that ∇' and ∇^2 commute, we also obtain

$$\nabla^2\nabla'\left(\frac{1}{|\mathbf{R} - \mathbf{R}'|}\right) = 0 \tag{5.19}$$

Consequently, the laplacian of the surface integral in (5.16) is zero:

$$\begin{aligned}\nabla^2 \iint_S \left[\frac{\nabla'\phi(\mathbf{R}')}{|\mathbf{R} - \mathbf{R}'|} - \phi(\mathbf{R}')\nabla'\left(\frac{1}{|\mathbf{R} - \mathbf{R}'|}\right)\right] \cdot \mathbf{n}' \, dS' \\ = \iint_S \left[\nabla'\phi(\mathbf{R}')\nabla^2\left(\frac{1}{|\mathbf{R} - \mathbf{R}'|}\right)\right. \\ \left. - \phi(\mathbf{R}')\nabla^2\nabla'\left(\frac{1}{|\mathbf{R} - \mathbf{R}'|}\right)\right] \cdot \mathbf{n}' \, dS' = 0\end{aligned}$$

This means that the laplacian of $-1/4\pi$ times the volume integral equals the laplacian of ϕ itself:

$$\nabla^2\phi(\mathbf{R}) = \nabla^2\left(\frac{-1}{4\pi} \iiint_D \frac{\nabla'^2\phi(\mathbf{R}')}{|\mathbf{R} - \mathbf{R}'|} dV'\right) \tag{5.20}$$

Rewriting $\nabla^2\phi$ as ρ, we conclude the following.

LEMMA 5.4 *The laplacian of*

$$\frac{-1}{4\pi} \iiint_D \frac{\rho(\mathbf{R}')}{|\mathbf{R} - \mathbf{R}'|} dV' \tag{5.21}$$

equals $\rho(\mathbf{R})$ (for $\mathbf{R}$ inside D).

(To be sure, we have proved this only for functions $\rho(\mathbf{R})$ that are known, *a priori,* to be laplacians of other functions; but the lemma can be proved for any ρ that is "Hölder continuous." If the domain D is unbounded, additional hypotheses have to be imposed to ensure that ρ approaches zero sufficiently rapidly at infinity that the integrals converge.)*

Lemma 5.4 is extremely useful in applications. It enables us to write down immediately a solution to *Poisson's equation*

$$\nabla^2\phi = \rho$$

---

*A good reference for the mathematical details is O. D. Kellogg's *Foundations of Potential Theory,* Dover Publications, New York (1953).

where $\rho(x,y,z)$ is a given scalar function; namely,

$$\phi(\mathbf{R}) = \frac{-1}{4\pi} \iiint_D \frac{\rho(\mathbf{R}')}{|\mathbf{R} - \mathbf{R}'|} dV' \tag{5.22}$$

Additional solutions to Poisson's equation are then generated by adding to ϕ solutions of the *homogeneous* form of Poisson's equation, *Laplace's equation*

$$\nabla^2\phi = 0$$

(recall exercise 4, section 3.6).

Solutions of Laplace's equation, which we have called *harmonic functions,* describe electrostatic, magnetostatic, and gravitational potentials in free space, incompressible fluid motions, steady-state temperature distributions, and equilibrium membrane deformations. The real and imaginary parts of complex analytic functions, as well, satisfy Laplace's equation. The search for solutions to Laplace's equation gave birth to Fourier series, Bessel functions, Legendre polynomials, and many techniques in the calculus of variations and functional analysis.

Poisson's equation describes the potentials in the presence of sources (charges or currents for electromagnetic, mass for gravitational); ρ denotes the source density.

The first important observation about Poisson's and Laplace's equations is that their solutions are completely determined throughout a region D by the values they take on the boundary S of D. To see this, consider two solutions $f(x,y,z)$ and $g(x,y,z)$ of, say, Poisson's equation, taking the same values on S. The difference $f - g$ obviously satisfies Laplace's equation. If we set $\phi = \psi = f - g$ in the first Green formula [eq. (5.9)], then we deduce

$$\iiint_D |\nabla(f - g)|^2 \, dV = 0$$

Thus the gradient of $f - g$ is identically zero throughout D, and f and g can differ only by a constant. Since they agree on the boundary, however, they must agree everywhere.

THEOREM 5.2 *Solutions of Laplace's and Poisson's equations are uniquely determined in a domain D by their values on the boundary of D.*

The third Green formula [eq. (5.15)], then, is a little misleading, in that it displays how a function is determined throughout a region by its laplacian, its values on the boundary, *and* the values of its normal derivative on the boundary. The latter data, according to the theorem, are redundant. To get a more precise statement we return to the second Green formula [eq. (5.10)] and, instead of replacing ψ by $1/R$, we replace it by something known as (what else?) a *Green's function.*

The Green's function $G(\mathbf{R})$ is chosen to have the form

$$G(\mathbf{R}) = \frac{1}{R} + \gamma(\mathbf{R}) \tag{5.23}$$

where γ is harmonic throughout D and equal to $-1/R$ on the surface S. (Finding such a function can be very difficult.) If we set $\psi = G$ in the second Green formula and divide by -4π, we get the right-hand side of eq. (5.15) from the $1/R$ term (as before), plus the contribution from γ:

$$\phi(0,0,0) = \frac{-1}{4\pi}\iiint_D \left(\frac{1}{R} + \gamma\right)\nabla^2\phi \, dV + \frac{1}{4\pi}\iint_S \left[\left(\frac{1}{R} + \gamma\right)\nabla\phi - \phi\nabla\left(\frac{1}{R} + \gamma\right)\right] \cdot \mathbf{n} \, dS \quad (5.24)$$

(exercise 4). Thus the coefficient of $\nabla\phi$ vanishes, and we derive the display

$$\phi(0,0,0) = \frac{-1}{4\pi}\iiint_D G\nabla^2\phi \, dV - \frac{1}{4\pi}\iint_S \phi\nabla G \cdot \mathbf{n} \, dS$$

which "relativizes" to the form

$$\phi(\mathbf{R}) = \frac{-1}{4\pi}\iiint_D G(\mathbf{R} - \mathbf{R}') \, \nabla'^2\phi(\mathbf{R}') \, dV' - \frac{1}{4\pi}\iint_S \phi(\mathbf{R}') \, \nabla' G(\mathbf{R} - \mathbf{R}') \cdot \mathbf{n}' \, dS' \quad (5.25)$$

(Some authors absorb the factor $-1/4\pi$ into the definition of G.)

Thus the Green's function, if it can be found, displays the solution to Poisson's (and hence Laplace's) equation in terms of its values on the boundary. (Recall that $\nabla^2\phi = \rho$ is a *given* part of Poisson's equation.)

In the interest of brevity we will consider only one Green's function in this book. It pertains to the special but important case where D is a sphere of radius b centered at (0,0,0). In this situation the constant function $\gamma = -1/b$ meets all the requirements, and we have

$$\phi(0,0,0) = \frac{-1}{4\pi}\iiint_D \left(\frac{1}{R} - \frac{1}{b}\right)\nabla^2\phi \, dV - \frac{1}{4\pi}\iint_S \phi\nabla\left(\frac{1}{R} - \frac{1}{b}\right) \cdot \mathbf{n} \, dS$$

which, with (5.13), becomes

$$\phi(0,0,0) = \frac{-1}{4\pi}\iiint_D \left(\frac{1}{R} - \frac{1}{b}\right)\nabla^2\phi \, dV + \frac{1}{4\pi b^2}\iint_S \phi \, dS \quad (5.26)$$

The final term is the average of ϕ on the surface of the sphere. Recall that in section 3.6 we promised to show that the laplacian $\nabla^2\phi$ provides a measure of the difference between the value of ϕ at a point and its average over surrounding points. Equation (5.26) is the fulfillment of this promise.

Notice in particular that, if ϕ is harmonic, then it satisfies the *mean value property* as follows.

THEOREM 5.3 *The value of a harmonic function at any point equals the average of its values on any sphere centered at that point.*

The mean value property explains why temperature distributions are harmonic at equilibrium: if every point is at the same temperature as its average surrounding, there is no reason for heat to flow. Advanced texts demonstrate that the mean value property is logically equivalent to harmonicity.

EXERCISES

1. Evaluate

$$\iint_S \left[\frac{1}{R}\nabla\phi - \phi\nabla\left(\frac{1}{R}\right)\right] \cdot d\mathbf{S}$$

over the surface of the sphere $(x - 3)^2 + y^2 + z^2 = 25$, where $\phi = xyz + 5$.

2. Evaluate

$$\iint_S \left[\phi\nabla\left(\frac{1}{R}\right) - \frac{1}{R}\nabla\phi\right] \cdot d\mathbf{S}$$

(a) over the surface of the ellipsoid $x^2/9 + y^2/16 + z^2/25 = 1$, where $\phi = x^2 + y^2 - 2z^2 + 4$.

(b) over the surface of the cylindrical pillbox bounded by $x^2 + y^2 = 25$ and $z = \pm 10$, where $\phi = x^2 - z^2 + 5$.

3. Show that eq. (5.16) follows by interpreting R as $|\mathbf{R} - \mathbf{R}'|$ in lemmas 5.1 through 5.3.

4. Derive eq. (5.24).

5. Prove that the solutions to Laplace's and Poisson's equations are uniquely determined in D by the values of their normal derivatives on S and the value of the solution at one (any) point of D.

6. Some authors bypass the auxiliary surface S' in the derivation of the third Green formula [eq. (5.15)] by claiming that

$$\nabla^2\left(\frac{1}{R}\right) = -4\pi\,\delta(\mathbf{R}) \tag{5.27}$$

Here the *Dirac delta function* $\delta(\mathbf{R})$ is defined to be zero everywhere except at $\mathbf{R} = \mathbf{0}$, where it is so "singular" that

$$\iiint_D \delta(\mathbf{R})\, dV = 1 \tag{5.28}$$

for any domain D containing the origin. Show how, with this artifice, eq. (5.15) follows from eq. (5.10) directly with the substitution $\psi = 1/R$.

7. The definition of the Green's function in the text actually describes the "Green's function of the first kind." The "Green's function of the second kind" takes the same form $1/R + \gamma(\mathbf{R})$ with γ harmonic throughout D, but with the boundary values of $\nabla G \cdot \mathbf{n}$ (rather than those of G itself) chosen so as to simplify the third Green formula [eq. (5.15)].

(a) What would result if we set $\psi = G$ in the second Green formula [eq. (5.10)] for a G making $\nabla G \cdot \mathbf{n} = 0$ on S?

(b) Derive the identity $\iint_S \nabla G \cdot \mathbf{n}\, dS = -4\pi$ and show why this makes the G described in part (a) infeasible.

(c) What would result if we set $\psi = G$ in eq. (5.10) for a G making $\nabla G \cdot \mathbf{n}$ a constant on S? What must the value of this constant be?

8. Let the domain D be bounded by the surface S as in the divergence theorem, and assume that all fields satisfy the appropriate differentiability conditions. Prove the identities

(a) $$\iiint_D \nabla\phi \cdot \nabla \times \mathbf{F}\, dV = \iint_S \mathbf{F} \times \nabla\phi \cdot d\mathbf{S}$$

(b) $$\iiint_D [(\nabla \times \mathbf{V}) \cdot (\nabla \times \mathbf{W}) - \mathbf{V} \cdot (\nabla \times \nabla \times \mathbf{W})]\, dV = \iint_S (\mathbf{V} \times \nabla \times \mathbf{W}) \cdot d\mathbf{S}$$

(c) $$\iiint_D [\mathbf{W} \cdot (\nabla \times \nabla \times \mathbf{V}) - \mathbf{V} \cdot (\nabla \times \nabla \times \mathbf{W})]\, dV = \iint_S [\mathbf{V} \times \nabla \times \mathbf{W} - \mathbf{W} \times \nabla \times \mathbf{V}] \cdot d\mathbf{S}$$

9. With D and S as in the previous exercise, suppose that $\nabla \cdot \mathbf{V} = 0$ and $\mathbf{W} = \nabla\phi$ with $\phi = 0$ on S. Prove that $\iiint_D \mathbf{V} \cdot \mathbf{W}\, dV = 0$.

10. What is the value of the surface integral in the third Green formula if ϕ is harmonic and the origin lies *outside* the closed surface S?

5.3 The Fundamental Theorem of Vector Analysis

We discovered in sections 4.4 and 4.5 that if $\nabla \times \mathbf{F} = 0$ then $\mathbf{F} = \nabla\phi$ and if $\nabla \cdot \mathbf{F} = 0$ then $\mathbf{F} = \nabla \times \mathbf{G}$ (in simply connected domains), and we derived integral formulas for the potentials ϕ and $\mathbf{G}$ in terms of the vector field $\mathbf{F}$. In many applications, however, $\mathbf{F}$ itself is unknown and it is desired to have formulas for $\mathbf{F}$ (or ϕ and $\mathbf{G}$) in terms of the "sources" of $\mathbf{F}$—its divergence and curl. Armed with lemma 5.4, we can now demonstrate the following:

(*i*) how to find an irrotational field $\mathbf{F}$ having a specified divergence ρ;
(*ii*) how to find a scalar potential ϕ for the field $\mathbf{F}$ in (*i*);
(*iii*) how to find a solenoidal field $\mathbf{F}$ having a specified curl $\mathbf{J}$;
(*iv*) how to find a vector potential $\mathbf{G}$ for the field $\mathbf{F}$ in (*iii*);
(*v*) how to find a vector field $\mathbf{F}$ with a specified divergence *and* a specified curl;

(*vi*) the extent to which a vector field **F** is determined by its divergence and curl; and

(*vii*) how to decompose any vector field **F** into a sum of an irrotational field and a solenoidal field.

Clearly these properties are related. For our analysis we shall assume that all fields encountered are sufficiently smooth that lemma 5.4, which states that the laplacian of

$$\phi(\mathbf{R}) = \frac{-1}{4\pi} \iiint_D \frac{\rho(\mathbf{R}')}{|\mathbf{R} - \mathbf{R}'|} \, dV' \tag{5.29}$$

equals ρ, can be invoked at will.*

We can dispose of (*i*) and (*ii*) very easily. Given ρ, we define the function ϕ by formula (5.29), and let $\mathbf{F} = \nabla\phi$. Then ϕ is a scalar potential for **F**, which is thus irrotational, and $\nabla \cdot \mathbf{F} = \nabla^2\phi = \rho$.

We can now use this tool—the facility to construct an irrotational field with any specified divergence—in solving (*iii*). We have already seen, in section 4.5, how to construct a vector field $\mathbf{F}_1$ having a specified curl **J** in a star-shaped domain *D*: with our present notation (exercise 1)

$$\mathbf{F}_1(\mathbf{R}) = \int_0^1 t \, \mathbf{J}[t(\mathbf{R} - \mathbf{R}_0)] \times [\mathbf{R} - \mathbf{R}_0] \, dt \tag{5.30}$$

where $\mathbf{R}_0$ is any point in *D*. However, as exercise 5 of section 4.5 demonstrates, the field $\mathbf{F}_1$ will not necessarily be solenoidal. But this is easy to repair; we simply subtract off an *irrotational* field with the same divergence as $\mathbf{F}_1$:

$$\mathbf{F}(\mathbf{R}) = \mathbf{F}_1(\mathbf{R}) - \nabla\phi(\mathbf{R})$$

where ϕ is computed from formula (5.29) with $\nabla \cdot \mathbf{F}_1$ in place of ρ. We thus have

$$\nabla \times \mathbf{F} = \nabla \times \mathbf{F}_1 - \nabla \times \nabla\phi = \mathbf{J} - 0 = \mathbf{J}$$

and

$$\nabla \cdot \mathbf{F} = \nabla \cdot \mathbf{F}_1 - \nabla^2\phi = \nabla \cdot \mathbf{F}_1 - \nabla \cdot \mathbf{F}_1 = 0$$

throughout the domain *D*. The assemblage of the final expression for **F** in terms of **J** is left as exercise 2.

Now since **F** is solenoidal, it has a vector potential **G**, which can be expressed in terms of **F** by the variant of formula (5.30):

$$\mathbf{G}(\mathbf{R}) = \int_0^1 t \, \mathbf{F}[t(\mathbf{R} - \mathbf{R}_0)] \times [\mathbf{R} - \mathbf{R}_0] \, dt \tag{5.31}$$

*Recall that the lemma can be established in full generality only if some additional mathematical details are hypothesized. Here we need to know that $\rho(\mathbf{R})$ is "Hölder continuous" and, if *D* is unbounded, goes to zero sufficiently rapidly at infinity. See Kellogg, *op. cit.*

Since **F** has been expressed in terms of **J**, eq. (5.31) generates a formula for **G** directly in terms of **J**, and items (*iii*) and (*iv*) are solved for arbitrary star-shaped domains.

The final expression for **G**, however, is so unwieldy that we shall derive an alternate vector potential that is easier to calculate. It does not require that D be star-shaped, but it does require an additional condition on **J**.

The vector potential **G** must satisfy the condition

$$\mathbf{J} = \nabla \times \mathbf{F} = \nabla \times (\nabla \times \mathbf{G}) = -\nabla^2\mathbf{G} + \nabla(\nabla \cdot \mathbf{G}) \tag{5.32}$$

Now by lemma 5.4 the expression

$$\mathbf{G}(\mathbf{R}) = \frac{1}{4\pi}\iiint_D \frac{\mathbf{J}(\mathbf{R}')}{|\mathbf{R} - \mathbf{R}'|}\, dV' \tag{5.33}$$

satisfies

$$-\nabla^2\mathbf{G} = \mathbf{J} \tag{5.34}$$

component-by-component. Thus it would be a solution to eq. (5.32) if we could also show that its divergence vanishes. Let us proceed; we have

$$\nabla \cdot \mathbf{G}(\mathbf{R}) = \frac{1}{4\pi}\nabla \cdot \iiint_D \frac{\mathbf{J}(\mathbf{R}')}{|\mathbf{R} - \mathbf{R}'|}\, dV' \tag{5.35}$$

Since the variable of integration is $\mathbf{R}'$ and ∇ operates on **R**, we move the operator inside the integral to obtain

$$\nabla \cdot \iiint_D \frac{\mathbf{J}(\mathbf{R}')}{|\mathbf{R} - \mathbf{R}'|}\, dV' = \iiint_D \mathbf{J}(\mathbf{R}') \cdot \nabla\left(\frac{1}{|\mathbf{R} - \mathbf{R}'|}\right) dV' \tag{5.36}$$

(Because the integrand is singular when $\mathbf{R}' = \mathbf{R}$, this step requires further justification; again we direct the reader to Kellogg's text.) If ∇' denotes differentiation with respect to the $\mathbf{R}'$ variables, then clearly

$$\mathbf{J}(\mathbf{R}') \cdot \nabla\left(\frac{1}{|\mathbf{R} - \mathbf{R}'|}\right) = -\mathbf{J}(\mathbf{R}') \cdot \nabla'\left(\frac{1}{|\mathbf{R} - \mathbf{R}'|}\right)$$

Moreover, by identity (3.28)

$$\mathbf{J}(\mathbf{R}') \cdot \nabla'\left(\frac{1}{|\mathbf{R} - \mathbf{R}'|}\right) + \frac{\nabla' \cdot \mathbf{J}(\mathbf{R}')}{|\mathbf{R} - \mathbf{R}'|} = \nabla' \cdot \left\{\frac{\mathbf{J}(\mathbf{R}')}{|\mathbf{R} - \mathbf{R}'|}\right\}$$

but the divergence of **J** must be zero since **J** is presumed to be a curl (of **F**). Therefore

$$\mathbf{J}(\mathbf{R}') \cdot \nabla\left\{\frac{1}{|\mathbf{R} - \mathbf{R}'|}\right\} = -\nabla' \cdot \left\{\frac{\mathbf{J}(\mathbf{R}')}{|\mathbf{R} - \mathbf{R}'|}\right\}$$

and if we insert this into eq. (5.36) and integrate over the volume **D**, the divergence theorem reduces it to

$$\nabla \cdot \iiint_D \frac{\mathbf{J}(\mathbf{R}')}{|\mathbf{R} - \mathbf{R}'|}\, dV' = -\iint_S \frac{\mathbf{J} \cdot \mathbf{n}'}{|\mathbf{R} - \mathbf{R}'|}\, dS'$$

For the divergence of the expression in eq. (5.33), then, we have

$$\nabla \cdot \mathbf{G} = \frac{1}{4\pi} \nabla \cdot \iiint_D \frac{\mathbf{J}(\mathbf{R}')}{|\mathbf{R} - \mathbf{R}'|}\, dV' = \frac{-1}{4\pi} \iint_S \frac{\mathbf{J} \cdot \mathbf{n}'}{|\mathbf{R} - \mathbf{R}'|}\, dS' \quad (5.37)$$

Finally, we see from eqs. (5.32) and (5.37) that the formula (5.33) *does* provide a solution to eq. (5.32), and thus to item (*iv*), ***whenever the surface integral in eq. (5.37) vanishes.*** (In practice one usually ensures this by taking the domain D large enough to encompass all the points where **J** is nonzero.)

We summarize: if

$$\phi(\mathbf{R}) = \frac{-1}{4\pi} \iiint_D \frac{\rho(\mathbf{R}')}{|\mathbf{R} - \mathbf{R}'|}\, dV'$$

and

$$\mathbf{G}(\mathbf{R}) = \frac{1}{4\pi} \iiint_D \frac{\mathbf{J}(\mathbf{R}')}{|\mathbf{R} - \mathbf{R}'|}\, dV'$$

then $\mathbf{F}_1 = \nabla\phi$ is irrotational and satisfies $\nabla \cdot \mathbf{F}_1 = \rho$ and $\mathbf{F}_2 = \nabla \times \mathbf{G}$ is solenoidal and satisfies $\nabla \times \mathbf{F}_2 = \mathbf{J}$. Both statements presume certain smoothness conditions for ρ and **J**, and that $\mathbf{J} \cdot \mathbf{n} = 0$ on the surface of D [or some other condition disposing of the surface integral in eq. (5.37)].

Item (*v*) is now trivial. To find a field possessing a given divergence and curl—$\rho(\mathbf{R})$ and $\mathbf{J}(\mathbf{R})$, respectively—we evaluate $\mathbf{F}_1$ and $\mathbf{F}_2$ according to the previous paragraph (or exercise 2 if necessary); the sum $\mathbf{F} = \mathbf{F}_1 + \mathbf{F}_2$ then has the stipulated properties.

Clearly such a field is not unique. For instance, we could add any constant vector **A** to **F** without altering its divergence or curl; or, we could add the gradient of any harmonic function (exercise 3). The following theorem, then, addresses item (*vi*); a vector field is completely determined by its divergence, its curl, *and its normal component on the surface surrounding D.*

THEOREM 5.4 *Suppose that* $\nabla \cdot \mathbf{F} = \nabla \cdot \mathbf{E}$ *and* $\nabla \times \mathbf{F} = \nabla \times \mathbf{E}$ *throughout a domain* D, *and that on the surface* S *bounding* D, $\mathbf{F} \cdot \mathbf{n} = \mathbf{E} \cdot \mathbf{n}$ *where* **n** *is the outward normal. Then* $\mathbf{F} = \mathbf{E}$ *throughout D.*

Proof The difference $\mathbf{F} - \mathbf{E}$ obviously has zero curl, so it is derivable from a scalar potential ϕ:

$$\mathbf{F} - \mathbf{E} = \nabla\phi$$

Also $\mathbf{F} - \mathbf{E}$ has zero divergence: thus $\nabla^2\phi = \nabla \cdot (\mathbf{F} - \mathbf{E}) = 0$. Finally, on the surface S

$$\nabla\phi \cdot \mathbf{n} = \mathbf{F} \cdot \mathbf{n} - \mathbf{E} \cdot \mathbf{n} = 0$$

When all this information is fed into the first Green formula [eq. (5.9)] with $\psi = \phi$,

$$\iiint_D \phi \, \nabla^2\phi \, dV = \iint_S \phi \, \nabla\phi \cdot \mathbf{n} \, dS - \iiint_D |\nabla\phi|^2 \, dV$$

we get

$$\iiint_D |\nabla\phi|^2 \, dV = 0 = \iiint_D |\mathbf{F} - \mathbf{E}|^2 \, dV$$

Therefore $\mathbf{F} = \mathbf{E}$ in D.

The validity of the final item (*vii*) is known as the *fundamental theorem of vector analysis:*

THEOREM 5.5 *Let* $\mathbf{F}(\mathbf{R})$ *be a continuously differentiable vector field defined in a star-shaped domain D. Then* $\mathbf{F}$ *can be written*

$$\mathbf{F} = \nabla\phi + \nabla \times \mathbf{G} \tag{5.38}$$

for some scalar field ϕ *and some vector field* $\mathbf{G}$.

Proof We define

$$\phi(\mathbf{R}) = \frac{-1}{4\pi} \iiint_D \frac{\nabla' \cdot \mathbf{F}(\mathbf{R}')}{|\mathbf{R} - \mathbf{R}'|} \, dV' \tag{5.39}$$

Then, according to lemma 5.4, $\nabla^2\phi = \nabla \cdot \mathbf{F}$. Thus the vector field $\mathbf{F} - \nabla\phi$ has zero divergence. Consequently (theorem 4.4), it is derivable from a vector potential $\mathbf{G}$:

$$\mathbf{F} - \nabla\phi = \nabla \times \mathbf{G}$$

Transposing $\nabla\phi$, we have the theorem.

An alternative proof would be to use the formula in exercise 2 to construct a solenoidal field with the same curl as $\mathbf{F}$; the difference would then be irrotational, and thus expressible as a gradient (exercise 4). The solenoidal field could be constructed through formula (5.33) if the curl of $\mathbf{F}$ satisfied the proper conditions on the boundary of D (exercise 5).

Despite its ostentatious title, the fundamental theorem is less useful in practice than the more modest potential theorems for the irrotational and solenoidal fields.

In closing we would like to alert the reader to the fact that *the highly plausible formula*

$$\mathbf{F}(\mathbf{R}) = \nabla\left\{\frac{-1}{4\pi}\iiint_D \frac{\nabla'\cdot\mathbf{F}(\mathbf{R}')}{|\mathbf{R}-\mathbf{R}'|}\,dV'\right\} + \nabla\times\left\{\frac{1}{4\pi}\iiint_D \frac{\nabla'\times\mathbf{F}(\mathbf{R}')}{|\mathbf{R}-\mathbf{R}'|}\,dV'\right\}$$

is incorrect in general (exercise 7)!

EXERCISES

1. Show that the formula (4.18) takes the form (5.30) when $\mathbf{J} = \nabla \times \mathbf{F}$.
2. Derive the expression

$$\mathbf{F}(\mathbf{R}) = \int_0^1 t\mathbf{J}[t(\mathbf{R}-\mathbf{R}_0)]\times[\mathbf{R}-\mathbf{R}_0]\,dt + \iiint_D \{(\mathbf{R}'-\mathbf{R}_0)\int_0^1 t^2\nabla\times\mathbf{J}[t(\mathbf{R}'-\mathbf{R}_0)]\,dt\}/4\pi|\mathbf{R}-\mathbf{R}'|\,dV'$$

   for a solenoidal field having $\mathbf{J}$ as its curl in a star-shaped domain D.
3. True or false: if $\nabla\cdot\mathbf{F} = 0$ and $\nabla\times\mathbf{F} = \mathbf{0}$ then $\mathbf{F}$ is constant.
4. Show that adding the gradient of a harmonic function to a vector field changes neither its divergence nor its curl.
5. Devise a proof of the fundamental theorem for star-shaped domains which begins by constructing a solenoidal field with the same curl as $\mathbf{F}$.
6. What conditions imposed at the surface S on $\nabla\times\mathbf{F}$ enable one to use formula (4.74) to obtain a solenoidal field with the same curl as $\mathbf{F}$? [See (5.37).]
7. Devise a counterexample to the final "identity" in the text. (*Hint:* There are trivial ones.)

5.4 Green's Theorem

This section is relatively elementary and is intended to provide some preparation for the rigorous proof of Stokes' theorem.

We shall work entirely in the xy plane. Let C denote a closed smooth arc in the plane (fig. 5.4). Consider the line integral of the vector field $\mathbf{F} = y\mathbf{i}$ around C:

$$\oint_C \mathbf{F}\cdot d\mathbf{R} = \int_C y\,dx$$

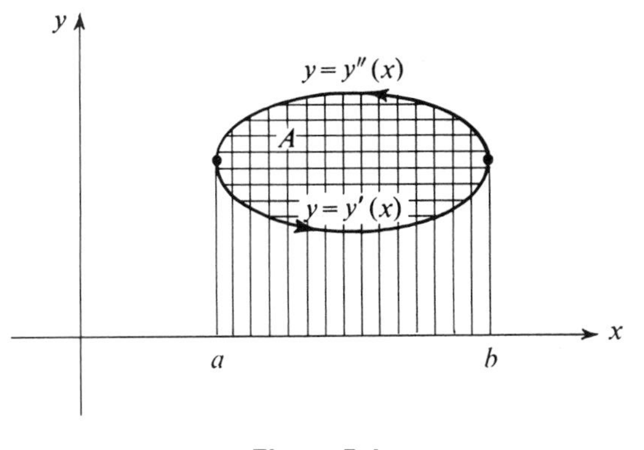

Figure 5.4

(Since it is conventional to orient closed curves in the xy plane so that $\mathbf{k}$ is the positive normal to the plane, we traverse C in a counterclockwise direction.) Then the line integral can be expressed as the sum of two ordinary integrals:

$$\int_C y\,dx = \int_a^b y'(x)\,dx + \int_b^a y''(x)\,dx \tag{5.40}$$

where the first integral is along the bottom portion of the curve and the second is along the top portion; the notation should be self-evident from the figure. (Note that the primes here do *not* denote derivatives.)

The first integral gives the area beneath the lower curve and above the x axis. The second integral equals

$$-\int_a^b y''(x)\,dx$$

and gives the negative of the area beneath the upper curve and above the x axis. Therefore the sum of the two integrals is $-A$, the negative of the area within C:

$$\int_C y\,dx = -A \tag{5.41}$$

Here we assumed C to be in the upper half-plane, but the reader can easily verify that eq. (5.41) also holds if C intersects the x axis or if C is beneath the x axis.

A similar argument shows that

$$\int_C x\,dy = A \tag{5.42}$$

Here we obtain A rather than $-A$, as is easily seen from figure 5.5.

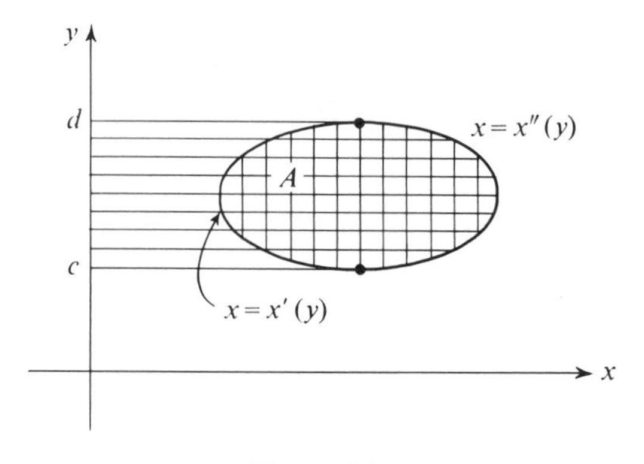

Figure 5.5

Now let us consider various other simple integrals about C. For instance, it is easy to verify that if the vector field is taken to be $\mathbf{F} = x\mathbf{i}$, then the line integral is

$$\int_C x\, dx = 0 \tag{5.43}$$

Indeed, $x\mathbf{i}$ is the gradient of the function $x^2/2$, so that the line integral of $x\, dx$ gives the change in $x^2/2$ as we move from initial point to final point, but for any closed curve these points coincide and hence the line integral is zero. In similar fashion, since $y\mathbf{j}$ is a gradient also, we have

$$\int_C y\, dy = 0 \tag{5.44}$$

Also, we have

$$\int_C dx = 0 \tag{5.45}$$

$$\int_C dy = 0 \tag{5.46}$$

It is entertaining, though not particularly instructive, to combine these line integrals in various ways. For instance, if x_0 is a constant, we derive

$$\int_C (x - x_0)\, dy = A \tag{5.47}$$

by using eqs. (5.42) and (5.46). Similarly,

$$\int_C (x - x_0)\, dx = 0 \tag{5.48}$$

by eqs. (5.43) and (5.45).

Somewhat more interesting is

$$\int_C \frac{1}{2}(x\,dy - y\,dx) = A \tag{5.49}$$

which we obtain by combining eqs. (5.41) and (5.42).

In view of the fact that the line integrals in eqs. (5.41), (5.42), (5.47), and (5.49) may be interpreted in terms of the area A within C, it is natural to ask whether there are any similar interpretations for the other line integrals. More generally, suppose we are given an arbitrary differential $F_1(x,y)\,dx + F_2(x,y)\,dy$, where F_1 and F_2 are continuous functions. Is there any connection between the line integral of this differential about C and the area within?

The answer is both yes and no. In general, there is no connection in the sense that we can draw a picture like that of figure 5.5 and interpret the integral in terms of areas. There is, however, a connection between the line integral about C and a double integral taken over the region within C. We will show that

$$\int_C F_1\,dx + F_2\,dy = \int\int_D \left(\frac{\partial F_2}{\partial x} - \frac{\partial F_1}{\partial y}\right) dx\,dy \tag{5.50}$$

where D is the domain within C (having area A). In the special case that the integrand in the double integral on the right side of eq. (5.50) is identically equal to 1, as in eq. (5.49), the right side of eq. (5.50) gives precisely A. If the integrand is zero, we get zero for the integral. In general, however, our result may not be related to A in any elementary manner and may be difficult to compute even with the help of eq. (5.50).

The reader may recognize eq. (5.50) as a special case of Stokes' theorem, discussed briefly in section 4.9. To see this, let $\mathbf{F} = F_1\mathbf{i} + F_2\mathbf{j}$ and $\mathbf{n} = \mathbf{k}$. The integral on the left is the line integral of the tangential component of $\mathbf{F}$ about C, and that on the right is the surface integral of the normal component of **curl F** over the surface enclosed by C.

This special case of Stokes' theorem is sometimes called *Green's theorem*. (Several other theorems are also called Green's theorem, incidentally.) The precise statement of the theorem is as follows.

THEOREM 5.6 *Let F_1 and F_2 be continuous functions of x and y for which the partial derivatives $\partial F_2/\partial x$ and $\partial F_1/\partial y$ exist and are continuous throughout a domain D in the xy plane. We require that D be bounded by a regular closed curve C, oriented by choosing* $\mathbf{k}$ *as the unit normal to the plane. We also require that any line passing through an interior point and parallel to either coordinate axis cut the boundary in exactly two points. Then eq.* (5.50) *is valid. More generally, eq.* (5.50) *is valid for regions in the plane that can be decomposed into finitely many domains having these properties.*

The proof of the theorem is similar to that of the divergence theorem and goes as follows.

Proof Let us first look at the right side of eq. (5.50). The integral can be broken up into two integrals, of which the first is

$$\iint_D \frac{\partial F_2}{\partial x} dx\, dy$$

Integrating first with respect to x, we have (with notation as in fig. 5.5)

$$\int_c^d \int_{x'(y)}^{x''(y)} \frac{\partial F_2}{\partial x} dx\, dy = \int_c^d [F_2(x'', y) - F_2(x', y)]\, dy = \int_C F_2\, dy$$

Similarly,

$$-\iint \frac{\partial F_1}{\partial y}\, dx\, dy = \int_a^b \int_{y'(x)}^{y''(x)} \frac{\partial F_1}{\partial y} dy\, dx$$

$$= \int_a^b [F_1(y') - F_1(y'')]\, dx = \int_C F_1\, dx$$

Adding these two gives the desired result. If D is a region that can be decomposed into finitely many domains having the stated properties, we simply sum the integrals involved over all the domains. The double integral then extends over all the parts, and the line integral extends over the entire boundary. If the boundaries of two parts have arcs in common, these arcs may be neglected, since the integrals will cancel (recall fig. 5.2 for a similar situation).

EXERCISES

1. Use Green's theorem to derive eq. (5.41).
2. Use Green's theorem to derive eq. (5.42).
3. Use Green's theorem to derive eq. (5.43).
4. Use Green's theorem to derive eq. (5.49).
5. Let $\mathbf{R} = x\mathbf{i} + y\mathbf{j}$ and $d\mathbf{R} = dx\,\mathbf{i} + dy\,\mathbf{j}$.
   (a) Compute the magnitude of the vector cross product $\mathbf{R} \times (\mathbf{R} + d\mathbf{R})$.
   (b) Thus, give a direct geometrical interpretation of the integrand of eq. (5.49). [*Hint:* Consider the triangle with the following vertices: $(0,0)$, (x,y), and $(x + dx, y + dy)$.]
   (c) Using figure 5.6 give an alternative derivation of eq. (5.49).
6. Let $\mathbf{F} = x\mathbf{i} + y\mathbf{j}$, and let C be an oriented closed curve enclosing an area A. What is $\int_C \mathbf{F} \cdot \mathbf{T}\, ds$? (As usual, $\mathbf{T}$ denotes the unit tangent to C in the positive direction.)
7. Let C denote the circle $x^2 + y^2 = 9$, and let $\mathbf{F} = y\mathbf{i} - 3x\mathbf{j}$. What is the line integral of the tangential component of $\mathbf{F}$ around C, taken in the usual counterclockwise direction?

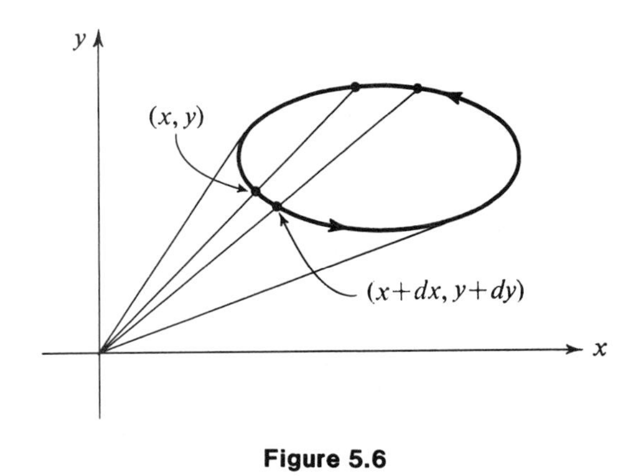

Figure 5.6

8. Let C denote the ellipse $x^2/4 + y^2/9 = 1$, and let $\mathbf{F} = (3y^2 - y)\mathbf{i} + (x^2 + 2)\mathbf{j}$.

(a) What is the area enclosed by C? (Don't integrate, for heaven's sake; we have already derived the area of an ellipse by using the area cosine principle.)

(b) Find the line integral of the tangential component of $\mathbf{F}$ around C, in the counterclockwise direction. [*Hint:* By Green's theorem, this resolves itself to a double integral, but no computation is necessary if you observe that the symmetry enables you to ignore certain terms. Just multiply the area by the average value of $(\partial F_2/\partial x) - (\partial F_1/\partial y)$.]

9. Compute $\int_C (4y^3\,dx - 2x^2\,dy)$ around the square bounded by the lines $x = \pm 1$ and $y = \pm 1$,

(a) directly, by performing the line integration.

(b) by using Green's theorem.

By symmetry, it is obvious that one of the terms in the integrand of the above line integral can be ignored. Which term?

10. Let $\mathbf{F} = 4z\mathbf{i} - 3x\mathbf{k}$. Compute the line integral of the tangential component of $\mathbf{F}$ about the circle $(x - 5)^2 + (z - 7)^2 = 4$ in the xz plane. Orient the plane by taking $\mathbf{j}$ to be unit normal. [*Careful:* If you just replace y by z in eq. (5.50), you will get the wrong orientation.]

11. In eq. (5.50), the functions F_1 and F_2 are fairly arbitrary functions of x and y (we require only that certain partial derivatives be continuous). It therefore appears that we can interchange F_1 and F_2 and also x and y to obtain the formula

$$\int_C (F_2\,dy + F_1\,dx) = \iint_D \left(\frac{\partial F_1}{\partial y} - \frac{\partial F_2}{\partial x}\right) dy\,dx$$

The left side of this equation is the same as the left side of eq. (5.50), but the right side has the opposite sign. It follows that this expression is incorrect. Give a clue, *in only one word,* to explain this paradox.

12. Use Green's theorem to find the area inside the loop of *Descartes' folium*

$$x = \frac{t}{1 + t^3} \qquad y = \frac{t^2}{1 + t^3} \qquad (0 \le t < \infty)$$

5.5 Stokes' Theorem

We are now in a position to give a rigorous proof of Stokes' theorem, which reduces certain surface integrals to line integrals. We are given a smooth oriented surface S in space, bounded by a piecewise smooth, closed curve C whose orientation is consistent with that of S (fig. 5.7). We assume that the surface can be parametrized by $\mathbf{R} = \mathbf{R}(u,v)$ in such a way that the coordinates x, y, and z are twice continuously differentiable functions of u and v (so that the mixed partials are equal in either order), with $(\partial\mathbf{R}/\partial u \times \partial\mathbf{R}/\partial v)$ pointing in the direction of the normal.

The set of values (u,v) in the uv plane that correspond to points on S will be denoted Σ (fig. 5.8). We assume that distinct points of Σ correspond to distinct points on S [the mapping $\mathbf{R}(u,v)$ is one-to-one] and that the region Σ and its boundary Γ satisfy the hypothesis of Green's theorem. (Notice that these assumptions imply that the positive orientation on C corresponds to the correct orientation on Γ; see exercise 4.) Then we can use this parametrization to derive the result.

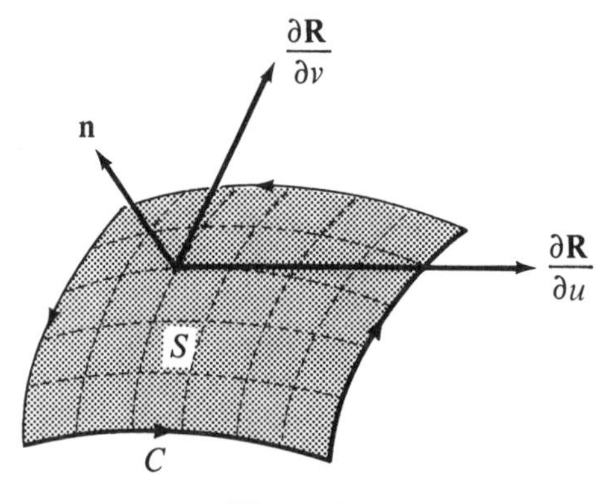

Figure 5.7

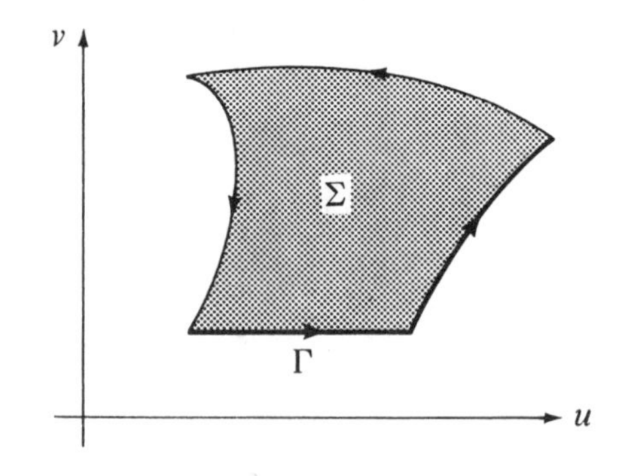

Figure 5.8

THEOREM 5.7 *Let S and C be as described above, and let* **F** *be a continuously differentiable vector field. Then*

$$\int_C \mathbf{F}\cdot d\mathbf{R} = \iint_S \mathbf{curl}\ \mathbf{F}\cdot d\mathbf{S} \tag{5.51}$$

In words: the circulation of **F** *around C equals the flux of its curl through S.*

Observe that we are using the notation

$$d\mathbf{S} = \frac{\partial \mathbf{R}}{\partial u}du \times \frac{\partial \mathbf{R}}{\partial v}dv \tag{5.52}$$

introduced in section 4.6. In the derivation we will use the identities

$$d\mathbf{R} = \frac{\partial \mathbf{R}}{\partial u}du + \frac{\partial \mathbf{R}}{\partial v}dv \tag{5.53}$$

and

$$\frac{\partial}{\partial u} = \frac{\partial \mathbf{R}}{\partial u}\cdot\nabla \tag{5.54}$$

$$\frac{\partial}{\partial v} = \frac{\partial \mathbf{R}}{\partial v}\cdot\nabla \tag{5.55}$$

Before going through the derivation, the reader may wish to review section 4.6, and also the operator convention (first paragraph of section 3.8). Thus, to derive eq. (5.54), we simply use the chain rule in operator form:

$$\frac{\partial}{\partial u} = \frac{\partial x}{\partial u}\frac{\partial}{\partial x} + \frac{\partial y}{\partial u}\frac{\partial}{\partial y} + \frac{\partial z}{\partial u}\frac{\partial}{\partial z} = \frac{\partial \mathbf{R}}{\partial u}\cdot\nabla \tag{5.56}$$

and eq. (5.55) is derived similarly. We will also use

$$\left(\frac{\partial \mathbf{R}}{\partial u}\times\frac{\partial \mathbf{R}}{\partial v}\right)\times\nabla = \frac{\partial \mathbf{R}}{\partial v}\left(\frac{\partial \mathbf{R}}{\partial u}\cdot\nabla\right) - \frac{\partial \mathbf{R}}{\partial u}\left(\frac{\partial \mathbf{R}}{\partial v}\cdot\nabla\right) \tag{5.57}$$

which is obtained by expanding the triple vector product and by using the operator convention. Similarly, the interchange of the $\cdot$ and $\times$ in

$$\mathbf{A}\cdot\nabla\times\mathbf{B} = \mathbf{A}\times\nabla\cdot\mathbf{B} \tag{5.58}$$

is easily verified.

Do not let these formalities obscure the basic idea, which is that the position vector **R**, for points on S, and also **F** itself at these points, can be written as functions of the parameters u and v, so the integrals in eq. (5.51) can be written in terms of u and v. After we have done this, we have carried the work down to the uv plane, and the proof simply amounts to some inspired juggling of vector identities, aided by Green's theorem.

Proof We write

$$\int_C \mathbf{F}\cdot d\mathbf{R} = \int_\Gamma \left[\left(\mathbf{F}\cdot\frac{\partial \mathbf{R}}{\partial u}\right)du + \left(\mathbf{F}\cdot\frac{\partial \mathbf{R}}{\partial v}\right)dv\right] \qquad \text{[by eq. (5.53)]}$$

$$= \iint_\Sigma \left[\frac{\partial}{\partial u}\left(\mathbf{F}\cdot\frac{\partial \mathbf{R}}{\partial v}\right) - \frac{\partial}{\partial v}\left(\mathbf{F}\cdot\frac{\partial \mathbf{R}}{\partial u}\right)\right] du\, dv \text{ (Green's theorem)}$$

$$= \iint_\Sigma \left(\frac{\partial \mathbf{F}}{\partial u}\cdot\frac{\partial \mathbf{R}}{\partial v} + \mathbf{F}\cdot\frac{\partial^2 \mathbf{R}}{\partial u\,\partial v} - \mathbf{F}\cdot\frac{\partial^2 \mathbf{R}}{\partial v\,\partial u} - \frac{\partial \mathbf{F}}{\partial v}\cdot\frac{\partial \mathbf{R}}{\partial u}\right) du\, dv$$

$$= \iint_\Sigma \left(\frac{\partial \mathbf{R}}{\partial v}\frac{\partial}{\partial u} - \frac{\partial \mathbf{R}}{\partial u}\frac{\partial}{\partial v}\right)\cdot \mathbf{F}\, du\, dv$$

$$= \iint_\Sigma \left[\frac{\partial \mathbf{R}}{\partial v}\left(\frac{\partial \mathbf{R}}{\partial u}\cdot\nabla\right) - \frac{\partial \mathbf{R}}{\partial u}\left(\frac{\partial \mathbf{R}}{\partial v}\cdot\nabla\right)\right]\cdot \mathbf{F}\, du\, dv \qquad \text{[by eqs. (5.54) and (5.55)]}$$

$$= \iint_\Sigma \left(\frac{\partial \mathbf{R}}{\partial u}\times\frac{\partial \mathbf{R}}{\partial v}\right)\times\nabla\cdot\mathbf{F}\, du\, dv \text{ [by eq. (5.57)]}$$

$$= \iint_\Sigma \left(\frac{\partial \mathbf{R}}{\partial u}\times\frac{\partial \mathbf{R}}{\partial v}\right)\cdot\nabla\times\mathbf{F}\, du\, dv \text{ [by eq. (5.58)]}$$

$$= \iint_\Sigma (\mathbf{curl\ F})\cdot\left(\frac{\partial \mathbf{R}}{\partial u}\times\frac{\partial \mathbf{R}}{\partial v}\right) du\, dv$$

$$= \iint_S \mathbf{curl\ F}\cdot d\mathbf{S} \text{ [by eq. (5.52)]}$$

which completes the derivation.

This proof avoids the pitfalls of the argument in section 4.9 only by devoiding itself of physical content. That is the value (and the liability) of relying on parametrizations.

This analytic proof justifies our heuristic derivations of the expressions for the curl in curvilinear coordinates (sections 3.10, 3.11). Recall that in section 3.4 we characterized the curl as the vector whose component in any given direction was the "swirl" of the vector field per unit area, in the limit of vanishing areas. In chapter 4 we identified this "swirl" as the line integral around the boundary of the area. For our formal definition, however, we adopted the formula

$$\mathbf{curl\ F} = \begin{vmatrix} \mathbf{i} & \mathbf{j} & \mathbf{k} \\ \dfrac{\partial}{\partial x} & \dfrac{\partial}{\partial y} & \dfrac{\partial}{\partial z} \\ F_1 & F_2 & F_3 \end{vmatrix}, \tag{5.59}$$

which was motivated by restricting the swirl characterization to rectangles with sides parallel to the coordinate axes.

Now with the analytic proof of Stokes' theorem, premised only on (5.59) and not on the heuristic picture, we have assurance that (5.59) *is* consistent with the swirl interpretation, *for areas of arbitrary shape:* the limiting value of the swirl per unit area is given by

$$\lim_{A\to 0} \frac{1}{A} \int_C \mathbf{F} \cdot d\mathbf{R} = \lim_{A\to 0} \frac{1}{A} \iint_A (\mathbf{curl\ F}) \cdot \mathbf{n}\, ds = (\mathbf{curl\ F}) \cdot \mathbf{n},$$

the component of **curl F** normal to the area. The curl is thus a coordinate-free concept; in physicists' terms, it is "covariant."

EXERCISES

1. Given the vector field $\mathbf{F} = 3y\mathbf{i} + (5 - 2x)\mathbf{j} + (z^2 - 2)\mathbf{k}$, find
   (a) div **F**.
   (b) **curl F**.
   (c) the surface integral of the normal component of **curl F** over the open hemispherical surface $x^2 + y^2 + z^2 = 4$ above the xy plane.
   [*Hint:* By a double application of Stokes' theorem, part (c) can be reduced to a triviality.]
2. Given that $\mathbf{curl\ F} = 2y\mathbf{i} - 2z\mathbf{j} + 3\mathbf{k}$, find the surface integral of the normal component of **curl F** (*not* **F**) over
   (a) the open hemispherical surface $x^2 + y^2 + z^2 = 9$, $z > 0$.
   (b) the sphere $x^2 + y^2 + z^2 = 9$.
   (In both parts, you should be able to write the answer down by inspection.)
3. Prove $\iint_S \nabla\phi \times \nabla\psi \cdot d\mathbf{S} = \oint_C \phi\nabla\psi \cdot d\mathbf{R}$.
4. Show why the positive orientations on C and Γ, in the proof of Stokes' theorem, correspond. (*Hint:* Reread the beginning of section 4.6.)
5. Be a bit fanciful, and imagine that S is the surface of a laundry bag with a drawstring forming the boundary C. The Stokes' theorem states that the surface integral of the normal component of **curl F** over the laundry bag equals the line integral of its tangential component around the drawstring. Now suppose that we close the bag by pulling the drawstring; the effective length of the drawstring becomes zero and the line integral is therefore zero. S has become a closed surface.
   (a) What is the surface integral of the normal component of **curl F** over a closed surface?
   We now apply the divergence theorem, which says that the volume integral of the divergence of a vector field through the interior of a closed laundry bag equals the surface integral of the normal component of the field over its surface. Let the vector field be **curl F.**
   (b) What is the volume integral of the divergence of **curl F** over a domain?
   If the laundry bag is very, very small, the divergence of **curl F** will be approximately constant throughout, and the volume integral of div (**curl F**) will be approximately div (**curl F**) at a point within the laundry bag times the volume the bag encloses.
   (c) What is div (**curl F**) at any point P?
   (d) To which of the identities in section 3.8 is this related?

6. This is very similar to exercise 5, but the point of view is somewhat different. Let S be the surface of a sphere, and let us imagine the sphere divided into two parts, an upper hemisphere and a lower hemisphere, by a plane parallel to the xy plane passing through its center. (Draw a diagram.) Let **F** be a vector field, and consider the surface integral of the normal component of **curl F** over the upper hemisphere. Relate this mentally to the line integral $\int_C \mathbf{F} \cdot d\mathbf{R}$, where C is the equator, oriented relative to the outward normal of the upper hemisphere (i.e., the positive direction is west to east). Now do the same thing for the lower hemisphere: the surface integral of (**curl F**) · **n** over the lower hemisphere equals the line integral over the equator with, however, an east-to-west direction of integration. Add the two.
   (a) What is the surface integral of the normal component of **curl F** over a sphere?
   (b) What is the volume integral of div (**curl F**) through the interior of a sphere?
   (c) Let the sphere shrink to a point; what does this say about div (**curl F**) at a point?
   (d) To which of the identities in section 3.8 is this related?
7. Suppose that $\mathbf{F} = \mathbf{grad}\ \phi$, so that the line integral of the tangential component of **F** along any curve is equal to the difference in the values of ϕ at the endpoints of the curve. In particular, if C is a closed curve, $\int_C \mathbf{F} \cdot d\mathbf{R} = 0$. Let S be a surface with boundary C.
   (a) What is the surface integral of the normal component of **curl** (**grad** ϕ) over a surface S?

   If S is a small element of surface, bounded by a closed curve C, **curl** (**grad** ϕ) will be approximately constant on S, and the surface integral of the normal component of **curl** (**grad** ϕ) will be approximately **n** · **curl** (**grad** ϕ) times the area of the surface.
   (b) For any unit vector **n**, and any point in space, what is **n** · **curl** (**grad** ϕ) at this point?
   (c) Since this result is independent of the direction of **n**, what can you say about **curl** (**grad** ϕ)?
   (d) To which of the identities in section 3.8 is this related?

5.6 The Transport Theorems

Let us reconsider the statement of the transport theorems given in section 4.10.

THEOREM 5.8 (Flux Transport Theorem) *Let $\mathbf{F}(\mathbf{R},t)$ be a vector field that changes with time, and let $S(t)$ be a smooth oriented surface, also changing with time, with the curve $C(t)$ as its boundary. Let $\mathbf{v}(\mathbf{R},t)$ be the velocity of the points of $S(t)$. Assume all these functions are continuously differentiable with respect to their arguments. Then*

$$\frac{d}{dt}\int\int_{S(t)} \mathbf{F}(\mathbf{R},t) \cdot d\,\mathbf{S} = \int\int_{S(t)} \left\{\frac{\partial \mathbf{F}}{\partial t} + (\nabla \cdot \mathbf{F})\mathbf{v}\right\} \cdot d\,\mathbf{S} + \int_{C(t)} \mathbf{F} \times \mathbf{v} \cdot d\,\mathbf{R} \qquad (5.60)$$

THEOREM 5.9 (Reynold's Transport Theorem) *Let $\eta(\mathbf{R},t)$ be a scalar field that changes with time, and let $\mathbf{D}(t)$ be a domain also changing with time, with the surface $\mathbf{S}(t)$ as its boundary. Let $\mathbf{v}(\mathbf{R},t)$ be the velocity of the points of $\mathbf{D}(t)$. Assume all these functions are continuously differentiable with respect to their arguments. Then*

$$\frac{d}{dt}\int\int\int_{D(t)} \eta(\mathbf{R},t)\ dV = \int\int\int_{D(t)} \left\{\frac{d\eta}{dt} + \eta\nabla\cdot\mathbf{v}\right\} dV \tag{5.61}$$

$$= \int\int\int_{D(t)} \left\{\frac{\partial\eta}{\partial t} + \nabla\cdot\eta\mathbf{v}\right\} dV \tag{5.62}$$

$$= \int\int\int_{D(t)} \frac{\partial\eta}{\partial t}\,dV + \int\int_{S(t)} \eta\mathbf{v}\cdot d\mathbf{S} \tag{5.63}$$

(The equivalence of the latter three formulations was demonstrated earlier.)

The main difficulty with the heuristic derivations given in section 4.10 is the vagueness of the velocity field concept. It presupposes some way of identifying, in a one-to-one fashion, points on the surface $S(t)$ with corresponding points on $S(t + dt)$, with $\mathbf{v}(t)dt$ describing their displacements. This is fine as long as $\mathbf{v}$ is derived from some sort of fluid motion, and the "corresponding points" are the physical trajectories of specific fluid particles. However, if $S(t)$ is only a mathematical surface, without physical substance, this correspondence between points is rather arbitrary. Hence, so also is the velocity field.

Thus we proceed with a more abstract approach. Let's start with a parametrization of the surface at some fixed time, say at $t = 0$:

$$\mathbf{R} = \mathbf{R}_0(u,v) \tag{5.64}$$

Here u, v range over a region Σ, bounded by the curve Γ in the uv plane (as in section 5.5). As time progresses, each point originally on S_0 traces out a curve, and we write

$$\mathbf{R} = \mathbf{R}(\mathbf{R}_0,t) \tag{5.65}$$

to describe the location at time t of the point originating from $\mathbf{R}_0$. Substituting eq. (5.64), we rewrite eq. (5.65) (and liberalize the notation):

$$\mathbf{R} = \mathbf{R}(\mathbf{R}_0,t) = \mathbf{R}(u,v,t) \tag{5.66}$$

Thus, fixing t and letting u, v roam over Σ, eq. (5.66) traces out the surface S_t, while holding u and v fixed and varying t produces a function that describes how a single point migrates from surface to surface. In this context it is clear that the velocity $\mathbf{v}$ of a point $\mathbf{R} = \mathbf{R}(u,v,t)$ on S_t is given by

$$\mathbf{v} = \frac{\partial\mathbf{R}}{\partial t}$$

We assume that the orientation of S_t, C_t, Σ, and Γ are all consistent with the parametrization equation (5.66), as in section 5.5. The rigorous derivation of (5.60) then proceeds as follows. We have

$$\Phi = \iint_{S_t} \mathbf{F} \cdot d\mathbf{S} = \iint_{\Sigma} \mathbf{F}[\mathbf{R}(u,v,t),t] \cdot \frac{\partial \mathbf{R}}{\partial u} \times \frac{\partial \mathbf{R}}{\partial v}\, du\, dv$$

Thus, since Σ is fixed,

$$\frac{d\Phi}{dt} = \iint_{\Sigma} \frac{d\mathbf{F}}{dt} \cdot \frac{\partial \mathbf{R}}{\partial u} \times \frac{\partial \mathbf{R}}{\partial v}\, du\, dv + \iint_{\Sigma} \mathbf{F} \cdot \frac{\partial}{\partial t}\left(\frac{\partial \mathbf{R}}{\partial u} \times \frac{\partial \mathbf{R}}{\partial v}\right) du\, dv \tag{5.67}$$

For the first term we have, by the chain rule,

$$\frac{d\mathbf{F}[\mathbf{R}(u,v,t),t]}{dt} = \frac{\partial \mathbf{F}}{\partial x}\frac{\partial x}{\partial t} + \frac{\partial \mathbf{F}}{\partial y}\frac{\partial y}{\partial t} + \frac{\partial \mathbf{F}}{\partial z}\frac{\partial z}{\partial t} + \frac{\partial \mathbf{F}}{\partial t} = (\mathbf{v} \cdot \nabla)\mathbf{F} + \frac{\partial \mathbf{F}}{\partial t} \tag{5.68}$$

The second term requires considerable labor, but nothing more profound than an inspired juggling of identities. Observe that

$$\mathbf{F} \cdot \frac{\partial}{\partial t}\left(\frac{\partial \mathbf{R}}{\partial u} \times \frac{\partial \mathbf{R}}{\partial v}\right) = \mathbf{F} \cdot \left[\frac{\partial}{\partial u}\left(\frac{\partial \mathbf{R}}{\partial t} \times \frac{\partial \mathbf{R}}{\partial v}\right) - \frac{\partial}{\partial v}\left(\frac{\partial \mathbf{R}}{\partial t} \times \frac{\partial \mathbf{R}}{\partial u}\right)\right] = \mathbf{F} \cdot \frac{\partial}{\partial u}\left(\mathbf{v} \times \frac{\partial \mathbf{R}}{\partial v}\right) - \mathbf{F} \cdot \frac{\partial}{\partial v}\left(\mathbf{v} \times \frac{\partial \mathbf{R}}{\partial u}\right)$$

and, moreover, that this equals

$$\frac{\partial}{\partial u}\left(\mathbf{F} \cdot \mathbf{v} \times \frac{\partial \mathbf{R}}{\partial v}\right) - \frac{\partial}{\partial v}\left(\mathbf{F} \cdot \mathbf{v} \times \frac{\partial \mathbf{R}}{\partial u}\right) - \frac{\partial \mathbf{F}}{\partial u} \cdot \left(\mathbf{v} \times \frac{\partial \mathbf{R}}{\partial v}\right) + \frac{\partial \mathbf{F}}{\partial v} \cdot \left(\mathbf{v} \times \frac{\partial \mathbf{R}}{\partial u}\right) \tag{5.69}$$

In the next paragraph we will show that

$$\frac{\partial \mathbf{F}}{\partial v} \cdot \mathbf{v} \times \frac{\partial \mathbf{R}}{\partial u} - \frac{\partial \mathbf{F}}{\partial u} \cdot \mathbf{v} \times \frac{\partial \mathbf{R}}{\partial v} = [(\nabla \cdot \mathbf{F})\mathbf{v} - (\mathbf{v} \cdot \nabla)\mathbf{F}] \cdot \left(\frac{\partial \mathbf{R}}{\partial u} \times \frac{\partial \mathbf{R}}{\partial v}\right) \tag{5.70}$$

Taking eq. (5.70) for granted for the moment, we insert it into eq. (5.69) and use the result in eq. (5.67) to derive

$$\frac{d\Phi}{dt} = \iint_\Sigma \left((\mathbf{v} \cdot \nabla)\mathbf{F} + \frac{\partial \mathbf{F}}{\partial t} + (\nabla \cdot \mathbf{F})\mathbf{v} - (\mathbf{v} \cdot \nabla)\mathbf{F} \right) \cdot \left(\frac{\partial \mathbf{R}}{\partial u} \times \frac{\partial \mathbf{R}}{\partial v} \right) du\, dv$$

$$+ \iint_\Sigma \left[\frac{\partial}{\partial u} \left(\mathbf{F} \cdot \mathbf{v} \times \frac{\partial \mathbf{R}}{\partial v} \right) - \frac{\partial}{\partial v} \left(\mathbf{F} \cdot \mathbf{v} \times \frac{\partial \mathbf{R}}{\partial u} \right) \right] du\, dv$$

In the last term we interchange the dot and cross products and apply Green's theorem:

$$\frac{d\Phi}{dt} = \iint_\Sigma \left((\nabla \cdot \mathbf{F})\mathbf{v} + \frac{\partial \mathbf{F}}{\partial t} \right) \cdot \left(\frac{\partial \mathbf{R}}{\partial u} \times \frac{\partial \mathbf{R}}{\partial v} \right) du\, dv$$

$$+ \oint_\Gamma \left(\mathbf{F} \times \mathbf{v} \cdot \frac{\partial \mathbf{R}}{\partial u} du + \mathbf{F} \times \mathbf{v} \cdot \frac{\partial \mathbf{R}}{\partial v} dv \right) \tag{5.71}$$

Identifying $d\mathbf{S}$ and

$$d\mathbf{R} = \frac{\partial \mathbf{R}}{\partial u} du + \frac{\partial \mathbf{R}}{\partial v} dv$$

we recover eq. (5.60).

The proof of the identity (5.70) is best handled with tensor notation. First, observe that, by the chain rule,

$$\frac{\partial \mathbf{F}}{\partial u} = \left(\frac{\partial \mathbf{R}}{\partial u} \cdot \nabla \right) \mathbf{F}$$

and similarly for v. Using the symbols x_i^u, x_i^v for the ith components of $\partial \mathbf{R}/\partial u$ and $\partial \mathbf{R}/\partial v$, respectively, we write the left-hand side of eq. (5.70) as

$$\begin{aligned}
&\epsilon_{ijk}(x_l^v \partial_l F_i) v_j x_k^u - \epsilon_{ijk}(x_l^u\, \partial_l F_i) v_j x_k^v \\
&= \epsilon_{ijk}(x_l^v x_k^u - x_l^u x_k^v) v_j \partial_l F_i \\
&= \epsilon_{ijk}(\delta_{ls}\delta_{kt} - \delta_{lt}\delta_{ks}) x_s^v x_t^u v_j \partial_l F_i \\
&= \epsilon_{ijk}\epsilon_{lkp}\epsilon_{stp} x_s^v x_t^u v_j \partial_l F_i \text{ [by eq. (1.40)]} \\
&= \epsilon_{ijk}\epsilon_{plk}\epsilon_{stp} x_s^v x_t^u v_j \partial_l F_i \text{ [by eq. (1.37)]} \\
&= (\delta_{ip}\delta_{jl} - \delta_{il}\delta_{jp})\epsilon_{stp} x_s^v x_t^u v_j \partial_l F_i \text{ [by eq. (1.40)]} \\
&= \epsilon_{sti} x_s^v x_t^u v_l \partial_l F_i - \epsilon_{stj} x_s^v x_t^u v_j \partial_i F_i \\
&= -\epsilon_{its} v_l (\partial_l F_i) x_t^u x_s^v + \epsilon_{jts} (\partial_i F_i) v_j x_t^u x_s^v \\
&= -[(\mathbf{v} \cdot \nabla)\mathbf{F}] \cdot \frac{\partial \mathbf{R}}{\partial u} \times \frac{\partial \mathbf{R}}{\partial v} + (\nabla \cdot \mathbf{F})\mathbf{v} \cdot \frac{\partial \mathbf{R}}{\partial u} \times \frac{\partial \mathbf{R}}{\partial v}
\end{aligned}$$

The identity is proved.

A rigorous proof of eq. (5.63) is based on the observation that any scalar field η can be written as the divergence of some vector field $\mathbf{F}$. This follows from lemma 5.4 (with the incumbent restrictions on ρ), which produces a scalar ϕ such that $\eta = \nabla^2\phi$; we simply set $\mathbf{F} = \nabla\phi$.

With this in hand we use the divergence theorem to express

$$\iiint_{V_t} \eta \, dV = \iint_{S_t} \mathbf{F} \cdot d\mathbf{S}$$

and apply eq. (5.60) to the (closed) surface S_t to derive

$$\frac{d}{dt}\iiint_{V_t} \eta \, dV = \iint_{S_t} \frac{\partial \mathbf{F}}{\partial t} \cdot d\mathbf{S} + \iint_{S_t} (\nabla \cdot \mathbf{F})\mathbf{v} \cdot d\mathbf{S}$$

One more application of the divergence theorem does it:

$$\begin{aligned}\frac{d}{dt}\iiint_{V_t} \eta \, dV &= \iiint_{V_t} \nabla \cdot \frac{\partial \mathbf{F}}{\partial t} dV + \iint_{S_t} (\nabla \cdot \mathbf{F})\mathbf{v} \cdot d\mathbf{S} \\ &= \iiint_{V_t} \frac{\partial \eta}{\partial t} dV + \iint_{S_t} \eta\mathbf{v} \cdot d\mathbf{S}\end{aligned}$$

in agreement with eq. (5.63).

EXERCISE

1. With a rigorous proof of Reynold's theorem in hand, revisit exercise 12 of section 3.3 by proving Euler's expansion formula:

$$\frac{d}{dt}\iiint_{D(t)} dV = \frac{d\,(\text{volume})}{dt} = \iiint_{D(t)} \nabla \cdot \mathbf{v} \, dV = \iint_{S(t)} \mathbf{v} \cdot \, d\mathbf{S}.$$

5.7 Matrix Techniques in Vector Analysis

In this section we are going to take a quick look at matrix theory, as it is used in vector analysis. Beginners should be aware that they are going to learn about a very restricted class of matrices. The subject covers much more ground than we will see here; indeed, most authors devote a whole book to it. However, the matrix calculus has some very nice interpretations when applied to three-dimensional vector analysis, and we shall exploit these features.

Readers who have already mastered linear algebra should, nonetheless, enjoy reading this section. They will have seen all the formulas before, but the point of view is quite different from the algebraic approach and offers some new insights.

A matrix is a rectangular array of numbers, like a bingo card or the box score of a baseball game. It can have any number of rows and columns. However, in vector analysis only three matrix "sizes" are commonly used: the 1-by-3 row matrix, for example,

$$[2, 4, 1] \tag{5.71}$$

the 3-by-1 column matrix, for example,

$$\begin{bmatrix} 2 \\ 1 \\ 5 \end{bmatrix} \tag{5.72}$$

and the 3-by-3 square matrix, for example,

$$\begin{bmatrix} 1 & -1 & 1 \\ 2 & 0 & 1 \\ 1 & 1 & 2 \end{bmatrix} \tag{5.73}$$

Notice that the dimensions of a matrix are stated "m-by-n," indicating m rows and n columns.

It is useful to think of the entries of a matrix as representing the components of a vector. Thus eq. (5.70) represents $2\mathbf{i} + 4\mathbf{j} + \mathbf{k}$ as a row and eq. (5.71) represents $2\mathbf{i} + \mathbf{j} + 5\mathbf{k}$ as a column. The square matrix of eq. (5.72) can be interpreted either as consisting of three rows, representing

$$\begin{aligned} \mathbf{A} &= \mathbf{i} - \mathbf{j} + \mathbf{k} \\ \mathbf{B} &= 2\mathbf{i} + \mathbf{k} \\ \mathbf{C} &= \mathbf{i} + \mathbf{j} + 2\mathbf{k} \end{aligned} \tag{5.74}$$

respectively, or as consisting of three columns, representing

$$\begin{aligned} \mathbf{D} &= \mathbf{i} + 2\mathbf{j} + \mathbf{k} \\ \mathbf{E} &= -\mathbf{i} + \mathbf{k} \\ \mathbf{F} &= \mathbf{i} + \mathbf{j} + 2\mathbf{k} \end{aligned} \tag{5.75}$$

respectively. Both interpretations are useful. To emphasize the point of view, we can write eq. (5.73) as

$$\begin{bmatrix} 1 & -1 & 1 \\ 2 & 0 & 1 \\ 1 & 1 & 2 \end{bmatrix} = \begin{bmatrix} \cdots & \mathbf{A} & \cdots \\ \cdots & \mathbf{B} & \cdots \\ \cdots & \mathbf{C} & \cdots \end{bmatrix}$$

if we are thinking of the rows as vectors, or as

$$\begin{bmatrix} 1 & -1 & 1 \\ 2 & 0 & 1 \\ 1 & 1 & 2 \end{bmatrix} = \begin{bmatrix} \vdots & \vdots & \vdots \\ \mathbf{D} & \mathbf{E} & \mathbf{F} \\ \vdots & \vdots & \vdots \end{bmatrix}$$

if the column-vector interpretation is appropriate.

When we want to address a single number in a matrix M, we use the notation m_{ij} for the entry in row number i and column number j, as illustrated:

$$[m_{11}, m_{12}, m_{13}] \quad \begin{bmatrix} m_{11} \\ m_{21} \\ m_{31} \end{bmatrix} \quad \begin{bmatrix} m_{11} & m_{12} & m_{13} \\ m_{21} & m_{22} & m_{23} \\ m_{31} & m_{32} & m_{33} \end{bmatrix} \tag{5.76}$$

Thus in matrix (5.71), $m_{13} = 1$; in eq. (5.72), $m_{21} = 1$; and in eq. (5.73), $m_{12} = -1$, $m_{33} = 2$, and $m_{23} = 1$. (Remember the order: m_{ij} refers to row i, column j. Think of the mnemonic $m_{row\ column}$, $m_{Roman\ Catholic}$.)

Matrix addition is performed by adding corresponding components:

$$[2 \quad 1 \quad 4] + [7 \quad 11 \quad 3] = [9 \quad 12 \quad 7]$$

$$\begin{bmatrix} -1 \\ 1 \\ 0 \end{bmatrix} + \begin{bmatrix} 1 \\ 2 \\ 3 \end{bmatrix} = \begin{bmatrix} 0 \\ 3 \\ 3 \end{bmatrix}$$

$$\begin{bmatrix} 1 & 2 & 3 \\ 4 & 5 & 6 \\ 7 & 8 & 9 \end{bmatrix} + \begin{bmatrix} 0 & 1 & -1 \\ 0 & 1 & -1 \\ 0 & 1 & -1 \end{bmatrix} = \begin{bmatrix} 1 & 3 & 2 \\ 4 & 6 & 5 \\ 7 & 9 & 8 \end{bmatrix}$$

This is consistent with the way we add vectors, whether interpreting them as rows or columns. Notice that we only add matrices of the same dimensions; we cannot add, for example, eqs. (5.71) to (5.72).

Scalar multiplication also proceeds by entries, in keeping with our vector interpretation:

$$2\,[2 \quad 1 \quad 4] = [4 \quad 2 \quad 8]$$

$$-\frac{1}{2}\begin{bmatrix} 1 & -1 & 1 \\ 2 & 0 & 1 \\ 1 & 1 & 2 \end{bmatrix} = \begin{bmatrix} -\frac{1}{2} & \frac{1}{2} & -\frac{1}{2} \\ -1 & 0 & -\frac{1}{2} \\ -\frac{1}{2} & -\frac{1}{2} & -1 \end{bmatrix}$$

It follows that matrix addition, like vector addition, has all the usual properties, that is, commutativity, associativity, and distributivity, with respect to scalar multiplication. In fact, matrix theory would be dull, indeed, if addition and scalar multiplication were the only operations.

The definition of matrix multiplication is what makes the subject useful and interesting. In general, the product $LR = P$ of two matrices L and R is defined only when the number of *columns* of the left factor L equals the number of *rows* of the right factor R. Denoting this common number by s, the entry in the ith row and jth column of P is then defined by

$$p_{ij} = \sum_{k=1}^{s} l_{ik} r_{kj} \tag{5.77}$$

From the formula (5.77), one can see that P has the same number of *rows* as L, and the same number of *columns* as R.

If we look at the possibilities for multiplication of the three types of matrices with which we are concerned, we discover that there are four interpretations of eq. (5.77):

(*i*) If L is a 1-by-3 row matrix and R is a 3-by-1 column matrix, then the product LR is the scalar product of the corresponding vectors:

$$[\cdots \quad \mathbf{A} \quad \cdots] \begin{bmatrix} \vdots \\ \mathbf{B} \\ \vdots \end{bmatrix} = \mathbf{A} \cdot \mathbf{B} \tag{5.78}$$

For example,

$$[2 \quad 4 \quad 1] \begin{bmatrix} 2 \\ 1 \\ 5 \end{bmatrix} = 4 + 4 + 5 = 13$$

(*ii*) If L is a 3-by-3 square matrix and R is a 3-by-1 column matrix, then the product LR is a 3-by-1 column matrix. The first entry is the scalar product of the first row of L with R; the second and third entries are the corresponding scalar products of the second and third rows, respectively, with R:

$$\begin{bmatrix} \cdots & \mathbf{A} & \cdots \\ \cdots & \mathbf{B} & \cdots \\ \cdots & \mathbf{C} & \cdots \end{bmatrix} \begin{bmatrix} \vdots \\ \mathbf{D} \\ \vdots \end{bmatrix} = \begin{bmatrix} \mathbf{A} \cdot \mathbf{D} \\ \mathbf{B} \cdot \mathbf{D} \\ \mathbf{C} \cdot \mathbf{D} \end{bmatrix} \tag{5.79}$$

For example,

$$\begin{bmatrix} 1 & -1 & 1 \\ 2 & 0 & 1 \\ 1 & 1 & 2 \end{bmatrix} \begin{bmatrix} 2 \\ 1 \\ 5 \end{bmatrix} = \begin{bmatrix} 6 \\ 9 \\ 13 \end{bmatrix} \tag{5.80}$$

(*iii*) If L is a 1-by-3 row matrix and R is a 3-by-3 square matrix, then the product LR is a 1-by-3 row matrix. The first entry is the scalar product of L with the first column of R; the second and third entries are the corresponding scalar products with the second and third columns respectively:

$$[\cdots \ \mathbf{A} \ \cdots] \begin{bmatrix} \vdots & \vdots & \vdots \\ \mathbf{B} & \mathbf{C} & \mathbf{D} \\ \vdots & \vdots & \vdots \end{bmatrix} = [\mathbf{A} \cdot \mathbf{B}, \quad \mathbf{A} \cdot \mathbf{C}, \quad \mathbf{A} \cdot \mathbf{D}] \tag{5.81}$$

For example,

$$[2 \quad 4 \quad 1] \begin{bmatrix} 1 & -1 & 1 \\ 2 & 0 & 1 \\ 1 & 1 & 2 \end{bmatrix} = [11 \quad -1 \quad 8] \tag{5.82}$$

(*iv*) If L and R are both 3-by-3 square matrices, then the product LR is also a 3-by-3 square matrix. *The (i, j)th entry of LR is the scalar product of row i of L with column j of R:*

$$\begin{bmatrix} \cdots & \mathbf{A} & \cdots \\ \cdots & \mathbf{B} & \cdots \\ \cdots & \mathbf{C} & \cdots \end{bmatrix} \begin{bmatrix} \vdots & \vdots & \vdots \\ \mathbf{D} & \mathbf{E} & \mathbf{F} \\ \vdots & \vdots & \vdots \end{bmatrix} = \begin{bmatrix} \mathbf{A} \cdot \mathbf{D} & \mathbf{A} \cdot \mathbf{E} & \mathbf{A} \cdot \mathbf{F} \\ \mathbf{B} \cdot \mathbf{D} & \mathbf{B} \cdot \mathbf{E} & \mathbf{B} \cdot \mathbf{F} \\ \mathbf{C} \cdot \mathbf{D} & \mathbf{C} \cdot \mathbf{E} & \mathbf{C} \cdot \mathbf{F} \end{bmatrix} \tag{5.83}$$

For example,

$$\begin{bmatrix} 1 & -1 & 1 \\ 2 & 0 & 1 \\ 1 & 1 & 2 \end{bmatrix} \begin{bmatrix} 1 & 2 & 3 \\ 3 & 2 & 1 \\ 2 & 2 & 2 \end{bmatrix} = \begin{bmatrix} 0 & 2 & 4 \\ 4 & 6 & 8 \\ 8 & 8 & 8 \end{bmatrix} \tag{5.84}$$

Notice that the italicized statement in (*iv*) actually covers cases (*i*) through (*iii*) also. Alert readers may observe that LR is also defined when L is a column and R is a row, but we have no use for such products here. (They are related to the dyadics mentioned in section 3.7.)

The most common use of matrices is in expressing a system of equations. Two matrices are equal when they have the same dimensions and their corresponding entries are equal.

Example 5.1 Express the system of equations

$$\begin{aligned} x - y + z &= 1 \\ 2x \qquad + z &= 2 \\ x + y + 2z &= 3 \end{aligned} \tag{5.85}$$

as a matrix equation.

Solution The reader can verify that

$$\begin{bmatrix} 1 & -1 & 1 \\ 2 & 0 & 1 \\ 1 & 1 & 2 \end{bmatrix} \begin{bmatrix} x \\ y \\ z \end{bmatrix} = \begin{bmatrix} 1 \\ 2 \\ 3 \end{bmatrix} \tag{5.86}$$

is equivalent to eq. (5.85). The square matrix is the same as eq. (5.73), and its rows **A, B,** and **C** are identified in eq. (5.74). The eq. (5.85) can be written as follows, with $\mathbf{R} = x\mathbf{i} + y\mathbf{j} + z\mathbf{k}$:

$$\mathbf{A} \cdot \mathbf{R} = 1$$
$$\mathbf{B} \cdot \mathbf{R} = 2$$
$$\mathbf{C} \cdot \mathbf{R} = 3$$

Hence **R** is the position vector to a point lying simultaneously in three planes, having normals **A, B,** and **C** respectively (recall section 1.10).

Since matrix products can be interpreted as arrays of scalar products, it should come as no surprise that matrix multiplication is linear in each factor; for any scalar s,

$$\begin{aligned} (sM + N)R &= sMR + NR \\ L(sM + N) &= sLM + LN \end{aligned} \tag{5.87}$$

assuming all products are defined. The proof is left as an exercise.

Since scalar products are commutative, one would expect that matrix multiplication is also. Surprisingly, this is not the case! Consider, for example, the matrices in eq. (5.84) multiplied in reverse order:

$$\begin{bmatrix} 1 & 2 & 3 \\ 3 & 2 & 1 \\ 2 & 2 & 2 \end{bmatrix} \begin{bmatrix} 1 & -1 & 1 \\ 2 & 0 & 1 \\ 1 & 1 & 2 \end{bmatrix} = \begin{bmatrix} 8 & 2 & 9 \\ 8 & -2 & 7 \\ 8 & 0 & 8 \end{bmatrix}$$

The answer is different! The reason is that when we form LR we use vectors from the *rows* of L and the *columns* of R; but when we form RL we use the *columns* of L and the *rows* of R. We are taking scalar products of different vectors in the two cases. (In fact, if the dimensions do not match, RL may be undefined.)

Let us consider the associative law. Is there any difference between $(LM)R$ and $L(MR)$? Notice that M must be square 3-by-3 if both products are defined, in the context of "vector analysis matrices."

One begins the computation of $(LM)R$ by regarding the *columns* of M as vectors. However, to compute $L(MR)$, one starts with the *row* vectors of M. Thus we would expect the products to be different. Astonishingly, they are the same!

Example 5.2 Verify the associative law

$$(LM)R = L(MR) \tag{5.88}$$

for L, M, and R given by eqs. (5.71), (5.73), and (5.72), respectively.

Solution We computed LM in eq. (5.82); hence

$$(LM)R = [11 \quad -1 \quad 8]\begin{bmatrix} 2 \\ 1 \\ 5 \end{bmatrix} = 22 - 1 + 40 = 61$$

MR was evaluated in eq. (5.80); therefore

$$L(MR) = [2 \quad 4 \quad 1]\begin{bmatrix} 6 \\ 9 \\ 13 \end{bmatrix} = 12 + 36 + 13 = 61$$

To prove eq. (5.88) in general, we have to abandon momentarily the vector interpretation of matrix multiplication, and work with the explicit formula (5.81) in terms of the entries.

We introduce the notation $N = LM$ and $Q = MR$. Then the (i, j)th entry of the left-hand side of eq. (5.88) is found by using eq. (5.63) twice:

$$\sum_{k=1}^{3} n_{ik}r_{kj} = \sum_{k=1}^{3}\left(\sum_{s=1}^{3} l_{is}m_{sk}\right) r_{kj} = \sum_{k=1}^{3}\sum_{s=1}^{3} l_{is}m_{sk}r_{kj} \tag{5.89}$$

The (i, j)th element of the right-hand side of eq. (5.88) is

$$\sum_{k=1}^{3} l_{ik}q_{kj} = \sum_{k=1}^{3} l_{ik}\left(\sum_{s=1}^{3} m_{ks}r_{sj}\right) = \sum_{k=1}^{3}\sum_{s=1}^{3} l_{ik}m_{ks}r_{sj} \tag{5.90}$$

Exactly the same terms appear in the final sums in eqs. (5.89) and (5.90), so they are equal and eq. (5.88) is proved.

One matrix plays a special role in linear algebra: it is the *identity matrix I*, defined by

$$I = \begin{bmatrix} 1 & 0 & 0 \\ 0 & 1 & 0 \\ 0 & 0 & 1 \end{bmatrix} = \begin{bmatrix} \cdots & \mathbf{i} & \cdots \\ \cdots & \mathbf{j} & \cdots \\ \cdots & \mathbf{k} & \cdots \end{bmatrix} = \begin{bmatrix} \vdots & \vdots & \vdots \\ \mathbf{i} & \mathbf{j} & \mathbf{k} \\ \vdots & \vdots & \vdots \end{bmatrix} \tag{5.91}$$

It is called the "identity" because multiplication by I, on the right or left, produces no change. For example,

$$\begin{bmatrix} 1 & 0 & 0 \\ 0 & 1 & 0 \\ 0 & 0 & 1 \end{bmatrix}\begin{bmatrix} 2 \\ 1 \\ 5 \end{bmatrix} = \begin{bmatrix} 2 \\ 1 \\ 5 \end{bmatrix}$$

$$[2 \quad 4 \quad 1]\begin{bmatrix} 1 & 0 & 0 \\ 0 & 1 & 0 \\ 0 & 0 & 1 \end{bmatrix} = [2 \quad 4 \quad 1]$$

Another way to see this is

$$\begin{bmatrix} \cdots & \mathbf{i} & \cdots \\ \cdots & \mathbf{j} & \cdots \\ \cdots & \mathbf{k} & \cdots \end{bmatrix}\begin{bmatrix} \vdots \\ \mathbf{A} \\ \vdots \end{bmatrix} = \begin{bmatrix} \mathbf{i}\cdot\mathbf{A} \\ \mathbf{j}\cdot\mathbf{A} \\ \mathbf{k}\cdot\mathbf{A} \end{bmatrix} = \begin{bmatrix} \vdots \\ \mathbf{A} \\ \vdots \end{bmatrix}$$

$$[\cdots \quad \mathbf{B} \quad \cdots]\begin{bmatrix} \vdots & \vdots & \vdots \\ \mathbf{i} & \mathbf{j} & \mathbf{k} \\ \vdots & \vdots & \vdots \end{bmatrix} = [\mathbf{B}\cdot\mathbf{i} \quad \mathbf{B}\cdot\mathbf{j} \quad \mathbf{B}\cdot\mathbf{k}] = [\cdots \quad \mathbf{B} \quad \cdots]$$

Given a square matrix M, we say R is a *right inverse* of M if $MR = I$. For instance, one should check that

$$\underset{M}{\begin{bmatrix} 1 & -1 & 1 \\ 2 & 0 & 1 \\ 1 & 1 & 2 \end{bmatrix}}\underset{R}{\begin{bmatrix} -\frac{1}{4} & \frac{3}{4} & -\frac{1}{4} \\ -\frac{3}{4} & \frac{1}{4} & \frac{1}{4} \\ \frac{1}{2} & -\frac{1}{2} & \frac{1}{2} \end{bmatrix}} = \underset{I}{\begin{bmatrix} 1 & 0 & 0 \\ 0 & 1 & 0 \\ 0 & 0 & 1 \end{bmatrix}} \tag{5.92}$$

so the second matrix is a right inverse of the first.

Notice the relation between the rows of M and the columns of R. If we express $MR = I$ as

$$\underset{M}{\begin{bmatrix} \cdots & \mathbf{A} & \cdots \\ \cdots & \mathbf{B} & \cdots \\ \cdots & \mathbf{C} & \cdots \end{bmatrix}}\underset{R}{\begin{bmatrix} \vdots & \vdots & \vdots \\ \mathbf{D} & \mathbf{E} & \mathbf{F} \\ \vdots & \vdots & \vdots \end{bmatrix}} = \underset{I}{\begin{bmatrix} 1 & 0 & 0 \\ 0 & 1 & 0 \\ 0 & 0 & 1 \end{bmatrix}}$$

we see that

$$\begin{gathered} \mathbf{A}\cdot\mathbf{D} = \mathbf{B}\cdot\mathbf{E} = \mathbf{C}\cdot\mathbf{F} = 1 \\ \mathbf{A}\cdot\mathbf{E} = \mathbf{A}\cdot\mathbf{F} = 0 \qquad \mathbf{B}\cdot\mathbf{D} = \mathbf{B}\cdot\mathbf{F} = 0 \qquad \mathbf{C}\cdot\mathbf{D} = \mathbf{C}\cdot\mathbf{E} = 0 \end{gathered} \tag{5.93}$$

Whenever two sets of vectors $\{\mathbf{A},\mathbf{B},\mathbf{C}\}$ and $\{\mathbf{D},\mathbf{E},\mathbf{F}\}$ satisfy relations (5.93), we say that one set of vectors is *reciprocal,* or *dual,* to the other set. The considerations of chapter 1 will enable us to find formulas for reciprocal vectors (and, as a result, a formula for the right inverse).

Example 5.3 Derive the formulas for the reciprocal vectors for the set $\{\mathbf{A},\mathbf{B},\mathbf{C}\}$.

Solution Examining eq. (5.93), we see that $\mathbf{D}$ must be perpendicular to $\mathbf{B}$ and $\mathbf{C}$. Let us try $\mathbf{D} = s\mathbf{B} \times \mathbf{C}$. To make $\mathbf{A} \cdot \mathbf{D} = 1$, we must choose s so that

$$s\mathbf{A} \cdot \mathbf{B} \times \mathbf{C} = s[\mathbf{A},\mathbf{B},\mathbf{C}] = 1$$

Consequently, if $[\mathbf{A},\mathbf{B},\mathbf{C}]$, which is the determinant of M, is not zero, we find

$$\mathbf{D} = \frac{\mathbf{B} \times \mathbf{C}}{[\mathbf{A},\mathbf{B},\mathbf{C}]} \tag{5.94}$$

Similarly,

$$\mathbf{E} = \frac{\mathbf{C} \times \mathbf{A}}{[\mathbf{A},\mathbf{B},\mathbf{C}]} \tag{5.95}$$

$$\mathbf{F} = \frac{\mathbf{A} \times \mathbf{B}}{[\mathbf{A},\mathbf{B},\mathbf{C}]} \tag{5.96}$$

On the other hand, if $[\mathbf{A},\mathbf{B},\mathbf{C}] = 0$, then $\mathbf{A}$, $\mathbf{B}$, and $\mathbf{C}$ are coplanar, and it is easy to see that no set satisfying eq. (5.93) exists (exercise 5).

This provides a formula for the right inverse: *if* $[\mathbf{A},\mathbf{B},\mathbf{C}] \neq 0$, *then a right inverse for the matrix*

$$M = \begin{bmatrix} \cdots & \mathbf{A} & \cdots \\ \cdots & \mathbf{B} & \cdots \\ \cdots & \mathbf{C} & \cdots \end{bmatrix}$$

is given by

$$R = \frac{1}{[\mathbf{A},\mathbf{B},\mathbf{C}]} \begin{bmatrix} \vdots & \vdots & \vdots \\ \mathbf{B} \times \mathbf{C} & \mathbf{C} \times \mathbf{A} & \mathbf{A} \times \mathbf{B} \\ \vdots & \vdots & \vdots \end{bmatrix} \tag{5.97}$$

If $[\mathbf{A},\mathbf{B},\mathbf{C}] = 0$, no right inverse exists. The reader should verify (exercise 7) that the right inverse displayed in eq. (5.92) is consistent with this formula.

Clearly, if we take the reciprocal set of vectors for the *columns* of M and arrange them in the *rows* of L, we obtain a *left inverse* of M; that is, $LM = I$. Hence, if we now let **A**,**B**,**C** represent the *columns* of M, we have

$$\frac{1}{[\mathbf{A},\mathbf{B},\mathbf{C}]}\begin{bmatrix} \cdots & \mathbf{B}\times\mathbf{C} & \cdots \\ \cdots & \mathbf{C}\times\mathbf{A} & \cdots \\ \cdots & \mathbf{A}\times\mathbf{B} & \cdots \end{bmatrix}\begin{bmatrix} \vdots & \vdots & \vdots \\ \mathbf{A} & \mathbf{B} & \mathbf{C} \\ \vdots & \vdots & \vdots \end{bmatrix} = I \tag{5.98}$$

A quick computation (exercise 6) reveals that the triple scalar product of the columns of M equals the triple scalar product of its rows, which in turn equals the determinant. So the existence of a left inverse also hinges on the condition $\det M \neq 0$.

In exercise 8 the reader is invited to compute the left inverse of the matrix M in eq. (5.92). Rather surprisingly, the left inverse L turns out to be equal to the right inverse R! To see this in general, we use the associative law, together with the equalities $LM = I = MR$, to derive

$$L = LI = L(MR) = (LM)R = IR = R$$

The fact of the matter, as the reader will show in exercise 9, is that if a matrix M has any inverse at all, it has only one; and this one is both a left and a right inverse. It can therefore be denoted M^{-1}, unambiguously. The condition for having an inverse is that the determinant of the matrix must be nonzero. The usefulness of an inverse in "undoing" a system of equations is illustrated in the next example.

Example 5.4 Solve the system of equations given in example 5.1.

Solution Consider the equations expressed in matrix form, eq. (5.86). If we multiply both sides of that equation on the left by the inverse of the matrix, given in eq. (5.92), we find the solution:

$$\begin{bmatrix} 1 & 0 & 0 \\ 0 & 1 & 0 \\ 0 & 0 & 1 \end{bmatrix}\begin{bmatrix} x \\ y \\ z \end{bmatrix} = \begin{bmatrix} x \\ y \\ z \end{bmatrix} = \begin{bmatrix} -\frac{1}{4} & \frac{3}{4} & -\frac{1}{4} \\ -\frac{3}{4} & \frac{1}{4} & \frac{1}{4} \\ \frac{1}{2} & -\frac{1}{2} & \frac{1}{2} \end{bmatrix}\begin{bmatrix} 1 \\ 2 \\ 3 \end{bmatrix} = \begin{bmatrix} \frac{1}{2} \\ \frac{1}{2} \\ 1 \end{bmatrix}$$

or $x = \frac{1}{2}$, $y = \frac{1}{2}$, $z = 1$.

Example 5.5 Derive a general formula for the solution of the system

$$\begin{bmatrix} m_{11} & m_{12} & m_{13} \\ m_{21} & m_{22} & m_{23} \\ m_{31} & m_{32} & m_{33} \end{bmatrix}\begin{bmatrix} x \\ y \\ z \end{bmatrix} = \begin{bmatrix} u \\ v \\ w \end{bmatrix} \tag{5.99}$$

where $\det M \neq 0$.

Solution First we form the left inverse of the square matrix. To achieve this, we have to regard it as made up of column vectors:

$$M = \begin{bmatrix} \vdots & \vdots & \vdots \\ \mathbf{D} & \mathbf{E} & \mathbf{F} \\ \vdots & \vdots & \vdots \end{bmatrix} \qquad M^{-1} = \frac{1}{[\mathbf{D},\mathbf{E},\mathbf{F}]} \begin{bmatrix} \cdots & \mathbf{E} \times \mathbf{F} & \cdots \\ \cdots & \mathbf{F} \times \mathbf{D} & \cdots \\ \cdots & \mathbf{D} \times \mathbf{E} & \cdots \end{bmatrix}$$

Next we multiply on the left by M^{-1} to obtain

$$I \begin{bmatrix} x \\ y \\ z \end{bmatrix} = \begin{bmatrix} x \\ y \\ z \end{bmatrix} = \frac{1}{[\mathbf{D},\mathbf{E},\mathbf{F}]} \begin{bmatrix} \cdots & \mathbf{E} \times \mathbf{F} & \cdots \\ \cdots & \mathbf{F} \times \mathbf{D} & \cdots \\ \cdots & \mathbf{D} \times \mathbf{E} & \cdots \end{bmatrix} \begin{bmatrix} u \\ v \\ w \end{bmatrix}$$

Looking at the first component, we find

$$x = \frac{\mathbf{E} \times \mathbf{F} \cdot (u\mathbf{i} + v\mathbf{j} + w\mathbf{k})}{[\mathbf{D},\mathbf{E},\mathbf{F}]} = \frac{\begin{vmatrix} u & v & w \\ \cdots & \mathbf{E} & \cdots \\ \cdots & \mathbf{F} & \cdots \end{vmatrix}}{\begin{vmatrix} \cdots & \mathbf{D} & \cdots \\ \cdots & \mathbf{E} & \cdots \\ \cdots & \mathbf{F} & \cdots \end{vmatrix}} = \frac{\begin{vmatrix} u & \vdots & \vdots \\ v & \mathbf{E} & \mathbf{F} \\ w & \vdots & \vdots \end{vmatrix}}{\begin{vmatrix} \vdots & \vdots & \vdots \\ \mathbf{D} & \mathbf{E} & \mathbf{F} \\ \vdots & \vdots & \vdots \end{vmatrix}}$$

Identifying the parts of the original matrix M, we have

$$x = \frac{\begin{vmatrix} u & m_{12} & m_{13} \\ v & m_{22} & m_{23} \\ w & m_{32} & m_{33} \end{vmatrix}}{\begin{vmatrix} m_{11} & m_{12} & m_{13} \\ m_{21} & m_{22} & m_{23} \\ m_{31} & m_{32} & m_{33} \end{vmatrix}} \tag{5.100}$$

Similarly,

$$y = \frac{\begin{vmatrix} m_{11} & u & m_{13} \\ m_{21} & v & m_{23} \\ m_{31} & w & m_{33} \end{vmatrix}}{\begin{vmatrix} m_{11} & m_{12} & m_{13} \\ m_{21} & m_{22} & m_{23} \\ m_{31} & m_{32} & m_{33} \end{vmatrix}} \qquad z = \frac{\begin{vmatrix} m_{11} & m_{12} & u \\ m_{21} & m_{22} & v \\ m_{31} & m_{32} & w \end{vmatrix}}{\begin{vmatrix} m_{11} & m_{12} & m_{13} \\ m_{21} & m_{22} & m_{23} \\ m_{31} & m_{32} & m_{33} \end{vmatrix}}$$

These formulas are collectively called *Cramer's rule.*

EXERCISES

1. Form the indicated products:

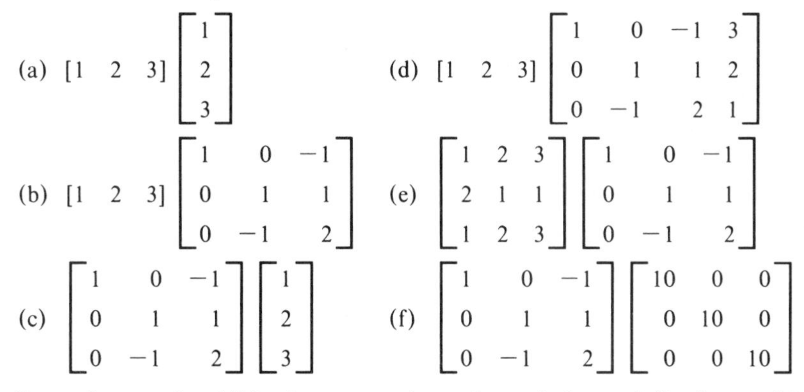

(a) $[1 \;\; 2 \;\; 3]\begin{bmatrix}1\\2\\3\end{bmatrix}$

(b) $[1 \;\; 2 \;\; 3]\begin{bmatrix}1 & 0 & -1\\0 & 1 & 1\\0 & -1 & 2\end{bmatrix}$

(c) $\begin{bmatrix}1 & 0 & -1\\0 & 1 & 1\\0 & -1 & 2\end{bmatrix}\begin{bmatrix}1\\2\\3\end{bmatrix}$

(d) $[1 \;\; 2 \;\; 3]\begin{bmatrix}1 & 0 & -1 & 3\\0 & 1 & 1 & 2\\0 & -1 & 2 & 1\end{bmatrix}$

(e) $\begin{bmatrix}1 & 2 & 3\\2 & 1 & 1\\1 & 2 & 3\end{bmatrix}\begin{bmatrix}1 & 0 & -1\\0 & 1 & 1\\0 & -1 & 2\end{bmatrix}$

(f) $\begin{bmatrix}1 & 0 & -1\\0 & 1 & 1\\0 & -1 & 2\end{bmatrix}\begin{bmatrix}10 & 0 & 0\\0 & 10 & 0\\0 & 0 & 10\end{bmatrix}$

2. Prove that matrix addition is commutative and associative, and distributes with scalar multiplication. (*Hint:* Exploit the vector interpretation.)

3. Prove the distributive laws in eq. (5.87).

4. Construct an example of two square matrices whose product *is* commutative.

5. Show that if **A, B,** and **C** are coplanar, then no reciprocal set of vectors exists. (Do not forget to consider the possibility that they are collinear.)

6. Show that the triple scalar product of the columns of a square matrix equals the triple scalar product of its rows.

7. Show that the right inverse in eq. (5.92) agrees with the formula (5.97).

8. Using eq. (5.98), compute the left inverse of the matrix in eq. (5.73), and compare with eq. (5.92).

9. Prove that if det $M \neq 0$, M has only one inverse.

10. Solve by computing the inverse:

$$\begin{aligned} 2x + y + 2z &= 2 \\ 3x \quad\;\; + 2z &= 4 \\ x + y + 2z &= 0 \end{aligned}$$

11. Solve exercise 10 by Cramer's rule.

12. Show that the inverse of a product equals the product of the inverses, in reverse order; that is, $(LR)^{-1} = R^{-1}L^{-1}$.

The transpose of a matrix M is a matrix M^T whose rows are the columns of the original matrix M.

13. Show that the (j, i)th entry in M^T equals the (i, j)th entry in M.

14. Show that the transpose of a product is the product of the transposes in reverse order; that is, $(LR)^T = R^TL^T$.

A symmetric matrix is a matrix that equals its transpose. An antisymmetric matrix is a matrix that equals the negative of its transpose. An orthogonal matrix is a matrix whose inverse equals its transpose.

15. Show that a symmetric matrix must have the form

$$\begin{bmatrix} m_{11} & m_{21} & m_{31} \\ m_{21} & m_{22} & m_{32} \\ m_{31} & m_{32} & m_{33} \end{bmatrix}$$

Is the product of two symmetric matrices symmetric?

16. Show that the columns of an orthogonal matrix are mutually orthogonal unit vectors. Show the same for the rows. What does this say about the reciprocal sets?

17. Show that if O is orthogonal and S is symmetric, then $O^{-1}SO$ is symmetric.

18. Show that if A is antisymmetric, it has the form

$$\begin{bmatrix} 0 & -m_{21} & m_{13} \\ m_{21} & 0 & -m_{32} \\ -m_{13} & m_{32} & 0 \end{bmatrix}$$

Then show if R is a column vector representing $\mathbf{v}$, AR represents

$$(m_{32}\,\mathbf{i} + m_{13}\,\mathbf{j} + m_{21}\,\mathbf{k}) \times \mathbf{v}$$

19. Construct examples of systems of three equations of the form of eq. (5.99) whose solutions constitute
    (a) a plane.
    (b) a straight line.
    (c) the empty set.
    (*Hint:* Remember the interpretation as the intersection of three planes.) What is the determinant of the matrix in these cases?

20. Consider the plane passing through the points located by the vectors $a\mathbf{A}$, $b\mathbf{B}$, and $c\mathbf{C}$ (presumed noncollinear). Let $\mathbf{D}$, $\mathbf{E}$, $\mathbf{F}$ be the reciprocal vectors to $\mathbf{A}$, $\mathbf{B}$, and $\mathbf{C}$. Show that $(1/a)\mathbf{D} + (1/b)\mathbf{E} + (1/c)\mathbf{F}$ is normal to this plane. Interpret this for the cases when $a = 0$ and $a = \infty$.

5.8 Linear Orthogonal Transformations

We now return to the study of different coordinate systems for describing scalar and vector fields. An important case is the utilization of another *cartesian* coordinate system (right-handed, of course), with axes labeled x', y', z' and their associated unit vectors $\mathbf{i}'$, $\mathbf{j}'$, $\mathbf{k}'$. This "new" coordinate system will have the same origin as the "old" x, y, z system (see fig. 5.9).

Consider a point in space. The point's coordinates in the old system are (x, y, z), and in the new system they are (x', y', z'). To see how these are related, draw the position vector $\mathbf{R}$ to the point (fig. 5.9). Then we have two descriptions of $\mathbf{R}$:

$$\mathbf{R} = x\mathbf{i} + y\mathbf{j} + z\mathbf{k} = x'\mathbf{i}' + y'\mathbf{j}' + z'\mathbf{k}' \tag{5.101}$$

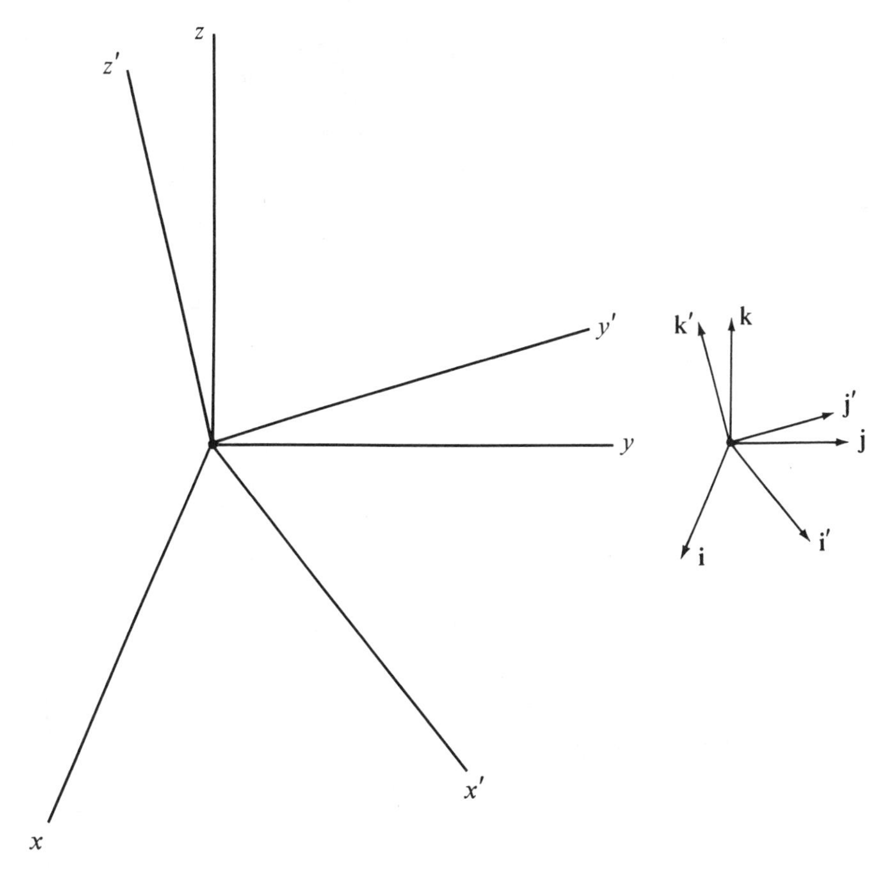

Figure 5.9

If we now take scalar products of eq. (5.101) with **i, j,** and **k** in turn, we find

$$\begin{aligned} x &= x'\mathbf{i}' \cdot \mathbf{i} + y'\mathbf{j}' \cdot \mathbf{i} + z'\mathbf{k}' \cdot \mathbf{i} \\ y &= x'\mathbf{i}' \cdot \mathbf{j} + y'\mathbf{j}' \cdot \mathbf{j} + z'\mathbf{k}' \cdot \mathbf{j} \\ z &= x'\mathbf{i}' \cdot \mathbf{k} + y'\mathbf{j}' \cdot \mathbf{k} + z'\mathbf{k}' \cdot \mathbf{k} \end{aligned} \tag{5.102}$$

Equation (5.102) can be compactly written

$$\begin{bmatrix} x \\ y \\ z \end{bmatrix} = J \begin{bmatrix} x' \\ y' \\ z' \end{bmatrix}$$

with the *transformation matrix J,* given by

$$J = \begin{bmatrix} \mathbf{i}' \cdot \mathbf{i} & \mathbf{j}' \cdot \mathbf{i} & \mathbf{k}' \cdot \mathbf{i} \\ \mathbf{i}' \cdot \mathbf{j} & \mathbf{j}' \cdot \mathbf{j} & \mathbf{k}' \cdot \mathbf{j} \\ \mathbf{i}' \cdot \mathbf{k} & \mathbf{j}' \cdot \mathbf{k} & \mathbf{k}' \cdot \mathbf{k} \end{bmatrix} \tag{5.103}$$

The matrix J has many useful and interesting properties. First of all, observe that the columns consist of the direction cosines of the vectors $\mathbf{i}', \mathbf{j}', \mathbf{k}'$ with respect to the old coordinate system (recall section 1.5). This is sometimes an aid in computing J.

Example 5.6 Suppose the new system is formed from the old by rotating through an angle θ about the z axis, as in figure 5.10. Compute the transformation matrix J.

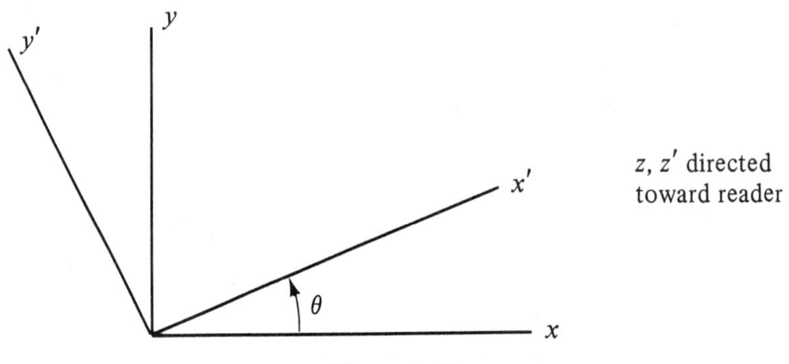

Figure 5.10

Solution The direction cosines of $\mathbf{i}'$ are $\cos\theta$, $\cos[(\pi/2) - \theta]$, and 0. For $\mathbf{j}'$, they are $\cos[(\pi/2) + \theta]$, $\cos\theta$, and 0. For $\mathbf{k}'$, they are 0, 0, and 1. Hence

$$J = \begin{bmatrix} \cos\theta & -\sin\theta & 0 \\ \sin\theta & \cos\theta & 0 \\ 0 & 0 & 1 \end{bmatrix} \tag{5.104}$$

From another point of view we can say that the columns of J represent the vectors $\mathbf{i}'$, $\mathbf{j}'$, and $\mathbf{k}'$ in the old coordinate system:

$$J = \begin{bmatrix} \vdots & \vdots & \vdots \\ \mathbf{i}' & \mathbf{j}' & \mathbf{k}' \\ \vdots & \vdots & \vdots \end{bmatrix} \tag{5.105}$$

This provides an easy prescription of the inverse of J. Since the set $\{\mathbf{i}', \mathbf{j}', \mathbf{k}'\}$ is self-reciprocal, by the analysis of section 5.7 we find

$$J^{-1} = \begin{bmatrix} \cdots & \mathbf{i}' & \cdots \\ \cdots & \mathbf{j}' & \cdots \\ \cdots & \mathbf{k}' & \cdots \end{bmatrix} = \begin{bmatrix} \mathbf{i}'\cdot\mathbf{i} & \mathbf{i}'\cdot\mathbf{j} & \mathbf{i}'\cdot\mathbf{k} \\ \mathbf{j}'\cdot\mathbf{i} & \mathbf{j}'\cdot\mathbf{j} & \mathbf{j}'\cdot\mathbf{k} \\ \mathbf{k}'\cdot\mathbf{i} & \mathbf{k}'\cdot\mathbf{j} & \mathbf{k}'\cdot\mathbf{k} \end{bmatrix} \tag{5.106}$$

The matrix J^{-1} is the *transpose of J;* that is, its rows are the columns of J. A matrix whose transpose equals its inverse is called *orthogonal* (this terminology was introduced in the previous set of exercises). We have shown that *the transformation that relates coordinates between two cartesian coordinate systems having the same origin is effected by a matrix multiplication* (a linear operation) *using an orthogonal matrix;* hence the name "linear orthogonal transformation."

We can use J^{-1} to get new coordinates in terms of old:

$$\begin{bmatrix} x' \\ y' \\ z' \end{bmatrix} = J^{-1} \begin{bmatrix} x \\ y \\ z \end{bmatrix} = \begin{bmatrix} \mathbf{i}' \cdot \mathbf{i} & \mathbf{i}' \cdot \mathbf{j} & \mathbf{i}' \cdot \mathbf{k} \\ \mathbf{j}' \cdot \mathbf{i} & \mathbf{j}' \cdot \mathbf{j} & \mathbf{j}' \cdot \mathbf{k} \\ \mathbf{k}' \cdot \mathbf{i} & \mathbf{k}' \cdot \mathbf{j} & \mathbf{k}' \cdot \mathbf{k} \end{bmatrix} \begin{bmatrix} x \\ y \\ z \end{bmatrix} \tag{5.107}$$

It is easy to see the reversal of roles: now the first column is the direction cosines of **i** with respect to the *new* system, etc. In fact, it is also easy to see (exercise 1) that the equations in (5.107) result from taking scalar products of eq. (5.101) with $\mathbf{i}', \mathbf{j}', \mathbf{k}'$ in turn.

Now that we know how coordinates transform, let us see how *vectors* transform. Obviously,

$$\begin{aligned} \mathbf{i} &= (\mathbf{i} \cdot \mathbf{i}')\mathbf{i}' + (\mathbf{i} \cdot \mathbf{j}')\mathbf{j}' + (\mathbf{i} \cdot \mathbf{k}')\mathbf{k}' \\ \mathbf{j} &= (\mathbf{j} \cdot \mathbf{i}')\mathbf{i}' + (\mathbf{j} \cdot \mathbf{j}')\mathbf{j}' + (\mathbf{j} \cdot \mathbf{k}')\mathbf{k}' \\ \mathbf{k} &= (\mathbf{k} \cdot \mathbf{i}')\mathbf{i}' + (\mathbf{k} \cdot \mathbf{j}')\mathbf{j}' + (\mathbf{k} \cdot \mathbf{k}')\mathbf{k}' \end{aligned} \tag{5.108}$$

An arbitrary vector **V** will have the two representations

$$\begin{aligned} \mathbf{V} &= \mathbf{V} \cdot \mathbf{i}\,\mathbf{i} + \mathbf{V} \cdot \mathbf{j}\,\mathbf{j} + \mathbf{V} \cdot \mathbf{k}\,\mathbf{k} = V_1\mathbf{i} + V_2\mathbf{j} + V_3\mathbf{k} \\ &= \mathbf{V} \cdot \mathbf{i}'\,\mathbf{i}' + \mathbf{V} \cdot \mathbf{j}'\,\mathbf{j}' + \mathbf{V} \cdot \mathbf{k}'\,\mathbf{k}' = V'_1\mathbf{i}' + V'_2\mathbf{j}' + V'_3\mathbf{k}' \end{aligned} \tag{5.109}$$

If we now take scalar products of eq. (5.109) with **i**, **j**, and **k**, we see that *the components of* **V** *transform just like the coordinates of a point:*

$$\begin{bmatrix} V_1 \\ V_2 \\ V_3 \end{bmatrix} = J \begin{bmatrix} V'_1 \\ V'_2 \\ V'_3 \end{bmatrix} \qquad \begin{bmatrix} V'_1 \\ V'_2 \\ V'_3 \end{bmatrix} = J^{-1} \begin{bmatrix} V_1 \\ V_2 \\ V_3 \end{bmatrix} \tag{5.110}$$

What happens to a scalar field $f(x, y, z)$ under the transformation? If we think of f as representing, say, temperature, it is clear that we do not change the value of f at a given point merely by changing coordinate systems. However, the three numbers describing the point's coordinates *do* change; so the *formula* for f, or its functional form, will be different. Let us consider an example before attempting the general formulation.

Example 5.7 Suppose $f(x,y,z) = y^2 - x^2$ in the old coordinate system. The linear orthogonal transformation of example 5.6 is performed, with $\theta = \pi/6$. What is the value of f at the point whose coordinates in the new system are (1,1,0)?

Solution Before we can apply the formula $y^2 - x^2$ for f, we must find x and y; that is, the *original* coordinates of (1,1,0). Using the transformation of eq. (5.103) with J given by eq. (5.104) and $\theta = \pi/6$, we find

$$x = x' \cos\theta - y' \sin\theta = 1\left(\frac{\sqrt{3}}{2}\right) - 1\left(\frac{1}{2}\right) = \frac{\sqrt{3} - 1}{2}$$

$$y = x' \sin\theta + y' \cos\theta = 1\left(\frac{1}{2}\right) + 1\left(\frac{\sqrt{3}}{2}\right) = \frac{\sqrt{3} + 1}{2}$$

Hence $y^2 - x^2 = \sqrt{3}$, the value of f.

Notice that the value of f could not be computed by taking $f(x',y',z') = y'^2 - x'^2 = 1 - 1 = 0$. First we have to express the original coordinates in terms of the new, and then apply the formula for f to the original coordinates. In other words, to express a function $f(x,y,z)$ in a new coordinate system, $f'(x',y',z')$, we first express the old coordinates in terms of the new:

$$\begin{aligned} x &= x(x',y',z') \\ y &= y(x',y',z') \\ z &= z(x',y',z') \end{aligned} \tag{5.111}$$

[as in eq. (5.102)], then plug into the original functional form for f:

$$f'(x',y',z') = f(x(x',y',z'),y(x',y',z'),z(x',y',z')) \tag{5.112}$$

In particular, the transformed function should not be written as $f(x',y',z')$, as this is literally incorrect. (However, this notation is often used for abbreviation when there is no possibility of misinterpretation.)

In example 5.7, the function $f(x,y,z) = y^2 - x^2$ is computed in the new system by

$$\begin{aligned} f'(x',y',z') &= (x' \sin\theta + y' \cos\theta)^2 - (x' \cos\theta - y' \sin\theta)^2 \\ &= (y'^2 - x'^2) \cos 2\theta + 2x'y' \sin 2\theta \end{aligned} \tag{5.113}$$

The transformation of vector fields is doubly complicated; one must get new *components* in terms of old through the transformation of eq. (5.110), but *since the components are functions of position,* the rule of eq. (5.112) must be employed on each component. Thus, first we use eq. (5.112) to express the old (**i**,**j**,**k**) components of **V** in terms of the new (x',y',z') coordinates of the point in space:

$$\begin{aligned} \mathbf{V} = {} & V_1(x(x',y',z'),y(x',y',z'),z(x',y',z'))\mathbf{i} \\ & + V_2(x(x' \cdots))\mathbf{j} + V_3(x(x' \cdots))\mathbf{k} \end{aligned}$$

Then we use J^{-1} to get the new components of **V**. To summarize, we state the rule for transforming a vector field: *if* **V** *is given in the old system by*

$$\mathbf{V} = V_1(x,y,z)\mathbf{i} + V_2(x,y,z)\mathbf{j} + V_3(x,y,z)\mathbf{k} \tag{5.114}$$

and in the new system by

$$\mathbf{V} = V_1'(x',y',z')\mathbf{i}' + V_2'(x',y',z')\mathbf{j}' + V_3'(x',y',z')\mathbf{k}' \tag{5.115}$$

then the components are related by

$$\begin{bmatrix} V_1'(x',y',z') \\ V_2'(x',y',z') \\ V_3'(x',y',z') \end{bmatrix} = J^{-1} \begin{bmatrix} V_1(x(x',y',z'),y(x',y',z'),z(x',y',z')) \\ V_2(x(x',y',z'),y(x',y',z'),z(x',y',z')) \\ V_3(x(x',y',z'),y(x',y',z'),z(x',y',z')) \end{bmatrix} \tag{5.116}$$

Example 5.8 Express the vector field

$$\mathbf{V} = \mathbf{i} + (yz)\mathbf{j} + (x^2 + y^2)\mathbf{k} \tag{5.117}$$

in the new coordinate system of example 5.6.

Solution We use the matrix J in eq. (5.104) to transform coordinates via eq. (5.103), and we use its transpose for J^{-1}; hence

$$\begin{bmatrix} V_1'(x',y',z') \\ V_2'(x',y',z') \\ V_3'(x',y',z') \end{bmatrix} = \begin{bmatrix} \cos\theta & \sin\theta & 0 \\ -\sin\theta & \cos\theta & 0 \\ 0 & 0 & 1 \end{bmatrix} \begin{bmatrix} 1 \\ (x'\sin\theta + y'\cos\theta)z' \\ (x'\cos\theta - y'\sin\theta)^2 + (x'\sin\theta + y'\cos\theta)^2 \end{bmatrix}$$

$$= \begin{bmatrix} \cos\theta + x'z'\sin^2\theta + y'z'\sin\theta\cos\theta \\ -\sin\theta + x'z'\sin\theta\cos\theta + y'z'\cos^2\theta \\ x'^2 + y'^2 \end{bmatrix}$$

and

$$\begin{aligned} \mathbf{V} = {} & (\cos\theta + x'z'\sin^2\theta + y'z'\sin\theta\cos\theta)\mathbf{i}' \\ & + (-\sin\theta + x'z'\sin\theta\cos\theta + y'z'\cos^2\theta)\mathbf{j}' \\ & + (x'^2 + y'^2)\mathbf{k}' \end{aligned} \tag{5.118}$$

Now we turn to the question of what happens to the vector operators **grad,** div, and **curl** when we transform coordinates. As a preliminary experiment, consider the following example.

Example 5.9 Let $f(x,y,z) = y^2 - x^2$. Apply the linear orthogonal transformation of example 5.6. Compare the expression

$$\frac{\partial f}{\partial x}\mathbf{i} + \frac{\partial f}{\partial y}\mathbf{j} + \frac{\partial f}{\partial z}\mathbf{k} \tag{5.119}$$

with the analogous expression in the new coordinates

$$\frac{\partial f'}{\partial x'}\mathbf{i}' + \frac{\partial f'}{\partial y'}\mathbf{j}' + \frac{\partial f'}{\partial z'}\mathbf{k}' \tag{5.120}$$

Solution The transformed function f' was computed in eq. (5.99), and eq. (5.106) becomes

$$(-2x' \cos 2\theta + 2y' \sin 2\theta)\mathbf{i}' + (2x' \sin 2\theta + 2y' \cos 2\theta)\mathbf{j}'$$

Now let us compute eq. (5.119) and then transform the resulting vector field to the new system. First,

$$\frac{\partial f}{\partial x}\mathbf{i} + \frac{\partial f}{\partial y}\mathbf{j} + \frac{\partial f}{\partial z}\mathbf{k} = -2x\mathbf{i} + 2y\mathbf{j}$$

Applying eq. (5.116),

$$J^{-1}\begin{bmatrix} -2x(x',y',z') \\ 2y(x',y',z') \\ 0 \end{bmatrix} = \begin{bmatrix} \cos\theta & \sin\theta & 0 \\ -\sin\theta & \cos\theta & 0 \\ 0 & 0 & 1 \end{bmatrix}\begin{bmatrix} -2x'\cos\theta + 2y'\sin\theta \\ 2x'\sin\theta + 2y'\cos\theta \\ 0 \end{bmatrix}$$
$$= \begin{bmatrix} -2x'\cos 2\theta + 2y'\sin 2\theta \\ 2x'\sin 2\theta + 2y'\cos 2\theta \\ 0 \end{bmatrix}$$

Hence eq. (5.119) transforms into eq. (5.120) and they describe the same vector field.

Of course, both expressions in the last example represent ∇f, computed in different coordinate systems. The reason they resulted in the same vector field is that **grad** f can be characterized without reference to any coordinate system; it is a vector pointing in the direction of maximum rate of change of f with respect to distance, and having length equal to this maximum rate of change. The same argument that showed **grad** f could be computed by eq. (5.119) in the old coordinate system also shows it can be computed by eq. (5.120) in the new coordinate system.

Similarly, the divergence of a vector field can be defined in a coordinate-free manner, thanks to the divergence theorem (section 5.1). If we compute div $\mathbf{V}$ as the flux per unit volume out of a box with sides perpendicular to the x,y,z axes, we get

$$\frac{\partial V_1}{\partial x} + \frac{\partial V_2}{\partial y} + \frac{\partial V_3}{\partial z} \tag{5.121}$$

If we use a box with sides perpendicular to the x',y',z' axes, we get

$$\frac{\partial V_1'}{\partial x'} + \frac{\partial V_2'}{\partial y'} + \frac{\partial V_3'}{\partial z'} \tag{5.122}$$

Since they both represent div $\mathbf{V}$, eqs. (5.121) and (5.122) are the same scalar field, expressed in different coordinates.

Example 5.10 Verify this statement for **V** as in example 5.8.

Solution Applying eq. (5.121) to eq. (5.117), we obtain $\nabla \cdot \mathbf{V} = z$. Applying eq. (5.122) to eq. (5.118), we obtain $z' \sin^2 \theta + z' \cos^2 \theta = z'$. Since $z' = z$ for this transformation, these are the same.

Of course, we can say the same about **curl V.** Because of Stokes' theorem, the component of **curl V** in any direction is the "swirl" of **V** in that direction, a concept defined without reference to a coordinate system (recall section 5.5). Hence the vector field

$$\begin{vmatrix} \mathbf{i} & \mathbf{j} & \mathbf{k} \\ \dfrac{\partial}{\partial x} & \dfrac{\partial}{\partial y} & \dfrac{\partial}{\partial z} \\ V_1 & V_2 & V_3 \end{vmatrix} \tag{5.123}$$

transforms to the vector field

$$\begin{vmatrix} \mathbf{i}' & \mathbf{j}' & \mathbf{k}' \\ \dfrac{\partial}{\partial x'} & \dfrac{\partial}{\partial y'} & \dfrac{\partial}{\partial z'} \\ V_1' & V_2' & V_3' \end{vmatrix} \tag{5.124}$$

and they both equal **curl V.** Exercise 7 asks the reader to verify this for the vector field of eq. (5.117).

The following example reveals another interpretation of the matrix J in eq. (5.103).

Example 5.11 Verify that eqs. (5.119) and (5.120) describe the same vector field for a *general* linear orthogonal transformation.

Solution The functional forms f and f' are related by eq. (5.112); hence

$$\begin{aligned} \frac{\partial f'}{\partial x'} &= \frac{\partial f(x(x',y',z'),y(x',y',z'),z(x',y',z'))}{\partial x'} \\ &= \frac{\partial f}{\partial x}\frac{\partial x}{\partial x'} + \frac{\partial f}{\partial y}\frac{\partial y}{\partial x'} + \frac{\partial f}{\partial z}\frac{\partial z}{\partial x'} \end{aligned} \tag{5.125}$$

by the chain rule. Similar equations hold for $\partial f'/\partial y'$ and $\partial f'/\partial z'$.

From the eqs. (5.102), we find

$$\frac{\partial x}{\partial x'} = \mathbf{i}' \cdot \mathbf{i} \qquad \frac{\partial x}{\partial y'} = \mathbf{j}' \cdot \mathbf{i} \qquad \frac{\partial x}{\partial z'} = \mathbf{k}' \cdot \mathbf{i} \qquad \frac{\partial y}{\partial x'} = \mathbf{i}' \cdot \mathbf{j} \qquad (etc.)$$

In other words, the partial derivatives of the old coordinates with respect to the new coordinates are the entries of the matrix J:

$$J = \begin{bmatrix} \dfrac{\partial x}{\partial x'} & \dfrac{\partial x}{\partial y'} & \dfrac{\partial x}{\partial z'} \\ \dfrac{\partial y}{\partial x'} & \dfrac{\partial y}{\partial y'} & \dfrac{\partial y}{\partial z'} \\ \dfrac{\partial z}{\partial x'} & \dfrac{\partial z}{\partial y'} & \dfrac{\partial z}{\partial z'} \end{bmatrix} \tag{5.126}$$

Hence we can see that eq. (5.125), and the corresponding y' and z' equations, can be used to express the vector eq. (5.120) by way of the following matrix product:

$$\begin{bmatrix} \dfrac{\partial f'}{\partial x'} & \dfrac{\partial f'}{\partial y'} & \dfrac{\partial f'}{\partial z'} \end{bmatrix} = \begin{bmatrix} \dfrac{\partial f}{\partial x} & \dfrac{\partial f}{\partial y} & \dfrac{\partial f}{\partial z} \end{bmatrix} J$$

Taking transposes and remembering that J is orthogonal, we find

$$\begin{bmatrix} \dfrac{\partial f'}{\partial x'} \\ \dfrac{\partial f'}{\partial y'} \\ \dfrac{\partial f'}{\partial z'} \end{bmatrix} = J^{-1} \begin{bmatrix} \dfrac{\partial f}{\partial x} \\ \dfrac{\partial f}{\partial y} \\ \dfrac{\partial f}{\partial z} \end{bmatrix} \tag{5.127}$$

But the right-hand side of eq. (5.127) is precisely what we obtain by applying the transformation rule of eq. (5.116) to the vector of eq. (5.119)! Hence eqs. (5.119) and (5.120) describe the same vector field.

The analogous verifications of the "invariance" of the formulas for divergence and curl are left to the exercises.

The fact that J is the matrix of partial derivatives is reflected by the following, alternative notation:

$$J = \frac{\partial(x,y,z)}{\partial(x',y',z')}$$

The terminology "Jacobian matrix" is often used when one refers to the matrix of partial derivatives; hence the symbol J.

Finally, we would like to extend the analysis of this section to a slightly more general type of transformation. Suppose that the origin of the new coordinate system does not coincide with the origin of the first; it is located at, say, the point with coordinates (x_0, y_0, z_0) in the old system (see fig. 5.11). Then if a point P has old

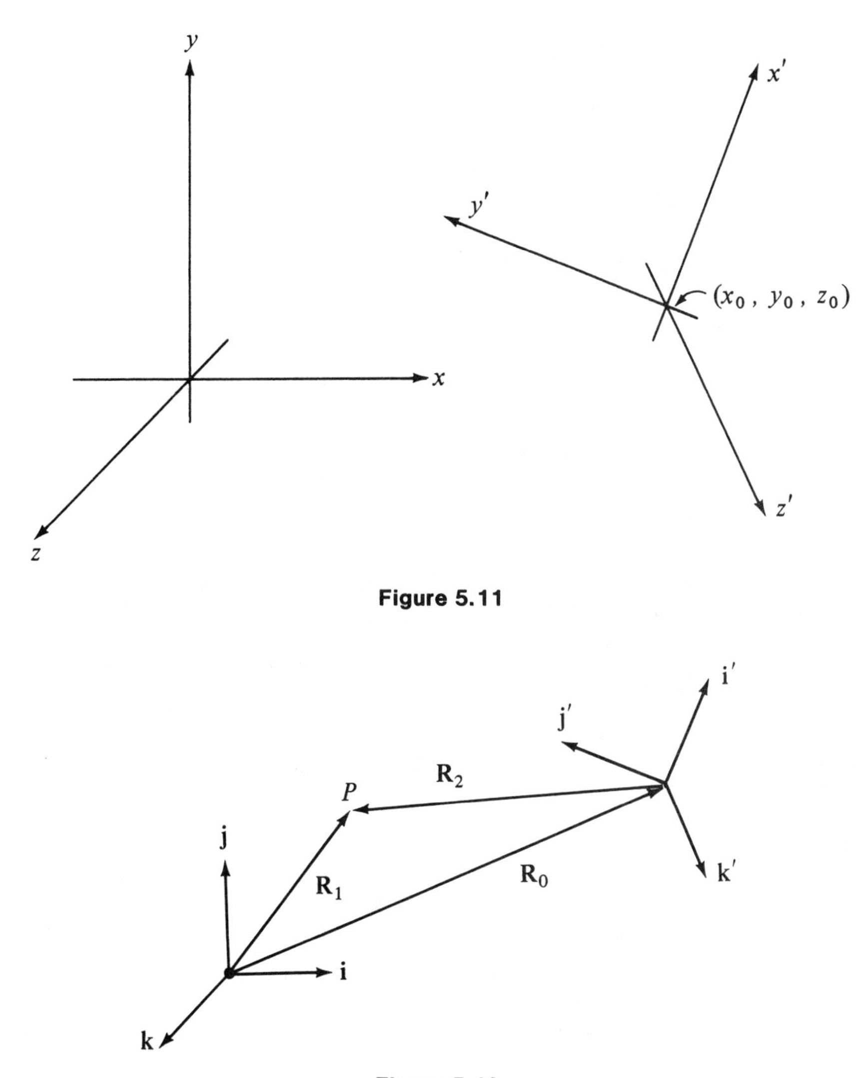

Figure 5.11

Figure 5.12

coordinates (x,y,z) and new coordinates (x',y',z'), the vectors $\mathbf{R}_1$ and $\mathbf{R}_2$ in figure 5.12 are related by

$$\mathbf{R}_1 = \mathbf{R}_0 + \mathbf{R}_2$$

$$x\mathbf{i} + y\mathbf{j} + z\mathbf{k} = x_0\mathbf{i} + y_0\mathbf{j} + z_0\mathbf{k} + x'\mathbf{i}' + y'\mathbf{j}' + z'\mathbf{k}' \qquad (5.128)$$

Taking scalar products of (5.128) with **i, j,** and **k** in turn leads to

$$\begin{bmatrix} x \\ y \\ z \end{bmatrix} = \begin{bmatrix} x_0 \\ y_0 \\ z_0 \end{bmatrix} + J \begin{bmatrix} x' \\ y' \\ z' \end{bmatrix} \tag{5.129}$$

with J defined as before, from eq. (5.103). Thus coordinates of a point do not transform by a linear orthogonal transformation in this case; eq. (5.129) describes an *affine transformation.*

However, vectors still transform according to the old rules of eqs. (5.109) and (5.110) because they are free; shifting the origin makes no difference to them.

Moreover, the vector operators **grad,** div, and **curl** retain their "invariant" formulas, again because they can be defined without reference to a coordinate system. Even the direct proofs of their invariance, as exemplified by example 5.11, can be carried out unchanged because the matrix of eq. (5.103) is still the matrix of partial derivatives. The reader should review example 5.11 to be sure this point is understood.

Consequently, most of the analysis we have developed for transformations between cartesian coordinate systems having the same origin holds for transformations between cartesian coordinate systems located at different origins. This includes the orthogonality of the matrix J, the rules of eq. (5.110) for transforming vectors, the rules of eqs. (5.112) and (5.116) for transforming scalar and vector fields, and the expressions of eqs. (5.120), (5.122), and (5.124) for **grad,** div, and **curl.** The only difference is the appearance of the nonhomogeneous "shifts" in eq. (5.129) relating point coordinates.

EXERCISES

1. Show that eq. (5.107) can be derived by taking scalar products of eq. (5.101) with **i**′, **j**′, and **k**′ in turn.
2. (a) Derive the matrix for the transformation generated by rotating the x,y,z system about the x axis through an angle ϕ.
   (b) Repeat part (a) if the x,y,z system is rotated about the y axis through an angle ψ.
3. Verify that the transpose of J in eq. (5.104) equals its inverse.
4. If $\mathbf{V} = 3\mathbf{i} + 4\mathbf{j} + \mathbf{k}$ and $\mathbf{W} = 2\mathbf{i} - \mathbf{j} - \mathbf{k}$, compute **V** and **W** in the new coordinate system of example 5.6. Verify that the scalar product $\mathbf{V} \cdot \mathbf{W}$ remains the same.

5. Consider the scalar and vector fields

$$f(x,y,z) = xyz$$
$$\mathbf{V}(x,y,z) = xz\mathbf{i} + \mathbf{j} + xyz\mathbf{k}$$

If the coordinate transformation of example 5.6 is performed with $\theta = \pi/6$, express the following fields in the new coordinate system:
(a) the scalar field f.
(b) the vector field $\mathbf{V}$.
(c) $\mathbf{grad}\, f$.
(d) div $\mathbf{V}$.
(e) $\mathbf{curl}\ \mathbf{V}$.

6. Repeat exercise 5 for the fields

$$f = x^2 + y^2$$
$$\mathbf{V} = x\mathbf{i} + y\mathbf{j} + z\mathbf{k}$$

Interpret the results.

7. Verify that for the vector field in eq. (5.117), the computation of **curl V** via eq. (5.123) in the old system leads to the same vector field as the computation of **curl V** via eq. (5.124) in the new system, for the transformation of example 5.6.

8. Modeling example 5.11, give a direct proof of the "invariance" under a general linear orthogonal transformation of eq. (5.102), of
(a) the divergence of an arbitrary vector field.
(b) the curl of an arbitrary vector field.

9. Repeat exercise 8 for the linear orthogonal transformation plus shift of eq. (5.129).

10. Is the laplacian of a scalar "invariant" under the transformations considered in this section?

11. What is the element of arc length, $ds = (dx^2 + dy^2 + dz^2)^{1/2}$, in the new coordinate system?

12. What is the volume element, $dV = dx\,dy\,dz$, in the new coordinate system?

13. Suppose a new x', y', z' coordinate system is related to the original system by a linear orthogonal transformation described by a matrix J, as in eq. (5.103), and a "still newer" x'', y'', z'' coordinate system is related to the "new" system by a linear orthogonal transformation generated by a matrix K. Show that the x'', y'', z'' system is related to the original x, y, z system by a linear orthogonal transformation, with the associated matrix given by JK. Prove directly that the product of two orthogonal matrices is orthogonal. What interpretation can you give to the columns of JK?

14. Compute the matrix J associated with the following sequences of operations, taking the x, y, z axes into the corresponding x' y', z' axes:
(a) First rotate through an angle $\pi/4$ about the z axis.
(b) Then rotate through an angle $\pi/2$ about the "current" y axis.
(c) Finally rotate through an angle $(-\pi/4)$ about the "current" x axis.

15. Show directly that scalar products are preserved under the general linear orthogonal transformation of eq. (5.110); that is, show that $V_1W_1 + V_2W_2 + V_3W_3 = V_1' W_1' + V_2' W_2' + V_3' W_3'$ when the components of **V** and **W** are related by eq. (5.110). This provides verification of the (obvious) fact that lengths and angles are preserved under these transformations.

16. Show that under the linear orthogonal transformation of eq. (5.102) there is a straight line through the origin, all of whose points have the same coordinates before and after the transformation. That is, for all points on this line, $x = x'$, $y = y'$, and $z = z'$. [*Hint:* If **R** is the position vector of a point on the line, then $\mathbf{R} \cdot \mathbf{i} = \mathbf{R} \cdot \mathbf{i}'$; hence $\mathbf{R} \cdot (\mathbf{i} - \mathbf{i}') = 0$. Similarly, $\mathbf{R} \cdot (\mathbf{j} - \mathbf{j}') = \mathbf{R} \cdot (\mathbf{k} - \mathbf{k}') = 0$. If $\mathbf{i} = \mathbf{i}'$ or $\mathbf{j} = \mathbf{j}'$, there is nothing to prove, so try $\mathbf{R} = (\mathbf{i} - \mathbf{i}') \times (\mathbf{j} - \mathbf{j}')$.]

17. Based on the last two exercises, can you derive *Euler's theorem:* every transformation of the form of eq. (5.102) can be described as a rotation of the coordinate system about some straight line through the origin?

18. Euler's theorem (see exercise 17) implies that the sequence of operations in exercise 14 is equivalent to a single rotation about some straight line. Find the line, and the angles of rotation.

19. As a partial converse to the theory developed in this section, suppose we *begin* with a transformation of coordinates defined by

$$\begin{bmatrix} x \\ y \\ z \end{bmatrix} = J \begin{bmatrix} x' \\ y' \\ z' \end{bmatrix} \tag{5.130}$$

and we know only that J is an orthogonal matrix.

(a) By examining the points with new coordinates (1,0,0), (0,1,0), and (0,0,1) in turn, show that the columns of J, interpreted as vectors expressed in the old system, point along the new x', y', and z' axes.

(b) Exploit the orthogonality of J to prove that the new axes are mutually orthogonal and that $(x'^2 + y'^2 + z'^2)^{1/2}$ equals the distance of (x',y',z') from the origin.

(c) From (a) and (b) we may conclude that the new system is a bona fide cartesian coordinate system. However, it may be left-handed, as the simple example

$$\begin{bmatrix} x \\ y \\ z \end{bmatrix} = \begin{bmatrix} 1 & 0 & 0 \\ 0 & 1 & 0 \\ 0 & 0 & -1 \end{bmatrix} \begin{bmatrix} x' \\ y' \\ z' \end{bmatrix}$$

shows. What modifications of rules in eqs. (5.110), (5.112), (5.116), (5.120), (5.122), and (5.124) have to be made in transforming to a left-handed system?

(d) How can you determine, from the matrix J, whether or not the new system is right-handed?

20. Generalize exercise 18: given an arbitrary orthogonal matrix J that generates a transformation via eq. (5.130), and given that the new system is right-handed, show how to compute the angle and axis of rotation in terms of J.

21. Derive the matrix for the transformation generated by rotating the x, y, z system through an angle $\pi/2$ about the straight line through the origin parallel to $\mathbf{i} + \mathbf{j} + \mathbf{k}$.

22. Generalize the previous exercise: derive the matrix for a rotation of the x, y, z system through an angle θ about a straight line through the origin parallel to $\mathbf{n}$.

23. (*Significance of the Jacobian*) Obviously the transformation equations (3.69) for orthogonal coordinates are, in general, nonlinear. However, the relation

$$d\mathbf{R} = \frac{\partial \mathbf{R}}{\partial u_1} du_1 + \frac{\partial \mathbf{R}}{\partial u_2} du_2 + \frac{\partial \mathbf{R}}{\partial u_3} du_3 \tag{5.131}$$

between the differentials can be viewed as a "local linearization" of eq. (3.69).

(a) Show that eq. (5.131) can be expressed

$$\begin{bmatrix} dx \\ dy \\ dz \end{bmatrix} = \begin{bmatrix} \frac{\partial x}{\partial u_1} & \frac{\partial x}{\partial u_2} & \frac{\partial x}{\partial u_3} \\ \frac{\partial y}{\partial u_1} & \frac{\partial y}{\partial u_2} & \frac{\partial y}{\partial u_3} \\ \frac{\partial z}{\partial u_1} & \frac{\partial z}{\partial u_2} & \frac{\partial z}{\partial u_3} \end{bmatrix} \begin{bmatrix} du_1 \\ du_2 \\ du_3 \end{bmatrix} \tag{5.132}$$

Recall that the matrix of partial derivatives in eq. (5.132) was identified in this section as the Jacobian of the transformation in eq. (3.69). It was abbreviated

$$J = \frac{\partial(x,y,z)}{\partial(u_1,u_2,u_3)}$$

(b) Show that the chain rule implies that the inverse of the Jacobian is

$$J^{-1} = \frac{\partial(u_1,u_2,u_3)}{\partial(x,y,z)} = \begin{bmatrix} \frac{\partial u_1}{\partial x} & \frac{\partial u_1}{\partial y} & \frac{\partial u_1}{\partial z} \\ \frac{\partial u_2}{\partial x} & \frac{\partial u_2}{\partial y} & \frac{\partial u_2}{\partial z} \\ \frac{\partial u_3}{\partial x} & \frac{\partial u_3}{\partial y} & \frac{\partial u_3}{\partial z} \end{bmatrix} \tag{5.133}$$

(c) Show that the requirement of (u_1,u_2,u_3) forming orthogonal curvilinear coordinates forces the rows of J to be orthogonal. Nonetheless, J is not an orthogonal matrix. Why?

(d) Show that the determinant of J is $h_1h_2h_3$, the factor appearing in the volume element of eq. (3.81). This prompts the mnemonic

$$dx\,dy\,dz = \left| \frac{\partial(x,y,z)}{\partial(u_1,u_2,u_3)} \right| du_1\,du_2\,du_3$$

APPENDIX A

Historical Notes

It is not really possible to appreciate the history of vector analysis without knowing something of the history of mathematics in general, and this is too broad a topic for us to discuss here. We shall confine our remarks to certain specific topics, and let the interested reader pursue the subject further elsewhere.

The word "vector" comes from a Latin word meaning "to carry" and is still sometimes used to mean "that which carries." For example, one says "the mosquito is the vector of malaria." The word entered mathematics via astronomy, where it was originally used with a somewhat different meaning. The notion of vector addition was arrived at independently by Möbius and others in the early part of the nineteenth century, thus giving rise to vector *algebra.* Vector *analysis* is somewhat more recent. For example, the notion of *curl* apparently was introduced by J. C. Maxwell in his *Treatise on Electricity and Magnetism* (1873).

The notation used in this book is essentially due to J. Willard Gibbs, whose book on vector analysis was printed privately in the early 1880s, and Oliver Heaviside, whose book *Electromagnetic Theory* (1893) makes hilarious reading because of its jibes at mathematicians.

One of the most interesting events in the history of vector analysis is the controversy that once existed between exponents of vector analysis and a few other mathematicians who felt that *quaternions* were more suitable for solving problems in physics. Before proceeding, let us briefly discuss the algebra of quaternions. Quaternions are formally similar to complex numbers, so let us first consider the background of the idea of a complex number.

As long ago as 1545, a mathematician (Cardan) "solved" a problem in algebra that has no real solutions. The problem is to find two numbers whose sum is 10 and whose product is 40. Cardan gave a formal solution, involving the square root of a negative number, and verified by substitution that these "fictitious numbers" have the required properties. As early as 1629, Girard suggested that such "impossible solutions" should be considered for three reasons: (*i*) one can give a general rule for finding roots of certain equations, (*ii*) these solutions supply the lack of other solutions, and (*iii*) they may in any event have their own usefulness. In 1673, Wallis pointed out that numbers such as $\sqrt{-1}$ should be just as legitimate in mathematics as negative numbers. One cannot have $\sqrt{-1}$ eggs in a basket, but then neither can one have -7 eggs in a basket. Wallis came very close to giving the usual geometrical interpretation of complex numbers. It remained for a Norwegian surveyor named Wessel to do this in 1797. (Argand did it independently in 1806, which is why the term "Argand diagram" is used. Wessel published his work in an obscure journal and did not receive credit during his lifetime.)

It was not until 1831 that Gauss put complex numbers on a respectable basis. Since some readers of this book may have learned complex numbers from a viewpoint that predates 1831, let us briefly review complex numbers. A complex number is an ordered pair (x,y) of real numbers. They are added and multiplied by (real) scalars as if they were row matrices. However, the product $(x_1,y_1)(x_2,y_2)$ of two row matrices is not defined in matrix algebra. We define the product of two complex numbers according to

$$(x_1,y_1)(x_2,y_2) = (x_1x_2 - y_1y_2, x_1y_2 + x_2y_1)$$

If we identify (1,0) with the real number 1, and let i denote (0,1), then

$$(x,y) = (x,0) + (0,y) = x(1,0) + y(0,1) = x + yi$$

which is the usual notation for a complex number (except for electrical engineers, who use j instead of i). Moreover, we have $i^2 = (0,1)(0,1) = (-1,0) = -1$, so it is now possible to square a number and obtain a negative number.

Now we recall that multiplication by $\cos\theta + i\sin\theta$ has the effect of rotating a complex number through an angle θ. Hence rotations in a plane can be obtained by identifying the plane with the Argand diagram, and the rotation with the operation of multiplying by $\cos\theta + i\sin\theta$. This suggested to W. R. Hamilton that rotations in space might be similarly obtained, if there were some way to multiply *triples* of numbers to obtain a system of "hypercomplex" numbers that would provide a three-dimensional analog of the complex number system. Apparently this problem troubled him for a period of fifteen years. This is not too surprising when one considers that, up to the time of Hamilton, it was generally assumed that the commutative law $xy = yx$ was a necessary condition for the consistency of the rules of algebra. Hamilton is credited with the realization that this is not the case; actually, Gauss had the same idea earlier but did not publish his work.

The realization that a noncommutative algebra is needed is still not enough. Hamilton was still trying to do the impossible. It was proved later, by Frobenius in

1878, that it is impossible to multiply ordered triples in such a manner that the resulting algebraic system will have all of the properties Hamilton desired. Evidently Hamilton suspected this himself. It was on a famous day, October 16, 1843, when he was out walking with his wife, that Hamilton, in a great flash of insight, conceived of the quaternions. It is said that he carved the fundamental formulas of this new algebra in the stone of Brougham Bridge, on which he happened to be at the moment. He immediately recognized the importance of his discovery (some might say invention) and devoted the remainder of his life to quaternions.

At first glance, a quaternion looks like a cross between a complex number and a vector. The usual form for writing a quaternion is

$$x = x_0 + x_1 i + x_2 j + x_3 k$$

Quaternions are added and multiplied by real numbers in the obvious manner. The product of two quaternions is defined by formally multiplying them out according to the usual rules of algebra (except that we must be careful to preserve the order), and then simplifying the resulting expression by using the following rules:

$$\begin{array}{lll} i^2 = -1 & j^2 = -1 & k^2 = -1 \\ ij = k & jk = i & ki = j \\ ji = -k & kj = -i & ik = -j \end{array}$$

An example will illustrate the procedure. If, for instance, $x = 3 - i + 2j + k$ and $y = 3j - 2k$, we have

$$\begin{aligned} xy &= (3 - i + 2j + k)(3j - 2k) \\ &= 9j - 6k - 3ij + 2ik + 6j^2 - 4jk + 3kj - 2k^2 \\ &= 9j - 6k - 3k - 2j - 6 - 4i - 3i + 2 \\ &= -4 - 7i + 7j - 9k \end{aligned}$$

It can be shown that the quaternions constitute a division algebra. That is, to each quaternion $x \neq 0$, there is a quaternion x^{-1} such that $xx^{-1} = x^{-1}x = 1$. We shall not digress to show how an inverse is computed. It is important to note, however, that we cannot write y/x, since this would be ambiguous. We must write either $x^{-1}y$ or yx^{-1} and, since multiplication of quaternions is not commutative, these two expressions may not be equal.

The *real part* of a quaternion $x_0 + x_1 i + x_2 j + x_3 k$ is the number x_0. If the real part of a quaternion is zero, the quaternion is called a *pure quaternion.* In applying quaternions to problems in physics or geometry, pure quaternions are identified with ordinary vectors in three-dimensional space, as the notation suggests.

If x and y are pure quaternions, the real part of xy turns out to be the negative of the scalar product $x \cdot y$ (computed by the usual formula), and the pure quaternionic part represents the vector product $x \times y$. Thus it is possible to do with quaternions many of the things one ordinarily does in vector analysis by using scalar and vector products.

Although the quaternions represent a four-dimensional division algebra, rather than a three-dimensional one, it turned out that they fulfilled the needs envisaged by Hamilton. It is possible to represent rotations by the use of quaternions, although not so simply as one might have wished, and in general there is a certain awkwardness in the use of quaternions. After working for ten years, Hamilton published his *Lectures on Quaternions* (1853); his *Elements of Quaternions* appeared in 1866, the year after his death. Incidentally, the earliest use of the word "vector" (in the mathematical sense), according to the Oxford dictionary, is in this work.

Hamilton had one devoted disciple, P. G. Tait, who mastered all the tricks of quaternions, and devoted himself to the cause of convincing one and all that quaternions were the ultimate tool for geometers and physicists. There were others who disagreed.

At about the same time Hamilton made his remarkable discovery, H. G. Grassmann published a work called *Ausdehnungslehre,* or *Theory of Extension.* In this remarkable book, both matrix theory and tensor algebra are developed implicitly, but because Grassmann filled the book with philosophical abstractions, and because of its difficulty, the book was essentially ignored by mathematicians. A second edition was published in 1862, but the work was not much appreciated until the twentieth century.

The vector analysis of Gibbs and Heaviside, and the various generalities in this chapter, are more closely related to *Ausdehnungslehre* than to anything Hamilton did. Grassmann introduced various types of "products" of vectors, set things up for Gibbs to invent dyadics, and discussed linear transformations in general. The notion of a linear associative algebra was developed by Benjamin Peirce in the 1860s. The only other name we shall mention is that of Cayley, who was eminent for (among other things) conceiving of n-dimensional space (as did Grassmann) and who published *Memoir on the Theory of Matrices* in 1858.

A delightful controversy took place between Gibbs and Tait concerning the merits of the use of quaternions in solving problems in geometry and physics. There is a certain beauty and mathematical elegance in the quaternions, but they are not very well adapted to practical use. Tait viewed vector analysis as a "hermaphroditic monster" and did not hesitate to express this view in print. The replies of Gibbs can be found in his collected works, available in any library, and they are both entertaining and instructive to read. By the beginning of the twentieth century, vector analysis was well established, and it was amply demonstrated that Hamilton and Tait were overly optimistic in their thought that quaternions would be as revolutionary to mathematics as was the invention of calculus. The revolutionary idea contributed by Hamilton was simply that it is possible to have a self-consistent algebra in which multiplication is not commutative.

EXERCISES

1. Show that if u and v are pure quaternions,

$$uv = -\mathbf{u}\cdot\mathbf{v} + \mathbf{u}\times\mathbf{v}$$

2. Show that if the vectors $\mathbf{u}$ and $\mathbf{v}$ are identified with pure quaternions,

$$\mathbf{u}\cdot\mathbf{v} = -\frac{uv + vu}{2}$$

and

$$\mathbf{u}\times\mathbf{v} = \frac{uv - vu}{2}$$

3. (a) Let $\mathbf{n}$ denote a unit vector that is perpendicular to a plane P. Thinking of P as a mirror, show that the reflected image of a vector $\mathbf{v}$ in the mirror is given by

$$\mathbf{v}' = \mathbf{v} - 2(\mathbf{v}\cdot\mathbf{n})\mathbf{n}$$

(b) Show that this can be written in quaternionic form as

$$v' = nvn$$

4. Let P and P' be two planes intersecting in a line L and let $\theta/2$ be the angle between the two planes. Choose unit normals $\mathbf{n}$ and $\mathbf{n}'$, respectively, so that the angle between $\mathbf{n}$ and $\mathbf{n}'$ is $\theta/2$ and let $\mathbf{u}$ be a unit vector along L so chosen that $\mathbf{n}\times\mathbf{n}' = \sin(\theta/2)\mathbf{u}$.
   (a) Letting $\mathbf{v}'$ denote the reflected image of $\mathbf{v}$ in P and $\mathbf{v}''$ denote the reflected image of $\mathbf{v}'$ in P', show that $\mathbf{v}''$ is the vector obtained by rotating $\mathbf{v}$ through an angle θ about the axis L. (The positive sense of rotation is related to the direction of $\mathbf{u}$ by the right-hand rule.)
   (b) Derive the quaternionic relation

$$v'' = n'nvnn' \tag{A.1}$$

   (c) Writing (A.1) in the form $v'' = (-n'n)v(-nn')$, derive the relation

$$v'' = (\cos\tfrac{1}{2}\theta + \sin\tfrac{1}{2}\theta\, u)v(\cos\tfrac{1}{2}\theta - \sin\tfrac{1}{2}\theta\, u) \tag{A.2}$$

5. If z is a complex number, the exponential e^z is defined by the infinite series

$$e^z = \sum_{n=0}^{\infty}\frac{z^n}{n!}$$

Using the same expression to define e^z when z is a quaternion, let $z = \phi u$, where u is a pure quaternion representing a unit vector and ϕ is an angle, and derive

$$e^{\phi u} = -\cos\phi + \sin\phi\, u$$

6. Rewrite (A.2) in exponential notation. (This is the formula for rotations that Hamilton was seeking when he developed the algebra of quaternions.)

APPENDIX B

Two Theorems of Advanced Calculus

In this appendix we prove two theorems of advanced calculus that are important to vector analysis.

THEOREM B.1 *Let $f(x,y,z)$ be a scalar function possessing continuous first partial derivatives $\partial f/\partial x$, $\partial f/\partial y$, $\partial f/\partial z$ in some domain D. Also let $\mathbf{h}$ be a unit vector with components (h_1,h_2,h_3). Then the directional derivative of f in the direction $\mathbf{h}$ exists in D and is given by*

$$\frac{df}{ds} = \frac{\partial f}{\partial x}h_1 + \frac{\partial f}{\partial y}h_2 + \frac{\partial f}{\partial z}h_3$$

Proof The directional derivative, if it exists, is given by the limit of

$$\frac{f(x + sh_1,\ y + sh_2,\ z + sh_3) - f(x,y,z)}{s} \tag{B.1}$$

as s approaches zero.

By adding and subtracting equal terms, we rewrite (B.1) as

$$\frac{f(x + sh_1, y + sh_2, z + sh_3) - f(x + sh_1, y + sh_2, z)}{s} + \frac{f(x + sh_1, y + sh_2, z) - f(x + sh_1, y, z)}{s} + \frac{f(x + sh_1, y, z) - f(x,y,z)}{s} \tag{B.2}$$

Each of the terms of (B.2) involves differences of values of f when *only one* coordinate is changed. Hence we can use the powerful tools of the ordinary calculus of one variable; in particular, the mean value theorem applies to each term (since the partial derivatives are continuous). For the first term in (B.2), we conclude that there is a number α between 0 and 1 such that

$$f(x + sh_1, y + sh_2, z + sh_3) - f(x + sh_1, y + sh_2, z) = sh_3\frac{\partial f}{\partial z}(x + sh_1, y + sh_2, z + \alpha sh_3)$$

Analyzing the other terms similarly, we find numbers β and γ also between 0 and 1 such that the expression (B.1) is equal to

$$\frac{sh_3\dfrac{\partial f}{\partial z}(x + sh_1, y + sh_2, z + \alpha sh_3)}{s} + \frac{sh_2\dfrac{\partial f}{\partial y}(x + sh_1, y + \beta sh_2, z)}{s} + \frac{sh_1\dfrac{\partial f}{\partial x}(x + \gamma sh_1, y, z)}{s} \tag{B.3}$$

Now we let s approach zero. The numbers α, β, and γ are always between 0 and 1, and since the partials are continuous, we conclude that the limit exists and is given by

$$\frac{\partial f}{\partial x}h_1 + \frac{\partial f}{\partial y}h_2 + \frac{\partial f}{\partial z}h_3$$

In theorem B.2 the notation is rather confusing. We advise the reader that the following symbols mean the same thing:

$$\partial^2 f/\partial y\, \partial x = \frac{\partial^2 f}{\partial y\, \partial x} = \frac{\partial}{\partial y}\left(\frac{\partial f}{\partial x}\right)$$

THEOREM B.2 *Let $f(x, y)$ be a scalar function possessing continuous first partial derivatives $\partial f/\partial x$ and $\partial f/\partial y$ in some domain D. Furthermore, let the second derivative $\partial^2 f/\partial y\, \partial x$ exist and be continuous in D. Then the second derivative $\partial^2 f/\partial x\, \partial y$ also exists in D and*

$$\frac{\partial^2 f}{\partial x\, \partial y} = \frac{\partial^2 f}{\partial y\, \partial x}$$

Proof The second derivative $\partial^2 f/\partial x\, \partial y$, if it exists, is the limit as s goes to zero of

$$\frac{\dfrac{\partial f}{\partial y}(x + s, y) - \dfrac{\partial f}{\partial y}(x, y)}{s} \tag{B.4}$$

On the other hand, the y derivatives in (B.4) can also be expressed as limits; (B.4) is equal to

$$\frac{1}{s}\left(\lim_{t\to 0} \frac{f(x + s, y + t) - f(x + s, y)}{t} - \lim_{t\to 0} \frac{f(x, y + t) - f(x, y)}{t}\right) \tag{B.5}$$

Since limits and sums are interchangeable, we can write (B.5) as

$$\lim_{t\to 0} \frac{[f(x + s, y + t) - f(x + s, y)] - [f(x, y + t) - f(x, y)]}{st} \tag{B.6}$$

The numerator in (B.6) can be regarded as the difference between the values of a function $F(u)$ evaluated at $u = x + s$ and $u = s$; this complicated function $F(u)$ is defined by

$$F(u) = f(u, y + t) - f(u, y)$$

As a function of u, F has a continuous derivative given by

$$F'(u) = \frac{\partial f}{\partial u}(u, y + t) - \frac{\partial f}{\partial u}(u, y)$$

and is thus vulnerable to the mean value theorem of one-dimensional calculus. We conclude that there is a number α between 0 and 1 such that

$$F(x + s) - F(x) = sF'(x + \alpha s)$$

Thus the expression (B.6) is equal to

$$\lim_{t\to 0} \frac{s\left(\dfrac{\partial f}{\partial x}(x + \alpha s, y + t) - \dfrac{\partial f}{\partial x}(x + \alpha s, y)\right)}{st} \tag{B.7}$$

We can now apply the mean value theorem to the function $\partial f/\partial x$ in (B.7), since only the second argument is changing; by hypothesis, $\partial f/\partial x$ has a continuous derivative with respect to its second argument, namely, $\partial^2 f/\partial y\, \partial x$! We conclude that there is a number β between 0 and 1 such that expression (B.7) is equal to

$$\lim_{t\to 0} \frac{st\dfrac{\partial^2 f}{\partial y\, \partial x}(x + \alpha s, y + \beta t)}{st} \tag{B.8}$$

Since $\partial^2 f/\partial y\, \partial x$ is continuous, the expression (B.8) equals

$$\frac{\partial^2 f}{\partial y\, \partial x}(x + \alpha s, y) \tag{B.9}$$

We have shown that (B.4) equals (B.9). Now taking limits as s goes to zero, again invoking the continuity of $\partial^2 f/\partial y\, \partial x$, we see that (B.4) does indeed have a limit, and that it is given by

$$\frac{\partial^2 f}{\partial y\, \partial x}(x, y)$$

APPENDIX C

The Vector Equations of Classical Mechanics

C.1 Mechanics of Particles and Systems of Particles

The basic equation of classical mechanics is expressed by Newton's second law: *force equals mass times acceleration.* More explicitly, if a particle, or a "point mass," is located at the position $\mathbf{R}$ in an inertial coordinate system, and if it has mass m, then its motion will be governed by the equation

$$\mathbf{F} = m\frac{d^2\mathbf{R}}{dt^2} = m\frac{d\mathbf{v}}{dt} = m\mathbf{a} \tag{C.1}$$

Here $\mathbf{F}$ is the *vector* sum of all the forces acting on the particle, and $\mathbf{v}$ and $\mathbf{a}$ are velocity and acceleration, respectively (t, of course, is time). All of the equations in this section will be derived from (C.1).

The *momentum* **p** of the particle is defined by

$$\mathbf{p} = m\mathbf{v} = m\frac{d\mathbf{R}}{dt} \tag{C.2}$$

Its *angular momentum* $\boldsymbol{\ell}$ is defined by

$$\boldsymbol{\ell} = \mathbf{R} \times \mathbf{p} = m\mathbf{R} \times \mathbf{v} \tag{C.3}$$

The *torque* **T** of a force **F** exerted at the point **R** is defined by

$$\mathbf{T} = \mathbf{R} \times \mathbf{F} \tag{C.4}$$

Combining these equations, we find that *the torque on a particle equals its rate of change of angular momentum:*

$$\frac{d\boldsymbol{\ell}}{dt} = \frac{d}{dt}\left(\mathbf{R} \times m\frac{d\mathbf{R}}{dt}\right) = \frac{d\mathbf{R}}{dt} \times m\frac{d\mathbf{R}}{dt} + \mathbf{R} \times m\frac{d^2\mathbf{R}}{dt^2} = \mathbf{R} \times \mathbf{F} = \mathbf{T} \tag{C.5}$$

The *kinetic energy* K of the particle is defined by

$$K = \frac{1}{2}m|\mathbf{v}|^2 = \frac{1}{2}m\left|\frac{d\mathbf{R}}{dt}\right|^2 = \frac{|\mathbf{p}|^2}{2m} \tag{C.6}$$

It obeys the equation

$$\frac{dK}{dt} = m\mathbf{v} \cdot \frac{d\mathbf{v}}{dt} = m\mathbf{a} \cdot \mathbf{v} = \mathbf{F} \cdot \mathbf{v} \tag{C.7}$$

The quantity $\mathbf{F} \cdot \mathbf{v}$ is called the *power* delivered by the force **F**, and it equals the rate of change of kinetic energy.

If a particular force acting on the particle can be expressed as a function of the position **R**, so that **F** is a vector field, it is natural to ask if this vector field is conservative (section 4.3). If so, we introduce the *potential energy* $V(\mathbf{R})$, satisfying

$$\mathbf{F}^{(\text{cons})}(\mathbf{R}) = -\nabla V(\mathbf{R}) \tag{C.8}$$

[Notice that a minus sign is incorporated in the definition so that $V(\mathbf{R}) = -\phi(\mathbf{R})$, as in section 4.3.] The total force on the particle is then expressed as the sum of all the conservative and the nonconservative forces:

$$\begin{aligned}
\mathbf{F} &= \sum_{s} \mathbf{F}_{s}^{(\text{cons})} + \mathbf{F}^{(\text{noncons})} \\
&= -\sum_{s} \nabla V_s(\mathbf{R}) + \mathbf{F}^{(\text{noncons})} \\
&= -\nabla V(\mathbf{R}) + \mathbf{F}^{(\text{noncons})}
\end{aligned}$$

where $V = \Sigma_s V_s$ is the total potential energy. Observe that the power generated by the conservative forces can be written as the time derivative of $-V(\mathbf{R})$ taken along the trajectory $\mathbf{R}(t)$:

$$\begin{aligned}\sum_s \mathbf{F}_s^{(\text{cons})} \cdot \mathbf{v} &= -\nabla V(\mathbf{R}) \cdot \frac{d\mathbf{R}}{dt} \\ &= -\frac{\partial V}{\partial x}\frac{dx}{dt} - \frac{\partial V}{\partial y}\frac{dy}{dt} - \frac{\partial V}{\partial z}\frac{dz}{dt} \\ &= -\frac{d}{dt}V(\mathbf{R})\end{aligned}$$

Consequently, eq. (C.7) can be expressed

$$\frac{dK}{dt} = -\frac{dV(\mathbf{R}(t))}{dt} + \mathbf{F}^{(\text{noncons})} \cdot \mathbf{v}$$

or introducing the *total energy* $K + V = E$, we find

$$\frac{dE}{dt} = \mathbf{F}^{(\text{noncons})} \cdot \mathbf{v} \tag{C.9}$$

In the important case when all of the forces are conservative, we see that the total energy of the particle is constant along the trajectory.

Now let us derive some of the basic laws governing the motion of a *system* of N particles located at the positions $\mathbf{R}_\alpha$ and having masses m_α ($\alpha = 1, 2, \ldots, N$); the total force on the αth particle will be denoted $\mathbf{F}_\alpha$. We introduce the total mass $M = \Sigma_\alpha m_\alpha$, the total momentum $\mathbf{P} = \Sigma_\alpha \mathbf{P}_\alpha$, the total angular momentum $\mathbf{L} = \Sigma_\alpha \boldsymbol{\ell}_\alpha$, and the *position vector* $\mathbf{R}_{\text{cm}}$ *of the center of mass,* defined by

$$\mathbf{R}_{\text{cm}} = \frac{\Sigma_\alpha m_\alpha \mathbf{R}_\alpha}{M} \tag{C.10}$$

Several useful theorems emerge from these definitions.

THEOREM C.1 *The total momentum* $\mathbf{P}$ *of the system equals the momentum of a single particle of mass* M *moving with the center of mass.*

Proof

$$M\frac{d\mathbf{R}_{\text{cm}}}{dt} = \sum_\alpha m_\alpha \frac{d\mathbf{R}_\alpha}{dt} = \mathbf{P} \tag{C.11}$$

THEOREM C.2 *The motion of the center of mass* $\mathbf{R}_{\text{cm}}$ *is the same as that of a single particle of mass* M *subjected to all the forces on all the particles simultaneously.*

Proof

$$M\frac{d^2\mathbf{R}_{cm}}{dt^2} = \sum_\alpha m_\alpha \frac{d^2\mathbf{R}_\alpha}{dt^2} = \sum_\alpha \mathbf{F}_\alpha$$

It is sometimes convenient to refer to certain vector quantities (position, velocity) *measured relative to the center of mass,* or measured "in the center of mass system." The center of mass system (hereafter known as the "CM system") is a moving coordinate system whose origin is located at $\mathbf{R}_{cm}$ and whose axes remain parallel to the corresponding axes of the inertial system. Clearly, the position vector of the αth particle in the CM system is $\mathbf{R}_\alpha - \mathbf{R}_{cm}$; and its velocity in the CM system, *defined as the derivative of this vector,* is the difference of the inertial velocities:

$$\frac{d}{dt}(\mathbf{R}_\alpha - \mathbf{R}_{cm}) = \frac{d\mathbf{R}_\alpha}{dt} - \frac{d\mathbf{R}_{cm}}{dt} = \mathbf{v}_\alpha - \frac{d\mathbf{R}_{cm}}{dt}$$

Its angular momentum is

$$\boldsymbol{\ell}_\alpha^{cm} = (\mathbf{R}_\alpha - \mathbf{R}_{cm}) \times m_\alpha\left(\mathbf{v}_\alpha - \frac{d\mathbf{R}_{cm}}{dt}\right)$$

THEOREM C.3 *The total angular momentum of the system equals the sum of the angular momenta of the individual particles in the CM system, plus the angular momentum of a single particle of mass M moving with the center of mass.*

Proof We have to show that

$$\mathbf{L} = \sum_\alpha \mathbf{R}_\alpha \times m_\alpha \frac{d\mathbf{R}_\alpha}{dt} = \sum_\alpha (\mathbf{R}_\alpha - \mathbf{R}_{cm}) \times m_\alpha\left(\frac{d\mathbf{R}_\alpha}{dt} - \frac{d\mathbf{R}_{cm}}{dt}\right) + \mathbf{R}_{cm} \times M\frac{d\mathbf{R}_{cm}}{dt} \tag{C.12}$$

The *sum* on the right equals

$$\sum_\alpha \mathbf{R}_\alpha \times m_\alpha \frac{d\mathbf{R}_\alpha}{dt} - \mathbf{R}_{cm} \times \left(\sum_\alpha m_\alpha \frac{d\mathbf{R}_\alpha}{dt}\right) - \left(\sum_\alpha m_\alpha \mathbf{R}_\alpha\right) \times \frac{d\mathbf{R}_{cm}}{dt} + \mathbf{R}_{cm} \times \sum_\alpha m_\alpha \frac{d\mathbf{R}_{cm}}{dt}$$

Using eq. (C.10), we express this as

$$\mathbf{L} - \mathbf{R}_{cm} \times M\frac{d\mathbf{R}_{cm}}{dt} - M\mathbf{R}_{cm} \times \frac{d\mathbf{R}_{cm}}{dt} + \mathbf{R}_{cm} \times M\frac{d\mathbf{R}_{cm}}{dt}$$
$$= \mathbf{L} - \mathbf{R}_{cm} \times M\frac{d\mathbf{R}_{cm}}{dt}$$

and inserting this back into eq. (C.12) verifies the identity.

We invite the reader to supply the proof of the following, similar result.

THEOREM C.4 *The total kinetic energy of the system equals the sum of the kinetic energies of the individual particles in the CM system, plus the kinetic energy of a single particle of mass M moving with the center of mass:*

$$\sum_\alpha \frac{1}{2} m_\alpha |\mathbf{v}_\alpha|^2 = \sum_\alpha \frac{1}{2} m_\alpha \left| \mathbf{v}_\alpha - \frac{d\mathbf{R}_{cm}}{dt} \right|^2 + \frac{1}{2} M \left| \frac{d\mathbf{R}_{cm}}{dt} \right|^2$$

Usually it is convenient to separate the forces on the αth particle into two categories: *internal* and *external* forces. The internal forces are produced by interaction with the other particles, while the external forces are produced outside the system. The advantages of this separation accrue from the following facts:

(*i*) Most internal forces between particles are *two-particle interactions,* so that if the force on the αth particle due to the βth particle is $\mathbf{F}_\alpha^{(\beta)}$, the total internal force on the αth particle is

$$\mathbf{F}_\alpha^{\text{int}} = \sum_{\beta \neq \alpha} \mathbf{F}_\alpha^{(\beta)}$$

(*ii*) Most two-particle interactions obey Newton's third law (NTL)—*every action is accompanied by an equal and opposite reaction*—which is interpreted to mean that

$$\mathbf{F}_\alpha^{(\beta)} = -\mathbf{F}_\beta^{(\alpha)} \tag{C.13}$$

and that both $\mathbf{F}_\alpha^{(\beta)}$ and $\mathbf{F}_\beta^{(\alpha)}$ are directed along the line connecting particles α and β:

$$(\mathbf{R}_\beta - \mathbf{R}_\alpha) \times \mathbf{F}_\alpha^{(\beta)} = (\mathbf{R}_\beta - \mathbf{R}_\alpha) \times \mathbf{F}_\beta^{(\alpha)} = \mathbf{0} \tag{C.14}$$

Systems having these properties will be said to "satisfy NTL." (Observe that if a system has some internal forces that violate NTL, one can use the artifice of categorizing these as *external* forces and interpreting the theorems accordingly.)

Clearly, if a system satisfies NTL, then one can omit the internal forces from the sum in theorem C.2, because they cancel pairwise. A more profound result is the following.

THEOREM C.5 *If a system satisfies NTL, then the rate of change of its total angular momentum measured in the CM system equals the sum of the torques of the external forces, measured relative to the center of mass.*

Proof The theorem states that

$$\begin{aligned} \frac{d\mathbf{L}^{\text{cm}}}{dt} &\equiv \frac{d}{dt} \sum_\alpha (\mathbf{R}_\alpha - \mathbf{R}_{\text{cm}}) \times m_\alpha \left(\mathbf{v}_\alpha - \frac{d\mathbf{R}_{\text{cm}}}{dt} \right) \\ &= \sum_\alpha (\mathbf{R}_\alpha - \mathbf{R}_{\text{cm}}) \times \mathbf{F}_\alpha^{\text{ext}} \end{aligned} \tag{C.15}$$

where the force on the αth particle satisfies

$$\mathbf{F}_\alpha = \mathbf{F}_\alpha^{\text{ext}} + \mathbf{F}_\alpha^{\text{int}} = \mathbf{F}_\alpha^{\text{ext}} + \sum_{\beta \neq \alpha} \mathbf{F}_\alpha^{(\beta)}$$

To derive this, we sum eq. (C.5) over all particles:

$$\frac{d\mathbf{L}}{dt} = \frac{d}{dt}\sum_\alpha \mathbf{R}_\alpha \times m_\alpha \frac{d\mathbf{R}_\alpha}{dt} = \sum_\alpha \mathbf{R}_\alpha \times \mathbf{F}_\alpha \tag{C.16}$$

By eq. (C.12), the left-hand side can be written

$$\frac{d\mathbf{L}}{dt} = \frac{d\mathbf{L}^{\text{cm}}}{dt} + \frac{d}{dt}\left(\mathbf{R}_{\text{cm}} \times M\frac{d\mathbf{R}_{\text{cm}}}{dt}\right) = \frac{d\mathbf{L}^{\text{cm}}}{dt} + \mathbf{R}_{\text{cm}} \times M\frac{d^2\mathbf{R}_{\text{cm}}}{dt^2}$$

and theorem C.2 for NTL systems reduces this to

$$\frac{d\mathbf{L}}{dt} = \frac{d\mathbf{L}^{\text{cm}}}{dt} + \mathbf{R}_{\text{cm}} \times \sum_\alpha \mathbf{F}_\alpha^{\text{ext}} \tag{C.17}$$

The right-hand side of (C.16) can also be reduced:

$$\sum_\alpha \mathbf{R}_\alpha \times \mathbf{F}_\alpha = \sum_\alpha \mathbf{R}_\alpha \times (\mathbf{F}_\alpha^{\text{ext}} + \mathbf{F}_\alpha^{\text{int}}) = \sum_\alpha \mathbf{R}_\alpha \times \mathbf{F}_\alpha^{\text{ext}} + \sum_\alpha \sum_{\beta \neq \alpha} \mathbf{R}_\alpha \times \mathbf{F}_\alpha^{(\beta)}$$

The terms in the second sum cancel pairwise by eqs. (C.13) and (C.14):

$$\mathbf{R}_\alpha \times \mathbf{F}_\alpha^{(\beta)} + \mathbf{R}_\beta \times \mathbf{F}_\beta^{(\alpha)} = (\mathbf{R}_\alpha - \mathbf{R}_\beta) \times \mathbf{F}_\alpha^{(\beta)} = \mathbf{0}$$

while the other sum splits into

$$\sum_\alpha (\mathbf{R}_\alpha - \mathbf{R}_{\text{cm}}) \times \mathbf{F}_\alpha^{\text{ext}} + \mathbf{R}_{\text{cm}} \times \sum_\alpha \mathbf{F}_\alpha^{\text{ext}} \tag{C.18}$$

Equating eqs. (C.17) and (C.18) and dropping the common term, we arrive at eq. (C.15).

Theorems C.2 and C.5 are very useful for NTL systems with many particles and complicated internal forces. They frequently permit a general overall description of the motion (position of CM, and angular momentum about CM) in terms of the external forces only. As we shall see, these data provide a complete characterization of the motion of rigid bodies.

The concept of potential energy can be applied to *systems* as well as to single-particle motions. If any of the forces on the αth particle are functions of $\mathbf{R}_\alpha$ only, and if the field $\mathbf{F}_\alpha(\mathbf{R}_\alpha)$ is conservative, we introduce the corresponding potential energy $V_\alpha(\mathbf{R}_\alpha)$ as before:

$$\mathbf{F}_\alpha^{(\text{cons})}(\mathbf{R}_\alpha) = -\nabla^{(\alpha)} V_\alpha(\mathbf{R}_\alpha)$$

with, obviously,

$$\nabla^{(\alpha)} = \mathbf{i}\frac{\partial}{\partial x_\alpha} + \mathbf{j}\frac{\partial}{\partial y_\alpha} + \mathbf{k}\frac{\partial}{\partial z_\alpha}$$

The two-particle interaction forces, however, generally depend on the coordinates of *both* particles:

$$\mathbf{F}_1^{(2)} = \mathbf{F}_1^{(2)}(\mathbf{R}_1,\mathbf{R}_2)$$

and even worse situations are imaginable.

To get a convenient theory for systems, therefore, *we shall classify as conservative those forces* $\mathbf{F}_\alpha$ *that are derivable from a single potential energy function* $V(\mathbf{R}_1, \mathbf{R}_2, \ldots, \mathbf{R}_N)$ *via*

$$\mathbf{F}_\alpha^{(\text{cons})} = -\nabla^{(\alpha)}V(\mathbf{R}_1, \mathbf{R}_2, \ldots, \mathbf{R}_N) \qquad (\alpha = 1, 2, \ldots, N) \qquad \text{(C.19)}$$

Here one regards $\mathbf{R}_\beta$ as constant when we apply $\nabla^{(\alpha)}$, if $\beta \neq \alpha$. Fortunately, many physical forces are conservative in the sense of eq. (C.19).

The classification of conservative and nonconservative system forces leads to a generalization of eq. (C.9). If E denotes the total system energy, defined by

$$E = \sum_\alpha \frac{1}{2}m_\alpha|\mathbf{v}_\alpha|^2 + V(\mathbf{R}_1, \mathbf{R}_2, \ldots, \mathbf{R}_N)$$

then *the rate of change of energy along the system trajectory equals the power generated by the nonconservative forces:*

$$\begin{aligned}\frac{dE}{dt} &= \sum_\alpha (\mathbf{F}_\alpha^{(\text{cons})} + \mathbf{F}_\alpha^{(\text{noncons})}) \cdot \frac{d\mathbf{R}_\alpha}{dt} + \sum_\alpha \nabla^{(\alpha)}V \cdot \frac{d\mathbf{R}_\alpha}{dt} \\ &= \sum_\alpha \mathbf{F}_\alpha^{(\text{noncons})} \cdot \frac{d\mathbf{R}_\alpha}{dt} \qquad \text{(C.20)}\end{aligned}$$

C.2 Mechanics of Rigid Bodies

A *rigid body* is a system of particles whose internal forces are so strong that they hold the interparticle distances fixed:

$$|\mathbf{R}_\alpha - \mathbf{R}_\beta| = \textit{constant} \qquad \text{(C.21)}$$

Rigid bodies are theoretical idealizations, but their motions provide very accurate descriptions of the mechanics of real physical solids. The conditions of rigidity eq. (C.21) can be enforced, in theory, by forces obeying eqs. (C.13) and (C.14), so we will treat rigid bodies as NTL systems.

Observe that eq. (C.21) implies that the *angles* between the interparticle vectors $(\mathbf{R}_\alpha - \mathbf{R}_\beta)$ are also fixed, because the angles can be expressed in terms of the lengths via the law of cosines. Hence, also, the dot products stay constant.

Furthermore, *the center of mass of a rigid body stays fixed in the body.* To see that $\mathbf{R}_{cm}$ stays a constant distance from, say, $\mathbf{R}_5$, we express $|\mathbf{R}_{cm} - \mathbf{R}_5|$ in terms of these fixed dot products:

$$|\mathbf{R}_{cm} - \mathbf{R}_5|^2 = \left|\frac{\Sigma_\alpha m_\alpha \mathbf{R}_\alpha}{M} - \mathbf{R}_5\right|^2 = \left|\frac{\Sigma_\alpha m_\alpha \mathbf{R}_\alpha}{M} - \frac{\Sigma_\alpha m_\alpha \mathbf{R}_5}{M}\right|^2$$

$$= \left|\frac{\Sigma_\alpha m_\alpha(\mathbf{R}_\alpha - \mathbf{R}_5)}{M}\right|^2 = \frac{\Sigma_\alpha \Sigma_\beta\, m_\alpha(\mathbf{R}_\alpha - \mathbf{R}_5)\cdot m_\beta(\mathbf{R}_\beta - \mathbf{R}_5)}{M^2}$$

Hence the distance from $\mathbf{R}_{cm}$ to each of the particles remains constant, and $\mathbf{R}_{cm}$ is fixed in the body.

Let us first consider the motion of a rigid body that moves *with one point held in place.* In this case it is convenient to use an inertial coordinate system with its origin at the stationary point. Also, we define an auxiliary coordinate system with the same origin, *but with its axes fixed in the body.* The position of the body is then completely specified by determining the position of the body-fixed axes.

In the subsequent developments we shall draw freely upon the techniques discussed in sections 5.7 and 5.8, and the corresponding exercises.

If we let $\mathbf{i}$, $\mathbf{j}$, and $\mathbf{k}$ be unit vectors along the inertial axes and $\mathbf{i}^b$, $\mathbf{j}^b$, and $\mathbf{k}^b$ be unit vectors along the body-fixed axes, then any vector $\mathbf{h}$ has the two representations

$$\mathbf{h} = h_1\mathbf{i} + h_2\mathbf{j} + h_3\mathbf{k}$$
$$= h_1^b\mathbf{i}^b + h_2^b\mathbf{j}^b + h_3^b\mathbf{k}^b$$

where the h_i, h_j^b are related by an *orthogonal transformation* expressed through the orthogonal 3-by-3 matrix O:

$$\begin{bmatrix} h_1 \\ h_2 \\ h_3 \end{bmatrix} = O \begin{bmatrix} h_1^b \\ h_2^b \\ h_3^b \end{bmatrix} \tag{C.22}$$

If $\mathbf{h}$ changes with time, then we have the two expressions for its derivative

$$\frac{d\mathbf{h}}{dt} = \frac{dh_1}{dt}\mathbf{i} + \frac{dh_2}{dt}\mathbf{j} + \frac{dh_3}{dt}\mathbf{k}$$

$$= \frac{dh_1^b}{dt}\mathbf{i}^b + h_1^b\frac{d\mathbf{i}^b}{dt} + \frac{dh_2^b}{dt}\mathbf{j}^b + h_2^b\frac{d\mathbf{j}^b}{dt} + \frac{dh_3^b}{dt}\mathbf{k}^b + h_3^b\frac{d\mathbf{k}^b}{dt}$$

(Of course, the inertial axes stay fixed.)

Now suppose an observer *turns with the body* and computes the rate of change of **h**. This person is unaware that the body axes are turning, and the only way to detect a change in **h** is *if its body-fixed components change.* Hence the observer computes

$$\left(\frac{d\mathbf{h}}{dt}\right)^b = \frac{dh_1^b}{dt}\mathbf{i}^b + \frac{dh_2^b}{dt}\mathbf{j}^b + \frac{dh_3^b}{dt}\mathbf{k}^b \tag{C.23}$$

What is the relation between $d\mathbf{h}/dt$ and $(d\mathbf{h}/dt)^b$? To answer this question, we turn to eq. (C.22). The components of $d\mathbf{h}/dt$ in the inertial system satisfy

$$\frac{d}{dt}\begin{bmatrix} h_1 \\ h_2 \\ h_3 \end{bmatrix} = O\frac{d}{dt}\begin{bmatrix} h_1^b \\ h_2^b \\ h_3^b \end{bmatrix} + \frac{dO}{dt}\begin{bmatrix} h_1^b \\ h_2^b \\ h_3^b \end{bmatrix} \tag{C.24}$$

(Here dO/dt has the obvious interpretation as the 3-by-3 matrix whose elements are the derivatives of the corresponding elements of O.) The first term on the right gives the components, in the inertial system, of the vector $(d\mathbf{h}/dt)^b$ in eq. (C.23). To unravel the meaning of the second term, we first rewrite it as

$$\frac{dO}{dt}O^TO\begin{bmatrix} h_1^b \\ h_2^b \\ h_3^b \end{bmatrix} \tag{C.25}$$

keeping in mind that $O^TO = I = OO^T$. Differentiating this last relationship reveals that

$$\frac{dO}{dt}O^T + O\frac{dO^T}{dt} = 0 \tag{C.26}$$

but since

$$\left(\frac{dO}{dt}O^T\right)^T = O\frac{dO^T}{dt}$$

eq. (C.26) says that $(dO/dt)O^T$ is antisymmetric. Thus we can write it as

$$\frac{dO}{dt}O^T = \begin{bmatrix} 0 & -\omega_3 & \omega_2 \\ \omega_3 & 0 & -\omega_1 \\ -\omega_2 & \omega_1 & 0 \end{bmatrix} \tag{C.27}$$

Now insert eqs. (C.27) and (C.22) into eq. (C.25). According to exercise 18, section 5.7, the resulting expression gives the inertial components of $\boldsymbol{\omega} \times \mathbf{h}$, where

$$\boldsymbol{\omega} = \omega_1\mathbf{i} + \omega_2\mathbf{j} + \omega_3\mathbf{k}$$

Hence we understand eq. (C.24) to say

$$\frac{d\mathbf{h}}{dt} = \left(\frac{d\mathbf{h}}{dt}\right)^b + \boldsymbol{\omega} \times \mathbf{h} \tag{C.28}$$

To determine the meaning of the vector $\boldsymbol{\omega}$, we let $\mathbf{h}$ be the vector from the origin to the αth particle: $\mathbf{h} = \mathbf{R}_\alpha$. Then $\mathbf{h}$ is fixed in the body, so we have

$$\frac{d\mathbf{R}_\alpha}{dt} = \mathbf{0} + \boldsymbol{\omega} \times \mathbf{R}_\alpha \tag{C.29}$$

But the left-hand side is the velocity, in the inertial system, of the particle at $\mathbf{R}_\alpha$. Hence, eq. (C.29) is the same as eq. (1.22), and we identify $\boldsymbol{\omega}$ as *the angular velocity of the body-fixed system with respect to the inertial system.* The result (C.28) is known as *Coriolis' law,* and the formula (C.27) tells how to compute the angular velocity vector $\boldsymbol{\omega}$ from the transformation matrix.

Let us use the angular velocity vector to express the angular momentum. We have

$$\begin{aligned}\mathbf{L} &= \sum_\alpha \mathbf{R}_\alpha \times m_\alpha \mathbf{v}_\alpha = \sum_\alpha \mathbf{R}_\alpha \times m_\alpha(\boldsymbol{\omega} \times \mathbf{R}_\alpha) \\ &= \sum_\alpha m_\alpha |\mathbf{R}_\alpha|^2 \boldsymbol{\omega} - \sum_\alpha m_\alpha \mathbf{R}_\alpha \, \mathbf{R}_\alpha \cdot \boldsymbol{\omega}\end{aligned}$$

by eq. (1.30). To study this, we must use tensor notation. If we let $R_i^{(\alpha)}$ denote the ith component of R_α, then the $\mathbf{i}$th component of $\mathbf{L}$ can be written

$$\begin{aligned}L_i &= \sum_\alpha m_\alpha R_l^{(\alpha)2} \omega_i - \sum_\alpha m_\alpha R_i^{(\alpha)} R_j^{(\alpha)} \omega_j \\ &= I_{ij}\omega_j \end{aligned} \tag{C.30}$$

(recall the summation convention.) Here the I_{ij} are the components of a tensor of rank two called the *inertia tensor:*

$$I_{ij} = \sum_\alpha m_\alpha (R_l^{(\alpha)2} \delta_{ij} - R_i^{(\alpha)} R_j^{(\alpha)})$$

Observe that I is symmetric: $I_{ij} = I_{ji}$.

Now that we have an expression for the angular momentum, we can write the equations of motion, derived by summing eq. (C.5) over all the particles: *the total torque equals the rate of change of the total angular momentum.* Hence, in tensor form,

$$T_i = \frac{d}{dt}(I_{ij}\omega_j)$$

Notice that since the inertial components of $\mathbf{R}^{(\alpha)}$ change as the body moves, I_{ij} is a function of time. However, if we work in the body-fixed coordinate system, all the $R_i^{(\alpha)}$ are fixed, and I_{ij} is constant! Hence, acknowledging Coriolis' law, we find that in the body-fixed system the equations of motion are

$$T_i = I_{ij}\frac{d\omega_j}{dt} + \epsilon_{ijk}\omega_j I_{kl}\omega_l \tag{C.31}$$

Furthermore, it is well known that because I is a symmetric tensor, there exist body-fixed axis systems wherein the off-diagonal components of I are zero; in such *principal axis systems,*

$$I = \begin{bmatrix} I_{11} & 0 & 0 \\ 0 & I_{22} & 0 \\ 0 & 0 & I_{33} \end{bmatrix}$$

Writing eq. (C.31) accordingly, we derive *Euler's equations of motion* for a body-fixed principal axis system:

$$\begin{aligned} T_1 &= I_{11}\frac{d\omega_1}{dt} - (I_{22} - I_{33})\omega_2\omega_3 \\ T_2 &= I_{22}\frac{d\omega_2}{dt} - (I_{33} - I_{11})\omega_3\omega_1 \\ T_3 &= I_{33}\frac{d\omega_3}{dt} - (I_{11} - I_{22})\omega_1\omega_2 \end{aligned} \tag{C.32}$$

In principle, one determines the motion by solving the Euler equations for $\boldsymbol{\omega}$ and then finding the transformation O from eq. (C.27). The latter task is made somewhat easier by choosing a convenient parametrization for O in terms of "Eulerian angles."

The Eulerian angles are discussed in the references below. Suffice it to say that the rotation described by O can be decomposed into a sequence of three successive rotations about the z, x, and z (again) body axes through the angles ϕ, θ, and ψ respectively; hence

$$O = \begin{bmatrix} \cos\phi & -\sin\phi & 0 \\ \sin\phi & \cos\phi & 0 \\ 0 & 0 & 1 \end{bmatrix} \begin{bmatrix} 1 & 0 & 0 \\ 0 & \cos\theta & -\sin\theta \\ 0 & \sin\theta & \cos\theta \end{bmatrix} \begin{bmatrix} \cos\psi & -\sin\psi & 0 \\ \sin\psi & \cos\psi & 0 \\ 0 & 0 & 1 \end{bmatrix}$$

$$= \begin{bmatrix} \cos\psi\cos\phi - \cos\theta\sin\phi\sin\psi & -\sin\psi\cos\phi - \cos\theta\sin\phi\cos\psi & \sin\theta\sin\phi \\ \cos\psi\sin\phi + \cos\theta\cos\phi\sin\psi & -\sin\psi\sin\phi + \cos\theta\cos\phi\cos\psi & -\sin\theta\cos\phi \\ \sin\theta\sin\psi & \sin\theta\cos\psi & \cos\theta \end{bmatrix}$$

and eq. (C.27) becomes, in the body-fixed system,

$$\omega_1 = \dot{\phi} \sin\theta \sin\psi + \dot{\theta} \cos\psi$$

$$\omega_2 = \dot{\phi} \sin\theta \cos\psi - \dot{\theta} \sin\psi$$

$$\omega_3 = \dot{\phi} \cos\theta + \dot{\psi} \qquad \text{(C.33)}$$

Once ω is determined from the Euler equations, the Eulerian angles ϕ, θ, and ψ are determined by solving eq. (C.33). Then O can be computed.

It should be mentioned that other conventions are sometimes used to define the Eulerian angles.

It is at times useful to express the kinetic energy of this motion in terms of the inertia tensor. We have

$$K = \frac{1}{2}\sum_{\alpha} m_\alpha |\mathbf{v}_\alpha|^2 = \frac{1}{2}\sum_{\alpha} m_\alpha |\omega \times \mathbf{R}_\alpha|^2$$

$$= \frac{1}{2}\sum_{\alpha} m_\alpha (\omega \times \mathbf{R}_\alpha) \cdot (\omega \times \mathbf{R}_\alpha)$$

$$= \frac{1}{2}\sum_{\alpha} m_\alpha \big[|\mathbf{R}_\alpha|^2 |\omega|^2 - (\mathbf{R}_\alpha \cdot \omega)^2 \big]$$

by eq. (1.33). In tensor notation we find

$$K = \frac{1}{2}\sum_{\alpha} m_\alpha (R_l^{(\alpha)2} \omega_j^2 - R_j^{(\alpha)} \omega_j R_i^{(\alpha)} \omega_i)$$

$$= \frac{1}{2}\omega_i \sum_{\alpha} m_\alpha (R_l^{(\alpha)2} \delta_{ij} - R_i^{(\alpha)} R_j^{(\alpha)}) \omega_j$$

$$= \frac{1}{2}\omega_i I_{ij} \omega_j$$

If we let $\omega = \omega\mathbf{n}$, where $\mathbf{n}$ is a unit vector, then $K = \frac{1}{2}(n_i I_{ij} n_j)\omega^2$. The quantity $n_i I_{ij} n_j$ is called the *moment of inertia* about the $\mathbf{n}$ direction. If $\mathbf{n} = \mathbf{k}^b$, we have $n_i = \delta_{i3}$ and the moment of inertia about the z axis becomes

$$\delta_{i3} I_{ij} \delta_{j3} = I_{33} = \sum_{\alpha} m_\alpha (R_l^{(\alpha)2} - R_3^{(\alpha)2}) = \sum_{\alpha} m_\alpha (x^{(\alpha)2} + y^{(\alpha)2})$$

— a formula familiar to most calculus students.

We now turn our attention to the general motion of the rigid body, dropping the assumption about the stationary point. Theorem C.2 allows us to write the equations of motion for the center of mass:

$$\sum_{\alpha} \mathbf{F}_{\alpha}^{\text{ext}} = M\frac{d^2\mathbf{R}_{\text{cm}}}{dt^2} \tag{C.34}$$

(invoking NTL for the internal forces). Also, theorem C.5 says that the external torque, measured about the center of mass, equals the rate of change of angular momentum measured in the CM system. However, since the center of mass is a body-fixed point, and relativizing all quantities to the center of mass is equivalent to treating the center of mass as stationary, the theory we just discussed can be used to analyze this motion. In particular, if I denotes the inertia tensor *with components computed in a body-fixed principal axis system with its origin at the center of mass,* the three Euler equations (C.32) describe the angular velocity of the body system with respect to the CM system. Hence the six scalar equations (C.34) and (C.32) determine the three components of $\mathbf{R}_{\text{cm}}$ and the three Eulerian angles, and the position of the rigid body can thus be completely specified.

EXERCISES

1. As a rule, the angular momentum $\mathbf{L}$ is *not* parallel to the angular velocity ω; but if ω is directed along the ith principal axis of I, $\mathbf{L} = I_{ii}\omega$ (not summed). Demonstrate this.
2. Prove: the moment of inertia I_0 in the direction $\mathbf{n}$ measured about the point $\mathbf{R}_0$ is related to the corresponding moment of inertia I_{cm} measured about $\mathbf{R}_{\text{cm}}$ by the formula

$$I_0 = I_{\text{cm}} + M|(\mathbf{R}_{\text{cm}} - \mathbf{R}_0) \times \mathbf{n}|^2$$

3. Prove theorem C.4.

References

1. GOLDSTEIN, H. *Classical Mechanics.* Reading, Mass.: Addison-Wesley, 1959.
2. NAGEL, E. *The Structure of Science.* New York: Harcourt, Brace, and World, 1961.

APPENDIX D

The Vector Equations of Electromagnetism

D.1 Electrostatics

It is well known that two electrically charged particles at rest will exert forces on each other. The forces vary inversely with the square of the distance separating them (the "inverse square law"), and directly with the charge on each particle. They are attractive or repulsive accordingly as they are of opposite or of equal polarity. This body of facts is known as "Coulomb's law," and it can be formulated vectorially as follows.

Let q_1 and q_2 denote the charge on each particle (signed according to polarity), and let $\mathbf{R}_1$ and $\mathbf{R}_2$ designate the respective position vectors. Then the force on particle 1 due to particle 2 equals

$$\mathbf{F}_1^{(2)} = \frac{kq_1q_2}{|\mathbf{R}_1 - \mathbf{R}_2|^2} \frac{(\mathbf{R}_1 - \mathbf{R}_2)}{|\mathbf{R}_1 - \mathbf{R}_2|} = kq_1q_2 \frac{\mathbf{R}_1 - \mathbf{R}_2}{|\mathbf{R}_1 - \mathbf{R}_2|^3}$$

where the (positive) constant k depends on the system of units. In SI units $k = 1/4\pi\epsilon_0$, where the *permittivity of free* space $\epsilon_0 = 8.854 \times 10^{-12}$ Farads per meter.

If there are N stationary particles with charges q_i, then the forces they exert on a "test charge" q located at $\mathbf{R}$ add vectorially:

$$\mathbf{F} = \sum_{i=1}^{N} kqq_i \frac{\mathbf{R} - \mathbf{R}_i}{|\mathbf{R} - \mathbf{R}_i|^3} \tag{D.1}$$

Historically, it has proved convenient to interpret eq. (D.1) as

$$\mathbf{F} = q\mathbf{E}(\mathbf{R}) \tag{D.2}$$

where the *electric field* $\mathbf{E}(\mathbf{R})$ is the force, per unit charge, that would be exerted by the charged particles 1 through N on a charged particle located at $\mathbf{R}$. Hence $\mathbf{E}(\mathbf{R})$ is a vector field given by

$$\mathbf{E}(\mathbf{R}) = k \sum_{i=1}^{N} q_i \frac{\mathbf{R} - \mathbf{R}_i}{|\mathbf{R} - \mathbf{R}_i|^3} \tag{D.3}$$

The electric field is irrotational for $\mathbf{R} \neq \mathbf{R}_i$, as the following computation shows:

$$\begin{aligned}
\nabla \times \mathbf{E} &= \sum kq_i \nabla \times \frac{(\mathbf{R} - \mathbf{R}_i)}{|\mathbf{R} - \mathbf{R}_i|^3} \\
&= \sum kq_i \left[\frac{\nabla \times (\mathbf{R} - \mathbf{R}_i)}{|\mathbf{R} - \mathbf{R}_i|^3} + \nabla \left(\frac{1}{|\mathbf{R} - \mathbf{R}_i|^3} \right) \times (\mathbf{R} - \mathbf{R}_i) \right] \\
&= \sum kq_i \left[\mathbf{0} - \frac{3(\mathbf{R} - \mathbf{R}_i) \times (\mathbf{R} - \mathbf{R}_i)}{|\mathbf{R} - \mathbf{R}_i|^5} \right] \\
&= \mathbf{0}
\end{aligned} \tag{D.4}$$

using identities (3.29), (3.32), and (3.33). In fact, $\mathbf{E}$ is the negative gradient of the *electrostatic potential* $V(\mathbf{r})$, given by

$$V(\mathbf{R}) = \sum kq_i / |\mathbf{R} - \mathbf{R}_i| \tag{D.5}$$

The equation

$$\mathbf{E} = -\nabla V \tag{D.6}$$

follows from identity (3.33).

The divergence of the electric field, for $\mathbf{R} \neq \mathbf{R}_i$, is computed by using eqs. (3.28), (3.31), and (3.33):

$$\nabla \cdot \mathbf{E} = k \sum q_i \left[\frac{3}{|\mathbf{R} - \mathbf{R}_i|^3} - \frac{3(\mathbf{R} - \mathbf{R}_i) \cdot (\mathbf{R} - \mathbf{R}_i)}{|\mathbf{R} - \mathbf{R}_i|^5} \right] = 0$$

However, $\nabla \cdot \mathbf{E}$ is undefined (infinite?) if the tip of $\mathbf{R}$ coincides with one of the point charges. To investigate this more fully, we consider the electric field due to a single

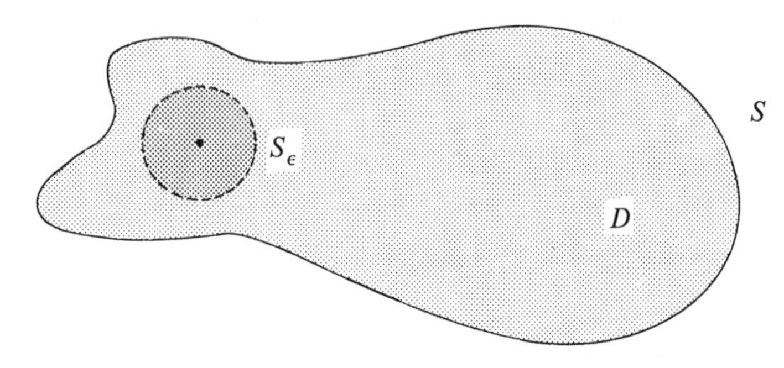

Figure D.1

point charge q_1 located at the origin $\mathbf{R}_1 = \mathbf{0}$, and compute its flux through a closed surface S:

$$\text{flux} = \iint_S \mathbf{E} \cdot d\mathbf{S} = kq_1 \iint \frac{\mathbf{R}}{R^3} \cdot d\mathbf{S} \tag{D.7}$$

(where, as usual, $R = |\mathbf{R}|$). If the domain D enclosed by S does not contain the origin, then $\nabla \cdot \mathbf{E} = 0$ throughout D and the net flux out of S is zero, by the divergence theorem. If the origin lies inside D, then we consider a sphere S_ϵ around the origin, whose radius ϵ is so small that the interior of S_ϵ lies inside D (see fig. D.1). Since $\nabla \cdot \mathbf{E} = 0$ in the intervening region bounded by S and S_ϵ, the flux out of S equals the flux of S_ϵ (compare section 5.2). Parametrizing S_ϵ by the spherical coordinates θ and ϕ (with $R = \epsilon = $ constant), we have

$$\begin{aligned} d\mathbf{S} &= \left(\frac{\partial \mathbf{R}}{\partial \phi} \times \frac{\partial \mathbf{R}}{\partial \theta}\right) d\phi\, d\theta \\ &= R^2 \sin\phi\, d\phi\, d\theta \frac{\mathbf{R}}{R} \end{aligned}$$

Thus the flux out of S_ϵ, and hence S, equals

$$kq_1 \int_0^{2\pi} \int_0^{\pi} \frac{\mathbf{R}}{R^3} \cdot \frac{\mathbf{R}}{R} R^2 \sin\phi\, d\phi\, d\theta = 4\pi k q_1$$

Shifting the charge q_1 to position $\mathbf{R}_1$, we conclude that

$$\iint_S \mathbf{E} \cdot d\mathbf{S} = \begin{cases} 4\pi k q_1 & \text{if } S \text{ encloses } \mathbf{R}_1 \\ 0 & \text{otherwise} \end{cases}$$

For N charges, we add all the individual contributions to the flux and derive *Gauss' law of electrostatics:*

$$\iint_S \mathbf{E} \cdot d\mathbf{S} = 4\pi k \sum (q_i \text{ or } 0) = 4\pi k (\text{total charge enclosed by } S) \tag{D.8}$$

[recall eq. (4.38)]. Here we are assuming that none of the charges q_i actually lies on the surface S.

Point charges do occur in nature as electrons, protons, and so forth, but in most macroscopic physical situations involving matter so many particles are present that it becomes necessary to use a continuum approximation, that is, to replace sums over point charges by integrals over charge densities. Thus, we introduce the *charge density function* $\rho(\mathbf{R})$ with units of charge per unit volume, so that the total charge q in a region D is given by

$$q = \iiint_D \rho(x,y,z)\, dx\, dy\, dz = \iiint \rho(\mathbf{R})\, dV \tag{D.9}$$

This leads us to consider the analogs of eqs. (D.3), (D.5), (D.6), and (D.7) for the continuous case:

$$\mathbf{E}(\mathbf{R}) = k \iiint \rho(\mathbf{R}') \frac{(\mathbf{R} - \mathbf{R}')}{|\mathbf{R} - \mathbf{R}'|^3}\, dV' \tag{D.10}$$

$$V(\mathbf{R}) = k \iiint \frac{\rho(\mathbf{R}')}{|\mathbf{R} - \mathbf{R}'|}\, dV' \tag{D.11}$$

$$\mathbf{E}(\mathbf{R}) = -\nabla V(\mathbf{R}) \tag{D.12}$$

$$\iint_S \mathbf{E} \cdot d\mathbf{S} = 4\pi k \iiint_{D \text{ (enclosed by } S)} \rho(\mathbf{R})\, dV \tag{D.13}$$

However, we now encounter some mathematical difficulties. The troubles arise from the appearance of zeros in the denominators. For the discrete-particle case, we simply excluded the points $\mathbf{R} = \mathbf{R}_i$ from consideration in equations like (D.3); these points were finite in number and we were willing to work around them. However, our continuum model will have whole regions of space where ρ is nonzero and the integrands in eqs. (D.10) and (D.11) will diverge if $\mathbf{E}$ or V is to be evaluated in such regions. Thus, the questions arise:

(*i*) Are the improper integrals in eqs. (D.10) and (D.11) well defined at points where $\rho(\mathbf{R}) \neq 0$?

(*ii*) If so, are eqs. (D.12) and (D.13) still valid?

These questions form the basis of *potential theory,* a subject that is treated in the textbook by Kellogg. Without going into details, we can get some insight by considering the following example. Suppose $f(\mathbf{R})$ is a continuous function and we wish to integrate $f(\mathbf{R})/R^p$ over some region containing the origin, where $\mathbf{R} = \mathbf{0}$. If we use spherical coordinates, we have

$$\begin{aligned}\iiint \frac{f(\mathbf{R})}{R^p}\, dV &= \iiint \frac{f(R,\phi,\theta)}{R^p} R^2 \sin\phi\, dR\, d\phi\, d\theta \\ &= \iiint [f(R,\phi,\theta) R^{2-p}\, dR] \sin\phi\, d\phi\, d\theta\end{aligned}$$

Since the improper integral $\int_0^a R^q \, dR$ converges for $q > -1$, we conclude that the inner integral over R will be finite for $p < 3$. The other integrals present no difficulty, so we propose the following rule of thumb for manipulating the improper integrals that arise due to the continuum approximation: *all the usual operations and theorems may be applied in a straightforward manner unless one encounters integrals of the form*

$$\iiint f(\mathbf{R}')|\mathbf{R} - \mathbf{R}'|^{-p} \, dV'$$

with $p \geq 3$. Of course such a glib statement is mathematically treacherous, but it roughly summarizes the results of the more rigorous potential theory.

Accordingly, we see that the potential $V(\mathbf{R})$ in eq. (D.11) is well defined at all points $\mathbf{R}$ where the charge density $\rho(\mathbf{R})$ is continuous. (In fact, jump discontinuities in ρ are permissible.) So also is the electric field $\mathbf{E}(\mathbf{R})$; keep in mind that the net component of $|\mathbf{R} - \mathbf{R}'|$ in the denominator is 2 in eq. (D.10). Furthermore, since eq. (D.10) is obtained from eq. (D.11) by formal differentiation, and all the integrals are convergent, the relation

$$\mathbf{E}(\mathbf{R}) = -\nabla V(\mathbf{R}) \tag{D.12}$$

is true. Thus $\mathbf{E}(\mathbf{R})$ is still irrotational and eq. (D.4) holds. Moreover, both integrals in Gauss' law eq. (D.13) are quite regular, and the law remains valid.

Now let us compute the divergence of $\mathbf{E}$, which by eq. (D.12) equals the negative laplacian of V. Modeling the computations we made earlier, we are tempted to proceed formally from eq. (D.10) to obtain

$$\begin{aligned}
\nabla \cdot \mathbf{E} &= k \iiint \rho(\mathbf{R}')\nabla \cdot \frac{\mathbf{R} - \mathbf{R}'}{|\mathbf{R} - \mathbf{R}'|^3} \, dV' \\
&= k \iiint \rho(\mathbf{R}') \left(\frac{3}{|\mathbf{R} - \mathbf{R}'|^3} - \frac{3(\mathbf{R} - \mathbf{R}') \cdot (\mathbf{R} - \mathbf{R}')}{|\mathbf{R} - \mathbf{R}'|^5} \right) dV' \\
&= 0 \qquad\qquad (\textit{WRONG!})
\end{aligned}$$

but, as the warning flag indicates, this calculation is suspect because of the exponents in the denominators; the rule of thumb is violated. The correct expression is obtained by applying the divergence theorem to the left-hand side of eq. (D.13), resulting in

$$\iiint_D \nabla \cdot \mathbf{E} \, dV = 4\pi k \iiint \rho(\mathbf{R}) \, dV$$

Since this equation holds for any domain D, we conclude

$$\nabla \cdot \mathbf{E} = 4\pi k \rho \tag{D.14}$$

—the "differential form of Gauss' law." Expressing **E** in terms of V, we have *Poisson's equation for the electrostatic potential*

$$\nabla^2 V = -4\pi k\rho \tag{D.15}$$

A typical situation in electrostatics is the following. One is explicitly given (*i*) the charge distribution ρ in a certain domain D and (*ii*) the values of the potential $V(\mathbf{R})$ on the boundary surface S of the domain; the problem is to find $V(\mathbf{R})$ everywhere inside the domain. Observe that one cannot simply use eq. (D.11), because the charge density function ρ is known only inside D. All one knows about the charges outside D is that they, together with the interior charges, give rise to the specified values of V on the surface S. (For example, the charges within an electric conductor always distribute themselves along its surface so that the conductor is an equipotential, in the electrostatic case.) Thus one has to solve Poisson's equation in D, with the specified boundary conditions, to find $V(\mathbf{R})$. Taking the gradient then gives the electric field $\mathbf{E}(\mathbf{R})$.

D.2 Magnetostatics

Just as stationary charges produce an electric field that can be detected as a force on a test charge, moving charges, or *currents*, produce *magnetic fields* that exert forces on "test currents." However, the geometric properties of these fields are somewhat more complicated.

A point charge q_1 located at $\mathbf{R}_1$, and moving with a velocity $\mathbf{v}_1$, produces a *magnetic induction field* **B** whose magnitude and direction at the point **R** are given (for nonrelativistic speeds) by the Biot-Savart law

$$\mathbf{B}(\mathbf{R}) = \gamma q_1 \frac{\mathbf{v}_1 \times (\mathbf{R} - \mathbf{R}_1)}{|\mathbf{R} - \mathbf{R}_1|^3} \tag{D.16}$$

where γ is a positive constant depending on the system of units. In SI units $\gamma = \mu_0/4\pi$, where the *permeability of free space* $\mu_0 = 4\pi \times 10^{-7}$ Henries per meter. The magnetic induction vector exerts a force on a particle with charge q and velocity $\mathbf{v}$ given by

$$\mathbf{F} = \eta q \mathbf{v} \times \mathbf{B} \tag{D.17}$$

Here again η is a dimensional constant. In SI units $\eta = 1$.

Note the dependence on velocity in these equations; a stationary particle produces no magnetic field, nor is it influenced by a magnetic field. A moving particle produces a field perpendicular to its velocity, and is forced by an external field in a direction perpendicular to both its velocity and the field.

Furthermore, the mutual interaction between two moving particles does not satisfy Newton's third law; the force on particle 2 $(q_2,\mathbf{R}_2,\mathbf{v}_2)$ resulting from particle 1 $(q_1,\mathbf{R}_1,\mathbf{v}_1)$ is

$$\mathbf{F}_2^{(1)} = \eta q_2\mathbf{v}_2 \times \frac{q_1\mathbf{v}_1 \times (\mathbf{R}_2 - \mathbf{R}_1)}{|\mathbf{R}_2 - \mathbf{R}_1|^3}$$

whereas

$$\mathbf{F}_1^{(2)} = \eta q_1\mathbf{v}_1 \times \frac{q_2\mathbf{v}_2 \times (\mathbf{R}_1 - \mathbf{R}_2)}{|\mathbf{R}_1 - \mathbf{R}_2|^3}$$

and a little experimentation will reveal that $\mathbf{F}_1^{(2)}$ is not equal to $\mathbf{F}_2^{(1)}$, nor are these forces directed along $\mathbf{R}_2 - \mathbf{R}_1$. (See exercise 31, section 1.13.)

Once again it is necessary to make a continuum approximation to solve most physical problems. Thus we shall consider a continuum of moving charges, with a charge density function $\rho_m(\mathbf{R})$ and a velocity field $\mathbf{v}(\mathbf{R})$. (In many situations one has moving charges flowing through a background of stationary charges; for example, in a current-carrying conductor the conducting electrons flow past the stationary ions. Thus we use a subscript to distinguish between the density of moving charges ρ_m and the total charge density ρ.)

These moving charges give rise to a current density $\mathbf{j}(\mathbf{R})$:

$$\mathbf{j}(\mathbf{R}) = \rho_m(\mathbf{R})\mathbf{v}(\mathbf{R}) \tag{D.18}$$

According to the discussion in section 3.3, the flux of $\mathbf{j}$ through a surface S equals the amount of charge crossing the surface per unit time, or, in other words, the current I through the surface:

$$I = \iint_S \mathbf{j} \cdot d\mathbf{S}$$

Moreover, the conservation of charge can be expressed, according to section 3.4, as

$$\nabla \cdot \mathbf{j} = -\frac{\partial \rho_m}{dt} = -\frac{\partial \rho}{dt} \tag{D.19}$$

(since the stationary charges do not change). For magnetostatics, $\partial\rho/\partial t = 0$, so $\mathbf{j}$ is solenoidal.

The total magnetic induction field due to a steady (i.e., time-independent) current distribution is obtained by superposition:

$$\begin{aligned}\mathbf{B}(\mathbf{R}) &= \gamma \iiint \rho_m(\mathbf{R}')\mathbf{v}(\mathbf{R}') \times \frac{(\mathbf{R} - \mathbf{R}')}{|\mathbf{R} - \mathbf{R}'|^3}\, dV' \\ &= \gamma \iiint \mathbf{j}(\mathbf{R}') \times \frac{(\mathbf{R} - \mathbf{R}')}{|\mathbf{R} - \mathbf{R}'|^3}\, dV' \end{aligned} \tag{D.20}$$

In some physical experiments the current producing the **B** field is carried by a wire filament. It is convenient to model this situation by letting the cross-sectional area of the wire go to zero, and the current density to infinity, in such a manner that the flux of **j** along the wire, that is, the current I, is constant. Then the wire becomes a space curve $\mathbf{R}' = \mathbf{R}'(s)$ carrying a current I. The volume integration in eq. (D.20) is regarded as an iterated integral, integrated first over the cross-sectional area, then along the length of the wire. In our model the first integral yields the current I, and eq. (D.20) becomes

$$\mathbf{B}(\mathbf{R}) = \gamma I \int \frac{d\mathbf{R}' \times (\mathbf{R} - \mathbf{R}')}{|\mathbf{R} - \mathbf{R}'|^3}$$

$$\left(= \gamma I \int \frac{d\mathbf{R}'(s)}{ds} \times \frac{\mathbf{R} - \mathbf{R}'(s)}{|\mathbf{R} - \mathbf{R}'(s)|^3}\, ds \right) \tag{D.21}$$

In magnetostatic situations, of course, the current is carried around a closed curve; otherwise, **j** could not be solenoidal. Thus the integral in eq. (D.21) could be more accurately written $\oint$.

It is very instructive to compute the total force exerted on one current loop by another. The effect of the field $\mathbf{B}(\mathbf{R})$ on the loop $\mathbf{R}_1(s)$ carrying current I_1 is given by

$$\mathbf{F}_1 = \eta \iiint \mathbf{j} \times \mathbf{B}\, dV$$

$$= \eta I_1 \oint d\mathbf{R}_1 \times \mathbf{B}(\mathbf{R}_1)$$

If **B** is produced by loop $\mathbf{R}_2(s)$ carrying current I_2, then the interaction force is

$$\mathbf{F}_1^{(2)} = \eta\gamma I_1 I_2 \oint_2 \oint_1 d\mathbf{R}_1 \times \frac{[d\mathbf{R}_2 \times (\mathbf{R}_1 - \mathbf{R}_2)]}{|\mathbf{R}_1 - \mathbf{R}_2|^3}$$

By identities (1.30) and (3.34)

$$\mathbf{F}_1^{(2)} = \eta\gamma I_1 I_2 \oint_2 \oint_1 \left(\frac{d\mathbf{R}_1 \cdot (\mathbf{R}_1 - \mathbf{R}_2)}{|\mathbf{R}_1 - \mathbf{R}_2|^3}\, d\mathbf{R}_2 - d\mathbf{R}_1 \cdot d\mathbf{R}_2 \frac{\mathbf{R}_1 - \mathbf{R}_2}{|\mathbf{R}_1 - \mathbf{R}_2|^3} \right)$$

$$= \eta\gamma I_1 I_2 \oint_2 d\mathbf{R}_2 \oint_1 \nabla_1 \left(\frac{-1}{|\mathbf{R}_1 - \mathbf{R}_2|} \right) \cdot d\mathbf{R}_1$$

$$- \eta\gamma I_1 I_2 \oint_2 \oint_1 \frac{\mathbf{R}_1 - \mathbf{R}_2}{|\mathbf{R}_1 - \mathbf{R}_2|^3}\, d\mathbf{R}_1 \cdot d\mathbf{R}_2$$

The first term vanishes by theorem 4.2, and the resulting expression,

$$\mathbf{F}_1^{(2)} = -\eta\gamma I_1 I_2 \oint_2 \oint_1 \frac{\mathbf{R}_1 - \mathbf{R}_2}{|\mathbf{R}_1 - \mathbf{R}_2|^3} d\mathbf{R}_1 \cdot d\mathbf{R}_2$$

is antisymmetric; that is, on interchanging indices, we find that $\mathbf{F}_1^{(2)} = -\mathbf{F}_2^{(1)}$. Therefore the total magnetic force between two current loops satisfies a form of Newton's third law, even though the interparticle forces do not!

It can be shown that a current loop does not exert a net force on itself, but that any given portion of the loop is, in general, subjected to a force. We do not present this analysis because the improper integrals are quite complicated.

Returning to eq. (D.20), we use eq. (3.33) to rewrite the equation for the magnetic induction:

$$\begin{aligned}\mathbf{B}(\mathbf{R}) &= -\gamma \iiint \mathbf{j}(\mathbf{R}') \times \nabla\left(\frac{1}{|\mathbf{R} - \mathbf{R}'|}\right) dV' \\ &= \gamma\nabla \times \iiint \frac{\mathbf{j}(\mathbf{R}')}{|\mathbf{R} - \mathbf{R}'|} dV' \end{aligned} \tag{D.22}$$

(since ∇ operates only on $\mathbf{R}$). It follows immediately that $\mathbf{B}$ is solenoidal:

$$\nabla \cdot \mathbf{B} = 0 \tag{D.23}$$

Hence $\mathbf{B}$ is derivable from a vector potential $\mathbf{A}$:

$$\mathbf{B} = \nabla \times \mathbf{A}$$

and, in fact, we can read off from eq. (D.22) what $\mathbf{A}$ should be:

$$\mathbf{A}(\mathbf{R}) = \gamma \iiint \frac{\mathbf{j}(\mathbf{R}')}{|\mathbf{R} - \mathbf{R}'|} dV' + \nabla\psi(\mathbf{R}) \tag{D.24}$$

where ψ is an arbitrary scalar function. (This degree of freedom is known as *gauge invariance.*)

To find $\nabla \times \mathbf{B} = \nabla \times (\nabla \times \mathbf{A})$ we refer the reader to section 5.3, where it was shown that

$$\begin{aligned}\nabla \times (\nabla \times \mathbf{A}) &= -\nabla^2\mathbf{A} + \nabla(\nabla \cdot \mathbf{A}) \\ &= 4\pi\gamma\mathbf{j} - \gamma\nabla \iint_S \frac{\mathbf{j} \cdot \mathbf{n}'}{|\mathbf{R} - \mathbf{R}'|} dS' \end{aligned} \tag{D.25}$$

Since we are integrating over a region containing all the current sources $\mathbf{j}(\mathbf{R})$, the enclosing surface has no currents; thus $\mathbf{j} = \mathbf{0}$ on the far right in eq. (D.25) and we ultimately have *Ampere's law:*

$$\nabla \times \mathbf{B}(\mathbf{R}) = 4\pi\gamma\mathbf{j}(\mathbf{R}) \tag{D.26}$$

Equations (D.23) and (D.26) are the basic laws of magnetostatics. It is sometimes convenient to apply Stokes' theorem to eq. (D.26), yielding the equation

$$\oint_C \mathbf{B} \cdot d\mathbf{R} = 4\pi\gamma \iint_S \mathbf{j} \cdot d\mathbf{S} = 4\pi\gamma I$$

where the surface S is bounded by the curve C and I is the total current crossing S.

The typical problem in magnetostatics involves solving eqs. (D.23) and (D.26) subject to certain boundary conditions between different media. We defer discussion of the latter for the moment.

D.3 Electrodynamics

In time-varying situations the previous equations must be modified. First of all Ampere's law [eq. (D.26)] becomes inconsistent with the equation of continuity eq. (D.19), because $\nabla \cdot \nabla \times \mathbf{B} = 0$ but $\nabla \cdot 4\pi\gamma\mathbf{j} = -4\pi\gamma\, \partial\rho/\partial t$. But note that

$$\frac{\partial \rho}{\partial t} = \frac{1}{4\pi k}\nabla \cdot \frac{\partial \mathbf{E}}{\partial t}$$

if Gauss's law [eq. (D.14)] continues to hold for electrodynamics. Thus one might suspect that Ampere's law is amended to read

$$\nabla \times \mathbf{B} = 4\pi\gamma\mathbf{j} + \frac{\gamma}{k}\frac{\partial \mathbf{E}}{\partial t} \tag{D.27}$$

for dynamic situations. Notice that eq. (D.27) implies that magnetic induction fields are produced not only by currents $\mathbf{j}$ but also by changing electric fields. Maxwell, the discoverer of this effect, called $(1/4\pi k)(\partial \mathbf{E}/\partial t)$ the "displacement current density."

Another necessary modification in the equations was discovered experimentally by Faraday. It involves the magnetic flux Φ across an oriented surface S:

$$\Phi = \iint_S \mathbf{B} \cdot d\mathbf{S}$$

Faraday observed that when the flux through S changes, an electric field is produced around the curve C forming the boundary of S, in accordance with

$$\frac{d\Phi}{dt} = \iint_S \frac{\partial \mathbf{B}}{\partial t} \cdot d\mathbf{S} = -\frac{1}{\alpha}\oint_C \mathbf{E} \cdot d\mathbf{R}$$

where α is another positive constant.

Applying Stokes' theorem we find

$$\alpha \iint_S \frac{\partial \mathbf{B}}{\partial t} \cdot d\mathbf{S} = -\iint_S \nabla \times \mathbf{E} \cdot d\mathbf{S}$$

and since this holds for arbitrary surfaces S we conclude that a changing magnetic induction field produces an electric field, so that eq. (D.4) must be modified for dynamic situations to read

$$\nabla \times \mathbf{E} = -\alpha \frac{\partial \mathbf{B}}{\partial t}$$

Actually, Faraday's law is more general than we have described. We have assumed that the flux Φ through S changed because the **B** field itself changed. In fact, Φ can also change if the surface S is moving or turning, in which case the flux transport theorem must be used to compute $d\Phi/dt$. Faraday observed that the same electric field is induced regardless of the mechanism that produces the change in Φ. This situation is relativistic and we refer the reader to the references for elaboration, but we remark that one consequence of the analysis is the identification of the constants α and η:

$$\alpha = \eta \tag{D.28}$$

and we incorporate this fact in our subsequent equations.

The four main equations that we have examined above,

$$\nabla \cdot \mathbf{E} = 4\pi k \rho \tag{D.14}$$

$$\nabla \cdot \mathbf{B} = 0 \tag{D.23}$$

$$\nabla \times \mathbf{E} = -\eta \frac{\partial \mathbf{B}}{\partial t} \tag{D.29}$$

$$\nabla \times \mathbf{B} = 4\pi\gamma \mathbf{j} + \frac{\gamma}{k} \frac{\partial \mathbf{E}}{\partial t} \tag{D.27}$$

are known as *Maxwell's equations,* and they can be used to find **E** and **B** when the charges ρ and currents **j** are known throughout all space. The charges, in turn, are subjected to the *Lorentz force,* obtained by adding eqs. (D.2) and (D.17):

$$\mathbf{F} = \iiint \rho(\mathbf{E} + \eta \mathbf{v} \times \mathbf{B})\, dV \tag{D.30}$$

ρ and **j**, of course, are constrained by the equation of continuity, (D.19). So, in general, the coupled system of Maxwell's equations, Lorentz's equation, and the equation of continuity describes how the charges produce, and are influenced by, the fields.

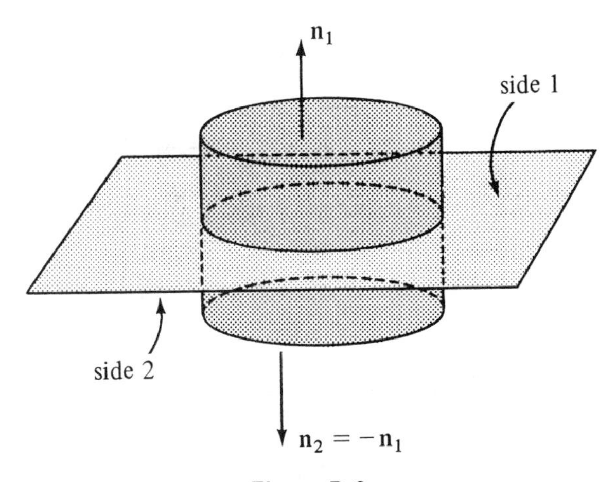

Figure D.2

If the charge sources are known only in a region D, Maxwell's equations must be supplemented with boundary conditions. These can be derived from the equations themselves, as follows. Suppose the region D is bounded by the smooth surface S. Consider an infinitesimal "Gaussian pillbox," that is, a very short circular cylinder with axis normal to S and with a face on either side of S, as in figure D.2.

Regarding the height of the cylinder as much shorter than the diameter of the faces, we apply Gauss' law eq. (D.8) to find

$$(\mathbf{E}_1 \cdot \mathbf{n}_1 + \mathbf{E}_2 \cdot \mathbf{n}_2)(\text{area of base}) = 4\pi k(\text{charge enclosed})$$

$$\mathbf{n}_1 \cdot (\mathbf{E}_1 - \mathbf{E}_2) = 4\pi k\left(\frac{\text{charge}}{\text{area}}\right) = 4\pi k(\text{surface charge density})$$

Thus the normal component of $\mathbf{E}$ jumps by an amount $4\pi k$ times the surface charge density as the surface is crossed.

Since $\nabla \cdot \mathbf{B} = 0$, the analogous argument shows that the normal component of $\mathbf{B}$ is continuous as the surface is crossed.

Now we consider an infinitesimal loop crossing the surface, as in figure D.3. We compute the line integral of $\mathbf{E}$ around this path, again treating the height ϵ as negligible compared to the length δ. If E_t denotes the relevant vector component of $\mathbf{E}$, we have

$$\oint \mathbf{E} \cdot d\mathbf{R} = (E_{t_1} - E_{t_2})\,\delta$$

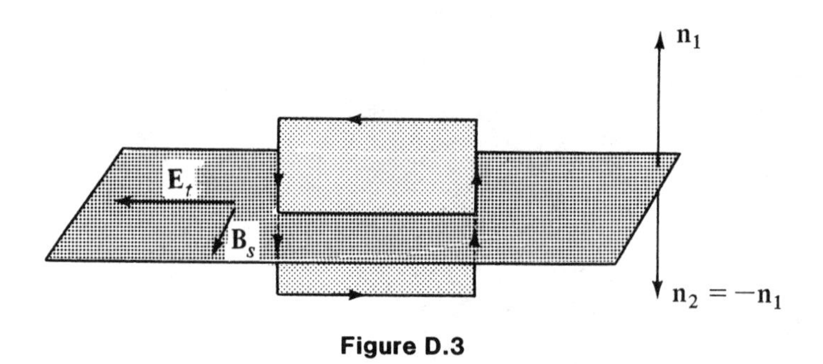

Figure D.3

Applying Stokes' theorem and eq. (D.29), we have

$$(E_{t_1} - E_{t_2})\,\delta = -\eta \iint \frac{\partial B}{\partial t} \cdot d\mathbf{S} = -\frac{\partial B_s}{\partial t}\,\delta\epsilon$$

where B_S is the indicated component of **B.** Since we are neglecting ϵ in comparison with δ, this equation implies $E_{t_1} = E_{t_2}$; that is, the tangential component of **E** is continuous as S is crossed.

If we integrate **B** around the same loop and use eq. (D.27) and Stokes' theorem, we find

$$(B_{t_1} - B_{t_2})\,\delta = 4\pi\gamma(j_S\,\epsilon)\,\delta + \frac{\gamma}{k}\frac{\partial E_s}{\partial t}\,\delta\epsilon$$

Again we neglect the last term; however, the term $j_s\epsilon$ might be appreciable if there is a *surface current density.* Checking the orientations, we conclude that the tangential component of **B** jumps by an amount $4\pi\gamma$ times the surface current density, in a direction perpendicular to the latter, as S is crossed.

Summarizing, we introduce σ as the surface charge density and **K** as the surface current density, and we say that the normal and tangential components of **E** and **B** jump by amounts

$$\begin{aligned}
\Delta E_{\text{normal}} &= 4\pi k\sigma \\
\Delta B_{\text{normal}} &= 0 \\
\Delta E_{\text{tangential}} &= 0 \\
\Delta B_{\text{tangential}} &= 4\pi\gamma\mathbf{K} \times \mathbf{n}
\end{aligned}$$

as we go from side 2 to side 1.

When there are material media inside the region of interest, the physics of the situation often makes it convenient to distinguish between free and bound charges, and free and bound currents. This is aided by splitting **E** into an electric displacement vector **D** and an electric polarization vector **P,** and by splitting **B** into a magnetic field vector **H** and a magnetization vector **M.** The details of these decompositions depend on the material properties, so we leave this matter to the references at the end of this appendix.

We wish to derive two results involving the interplay of the mechanical motions and the fields. The point charge q with mass m and velocity $\mathbf{v}$ has kinetic energy $\frac{1}{2}m|\mathbf{v}|^2$. According to eq. (C.7), the effect of a force $\mathbf{F}$ is to change the kinetic energy, at a rate $\mathbf{F} \cdot \mathbf{v}$. If we sum eq. (C.7) over the charges, call the total kinetic energy K, and replace $\mathbf{F}$ by the Lorentz force, we find

$$\frac{dK}{dt} = \sum q(\mathbf{E} + \eta\mathbf{v} \times \mathbf{B}) \cdot \mathbf{v} = \sum q\mathbf{E} \cdot \mathbf{v}$$

or, in continuous form,

$$\frac{dK}{dt} = \iiint \rho_m \mathbf{E} \cdot \mathbf{v}\, dV = \iiint \mathbf{E} \cdot \mathbf{j}\, dV$$

Using eq. (D.27) to eliminate $\mathbf{j}$ and then invoking eq. (3.36), we find

$$\begin{aligned}\frac{dK}{dt} &= \iiint \mathbf{E} \cdot \left(\frac{1}{4\pi\gamma}\nabla \times \mathbf{B} - \frac{1}{4\pi k}\frac{\partial \mathbf{E}}{\partial t}\right) dV \\ &= -\frac{1}{4\pi k}\iiint \mathbf{E} \cdot \frac{\partial \mathbf{E}}{\partial t}\, dV \\ &\quad + \frac{1}{4\pi\gamma}\iiint [\mathbf{B} \cdot \nabla \times \mathbf{E} + \nabla \cdot (\mathbf{B} \times \mathbf{E})]\, dV\end{aligned}$$

Applying eq. (D.29) and the divergence theorem,

$$\begin{aligned}\frac{dK}{dt} &= -\frac{1}{4\pi k}\iiint \mathbf{E} \cdot \frac{\partial \mathbf{E}}{\partial t} dV - \frac{\eta}{4\pi\gamma}\iiint \mathbf{B} \cdot \frac{\partial \mathbf{B}}{\partial t} dV \\ &\quad + \frac{1}{4\pi\gamma}\iint_S \mathbf{B} \times \mathbf{E} \cdot d\mathbf{S}\end{aligned}$$

where the surface integral is taken over the boundary of the region. Consequently,

$$\frac{d}{dt}\left[K + \iiint \left(\frac{|\mathbf{E}|^2}{8\pi k} + \frac{\eta|\mathbf{B}|^2}{8\pi\gamma}\right) dV\right] = -\iint_S \frac{\mathbf{E} \times \mathbf{B}}{4\pi\gamma} \cdot d\mathbf{S}$$

This equation leads one to postulate that the electromagnetic field itself has an energy distributed throughout space with a density

$$\frac{|\mathbf{E}|^2}{8\pi k} + \eta\frac{|\mathbf{B}|^2}{8\pi\gamma}$$

and that the energy of the electromechanical system is carried off by the field, with a flux density **P** known as the *Poynting vector:*

$$\mathbf{P} = \frac{\mathbf{E} \times \mathbf{B}}{4\pi\gamma}$$

A similar derivation can be carried out for momentum. If **P** denotes the total mechanical momentum, then

$$\begin{aligned} \frac{d\mathbf{P}}{dt} &= \sum \mathbf{F} = \iiint (\rho\mathbf{E} + \eta\mathbf{j} \times \mathbf{B})\, dV \\ &= \frac{1}{4\pi k}\iiint (\nabla \cdot \mathbf{E})\mathbf{E}\, dV + \frac{\eta}{4\pi\gamma}\iiint (\nabla \times \mathbf{B}) \times \mathbf{B}\, dV \\ &\quad - \frac{\eta}{4\pi k}\iiint \frac{\partial \mathbf{E}}{\partial t} \times \mathbf{B}\, dV \end{aligned} \tag{D.31}$$

from Maxwell's equations. Invoking the identity

$$\frac{\partial}{\partial t}(\mathbf{E} \times \mathbf{B}) = \frac{\partial \mathbf{E}}{\partial t} \times \mathbf{B} + \mathbf{E} \times \frac{\partial \mathbf{B}}{\partial t}$$

and using eqs. (D.29) and (D.23), we rewrite eq. (D.31) as

$$\begin{aligned} &\frac{d\mathbf{P}}{dt} + \frac{\eta}{4\pi k}\iiint \frac{\partial(\mathbf{E} \times \mathbf{B})}{\partial t}\, dV \\ &= \iiint \frac{1}{4\pi k}[(\nabla \times \mathbf{E}) \times \mathbf{E} + (\nabla \cdot \mathbf{E})\mathbf{E}]\, dV \\ &\quad + \iiint \frac{\eta}{4\pi\gamma}[(\nabla \times \mathbf{B}) \times \mathbf{B} + (\nabla \cdot \mathbf{B})\mathbf{B}]\, dV \end{aligned} \tag{D.32}$$

Employing tensor notation, we find

$$\begin{aligned} [(\nabla \times \mathbf{E}) \times \mathbf{E} + (\nabla \cdot \mathbf{E})\mathbf{E}]_i &= \epsilon_{ijk}\epsilon_{jlm}(\partial_l E_m)E_k + (\partial_l E_l)E_i \\ &= \epsilon_{kij}\epsilon_{lmj}(\partial_l E_m)E_k + (\partial_l E_l)E_i \\ &= (\delta_{kl}\delta_{im} - \delta_{km}\delta_{il})(\partial_l E_m)E_k + (\partial_l E_l)E_i \\ &= (\partial_l E_i)E_l - (\partial_i E_m)E_m + (\partial_l E_l)E_i \\ &= \partial_l(E_l E_i) - \partial_i\left(\frac{E_m^2}{2}\right) \\ &= \partial_l\left(E_l E_i - \frac{\delta_{il}|\mathbf{E}|^2}{2}\right) \end{aligned}$$

Hence, the ith component of eq. (D.32) can be expressed

$$\frac{dP_i}{dt} + \frac{\eta}{4\pi k}\iiint \frac{\partial(\mathbf{E}\times\mathbf{B})_i}{\partial t}dV = \iiint \partial_l T_{li}\, dV \tag{D.33}$$

where

$$T_{li} = \frac{E_l E_i}{4\pi k} + \eta\frac{B_l B_i}{4\pi\gamma} - \delta_{li}\left(\frac{|\mathbf{E}|^2}{8\pi k} + \frac{\eta|\mathbf{B}|^2}{8\pi\gamma}\right)$$

If we think of i as fixed, the right-hand side of eq. (D.33) looks like a divergence; therefore the momentum equation becomes

$$\frac{dP_i}{dt} + \iiint \frac{\partial}{\partial t}\frac{\eta(\mathbf{E}\times\mathbf{B})_i}{4\pi k}dV = \iint_S T_{li}n_l\, dS \tag{D.34}$$

where n_l represents the components of the outward unit normal $\mathbf{n}$. The interpretation of eq. (D.34) is to regard $\eta(\mathbf{E}\times\mathbf{B})/4\pi k$ as momentum stored in the field, and T_{li} as a "flux dyadic" or "stress tensor" giving, componentwise, the flow of momentum flux through the surface S. Elaboration of this *Maxwell stress tensor* will be found in the references at the end of this appendix.

We close out this appendix with a discussion of a procedure for solving Maxwell's equations in the case where the charge and current densities ρ and $\mathbf{j}$ are known throughout space, and where the time dependence is sinusoidal. Thus we assume that every quantity $\mathbf{E}$, $\mathbf{B}$, ρ, and $\mathbf{j}$ contains a factor $\mathbf{e}^{i\omega t}$, so that time derivatives are equivalent to multiplication by $i\omega$. This "frequency domain analysis" is a very useful approach in understanding radiation problems. Maxwell's equations take the form

$$\nabla\cdot\mathbf{E} = 4\pi k\rho \tag{D.35}$$

$$\nabla\cdot\mathbf{B} = 0 \tag{D.36}$$

$$\nabla\times\mathbf{E} = -\mathrm{i}\omega\eta\mathbf{B} \tag{D.37}$$

$$\nabla\times\mathbf{B} = 4\pi\gamma\mathbf{j} + \frac{i\omega\gamma}{k}\mathbf{E} \tag{D.38}$$

First of all note that eq. (D.35) is redundant; by taking the divergence of eq. (D.38)

$$\nabla\cdot\nabla\times\mathbf{B} = 0 = 4\pi\gamma\nabla\cdot\mathbf{j} + \frac{i\omega\gamma}{k}\nabla\cdot\mathbf{E}$$

and inserting the equation of continuity $\nabla\cdot\mathbf{j} = -i\omega\rho$ we derive

$$\frac{i\omega\gamma}{k}\nabla\cdot\mathbf{E} = \frac{i\omega\gamma}{k}4\pi k\rho$$

which is equivalent to eq. (D.35) if $\omega \neq 0$. (Observe that $\omega = 0$ is the *static* case, which we have analyzed already.)

Now eq. (D.36) is automatically satisfied if we take

$$\mathbf{B} = \nabla \times \mathbf{A} \tag{D.39}$$

for *any* vector field **A**. With this substitution in eq. (D.37) we learn

$$\nabla \times \{\mathbf{E} + i\omega\eta\mathbf{A}\} = 0$$

Thus eq. (D.37) will be satisfied if we take

$$\mathbf{E} + i\omega\eta\mathbf{A} = -\nabla\phi \qquad \text{or} \qquad \mathbf{E} = -i\omega\eta\mathbf{A} - \nabla\phi \tag{D.40}$$

for *any* scalar field ϕ.

As a result we only have to satisfy eq. (D.38) with the forms (D.39) and (D.40)—and we are free to choose ϕ and **A**, to enable us to do so.

Insertion of eqs. (D.39) and (D.40) into eq. (D.38) results in

$$\begin{aligned} \nabla \times (\nabla \times \mathbf{A}) &= -\nabla^2\mathbf{A} + \nabla(\nabla \cdot \mathbf{A}) \\ &= 4\pi\gamma\mathbf{j} + \frac{i\omega\gamma}{k}\mathbf{E} \\ &= 4\pi\gamma\mathbf{j} + \frac{\omega^2\gamma\eta}{k}\mathbf{A} - \frac{i\omega\gamma}{k}\nabla\phi \end{aligned}$$

or

$$\nabla\left(\nabla \cdot \mathbf{A} + \frac{i\omega\gamma}{k}\phi\right) = \nabla^2\mathbf{A} + \frac{\omega^2\gamma\eta}{k}\mathbf{A} + 4\pi\gamma\mathbf{j} \tag{D.41}$$

We shall solve eq. (D.41) by choosing a formula for **A** that makes the right-hand side zero, and then setting

$$\phi = \frac{ik}{\omega\gamma}\nabla \cdot \mathbf{A} \tag{D.42}$$

to make the left-hand side zero.

The formula for **A** is an extension of lemma 5.4, section 5.2:

LEMMA D.1 *The expression*

$$\begin{aligned} \mathbf{A}(\mathbf{R}) &= \gamma \iiint \frac{\mathbf{j}(\mathbf{R}')}{|\mathbf{R} - \mathbf{R}'|} e^{-i\lambda|\mathbf{R}-\mathbf{R}'|}\, dV' \\ \lambda &= \omega\sqrt{\left\{\frac{\gamma\eta}{k}\right\}} \end{aligned} \tag{D.43}$$

satisfies

$$\nabla^2\mathbf{A} + \frac{\omega^2\gamma\eta}{k}\mathbf{A} = \nabla^2\mathbf{A} + \lambda^2\mathbf{A} = -4\pi\gamma\mathbf{j} \tag{D.44}$$

Proof Because the variable **R** appears twice in the integrand we cannot apply lemma 5.4 directly. Also we must avoid powers of $|\mathbf{R} - \mathbf{R}'|$ higher than 2 in the denominator. We proceed cautiously.

It is easy to show that the product rule for the laplacian is

$$\nabla^2 f(\mathbf{R})g(\mathbf{R}) = g(\mathbf{R})\,\nabla^2 f(\mathbf{R}) + f(\mathbf{R})\,\nabla^2 g(\mathbf{R}) + 2\nabla f(\mathbf{R}) \cdot \nabla g(\mathbf{R}) \qquad \text{(D.45)}$$

One way of implementing this is to introduce the operators ∇_1 and ∇_2, which operate on the variables $\mathbf{R}_1$ and $\mathbf{R}_2$, respectively, and perform the operation

$$\nabla_1^2 f(\mathbf{R}_1)g(\mathbf{R}_2) + \nabla_2^2 f(\mathbf{R}_1)g(\mathbf{R}_2) + 2\nabla_1 \cdot \nabla_2 f(\mathbf{R}_1)g(\mathbf{R}_2) \qquad \text{(D.46)}$$

and finally set $\mathbf{R}_1 = \mathbf{R}_2 = \mathbf{R}$ in the result. Applied to eq. (D.43) this procedure leads to

$$\begin{aligned}\nabla^2 \iiint \frac{\mathbf{j}(\mathbf{R}')}{|\mathbf{R} - \mathbf{R}'|} e^{-i\lambda|\mathbf{R}-\mathbf{R}'|}\, dV' &= \nabla_1^2 \iiint \frac{\mathbf{j}(\mathbf{R}')}{|\mathbf{R}_1 - \mathbf{R}'|} e^{-i\lambda|\mathbf{R}_2-\mathbf{R}'|} dV' \\ &+ \nabla_2^2 \iiint \frac{\mathbf{j}(\mathbf{R}')}{|\mathbf{R}_1 - \mathbf{R}'|} e^{-i\lambda|\mathbf{R}_2-\mathbf{R}'|} dV' \\ &+ 2\nabla_1 \cdot \nabla_2 \iiint \frac{\mathbf{j}(\mathbf{R}')}{|\mathbf{R}_1 - \mathbf{R}'|} e^{-i\lambda|\mathbf{R}_2-\mathbf{R}'|} dV' \qquad \text{(D.47)}\end{aligned}$$

$(\mathbf{R}_1 = \mathbf{R}_2 = \mathbf{R})$.

Lemma 5.4 *can* be applied to the first term, since $\mathbf{R}_2$ is sensibly constant while the operator works on $\mathbf{R}_1$. It produces $4\pi\ \mathbf{j}(\mathbf{R}_1)e^{-i\lambda|\mathbf{R}_2-\mathbf{R}_1|}$, and with $\mathbf{R}_1 = \mathbf{R}_2 = \mathbf{R}$ this becomes

$$-4\pi\ \mathbf{j}(\mathbf{R}) \qquad \text{(D.48)}$$

The rest of the computation is straightforward, because the differential operators can be brought under the integral signs without creating divergent integrands. From the identities one can show (exercise 1)

$$\nabla_2^2 e^{-i\lambda|\mathbf{R}_2-\mathbf{R}'|} = \left(-\lambda^2 - \frac{i2\lambda}{|\mathbf{R}_2 - \mathbf{R}'|}\right) e^{-i\lambda|\mathbf{R}_2-\mathbf{R}'|},$$

$$2\nabla_1 \cdot \nabla_2 \frac{1}{|\mathbf{R}_1 - \mathbf{R}'|} e^{-i\lambda|\mathbf{R}_2-\mathbf{R}'|} = i2\lambda\ \frac{\mathbf{R}_1 - \mathbf{R}'}{|\mathbf{R}_1 - \mathbf{R}'|^3} \cdot \frac{\mathbf{R}_2 - \mathbf{R}'}{|\mathbf{R}_2 - \mathbf{R}'|}\ e^{-i\lambda|\mathbf{R}_2-\mathbf{R}'|} \qquad \text{(D.49)}$$

and insertion of eqs. (D.48) and (D.49) into the integrals in eq. (D.47) with $\mathbf{R}_1 = \mathbf{R}_2 = \mathbf{R}$ produces

$$\begin{aligned}
\nabla^2 \iiint \frac{\mathbf{j}(\mathbf{R}')}{|\mathbf{R}-\mathbf{R}'|} e^{-i\lambda|\mathbf{R}-\mathbf{R}'|}\, dV' \\
&= -4\pi\mathbf{j}(\mathbf{R}) \\
&\quad + \iiint \frac{\mathbf{j}(\mathbf{R}')}{|\mathbf{R}-\mathbf{R}'|}\left(-\lambda^2 - \frac{i2\lambda}{|\mathbf{R}-\mathbf{R}'|}\right) e^{-i\lambda|\mathbf{R}-\mathbf{R}'|}\, dV' \\
&\quad + \iiint \frac{\mathbf{j}(\mathbf{R}')}{|\mathbf{R}-\mathbf{R}'|^2}\, i2\lambda e^{-i\lambda|\mathbf{R}-\mathbf{R}'|}\, dV' \\
&= -4\pi\mathbf{j}(\mathbf{R}) - \lambda^2 \iiint \frac{\mathbf{j}(\mathbf{R}')}{|\mathbf{R}-\mathbf{R}'|} e^{-i\lambda|\mathbf{R}-\mathbf{R}'|}\, dV'
\end{aligned}$$

With the factor γ inserted and some rearrangement this becomes eq. (D.43).

As a result Maxwell's equations are solved for this situation by computing **A** from eq. (D.43), ϕ from eq. (D.42), **E** from eq. (D.40), and **B** from eq. (D.39).

EXERCISES

1. Verify the formulas (D.49).
2. In free space with $\rho = 0$, $\mathbf{j} = \mathbf{0}$, show that both **E** and **B** satisfy the *wave equation*

$$\nabla^2 \begin{pmatrix} \mathbf{E} \\ \mathbf{B} \end{pmatrix} = \frac{\eta\gamma}{k} \frac{\partial^2}{\partial t^2} \begin{pmatrix} \mathbf{E} \\ \mathbf{B} \end{pmatrix}$$

3. In many electric conductors the currents and fields obey an experimental law known as *Ohm's law:* $\mathbf{j} = \sigma\mathbf{E}$, where σ is a constant depending on the conductor, and is called the conductivity. If Ohm's law holds and $\rho = 0$, show that both **E** and **B** satisfy the *telegrapher's equation*

$$\nabla^2 \begin{pmatrix} \mathbf{E} \\ \mathbf{B} \end{pmatrix} = \frac{\eta\gamma}{k} \frac{\partial^2}{\partial t^2} \begin{pmatrix} \mathbf{E} \\ \mathbf{B} \end{pmatrix} + 4\pi\eta\gamma\sigma \frac{\partial}{\partial t} \begin{pmatrix} \mathbf{E} \\ \mathbf{B} \end{pmatrix}$$

4. How does the equation in exercise 3 change if the conductivity σ depends on position, $\sigma = \sigma(\mathbf{R})$?
5. If a wire loop is moved through a magnetic induction field $\mathbf{B}(\mathbf{R})$, the conduction electrons "feel" a force $\eta q \mathbf{v} \times \mathbf{B}$, where $\mathbf{v}$ is the velocity of the wire. However, an observer moving with the wire is unaware of any velocity and postulates that the source of this force is an electric field $\mathbf{E}$. Use the flux transport theorem to analyze this situation, and derive the relation (D.28) from Faraday's law.

References

1. KELLOGG, O. D. *Foundations of Potential Theory.* New York: Dover Publications, 1953.
2. JACKSON, J. D. *Classical Electrodynamics.* New York: John Wiley and Sons, 1962.
3. PANOFSKY, WOLFGANG K. H., and PHILLIPS, M. *Classical Electricity and Magnetism,* 2nd edition. Reading, Mass.: Addison-Wesley, 1962.
4. SADIKU, M. N. O. *Elements of Electromagnetics,* Saunders College Publishing, Fort Worth, Tex.: 1989.

APPENDIX E

Constrained Optimization

In this section we shall explore how the properties of the gradient, discussed in Section 3.1, can provide insight into the discipline known as mathematical optimization theory.

The fact that $\mathbf{grad}\, f \cdot \mathbf{u}$ gives the directional derivative of f in the direction of the unit vector $\mathbf{u}$ is expressed, in the language of finite differences, as

$$\lim_{s \to 0} \frac{f(\mathbf{R} + s\mathbf{u}) - f(\mathbf{R})}{s} = \mathbf{u} \cdot \mathbf{grad}\, f$$

It follows that if $\mathbf{u} \cdot \mathbf{grad}\, f > 0$, f must definitely increase for *some* distance in the direction of $\mathbf{u}$ (and decrease in the opposite direction). If f changes rapidly, it may start to decrease shortly thereafter, but there is no question that some increase must take place.

What does this say about a point where f takes a maximum value? If $\mathbf{R}$ is a local maximum for f, there *can be no direction* in which f increases. Therefore $\mathbf{grad}\, f$ can have "no direction" at such a point; that is, it must be the zero vector (see fig. E.1). ***Grad** f is zero at any local maximum (or minimum) of a continuously differentiable scalar field.* Thus the condition $\mathbf{grad}\, f = 0$ replaces the familiar necessary condition, $df/dx = 0$, for an extremum of a function of one variable.

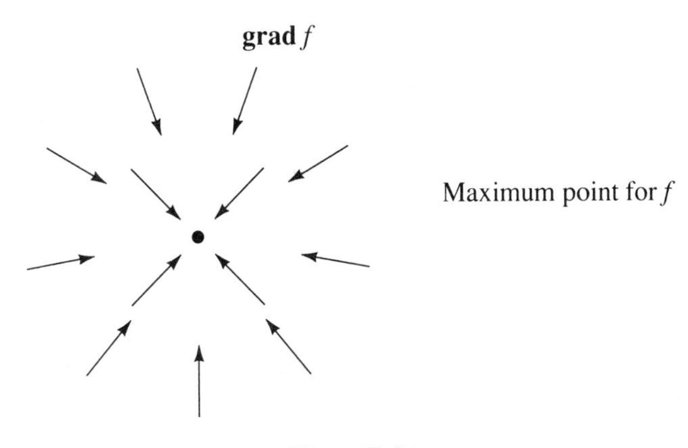

Figure E.1

Example E.1 Find the maximum value of $f(x,y,z) = -x^4 - 2y^2 - 4z^4 + 2z^2$.

Solution Clearly f approaches $-\infty$ as $|\mathbf{R}| = |x\mathbf{i} + y\mathbf{j} + z\mathbf{k}|$ grows, so the standard theorems of analysis show that f does, in fact, achieve a maximum value. At such a point $\mathbf{grad}\, f = 0$. Thus

$$-4x^3 = 0 \qquad -4y = 0 \qquad \text{and} \qquad -16z^3 + 4z = 0$$

The possible maximum points (remember that we have only derived *necessary* conditions) are $x = 0$, $y = 0$, and $z = 0, \pm\frac{1}{2}$. The corresponding values of f are 0, $\frac{1}{4}$, and $\frac{1}{4}$. Thus the maximum value of f is $\frac{1}{4}$ and it occurs at the *two* points $\mathbf{R} = \pm(\frac{1}{2})\mathbf{k}$.

A more interesting situation arises when the set of points over which we seek to maximize f is *constrained*. Thus suppose that $g(x,y,z)$ is another continuously differentiable scalar field and the surface $g(x,y,z) = 0$ divides all of space into regions where $g(x,y,z) < 0$ and $g(x,y,z) > 0$ (fig. E.2). Now we seek the point *in the region* $g \geq 0$ where f is maximal. Perhaps f takes still larger values outside this region, but such points are irrelevant to our quest. What are the necessary conditions on f at this constrained maximum point?

There are two possibilities. Either g is greater than zero at the maximum point of f, or it is equal to zero there. As figure E.2 suggests, points where $g > 0$ are called *interior* points, and $g = 0$ at the *boundary* points. [Together these are known as *feasible* points; if $g(x,y,z) < 0$, the point is called *infeasible*.] For technical reasons we make the assumption that $\mathbf{grad}\, g$ is continuous and nonzero on the boundary. (See exercise 12.)

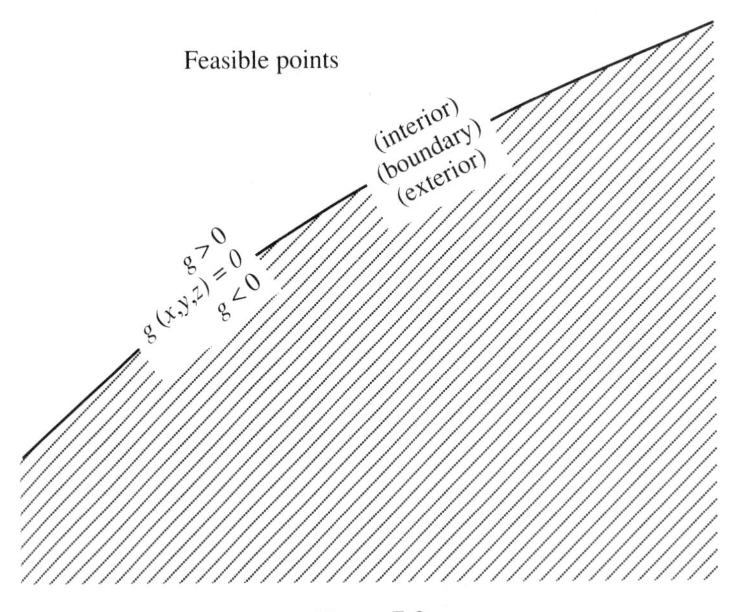

Figure E.2

If the maximum of f occurs at an interior point, then the situation as depicted in figure E.3 is similar to that in figure E.1, and the necessary condition **grad** f $= 0$ continues to hold. Since this is no different from the unconstrained case, we say the constraint $g \geq 0$ is *inactive*.

On the other hand, if the constrained maximum point for f lies on the boundary, it is not necessarily true that **grad** f equals zero there; it is only necessary that one is unable to increase f *without leaving the feasible region.* Figure E.4 tells us how to express this. In order that all *feasible* directions emanating from the maximal point be directed at least 90° away from **grad** f, clearly ***grad*** *f must be oppositely directed to* ***grad*** *g.* Consequently the necessary condition for a boundary point to be a constrained maximum for f is that, for some nonnegative (i.e., positive or zero) scalar λ,

$$\mathbf{grad}\, f = -\lambda\, \mathbf{grad}\, g \qquad \lambda \geq 0 \tag{E.1}$$

This condition is markedly different from the unconstrained criterion, and we say the constraint is *active* in this case. Notice, however, that both the active and inactive necessary conditions can be expressed by eq. (E.1), since $\lambda = 0$ is permitted.

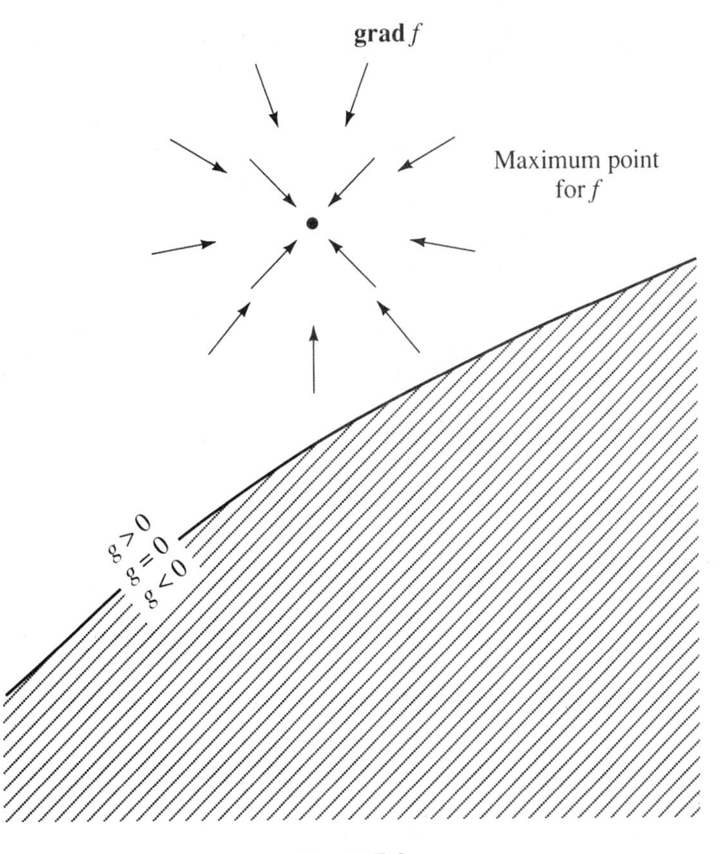

Figure E.3

Example E.2 Find the maximum of $f(x,y,z) = x^4 - 2y^2 - 4z^4 + 2z^2$ subject to the constraint $g(x,y,z) = x + y - 2 \geq 0$.

Solution Note that both *unconstrained* maximum points $(0,0,\frac{1}{2})$ and $(0,0,-\frac{1}{2})$ are now infeasible. At the maximum point the equations

$$-4x^3 = -\lambda \qquad -4y = -\lambda \qquad -16z^3 + 4z = 0 \tag{E.2}$$

must possess a common solution, with $\lambda \geq 0$ and $g(x,y,z) \geq 0$. Clearly the last equation implies z equals 0 or $\pm\frac{1}{2}$ as before.

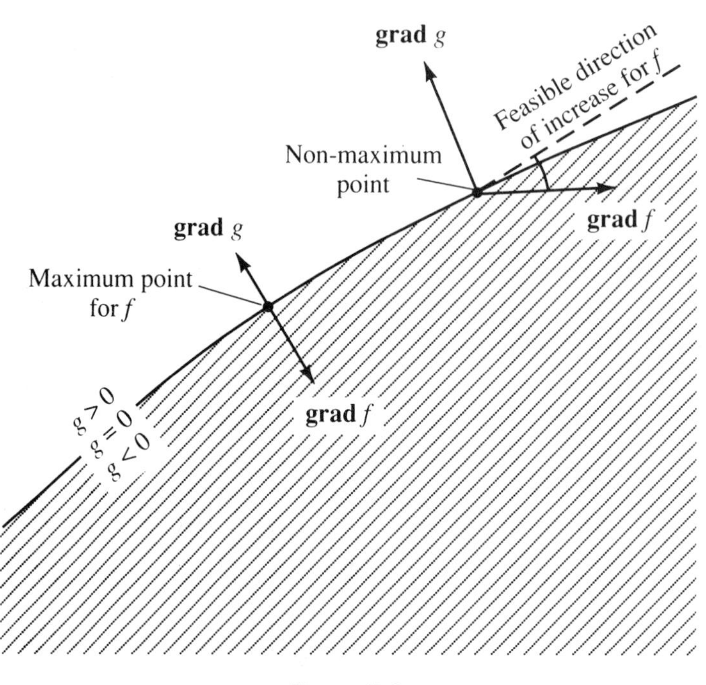

Figure E.4

(i) If the constraint is inactive then $\lambda = 0$ and $\mathbf{grad}\, f = 0$. The equations then imply $x = y = 0$; but this violates the constraint $x + y - 2 \geq 0$.
(ii) Thus the constraint is active, $\lambda > 0$, and $x + y + 2 = 0$. The only real solution of all these equations is $x = y = 1$, $\lambda = 4$ (see exercise 9).

The constrained maximum therefore lies at one of the points $(1,1,0)$, $(1,1,\frac{1}{2})$, or $(1,1,-\frac{1}{2})$. The corresponding values of f are -3 and $-\frac{11}{4}$ (twice), and the latter value is maximal.

How does this generalize to two constraints: $g(x,y,z) \geq 0$ and $h(x,y,z) \geq 0$? In figure E.5 we see that if the maximum of f occurs at an interior point then both constraints are inactive and $\mathbf{grad}\, f = 0$ there. If the maximum occurs at a point where $g = 0$ but $h > 0$ (fig. E.6) then one constraint is active and the other inactive; thus $\mathbf{grad}\, f = -\lambda\, \mathbf{grad}\, g$ with $\lambda \geq 0$. The analogous condition holds at points where $h = 0$ but $g > 0$.

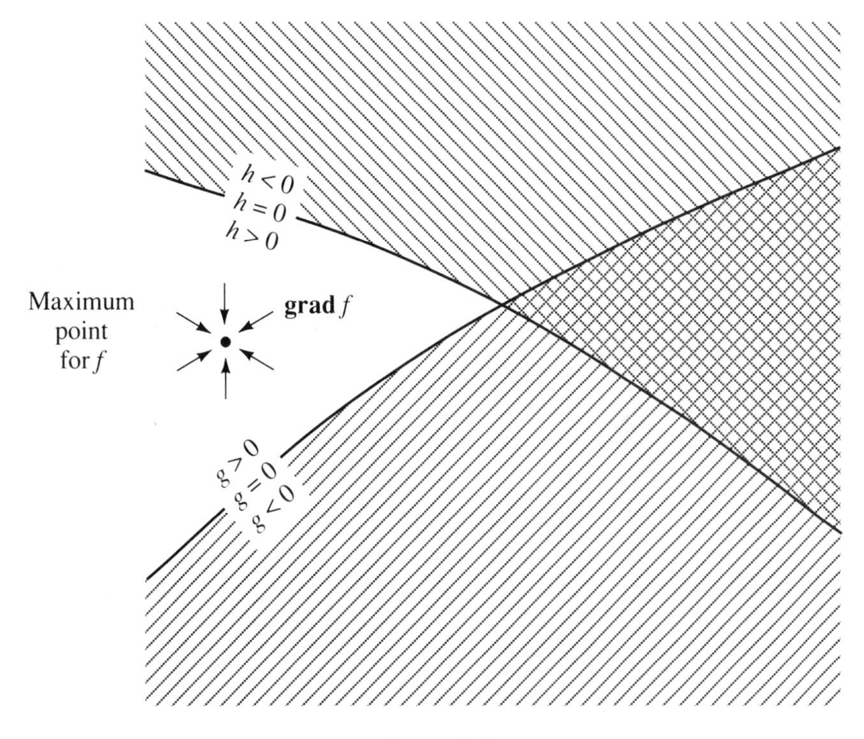

Figure E.5

Now at a point lying in both boundary surfaces, we must ensure that no feasible direction makes an angle of less than 90° with **grad** f. As figure E.7 demonstrates, this requires that **grad** f lie in the planar "wedge" (keep in mind that the situation depicted is three-dimensional) whose sides are rays parallel to $-$**grad** g and $-$**grad** h. This means that **grad** f is expressible as $-\lambda$ **grad** $g - \mu$ **grad** h, with nonnegative coefficients λ and μ. The necessary condition for a maximum point subject to two constraints becomes

$$\mathbf{grad}\, f = -\lambda\, \mathbf{grad}\, g - \mu\, \mathbf{grad}\, h \qquad \lambda \geq 0, \mu \geq 0$$

The generalization of all this is known as the *Kuhn-Tucker theorem of mathematical programming.* Roughly stated (that is, omitting the continuity hypotheses), it says that if **R** is a point at which f achieves a maximum subject to the constraints $g_1(x,y,z) \geq 0$, $g_2(x,y,z) \geq 0$, . . . , $g_n(x,y,z) \geq 0$, then the equation

$$\mathbf{grad}\, f = -\lambda_1\, \mathbf{grad}\, g_1 - \lambda_2\, \mathbf{grad}\, g_2 - \cdots - \lambda_n\, \mathbf{grad}\, g$$

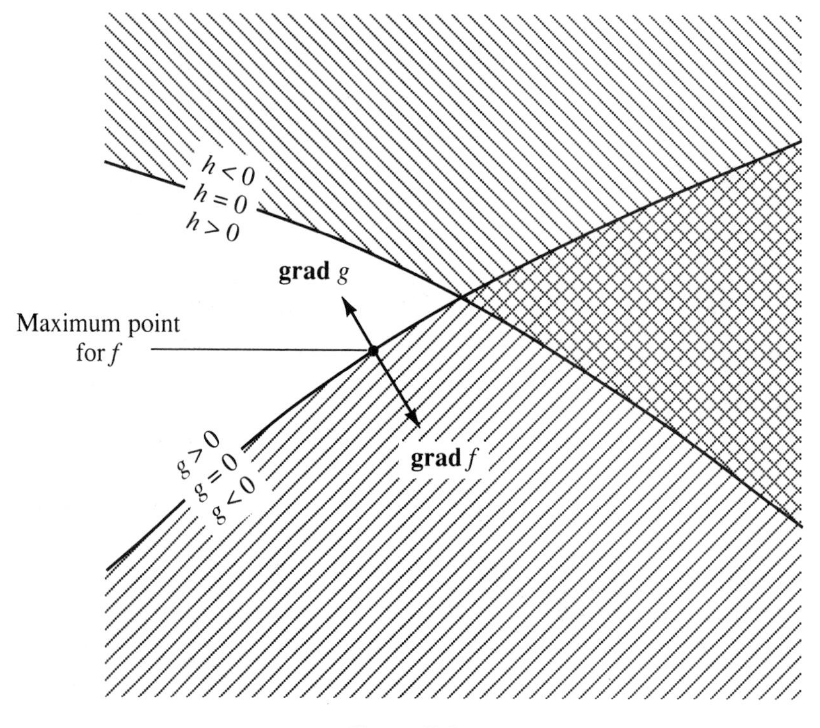

Figure E.6

must hold at **R**, with nonnegative coefficients $\lambda_1, \lambda_2, \ldots, \lambda_n$. These coefficients are known as *Lagrange multipliers.*

Example E.3 Find the maximum of $f(x,y,z) = -x^4 - 2y^2 - 4z^4 + 2z^2$ subject to the constraints $g(x,y,z) = x + y - 2 \geq 0$ and $h(x,y,z) = z^2 - 1 \geq 0$.

Solution The Kuhn-Tucker equations are

$$-4x^3 = -\lambda \qquad -4y = -\lambda \qquad \text{and} \qquad -16z^3 + 4z = -2\mu z$$

As before, $x = y = 1$ is the only real feasible solution to the first two equations. The last equation has solutions

$$z = 0 \qquad z = \pm\sqrt{[(4 + 2\mu)/16]}.$$

$z = 0$ is unfeasible. If $\mu = 0$, $z = \pm\frac{1}{2}$, neither of which meets the constraints. Thus $\mu > 0$ and the constraint is active: $z^2 - 1 = 0$. As a result the maximum occurs at $(1,1,\pm 1)$ and has the value -5.

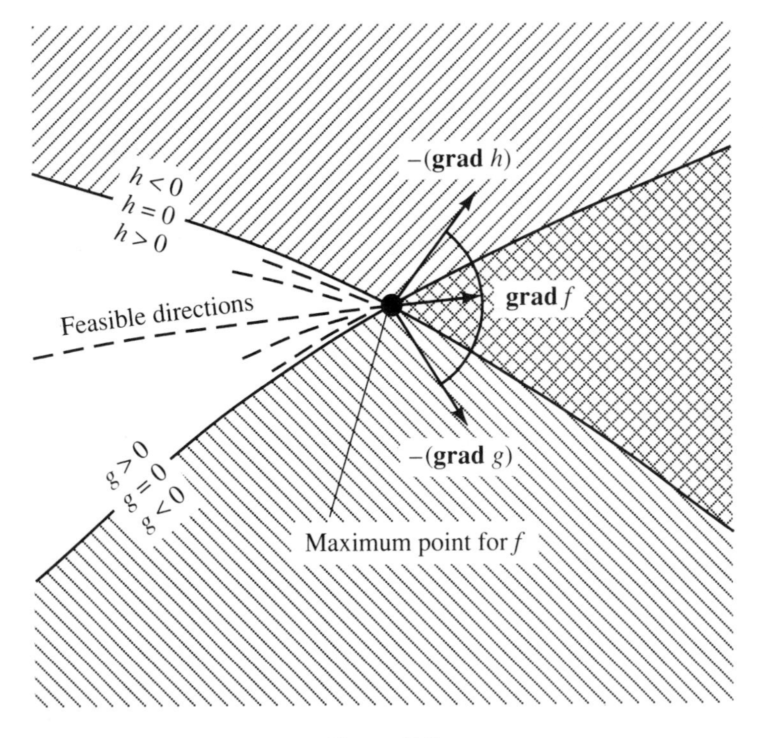

Figure E.7

EXERCISES

1. Find the maximum of $f(x,y,z) = -2x^2 - 3y^2 - 4z^2 + 8x + 12y + 24z + 15$.
2. (a) Restate the Kuhn-Tucker conditions for finding the *minimum* of a function f subject to the constraints $g_1 \geq 0, g_2 \geq 0, \ldots, g_n \geq 0$.
   (b) Restate the Kuhn-Tucker conditions for finding the minimum of f subject to $g_1 \leq 0, g_2 \leq 0, \ldots, g_n \leq 0$.
3. Find the minimum of $f(x,y) = x^2 - 4x + y^2 - 2y + 5$ subject to the constraint
   (a) $x \geq y$
   (b) $x \leq y$
   Interpret geometrically. [*Hint:* Complete the square for f.]
4. The equality constraint $g(x,y,z) = 0$ can be implemented as a *pair* of inequality constraints $g \geq 0$ and $g \leq 0$. State the Kuhn-Tucker conditions for a maximization problem with n equality constraints.

5. Use the Kuhn-Tucker conditions to test which of the following points could be the minimum point for a function $f(x,y)$ whose gradient is given by $(4x + 2y - 10)\mathbf{i} + (2x + 2y - 10)\mathbf{j}$, subject to the constraint $x^2 + y^2 \leq 5$:
   (a) (1,2)
   (b) (0,5)
   (c) (1,1)
6. Consider the problem of maximizing $y^2 - 2x - x^2$ subject to $x^2 + y^2 \leq 1$. Which of the following points satisfy the Kuhn-Tucker conditions? Which give(s) the maximum?
   (a) (1,0) (b) (−1,0) (c) $(-\frac{1}{2}, \frac{\sqrt{3}}{2})$ (d) $(-\frac{1}{2}, -\frac{\sqrt{3}}{2})$
7. Consider the problem of minimizing $f(x,y) = x$, subject to $(x - 3)^2 + (y - 2)^2 \geq 13$ and $(x - 4)^2 + y^2 \leq 16$. Find the three points satisfying the Kuhn-Tucker conditions, and find the minimum.
8. Minimize $x^2 + 2y$ subject to $x^2 + y^2 = 1$.
9. Prove the statement made in example E.2, that $x = y = 1$ is the only real feasible solution to the Kuhn-Tucker conditions. (*Hint:* The function $x^3 + x - 2$ has a positive slope everywhere, so it can cross the x axis only once.)
10. Draw a figure illustrating the possible directions for **grad** f at a maximum point lying on the boundary of *three* constraint surfaces. To make things easier, you might begin by taking the three constraints to be $x \geq 0$, $y \geq 0$, and $z \geq 0$.
11. How would figure E.7 look if **grad** g and **grad** h were parallel at the point S? What would the Kuhn-Tucker condition for **grad** f become?
12. The assumption was made in the text that the surface $g(x,y,z) = 0$ separates space into two regions, where $g < 0$ and $g > 0$ respectively. Of course, this would not occur if the *maximum* value of g happened to be 0, because then the region $g > 0$ would be empty. How is this related to the "technical assumption" that **grad** $g \neq 0$ on the boundary surface?

Selected Answers and Notes

Important: Not all the notes given here will be understood by a beginner. Some are intended for graduate students or teachers who may be teaching vector analysis for the first time.

In this book, vectors are represented by boldfaced letters such as **A, B, C**,. . . . Since you cannot conveniently imitate this, the authors suggest that you either underline the letter, $\underline{A}$, or put an arrow above it, $\vec{A}$. Be sure to distinguish between the number 0 and the vector **0**.

Section 1.1

Note: The reader who has studied modern algebra or logic will recognize that a vector is an *equivalence class* of directed line segments. Note that parallel vectors having the same length in feet will also have the same length in meters or centimeters. That is, vector equality is not a metric property; it does not depend on choice of unit of length.

Section 1.2

1. In problems of this kind, think of the vectors as displacements. The displacement **C** can be obtained by first moving backward along **F**, then moving along **E**, then upward in a direction opposite to **D**. Hence, $\mathbf{C} = -\mathbf{F} + \mathbf{E} - \mathbf{D}$.
2. $\mathbf{G} = -\mathbf{K} + \mathbf{C} + \mathbf{D} - \mathbf{E}$.
3. $\mathbf{x} = \mathbf{F} - \mathbf{B} = \mathbf{A}$.

4. $\mathbf{x} = \mathbf{D} - \mathbf{E} - \mathbf{H} = \mathbf{G}$.
5. Arrow extending from the same initial point and forming the diagonal of the parallelogram determined by the two vectors, as shown in figure 1.2.
6. Notice that $\mathbf{C} - \mathbf{A} = \mathbf{C} + (-\mathbf{A}) = (-\mathbf{A}) + \mathbf{C}$.
7. Yes, the statement is correct (the parallelogram may be "flat").
8. This is easy if you observe that a regular hexagon is composed of six equilateral triangles. (a) $\mathbf{B} - \mathbf{A}, -\mathbf{A}, -\mathbf{B}, \mathbf{A} - \mathbf{B}$, (b) the zero vector.

Note: This kind of addition was called *geometrical addition* when it was first introduced by Möbius and others over a century ago. Observe that the length of $\mathbf{A} + \mathbf{B}$ does not equal the length of $\mathbf{A}$ plus the length of $\mathbf{B}$. A student once announced happily that he had won a bet in a tavern by showing an instance in which three units added to four units produced five units (see exercise 4, section 1.4).

Section 1.3

1. No, length is never negative.
2. $|4\mathbf{A}| = 12$, $|-2\mathbf{A}| = 6$, $|s\mathbf{A}| \leq 6$.
3. $|s\mathbf{A}| = 1$, $|-s\mathbf{A}| = 1$.

Note: If s is a nonzero number and $\mathbf{A}$ is a vector, the vector $s^{-1}\mathbf{A}$ is sometimes said to be "$\mathbf{A}$ divided by s." Thus, if we divide a nonzero vector by its own length, we obtain a vector of unit magnitude. This is the point of the first part of exercise 3.

4. Equals the magnitude of $\mathbf{A}$.
5. No, $\mathbf{A}$ might be the zero vector.
6. Yes.
7. Not necessarily true, since the vectors may not point in the same direction.
8. Two. Think of the plane as the top of your desk. One of the vectors points upward and the other downward. Many students say, "There are infinitely many." This is incorrect, since we do not distinguish between vectors that are equal.
9. Infinitely many. Think of the line as perpendicular to the xy plane. The unit vector might make any angle θ with the x axis.
10. Two, pointing in opposite directions.
11. $\mathbf{C} = \frac{1}{2}(\mathbf{A} + \mathbf{B})$.
12. $|\mathbf{A}| = |\mathbf{A} - \mathbf{B} + \mathbf{B}| \leq |\mathbf{A} - \mathbf{B}| + |\mathbf{B}|$
    Hence $|\mathbf{A}| - |\mathbf{B}| \leq |\mathbf{A} - \mathbf{B}|$. If you prefer a less tricky method, draw a diagram and use a well-known theorem in geometry.
13. $a = -2$, $b = c = 1$ is one possible answer. There are others.

Section 1.4

Note: I think the only reason some students have trouble with some of these exercises is that they think more is expected of them than simply writing down the answer. When I work one of these problems by drawing a diagram and looking at it, students sometimes say, "Oh, is that all you want?" It is not necessary to use any equations or formulas in giving the answer to a trivial exercise.

1. 1.
2. 0.
3. $\sqrt{2}$.
4. 5.
5. $-\mathbf{i}, -\mathbf{j}, \frac{1}{2}\sqrt{2}\mathbf{i} + \frac{1}{2}\sqrt{2}\mathbf{j}$.
6. $\mathbf{A} = \mathbf{i} - 3\mathbf{j}$.
7. $A_1 = |\mathbf{A}| \cos \theta, A_2 = |\mathbf{A}| \sin \theta$.
8. $A_1 = 3\sqrt{3}, A_2 = 3$.
9. (a) $\frac{1}{2}\mathbf{i} + \frac{1}{2}\sqrt{3}\mathbf{j}$.
(b) $\frac{1}{2}\sqrt{3}\mathbf{i} - \frac{1}{2}\mathbf{j}$.
(c) $\frac{3}{5}\mathbf{i} + \frac{4}{5}\mathbf{j}$.
(d) $\frac{1}{2}\mathbf{i} + \frac{1}{2}\sqrt{3}\mathbf{j}, \frac{1}{2}\mathbf{i} - \frac{1}{2}\sqrt{3}\mathbf{j}$.
(e) $\pm(\frac{1}{2}\sqrt{2}\mathbf{i} + \frac{1}{2}\sqrt{2}\mathbf{j})$.
10. 10, 3, $\sqrt{1 + s^2}$, 1.
11. $2\mathbf{i} + 6\mathbf{j}$.

Section 1.5

1. 5, 3, 5.
2. $5\mathbf{i} + 6\mathbf{j} - \mathbf{k}, 4\mathbf{j} + 4\mathbf{k}$.
3. $4\sqrt{2}$
4. $\pm\frac{1}{3}$.
5. $\frac{3}{5}\mathbf{i} + \frac{4}{5}\mathbf{j}$.
6. (a) $4\sqrt{2}$. (b) *yz* plane.
7. $\cos \alpha = \frac{3}{5}$.
8. $\pm\mathbf{j}$.
9. $\sqrt{3}$.
10. $\mathbf{i} - 5\mathbf{j} - \mathbf{k}$.
11. $x\mathbf{i} + y\mathbf{j} + z\mathbf{k}$.
12. $s = 2, t = 3, r = -1$.
13. $\frac{2}{3}, -\frac{2}{3}, \frac{1}{3}$.
14. Use the pythagorean theorem.
15. Cone concentric with the positive *x* axis.
16. Two.
17. $-\cos \alpha, \cos \beta, \cos \gamma$.
18. $\pm\frac{1}{3}\sqrt{3}(\mathbf{i} + \mathbf{j} + \mathbf{k})$.

Section 1.6

1. $2\mathbf{i} - 5\mathbf{j} - 8\mathbf{k}$.
2. $\mathbf{i} + 2\mathbf{j} + 9\mathbf{k}$.
3. $32\mathbf{j} - 26\mathbf{k}$.
4. 10 miles.
5. 7 pounds.
7. $kq_1q_2(\mathbf{R}_1 - \mathbf{R}_2)/|\mathbf{R}_1 - \mathbf{R}_2|^3$.

Section 1.7

1. $\cos^{-1}(-\frac{2}{15})$.
2. $\cos^{-1}(\frac{1}{3}\sqrt{3})$.
3. $\cos^{-1}\frac{1}{41}\sqrt{1435}$, $\cos^{-1}\frac{1}{41}\sqrt{246}$.
4. $90° - \cos^{-1}\frac{1}{3}$.
5. $(\mathbf{i} + \mathbf{j} + \mathbf{k}) \cdot (x\mathbf{i} + y\mathbf{j} + z\mathbf{k}) = 0$ if and only if $x + y + z = 0$. Hence, $\theta = 90°$ if, and only if, $x + y + z = 0$.
7. If $\mathbf{A} + \mathbf{B} + \mathbf{C} + \mathbf{D} = \mathbf{0}$ and $\mathbf{A} = -\mathbf{C}$ then $\mathbf{B} = -\mathbf{D}$.
8. *Hint:* Let the sides be **A**, **B**, and $\mathbf{B} - \mathbf{A}$. If parallel to **A**, the line segment is $-\frac{1}{2}\mathbf{B} + \mathbf{A} + \frac{1}{2}(\mathbf{B} - \mathbf{A}) = \frac{1}{2}\mathbf{A}$.
9. Use the technique illustrated in example 1.3 of the text.
12. Show that the vector sum of the medians is zero.
13. True.
14. True.
15. True.
16. False (radius is 3).
17. $(x - 2)^2 + (y - 3)^2 + (z - 4)^2 = 9$.
18. $x^2 + y^2 = 4$.
19. Line.
20. y axis.
21. The single point $(2, -3, 4)$.
22. The three coordinate planes.
23. 5.
24. 8.
25. 1.
26. Cone of two sheets concentric with z axis.
27. Ellipsoid.

Section 1.8

1. $x = 3t, y = -2t, z = 7t$.
2. $x = 1, y = 2$.
3. $y = 2, z = 3$.
4. $\pm(\frac{3}{5}\mathbf{i} + \frac{4}{5}\mathbf{j})$.
5. $\pm(\frac{6}{7}\mathbf{i} + \frac{3}{7}\mathbf{j} + \frac{2}{7}\mathbf{k})$.
6. $\pm(\frac{3}{19}\sqrt{19}\mathbf{i} - \frac{3}{19}\sqrt{19}\mathbf{j} + \frac{1}{19}\sqrt{19}\mathbf{k})$.
7. $x = \frac{1}{4}y = -z$.
8. $x = 3, y = 4$.
9. $x - 1 = -\frac{1}{2}(y - 4) = \frac{1}{8}(z + 1)$. This may be written in other forms.
10. $\frac{1}{7}\sqrt{42}$.
11. $\cos^{-1}\frac{3}{70}\sqrt{42}$, about 74°.
14. (a) $0 < \lambda < \infty$; $-1 < \lambda < 0$; $-\infty < \lambda < -1$.
15. (a) (2,2,3).
    (b) The lines coincide.
    (c) No intersection (parallel lines).
    (d) No intersection.

Section 1.9

1. 19.
2. $8 + 27 - 12 = 23$.
3. 20.
4. $\cos^{-1} \frac{2}{15}$.
5. $\cos^{-1} \frac{3}{5}$.
6. -2.
7. $\frac{10}{3}$.
8. $\sqrt{2}$.
9. $\frac{15}{13}\sqrt{26}$.
10. $\sqrt{5}\mathbf{i} + \sqrt{5}\mathbf{j}$.
11. Nothing. But $\mathbf{A} = \mathbf{0}$.
12. (a) $-(\mathbf{i} + \mathbf{j} + \mathbf{k}) + (7\mathbf{i} - 2\mathbf{j} - 5\mathbf{k})$
    (b) $3(2\mathbf{i} - \mathbf{j} - 2\mathbf{k}) + \mathbf{0}$
    (c) $\mathbf{0} + (6\mathbf{i} - 3\mathbf{j} - 6\mathbf{k})$.
14. $-\frac{214}{49}\mathbf{i} + \frac{37}{49}\mathbf{j} - \frac{330}{49}\mathbf{k}$.
16. (a) Circle with diameter $|\mathbf{A}|$.
    (b) Sphere with diameter $|\mathbf{A}|$.
17. $|\sin \frac{1}{2}\theta|$.
20. Expand $(\mathbf{A} + \mathbf{B}) \cdot (\mathbf{A} + \mathbf{B}) + (\mathbf{A} - \mathbf{B}) \cdot (\mathbf{A} - \mathbf{B})$.

Section 1.10

Note: Quite often we speak of *the* equation of a plane where it would be better to speak of *an* equation, since distinct equations may represent the same plane. For example, $x + y + 2z = 3$ and $2x + 2y + 4z = 6$ both represent the same plane.

1. (a) $\pm(\frac{2}{3}\mathbf{i} + \frac{1}{3}\mathbf{j} + \frac{2}{3}\mathbf{k})$.
   (b) $\pm(\frac{1}{2}\sqrt{2}\mathbf{i} - \frac{1}{2}\sqrt{2}\mathbf{k})$.
   (c) $\pm(-\frac{1}{37}\sqrt{37}\mathbf{j} + \frac{6}{37}\sqrt{37}\mathbf{k})$.
   (d) $\pm\mathbf{i}$.
   (e) $\pm(\frac{1}{2}\sqrt{2}\mathbf{j} - \frac{1}{2}\sqrt{2}\mathbf{k})$.
   (f) $\pm(\frac{1}{2}\sqrt{2}\mathbf{i} - \frac{1}{2}\sqrt{2}\mathbf{j})$.
2. $x - 4y + z = 0$.
3. $2x - 2y + z + 3 = 0$.
4. $3x + y - z = 3$.
5. No.
6. $\frac{16}{3}$.
7. (a) $\sqrt{14}$. (b) $3\sqrt{2}$. (c) 2.
8. $\frac{1}{3}\sqrt{3}$.
9. $(\mathbf{i} + \mathbf{j} + 3\mathbf{k}) \cdot (2\mathbf{i} - 8\mathbf{j} + 2\mathbf{k}) = 0$.
11. $90° - \cos^{-1} \sqrt{35}/7$.
12. (a) Any multiple of $5\mathbf{i} - \mathbf{j} - \mathbf{k}$.
    (b) $8\sqrt{27}/27$.
13. $\sin^{-1} \frac{5}{9}\sqrt{3}$, about 74°.
14. $90° - \cos^{-1} \frac{5}{9}\sqrt{3}$.
15. $3x - y = C, z = 0$.
16. $\frac{5}{2}\sqrt{2}$.
17. $3x + 2y = 11, z = 0$.
18. $2x + 2y + z = 4$.
19. $\pm\frac{1}{3}$.
20. $7\ (= |AB|)$.
21. 6.
22. 30/7.
23. $3x + 2y + z = 18$.
24. (a) The point $(-2,1,5)$.
    (b) No intersection.
    (c) The line $x = y + 3 = -\frac{1}{2}z$
    (d) No intersection.

Section 1.11

Note to instructor: A k-dimensional vector space (or k-dimensional subspace) is oriented by selecting a linearly independent ordered set consisting of k vectors. Any other such linearly independent ordered set is said to have "positive" orientation if it can be obtained from the given set in the proper order by a linear transformation with positive determinant. If an n-dimensional space has been oriented, and if also an $(n - 1)$-dimensional subspace of the same space is oriented by an ordered set $\mathbf{A}_1, \mathbf{A}_2, \ldots, \mathbf{A}_{n-1}$, then the same orientation of the subspace can be prescribed just as well by selecting a single vector $\mathbf{C}$ not in the subspace, using the following convention: the ordered set $\mathbf{A}_1, \mathbf{A}_2, \ldots, \mathbf{A}_{n-1}, \mathbf{C}$ must have positive orientation.

1. Numerically they are equal to the areas of the projections of the area on the coordinate planes.

Section 1.12

1. (a) $2\mathbf{i} + 14\mathbf{j} + 4\mathbf{k}$.
(b) $-8\mathbf{i} + 23\mathbf{j} - \mathbf{k}$.
(c) $-11\mathbf{i} - 6\mathbf{j} + \mathbf{k}$.
(d) $\mathbf{k}$.
(e) $\mathbf{j} - \mathbf{i}$.

2. $\sqrt{26}$.

3. $\frac{1}{2}\sqrt{61}$.

4. $\mathbf{0}$. $\mathbf{A}$ and $\mathbf{B}$ are parallel.

5. $\pm(\frac{1}{11}\sqrt{11}\mathbf{i} - \frac{3}{11}\sqrt{11}\mathbf{j} + \frac{1}{11}\sqrt{11}\mathbf{k})$.

6. $\frac{1}{8}(x - 2) = -\frac{1}{13}(y - 3) = -\frac{1}{3}(z - 7)$.

7. $x = -\frac{1}{4}y = \frac{1}{3}z$.

8. $-64\mathbf{j} + 16\mathbf{k}$, $16\mathbf{i} - 16\mathbf{j} + 16\mathbf{k}$. No.

9. $17x - y + 9z = 43$.

10. $\pm\frac{1}{25}\sqrt{5}(5\mathbf{i} + 6\mathbf{j} - 8\mathbf{k})$.

11. $\sin(\psi - \theta) = \sin\psi\cos\theta - \cos\psi\sin\theta$.

13. $\sqrt{65}/\sqrt{26}$.

14. $\sqrt{34}$.

15. $\frac{x-4}{5} = \frac{y-2}{-8} = \frac{z-1}{-7}$.

16. (a) $8\sqrt{29}/29$.
(b) $8\sqrt{77}/77$.
(c) $5(6\mathbf{i} + 4\mathbf{j} + 5\mathbf{k})$.

17. $\pm\frac{5}{3}(2\mathbf{i} + \mathbf{j} - 2\mathbf{k})$.

18. $r = 3$, $s = -\frac{27}{2}$.

19. One of them is zero.

20. Yes.

22. $\pm 8\mathbf{i}$

23. (a) No.

(b) $\frac{1}{2}x - \frac{52}{7} = -\frac{1}{4}y + \frac{52}{21} = z - \frac{208}{21}$. (This answer can be written in many other ways, so don't be discouraged if your answer differs from this in appearance.)

(c) $4/\sqrt{21}$.

24. $x^2 + y^2 + z^2 - xy - yz - zx = 2$; a cylinder of radius $\frac{2}{3}\sqrt{3}$ ft.

31. No.

Section 1.13

1. (a) 30. (b) -13. (c) 5. (d) 1.

2. 5.

3. 0.

4. $\frac{2}{3}$.

5. 1.

6. $3x - 17y - 4z = 0$.

7. $3x - 7y + z = -20$.

8. (a) Their triple scalar product is zero. Alternatively, all three are perpendicular to $\mathbf{i} + \mathbf{j} + \mathbf{k}$.

(b) $x + y + z = 0$.

9. (a) $101/6$.

(b) $x - 11y - 14z = -43$.

(c) $1/\sqrt{319}$.

10. (a) $C_3 = 2$.

(c) Draw a diagram.

11. $\frac{2}{19}\sqrt{38}$.

12. (a) 13.

(b) $\mathbf{i} - 3\mathbf{j} - 4\mathbf{k}$.

(c) 3.

(d) -12.

13. (a) $\sqrt{26}/2$. (b) 2.

(c) $\cos^{-1}\frac{1}{9}$.

14. Yes.

15. They are coplanar.

16. (a) Compare $\mathbf{A} \cdot \mathbf{i}$ and $\mathbf{A} \cdot \mathbf{u}$.

(b) $\mathbf{A}$.

Section 1.14

5. $(\omega \cdot \mathbf{R})\omega - (\omega \cdot \omega)\mathbf{R}$. **6.** No. **7.** 0. **12.** $\mathbf{u} = \pm\mathbf{w}$ or $\mathbf{u}, \mathbf{w} \perp \mathbf{v}$

15. $\begin{vmatrix} \mathbf{u} \cdot \mathbf{u} & \mathbf{u} \cdot \mathbf{v} \\ \mathbf{u} \cdot \mathbf{v} & \mathbf{v} \cdot \mathbf{v} \end{vmatrix}$.

Section 1.15

These exercises appeared already in section 1.14.

Section 2.1

1. (a) $\cos t\,\mathbf{i} - \sin t\,\mathbf{j}$.
   (b) True since $\mathbf{k} \cdot \mathbf{F}'(t) = 0$.
   (c) $t = n\pi$ $(n = 0, \pm 1, \pm 2, \ldots)$.
   (d) Yes, $\sqrt{2}$.
   (e) Yes, 1.
   (f) $-\sin t\,\mathbf{i} - \cos t\,\mathbf{j}$.
2. (a) $3\mathbf{i} + 3t^2\mathbf{j}$.
   (b) $\cos t\,\mathbf{i} - e^{-t}\mathbf{j}$.
   (c) $-2t\mathbf{i} + (e^t + 5t^4)\mathbf{j} + (e^t - 3t^2)\mathbf{k}$.
   (d) $(\cos t + 3t^2)(\mathbf{i} + \mathbf{j} + 2\mathbf{k})$.
   (e) $\mathbf{0}$.
3. (a) $6t - 10t \sin t - 5t^2 \cos t$.
   (b) $(8t\sqrt{8t^2 + 1})/(8t^2 + 1)$.
   (c) $1 - 12t^3$.
4. Use theorem 2.4, noting that one term vanishes in this case.
5. (a) 7.
   (b) 33.
   (c) $8\mathbf{i} + 5\mathbf{j} - 6\mathbf{k}$.
   (d) 0.
   (e) -2.
   (f) $\frac{3}{5}\mathbf{i} + \frac{2}{5}\mathbf{j} + \frac{6}{5}\mathbf{k}$.
   (g) $-42\mathbf{i} + 66\mathbf{j} - \mathbf{k}$.
   (h) $\mathbf{B}$.
   (i) $\mathbf{B} \times \mathbf{C}$.

Section 2.2

1. $\mathbf{i}$.
2. (a) $2\sqrt{5}\pi^2$
   (b) $\frac{1}{5}\sqrt{5}(\sin t\,\mathbf{i} + \cos t\,\mathbf{j} + 2\mathbf{k})$.
   (c) $\frac{1}{5}\sqrt{5}(-\mathbf{j} + 2\mathbf{k})$.
3. $(\mathbf{i} + 2\pi\mathbf{j})/\sqrt{1 + 4\pi^2}$.
4. Along a straight line, $\mathbf{T}$ is constant.
5. (a) $\sqrt{2}(e - 1)$.
   (b) $x = \dfrac{s + \sqrt{2}}{\sqrt{2}} \cos \log \left(\dfrac{s + \sqrt{2}}{\sqrt{2}}\right)$, $y = \dfrac{s + \sqrt{2}}{\sqrt{2}} \sin \log \left(\dfrac{s + \sqrt{2}}{\sqrt{2}}\right)$, $z = 0$.
6. (a) $\int_0^1 \sqrt{14}\, dt = \sqrt{14}$.
   (b) Distance between points is $\sqrt{14}$, and the path is straight.
7. $x^2 - y^2 = 1$, $z = 0$.
9. At (0,0,0), corresponding to $t = 0$.
13. No, the tangent may be parallel to the y axis.

14. The arc $t^2\mathbf{i} + t^3\mathbf{j}$ has a cusp at $t = 0$. Physically, a particle can follow a curve with a sharp corner by decelerating to zero speed at the corner, then resuming with a different direction of velocity.

Section 2.3

1. (a) $\sqrt{2}e^t$.
(b) $a_t = \sqrt{2}e^t$, $a_n = \sqrt{2}e^t$.
(c) $\frac{1}{2}\sqrt{2}[(\cos t - \sin t)\mathbf{i} + (\sin t + \cos t)\mathbf{j}]$.
(d) $\frac{1}{2}\sqrt{2}e^{-t}$.

2. (a) $\sqrt{9t^2 + 25}$.
(b) $9t/\sqrt{9t^2 + 25}$, $[9(t^2 + 4) - 81t^2/(9t^2 + 25)]^{1/2}$.
(c) $\dfrac{3(\cos t - t \sin t)\mathbf{i} + 3(\sin t + t \cos t)\mathbf{j} + 4\mathbf{k}}{\sqrt{9t^2 + 25}}$.
(d) $3(9t^4 + 52t^2 + 100)^{1/2}/(9t^2 + 25)^{3/2}$.

3. (a) $\sqrt{3}e^t$.
(b) $a_t = \sqrt{3}e^t$, $a_n = \sqrt{2}e^t$.
(c) $\frac{1}{3}\sqrt{3}[(\cos t - \sin t)\mathbf{i} + (\sin t + \cos t)\mathbf{j} + \mathbf{k}]$.
(d) $\frac{1}{3}\sqrt{2}e^{-t}$.

4. (a) $10\sqrt{5}$.
(b) $a_t = 0$, $a_n = 80$.
(c) $\frac{1}{5}\sqrt{5}(2 \cos 4t\, \mathbf{i} - 2 \sin 4t\, \mathbf{j} + \mathbf{k})$.
(d) $\frac{4}{25}$.

5. (a) $v = \frac{3}{2}$.
(b) $\mathbf{a} = -\cos t\, (\mathbf{i} - \mathbf{j}) - \sin t\, (\mathbf{i} + \mathbf{j})$.
(c) $-\frac{2}{3} \sin t\, (\mathbf{i} - \mathbf{j}) + \frac{2}{3} \cos t\, (\mathbf{i} + \mathbf{j}) + \frac{1}{3}\mathbf{k}$.
(d) $k = 4\sqrt{2}/9$

6. $\dfrac{2(3t^4 + 2t^3 - 3t^2 - 2t + 2)^{1/2}}{3(2t^4 - 4t^3 + 10t^2 + 1)^{3/2}}$

12. $\mathbf{F} \times \dfrac{d\mathbf{F}}{dt} \cdot \dfrac{d^3\mathbf{F}}{dt^3}$.

13. $\frac{1}{2}$; $-\frac{1}{2}$.

14. 6, $\frac{1}{2}$, 0, circle of radius 2 in the plane $x = y$.

15. (a) 1.
(b) 0.
(c) $a_t = d^2s/dt^2$.
(d) 0.
(e) ds/dt.
(f) τ.
(g) 1.
(h) k.
(i) $-\tau\mathbf{N}$.

18. (a) False. (b) False. (c) True.

Section 2.4

1. $\mathbf{v} = 4b[(\sin\theta)\mathbf{u}_r + (1 - \cos\theta)\mathbf{u}_\theta]$
   $\mathbf{a} = 16b[(2\cos\theta - 1)\mathbf{u}_r + (2\sin\theta)\mathbf{u}_\theta]$.
2. $\mathbf{v} = b[(\cos t)\mathbf{u}_r - e^{-t}(1 + \sin t)\mathbf{u}_\theta]$
   $\mathbf{a} = b([-\sin t - e^{-2t}(1 + \sin t)]\mathbf{u}_r + e^{-t}[1 + \sin t - 2\cos t]\mathbf{u}_\theta)$.
5. Yes, except when its velocity is zero.
6. (a) The second term.
   (b) The second and third terms.
   (c) All are nonzero.
   (d) Many possibilities.
9. 24π, since $dr/dt = 3$ and $d\theta/dt = 4\pi$.
10. (a) $\pi^2 r$ cm/sec² (if r is in cm) directed towards the center. Note that 30 rev/min $= \pi$ rad/sec.
    (b) $4\pi\mathbf{u}_\theta$ cm/sec.
11. If the particle is moving parallel to the field no force will be exerted. (In elementary books it is sometimes stated that the force is proportional to the rate at which the particle "cuts" the lines of flow.)
12. $v/r|\mathbf{B}|$. [$qv\,|\mathbf{B}|$ must equal the component a_n discussed in section 2.3.]
14. $$\left[\frac{d^3r}{dt^3} - 3\frac{dr}{dt}\left(\frac{d\theta}{dt}\right)^2 - 3r\frac{d\theta}{dt}\frac{d^2\theta}{dt^2}\right]\mathbf{u}_r + \left[3\frac{d^2r}{dt^2}\frac{d\theta}{dt} + 3\frac{dr}{dt}\frac{d^2\theta}{dt^2} + r\frac{d^3\theta}{dt^3} - r\left(\frac{d\theta}{dt}\right)^3\right]\mathbf{u}_\theta.$$

Section 3.1

1. (a) $(\cos x + ye^{xy})\mathbf{i} + xe^{xy}\mathbf{j} + \mathbf{k}$. (b) $-\mathbf{R}/|\mathbf{R}|^3$. (c) $\mathbf{k}$.
2. yz plane, where $x = 0$.
3. f depends only on y.
4. $f(x,y,z) = x^2 + yz + C$.
6. Unit vector directed away from the z axis, except at points on the z axis, where it is not defined.
7. (a) 10.
   (b) The maximum rate of increase of R^2 is in the direction $\mathbf{R}$, where
   $$\frac{d}{ds}(R^2) = \frac{d}{dr}(R^2) = 2R$$
   which equals 10 at (3,0,4).
9. (a) 0. (b) $-\frac{4}{3}$.
10. (a) $\frac{5}{3}$. (b) $-\frac{2}{3}$. (c) $-\frac{28}{3}$. (d) $\frac{1}{42}\sqrt{14}$.

11. $150\sqrt{5}$. This function equals s^6, where s is the distance to the y axis. We have $(d/ds)(s^6) = 6s^5 = 150\sqrt{5}$ at this point.

13. (1,1,2). (At this point the tangent to the curve is perpendicular to **grad** ϕ. Of course, this can also be done without using vector methods, by observing that $\phi = t^2 - 6t + t^4$ has its minimum at $t = 1$.)

14. Any scalar multiple of $4\mathbf{i} + \mathbf{j} + \mathbf{k}$.

15. $2x + 4y - z = 21$.

16. (a) From your diagram you see that any scalar multiple of $\mathbf{i} + \mathbf{k}$ will do.
(b) $4\mathbf{i} + 4\mathbf{k}$.

17. $x + 2y - 8z = -28$.

18. $x = y, z = 0$.

19. $\pm\frac{1}{14}\sqrt{14}(3\mathbf{i} - \mathbf{j} + 2\mathbf{k})$.

20. $4x + 6y - z = 13$.

23. $\pm\frac{1}{2}\sqrt{2}(\mathbf{i} - \mathbf{j})$. In (c), let $\mathbf{R} = 2 \sin t\,\mathbf{i} + 2 \cos t\,\mathbf{j} + \sqrt{5}\mathbf{k}$.

24. $\cos^{-1} \frac{31}{32}$.

25. $\sin^{-1} \frac{2}{3}\sqrt{2}$.

27. $90° - \cos^{-1} 98/(\sqrt{157}\sqrt{65})$.

28. $\cos^{-1} 8/(3\sqrt{21})$.

30. If $\mathbf{T}$ is a unit tangent to the ellipse, $\nabla(|\mathbf{R}_1| + |\mathbf{R}_2|) \cdot \mathbf{T} = 0$ (why?). Also, $\nabla|\mathbf{R}_1|$ and $\nabla|\mathbf{R}_2|$ are unit vectors in the directions $\mathbf{R}_1$ and $\mathbf{R}_2$ respectively, so the cosine of the angle between $\nabla|\mathbf{R}_2|$ and $\mathbf{T}$ equals the cosine of the angle between $\nabla|\mathbf{R}_1|$ and $-\mathbf{T}$.

31. (2,4,8). (The vector extending from the center of the sphere to this point is perpendicular to the given plane.)

32. (1,1,1). [The gradient of $x^2 + 2y^2 + 3z^2$ is parallel to $\mathbf{i} + 2\mathbf{j} + 3\mathbf{k}$ at (1,1,1).]

33. (1,1,2). This is the preceding exercise in a different format.

36. (a)
$$\nabla w \cdot \nabla u \times \nabla v = (u\nabla v + v\nabla u) \cdot \nabla u \times \nabla v = u(\nabla v \cdot \nabla u \times \nabla v) + v(\nabla u \cdot \nabla u \times \nabla v) = 0$$
(b) At any point in space, the isotimic surfaces u = constant and v = constant intersect in a curve along which both u and v and, hence, w are constant. $\nabla u \times \nabla v$ is tangent to this curve and, hence, perpendicular to ∇w.

37. If u, v, and w are functionally related, $\nabla w \cdot \nabla u \times \nabla v = 0$.

Section 3.2

2. (a) $x(z + a) = -1, y(z + b) = -1$.
(b) $x(z - 3) = -1, y(z - 3) = -1$.

3. Half lines extending from the origin.

4. The gradient is normal to these surfaces:

Section 3.3

1. $ye^{xy} + x \cos xy - 2x \cos zx \sin zx$.
2. 3.
3. $6y^3z + 18x^2yz$.
4. Zero except at the origin, where the field is not defined. The magnitude of this field at any point is $1/R^2$, so this field can be thought of as the electric field intensity due to a charge of suitably chosen magnitude at the origin. A physicist or electrical engineer might say that the divergence is "infinity" at the origin, since the divergence of an electrostatic field is proportional to the charge density, and the charge density at a point charge is "infinite."
6. Let $\mathbf{F} \cdot \text{grad}\, \phi = 0$.
7. $\mathbf{F} = y\mathbf{i} + z\mathbf{j} + x\mathbf{k}$ is one example.
8. There are infinitely many possible answers, for example $\mathbf{F} = -x\mathbf{i}$.
9. Again there are infinitely many acceptable answers. Two of them are $e^x\mathbf{i}$ and $e^x\mathbf{i} + ye^x\mathbf{j}$.
10. Divergence is zero everywhere, since $\partial F_1/\partial x = 0$, $F_2 = 0$, and (we assume) $F_3 = 0$. Some students observe that $\mathbf{F} = Cy\mathbf{i}$ for some constant C, and then compute the answer using the formula for the divergence. This is clever, but not the point of the exercise.
11. Divergence is zero everywhere. For example, consider point P. Along the x axis, $F_1 = 0$, so $\partial F_1/\partial x = 0$ at P. As we move through P along the flow line indicated, F_2 takes on its maximum value $|\mathbf{F}|$, therefore $\partial F_2/\partial s = 0$ at P, where s is measured along the flow line. But at point P we are moving parallel to the y axis, so $\partial F_2/\partial y = \partial F_2/\partial s$ at P, hence is zero at this point. Another method: Conjecture that $\mathbf{F} = -y\mathbf{i} + x\mathbf{j}$ and use the formula.

Section 3.4

1. $x\mathbf{i} - y\mathbf{j} + y(1 - 2x)\mathbf{k}$.
2. $-z^2 \sin yz^2\mathbf{i} + (y \cos xy - xe^{xy})\mathbf{k}$.
3. $-(y^2 + z^2)\mathbf{i} + 2zx\mathbf{j}$.
4. (a) $1 + z^2 + x + y$.
   (b) $z\mathbf{i} + 2xz\mathbf{j} + y\mathbf{k}$.
5. The paddle wheel will not tend to rotate.
6. Think of the velocity field of a fluid swirling about the x axis. Assume constant angular velocity $\boldsymbol{\omega}$. Then $\mathbf{v} = \boldsymbol{\omega} \times \mathbf{R}$, and since $\textbf{curl } \mathbf{F} = 2\boldsymbol{\omega}$ as stated in the text (to be proved later) we have $\boldsymbol{\omega} = \mathbf{i}$ and

$$\mathbf{v} = \mathbf{i} \times \mathbf{R} = \mathbf{i} \times (x\mathbf{i} + y\mathbf{j} + z\mathbf{k}) = y\mathbf{k} - z\mathbf{j}$$

   This is one possible answer. Another is $2y\mathbf{k}$, which represents a shearing motion parallel to the xz plane.

7. No (fig. 3.15).
8. No.
12. 0 (by symmetry).

Section 3.5

1. 16
2. $12\mathbf{i} + 4\mathbf{j} + \mathbf{k}$.
3. 64.
4. (a) $2xy + 1$.
(b) $-2\mathbf{i} + \mathbf{j} - x^2\mathbf{k}$.
(c) $2y\mathbf{i} + 2x\mathbf{j}$.
5. Scalar field.
6. Vector field.
7. 3, $\mathbf{0}$.
8. $(x^2 + z^2)e^{xz}$.
9. Always $\mathbf{0}$.
10. Always 0.

Section 3.6

1. $20x^3yz^3 + 6x^5yz$.
2. 0 except at the origin.
3. $-2yz^2(y^2z^2 + 3x^2z^2 + 6x^2y^2)\mathbf{k}$.
4. (a) and (b). Also (c) provided that $p^2 = q^2$.
5. (a) Vector field.
(b) Scalar field.
(c) Vector field.
(d) Scalar field.
(e) Zero vector field.
(f) Meaningless.
(g) Vector field.
(h) Vector field.
(i) Meaningless.
(j) Vector field.
6. (b) $\dfrac{\sin x \sinh y}{\sinh 5} + \dfrac{\sin 2x \sinh 2y}{\sinh 10}$
7. (a) $4x\mathbf{i} + \mathbf{j}$.
(b) 3.
(c) 4.
(d) $z\mathbf{i} - 4xz\mathbf{j} + (4xy - x)\mathbf{k}$.
8. (a) $2\mathbf{i}$.
(b) $2\mathbf{k}$.
(c) -2.

Section 3.7

4. The Taylor series for $f(x,y,z)$.

Section 3.8

5. As written, the right side is symmetrical in **F** and **G**, but the left side is not, since $\mathbf{F} \times \mathbf{G} \neq \mathbf{G} \times \mathbf{F}$.
8. $40[(z - y)\mathbf{i} + (x - z)\mathbf{j} + (y - x)\mathbf{k}]$.
9. $2\mathbf{A}$.
10. (a) $2\mathbf{R} \cdot \mathbf{A}$.
    (b) $2\mathbf{R} \times \mathbf{A}$.
    (c) $2R^2\mathbf{A}$.
    (d) $4(\mathbf{A} \cdot \mathbf{R})^3\mathbf{A}$.
    (e) $(\mathbf{A} \cdot \mathbf{R})/R$.
    (f) $(\mathbf{A} \cdot \mathbf{R})\mathbf{A}$.
    (g) 0.
    (h) $2\mathbf{A}$.
    (i) 6.
11. Zero vector field except where $R = 0$; not defined where $R = 0$.
12. div (**curl F**) $= 0$. Hence $2 + C = 0$ and $C = -2$.

Section 3.10

6. See the end of section 3.10.
8. All zero. (Except, of course, at the origin.)
9. π.
10. $nr^{n-1}\mathbf{e}_r$.
11. $(-2 \cos \phi \, \mathbf{e}_r - \sin \phi \, \mathbf{e}_\phi)/r^3$.
12. $\nabla \cdot \mathbf{F} = \dfrac{2}{r} + \cot \phi - \dfrac{\sin \theta}{\sin \phi}$

    $\nabla \times \mathbf{F} = \cot \phi \cos \theta \, \mathbf{e}_r - 2 \cos \theta \, \mathbf{e}_\phi + 2\mathbf{e}_\theta$.
13. $n = -2$.
14. All n.
22. (a) $F_r = r^{m+1}/(m + 3)$.

Section 3.11

4. $(1/u_1) \, du_1 \, du_2 \, du_3$.
5. $(12u_1u_3 + 12u_2u_3 + 3)/4u_3$.
6. (a) Yes.
    (b) $x = (u_1 + u_2)/2$, $y = (u_1 - u_2)/2$, $z = u_3/2$.
    (c) $h_1 = \frac{1}{2}\sqrt{2}$, $h_2 = \frac{1}{2}\sqrt{2}$, $h_3 = \frac{1}{2}$.
    (d) $\nabla^2 f = 2\dfrac{\partial^2 f}{\partial u_1^2} + 2\dfrac{\partial^2 f}{\partial u_2^2} + 4\dfrac{\partial^2 f}{\partial u_3^2}$.
    (e) $\sqrt{2}\mathbf{u}_1 + \sqrt{2}\mathbf{u}_2 + 4\mathbf{u}_3$.

7. (a) $x = \frac{1}{3}(2u_1 + u_2), y = \frac{1}{3}(u_1 - u_2), z = \frac{1}{2}u_3.$
(c) This coordinate system is not orthogonal.

8. (b) $h_1 = h_2 = 2(u_1^2 + u_2^2)^{1/2}, h_3 = 1.$
(c) $\left(\dfrac{\partial^2 g}{\partial u_1^2} + \dfrac{\partial^2 g}{\partial u_2^2}\right) \Big/ 4(u_1^2 + u_2^2) + \dfrac{\partial^2 g}{\partial u_3^2}.$
(d) $\nabla \cdot \mathbf{F} = \dfrac{u_1(u_2 + u_3)}{2(u_1^2 + u_2^2)^{3/2}}$

$$\nabla \times \mathbf{F} = \mathbf{e}_1/2(u_1^2 + u_2^2)^{1/2} + \mathbf{e}_2 + (4u_1^2 + 2u_2^2 - 2u_2u_3)\mathbf{e}_3/4(u_1^2 + u_2^2)^{3/2}.$$

10. This coordinate system is not right-handed, hence the usual formula for curl does not apply.

11. (a) $1/uv.$
(b) $2w/uv.$

12. $h_u = h_v = \sqrt{u^2 + v^2}, h_z = 1.$

13. $(u^2 + v^2)\, du\, dv\, dz.$

14. (a) $\dfrac{1}{u^2 + v^2}\left(\dfrac{\partial}{\partial u}(\sqrt{u^2 + v^2}A_u) + \dfrac{\partial}{\partial v}(\sqrt{u^2 + v^2}A_v)\right) + \dfrac{\partial A_z}{\partial z}.$
(b) $\dfrac{\partial^2 \phi}{\partial u^2} + \dfrac{\partial^2 \phi}{\partial v^2} + (u^2 + v^2)\dfrac{\partial^2 \phi}{\partial z^2} = 0.$

Section 4.1

1. $\frac{41}{6}$.

3. (a) 8. (b) 8.

4. $\pm 8\pi$, depending on direction.

5. 40. (This can also be done by observing that $\mathbf{F} \cdot d\mathbf{R} = d\phi$ where $\phi = x^2y + zy$, so that the integral is $\phi(3,4,1) - \phi(1,0,2)$. See section 4.3 for further discussion of this "trick.")

6. Zero.

7. (a) Along this path, $\mathbf{F} = \sqrt{1 - x^2}\mathbf{i} - x\mathbf{j}$ and

$$d\mathbf{R} = dx\mathbf{i} - \frac{x\, dx}{\sqrt{1 - x^2}}\mathbf{j}$$

so

$$\mathbf{F} \cdot d\mathbf{R} = \frac{dx}{\sqrt{1 - x^2}} \quad \text{and} \quad \int \mathbf{F} \cdot d\mathbf{R} = \int_{-1}^{1} \frac{dx}{\sqrt{1 - x^2}} = \pi$$

(b) $\pi.$

8. $\mathbf{F} \cdot d\mathbf{R} = -d(\tan^{-1} y/x) = -d\theta.$

13. 36. (*Caution:* $\mathbf{R} \cdot d\mathbf{R} = s\, ds$ in this case because the points are collinear with the origin.)

14. 0. (**F** is perpendicular to $d\mathbf{R}$.)
15. (a) $\frac{1}{2}\sqrt{2}(\mathbf{i} + \mathbf{j})$.
(b) **i.**
(c) $-\mathbf{j}$.
16. (a) $\sqrt{2}\, dx$ or $\sqrt{2}\, dy$.
(b) dx.
(c) $-dy$.
17. (a) $d\mathbf{R} = dx\mathbf{i} + dy\mathbf{j} = dx\mathbf{i} + dx\mathbf{j} = (\frac{1}{2}\sqrt{2}\mathbf{i} + \frac{1}{2}\sqrt{2}\mathbf{j})\sqrt{2}\, dx = \mathbf{T}\, ds$.
(b) $d\mathbf{R} = dx\mathbf{i} + dy\mathbf{j} = \mathbf{i}\, dx = \mathbf{T}\, ds$.
(c) $d\mathbf{R} = dx\mathbf{i} + dy\mathbf{j} = dy\mathbf{j} = \mathbf{T}\, ds$.

Section 4.2

Note: In this book, any set of points is a *region* and a region is a *domain* if and only if it is open and connected. In some books other conventions are used; there is no standard agreement: for example, some books use *domain* to mean *domain of definition* and those domains that are open and connected are called regions.

1. Domain, not simply connected.
2. Simply connected domain.
3. Simply connected domain.
4. Not a domain. (Points on the plane $z = 0$ are not interior.)
5. Simply connected domain.
6. Domain, not simply connected.
7. Simply connected domain.
8. Not a domain (not connected).

Section 4.3

1. The integral over C equals that over C_1 minus that over C_2, so if the first of these is zero the other two are equal.
2. Many possibilities.
3. Many possibilities.
4. 2π or -2π, depending on which way the circle is oriented.
5. ϕ is a multiple-valued function, and hence not a scalar field as we have defined it.
6. $\phi = yx + \sin xz + C$.
7. $\mathbf{F} = \mathbf{grad}\, \phi$ where $\phi = x^2y + yz$.
8. They differ by a constant.

Section 4.4

1. (a) Conservative, $\phi = 6x^2y + xyz + C$.
   (b) Conservative, $\phi = e^{xz} + C$.
   (c) Conservative, $\phi = -\cos x + \frac{1}{3}y^3 + e^z + C$.
   (d) Not conservative.
   (e) Conservative $\phi = \ln(x^2 + y^2) + z^2$.
   The domain of definition is not simply connected in (e). You must explicitly construct ϕ.
2. Yes. $\phi + \psi$.
4. $\phi(1,2,3) = -\frac{1}{14}\sqrt{14}$ and $\phi(2,3,5) = -\frac{1}{38}\sqrt{38}$; hence the work done is

$$\phi(2,3,5) - \phi(1,2,3) = \tfrac{1}{14}\sqrt{14} - \tfrac{1}{38}\sqrt{38}$$

5. No, provided the path avoids the origin.

Note: Conservative fields are sometimes called *potential fields*. The term *irrotational* is also used. It is not possible for a flow line of such a field to be a closed curve, for the integral of a field about a closed flow line is nonzero, and this would contradict (*ii*). Therefore the flow lines either have no endpoints (e.g., if they "extend to infinity" in both directions) or perhaps they start at a point (called the "source") and perhaps end at another point (called the "sink"). For this reason, such fields are also called *source fields*. A simple example is the electrostatic field due to a positive point charge at the origin. The origin is the "source" and the flow lines extend radially away from the origin.

10. -1. **11.** No.

Section 4.5

2. $-\frac{1}{2}x^2\mathbf{k}$.
7. If $\mathbf{F} = \nabla\phi$, a vector potential is given by $\phi\mathbf{G}$.
8. $$\mathbf{G} = \left[-A_1\left(\frac{y^2}{3} + \frac{z^2}{2}\right) + A_2\frac{xy}{3} + A_3xz\right]\mathbf{i} + \left[-A_2\left(\frac{x^2}{3} + \frac{z^2}{2}\right) + A_1\frac{xy}{3} + A_3yz\right]\mathbf{j}.$$
9. (a) $\mathbf{G} = (A_1y - A_2x)\mathbf{k}$.
   (b) $\mathbf{G} = \dfrac{xy^2}{2}\mathbf{k}$.
   (c) $\mathbf{G} = -\ln\sqrt{x^2 + y^2}\,\mathbf{k}$.

Section 4.6

4. $\sqrt{4x^2 + 4y^2 + 1}\, dx\, dy$.
5. 11.
6. (a) $\frac{1}{3}\sqrt{3}(\mathbf{i} + \mathbf{j} + \mathbf{k})$.
(b) $\mathbf{k} \cdot \mathbf{n} = \frac{1}{3}\sqrt{3}$.
(c) $\displaystyle\int_0^1 \int_0^{1-y} \frac{dx\, dy}{|\cos \gamma|}$.
(d) $\frac{1}{2}\sqrt{3}$.

Section 4.7

1. 18π.
2. (a) 8.
(b) 16.
(c) 24.
(d) 0.
(e) 0.
(f) 0.
(g) 0.
3. $\int_0^1 \int_0^{2-2x} \frac{7}{2} \cdot \frac{6}{7} x\, dy\, dx = 1$.
4. Zero.
5. $3\pi a^2$.
6. -1.
7. (a) No.
(b) Yes.
(c) 4.
8. $13\frac{7}{8}$.
9. 60π.
10. 0.
11. $15\pi/2$.
12. 0.
13. $4/15$.
18. $|\mathbf{E}| = \lambda/2\pi\epsilon_0 r$.
19. (a) $T_r = T_a + \dfrac{1/r - 1/a}{1/b - 1/a}(T_b - T_a)$.
20. $\sqrt{2} + \log(1 + \sqrt{2})$
21. 2π

Section 4.8

3.

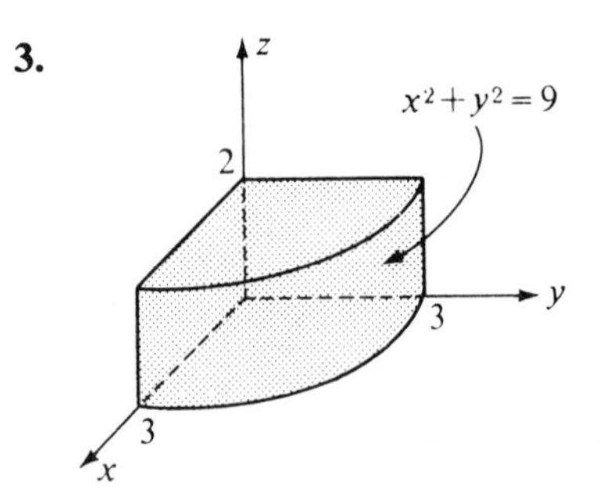

4. (a) 3.
(b) 3.
(e) This will be discussed in sec. 4.9.
5. $3v$.
6. $\pi(1 - e^{-1})$.

Section 4.9

6. 60. The z component of **F** can be ignored, so replace **F** by $\mathbf{G} = y\mathbf{i} + (x + 2)\mathbf{j}$. Div $\mathbf{G} = 0$ so (by the divergence theorem) the desired integral is the negative of the integral of **G** over the four *flat* faces of a certain five-sided closed surface. Only two faces contribute nonzero values to the integral. The average value of $x + 2$ over the face in the xz plane is $\frac{7}{2}$ and the average value of y over the face in the yz plane is $\frac{3}{2}$ so the arithmetic is $\frac{7}{2}(3)(4) + \frac{3}{2}(3)(4) = 60$.
9. (a) 0.
(b) -2.
(c) 4.
(d) 0.
(e) -1.
10. $\pm 2\pi$, depending on direction of integration.
11. 0.
17. 0.
18. (a) Use (3.20).
(b) $\frac{928}{3}\pi$.
[Exercise 18 part (a) does not apply in part (b) since this function is not harmonic.]
19. (a) 108π.
(b) 1944π.
21. The divergence is identically zero, so the desired integral equals the negative of the integral over the missing top, which, in this case, is trivial to compute.
22. The field is $\frac{1}{3}x^3\mathbf{i} + \frac{1}{3}y^3\mathbf{j}$.
24. (a) $4\pi b^4$.
(b) To avoid a triple integral, take $dV = 4\pi r^2\, dr$, so that the integral is

$$\int_0^b 16\pi r^3\, dr$$

25. 8π.
29. (a) 16.8 if volume is proportional to the cube of the minimal diameter.
(b) Yes.
30. $5\epsilon_0$.

31. $\iiint \nabla \cdot \mathbf{F}\, dV = \iint \mathbf{F} \cdot \mathbf{n}\, dS$ leads to $\mathbf{C} \cdot \iiint \nabla\phi\, dV = C \cdot \iint \phi_n\, dS$ and, since this is valid for every constant vector **C**, the identity follows.
32. In the divergence theorem, let $\mathbf{F} = \mathbf{A} \times \mathbf{C}$ where **C** is a constant vector field, and proceed as in the preceding problem.
33. (a) ∇f.
    (b) $\nabla \times \mathbf{F}$. (Make use of the two preceding problems.)

Section 5.1

1. In applying the fundamental theorem of calculus.
2. To ensure that the volume integral of div **F** over the bounded domain D exists.
5. $\cos \gamma = 0$, so the expression $dx\, dy/|\cos \gamma|$ is meaningless.
6. 30.
7. $5v$.
8. Yes.
9. (a) An outer sphere with **n** pointing away from the origin and an inner sphere with **n** pointing towards the origin.
   (b) Sum of two integrals.
   (c) They are equal.
   (d) No.
   (e) 4π.

Section 5.2

1. $4\pi\phi(0,0,0) = 20\pi$.
2. (a) $-4\pi\phi(0,1,0) = -20\pi$.
   (b) $-4\pi\phi(2,1,3) = 0$.

Section 5.3

3. False (for example $\mathbf{F} = y\mathbf{i} + x\mathbf{j}$).
7. $\mathbf{F} =$ constant.

Section 5.4

6. Zero.
7. -36π, since $\mathbf{curl\ F} \cdot \mathbf{k} = -4$ and area enclosed by C is 9π.
8. (a) 6π. (b) 6π.
9. (a) -16. (b) -16. (c) Second term.

10. 28π.
11. Orientation.

Section 5.5

1. (a) $2z$.
(b) $-5\mathbf{k}$.
(c) -20π.
2. (a) 27π.
(b) 0.
5. (a) Zero.
(b) Zero.
(c) Zero.
(d) div **curl F** = 0.

Note: The divergence of a vector field at a point is sometimes called the *source density* of the field at that point. This is because the divergence of the electric intensity of an electrostatic field is equal (within a factor) to the charge density, and electric charge is the "source" or "cause" of the field. The statement "a field has zero divergence in any region that is free of sources" has an intuitive appeal to many students. The above exercise can be worded: the curl of a vector field is another vector field that is free of sources.

6. (a) Zero. (b) Zero. (c) Zero. (d) div **curl F** = 0.
7. (a) Zero. (b) Zero. (c) Zero vector. (d) **curl grad** ϕ = **0.**

Note: The curl of a vector field at a point is sometimes called the *vortex density* of the field at that point. This is because, in some sense, the curl describes the "eddy" or "whirlpool" nature of the field. Note that vortex density is a vector quantity. Just as engineers sometimes think of a point source as a point where the divergence is "infinite," so also do they think of a *vortex filament* as a curve in space along which the magnitude of the curl is "infinite." The central part of a tornado provides an approximate idea. We leave to the reader the precise formulation of the definition. The intuitive content of exercise 7 is that any field that can be derived from a scalar potential must be vortex-free. It should be noted, however, that if we allow the scalar potential to be a multiple-valued function, it is sometimes possible to find a scalar potential for the velocity field of fluid swirling about a vortex filament. We heartily recommend the chapter on vector analysis in *Mathematics of Circuit Analysis,* by E. A. Guillemin (Wiley, 1949), in which these matters are taken up in greater detail.

Let us now briefly review and extend some of the earlier ideas. We consider only continuously differentiable vector fields.

If a vector field defined in a domain D has any one of the following properties, it has all of them:

(*i*) Its curl is zero at every point.
(*ii*) Its integral around any closed contour is zero, provided that there is a surface enclosed by the contour entirely within D.
(*iii*) It is the gradient of a scalar function, but this function may possibly be multiple-valued.

If the domain D is simply connected, we can omit the clauses starting "provided that . . ." and "but this . . ." from these properties. When D is simply connected, the following terms are used for these fields: conservative field, irrotational field, potential field, source field.

Similarly, any one of the following properties of a continuously differentiable vector field implies the others:

(*i*) Its divergence is zero at every point in D.
(*ii*) Its integral over every surface is zero, provided that we consider only closed surfaces enclosing points all of which are in D.
(*iii*) It is the curl of another (possibly multiple-valued) vector field.

These statements are not precise and should not be taken very seriously. Terms sometimes used for such fields are: solenoidal field, rotational field, turbulent field, source-free field, vortex field. The terminology is not standardized; in modern usage, the term "turbulent" has an altogether different meaning. In applications, vector fields that are discontinuous along a surface are of considerable importance. We have not discussed such fields because they arise more naturally in courses dealing with applications, where the motivation for studying them is more apparent.

Section 5.7

1. (a) 14. (b) $[1 \quad -1 \quad 7]$. (c) $\begin{bmatrix} -2 \\ 5 \\ 4 \end{bmatrix}$.

(d) $[1 \quad -1 \quad 7 \quad 10]$.

(e) $\begin{bmatrix} 1 & -1 & 7 \\ 2 & 0 & 1 \\ 1 & -1 & 7 \end{bmatrix}$ (f) $\begin{bmatrix} 10 & 0 & -10 \\ 0 & 10 & 10 \\ 0 & -10 & 20 \end{bmatrix}$.

10. Check your answer by substituting into the equations.
15. In general, no. (Yes, if they commute.)

Section 5.8

2. (b) $\begin{bmatrix} \cos\psi & 0 & \sin\psi \\ 0 & 1 & 0 \\ -\sin\psi & 0 & \cos\psi \end{bmatrix}$

6. $f' = x'^2 + y'^2 \qquad \mathbf{V}' = x'\mathbf{i}' + y'\mathbf{j}' + z'\mathbf{k}'$
   $\nabla f = 2x'\mathbf{i}' + 2y'\mathbf{j}' \qquad \nabla \cdot \mathbf{V}' = 3 \qquad \nabla \times \mathbf{V}' = \mathbf{0}.$
11. $ds = (dx'^2 + dy'^2 + dz'^2)^{1/2}$.
12. $dV' = dx'\, dy'\, dz'$.
14. $x = -y', y = z', z = -x'$.
18. The line is along $\mathbf{i} - \mathbf{j} - \mathbf{k}$. The angle around this axis is $-\pi/3$.
19. (d) Take the determinant.

20–22. Elaborate on the technique in exercise 18; or read "Coordinate-Free Rotation Formalism," by J. Matthews, *Amer. Jnl. of Physics, 44,* 1210 (1976).

23. (c) The transformation might be left-handed.

Appendix A

1. Simply multiply $(u_1 i + u_2 j + u_3 k)(v_1 i + v_2 j + v_3 k)$.
2. (a) $uv + vu = (-\mathbf{u} \cdot \mathbf{v} + \mathbf{u} \times \mathbf{v}) + (-\mathbf{v} \cdot \mathbf{u} + \mathbf{v} \times \mathbf{u})$. The cross products disappear.
3. (a) This is simple vector algebra.
   (b) $\mathbf{v}' = \mathbf{v} - 2(\mathbf{v} \cdot \mathbf{n})\mathbf{n} = v + (vn + nv)n = v + vnn + nvn$. But $\mathbf{n} \cdot \mathbf{n} = -\text{nn}$ and since $\mathbf{n}$ is a unit vector $nn = -1$. So $v' = v - v + nvn = nvn$.
4. $n'n = -(\mathbf{n}' \cdot \mathbf{n}) + \mathbf{n}' \times \mathbf{n} = -\cos \tfrac{1}{2}\theta - \mathbf{u} \sin \tfrac{1}{2}\theta$.
5. Since u is a unit vector, $u^2 = -1$, so $u^3 = -u$, $u^4 = 1$, $u^5 = u, \ldots$ and

$$1 + \phi u + \phi^2 u^2/2! + \phi^3 u^3/3! + \cdots$$
$$= (1 - \phi^2/2! + \phi^4/4! - \cdots) + (\phi - \phi^3/3! + \cdots)u$$

6. $v'' = e^{(\theta/2)u}\, v\, e^{-(\theta/2)u}$.

Appendix E

1. 71.
3. (a) 0. (b) $\frac{1}{2}$.
8. -2.

Index

IMPORTANT FORMULAS

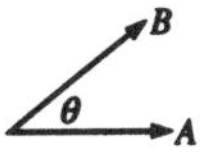

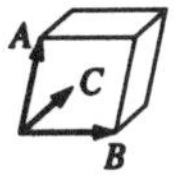

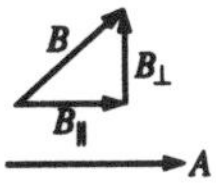

OPERATION	GEOMETRIC	ANALYTIC
Scalar (dot) product, $\mathbf{A}\cdot\mathbf{B}$	$\lvert\mathbf{A}\rvert\,\lvert\mathbf{B}\rvert\cos\theta$	$A_1B_1 + A_2B_2 + A_3B_3$
Vector or cross product, $\mathbf{A}\times\mathbf{B}$	$\mathbf{n}\,\lvert\mathbf{A}\rvert\,\lvert\mathbf{B}\rvert\sin\theta$	$\begin{vmatrix} \mathbf{i} & \mathbf{j} & \mathbf{k} \\ A_1 & A_2 & A_3 \\ B_1 & B_2 & B_3 \end{vmatrix}$
Triple scalar product, $[\mathbf{A}, \mathbf{B}, \mathbf{C}]$	Volume $\mathbf{A}\cdot\mathbf{B}\times\mathbf{C} = \mathbf{A}\times\mathbf{B}\cdot\mathbf{C}$	$\begin{vmatrix} A_1 & A_2 & A_3 \\ B_1 & B_2 & B_3 \\ C_1 & C_2 & C_3 \end{vmatrix}$
Parallel-perpendicular decomposition	$\mathbf{B}_\parallel$	$\dfrac{\mathbf{A}\cdot\mathbf{B}}{\mathbf{A}\cdot\mathbf{A}}\mathbf{A}$
	$\mathbf{B}_\perp$	$\dfrac{(\mathbf{A}\times\mathbf{B})\times\mathbf{A}}{\mathbf{A}\cdot\mathbf{A}}$

OPERATOR	INTERPRETATION	INTEGRAL THEOREM
grad $\phi = \nabla\phi$	Maximum rate of change of ϕ	$\int_P^Q \nabla\phi\cdot d\mathbf{R} = \phi(Q) - \phi(P)$
div $\mathbf{F} = \nabla\cdot\mathbf{F}$	Net outflux of $\mathbf{F}$ per unit volume	$\iiint_D \nabla\cdot\mathbf{F}\,dV = \iint_S \mathbf{F}\cdot\mathbf{n}\,dS$
curl $\mathbf{F} = \nabla\times\mathbf{F}$	Circulation of $\mathbf{F}$ per unit volume	$\iint_S \nabla\times\mathbf{F}\cdot\mathbf{n}\,dS = \int_C \mathbf{F}\cdot d\mathbf{R}$

VECTOR IDENTITY	ASSOCIATED POTENTIAL THEOREM
$\nabla\times\nabla\phi = 0$	$\nabla\times\mathbf{F} = 0$ implies $\mathbf{F} = \nabla\phi$
$\nabla\cdot\nabla\times\mathbf{G} = 0$	$\nabla\cdot\mathbf{F} = 0$ implies $\mathbf{F} = \nabla\times\mathbf{G}$
$\mathbf{A}\times(\mathbf{B}\times\mathbf{C}) = (\mathbf{A}\cdot\mathbf{C})\mathbf{B} - (\mathbf{A}\cdot\mathbf{B})\mathbf{C}$	$\nabla\times(\nabla\times\mathbf{F}) = \nabla(\nabla\cdot\mathbf{F}) - \nabla^2\mathbf{F}$

ORTHOGONAL CURVILINEAR COORDINATES (u_1, u_2, u_3)

$h_i = \lvert\partial\mathbf{R}/\partial u_i\rvert = 1/\lvert\nabla u_i\rvert$

Cylindrical: $h_\rho = 1,\ h_\theta = \rho,\ h_z = 1$

Spherical: $h_r = 1,\ h_\phi = r,\ h_\theta = r\sin\phi$

$dV = h_1h_2h_3\,du_1\,du_2\,du_3$

$d\mathbf{R} = h_1\,du_1\,\mathbf{e}_1 + h_2\,du_2\,\mathbf{e}_2 + h_3\,du_3\,\mathbf{e}_3$

$\nabla f = (\partial f/\partial u_1)\mathbf{e}_1/h_1 + (\partial f/\partial u_2)\mathbf{e}_2/h_2 + (\partial f/\partial u_3)\mathbf{e}_3/h_3$

$\nabla\cdot\mathbf{F} = [\partial(F_1h_2h_3)/\partial u_1 + \partial(F_2h_3h_1)/\partial u_2 + \partial(F_3h_1h_2)/\partial u_3]/h_1h_2h_3$

$$\nabla^2 f = \frac{1}{h_1h_2h_3}\left[\frac{\partial}{\partial u_1}\left(\frac{h_2h_3}{h_1}\frac{\partial f}{\partial u_1}\right) + \frac{\partial}{\partial u_2}\left(\frac{h_3h_1}{h_2}\frac{\partial f}{\partial u_2}\right) + \frac{\partial}{\partial u_3}\left(\frac{h_1h_2}{h_3}\frac{\partial f}{\partial u_3}\right)\right]$$

$$\nabla\times\mathbf{F} = \frac{1}{h_1h_2h_3}\begin{vmatrix} h_1\mathbf{e}_1 & h_2\mathbf{e}_2 & h_3\mathbf{e}_3 \\ \dfrac{\partial}{\partial u_1} & \dfrac{\partial}{\partial u_2} & \dfrac{\partial}{\partial u_3} \\ F_1h_1 & F_2h_2 & F_3h_3 \end{vmatrix}$$

Hawkes Publishing

Introduction to Vector Analysis
Seventh Edition

Harry F. Davis
Arthur David Snyder